AI SCRUM:
THE AGILE INVERSION

A PROJECT LEADERSHIP FRAMEWORK FOR EVERY TEAM, EVERY DISCIPLINE

JAMES J. BAKER

Published in the United States of America

AI SCRUM: THE AGILE INVERSION
A Project Leadership Framework for Every Team, Every Discipline

ISBN: 979-8-9950438-0-5 (Paperback / eBook)

Warning and Disclaimer
Every effort has been made to make this book as complete and accurate as possible, but no warranty or fitness is implied. The information provided is on an "as is" basis. The author and the publisher shall have neither liability nor responsibility to any person or entity with respect to any loss or damages arising from information contained in this book.

AI Involvement Disclaimer
The use of artificial intelligence in the creation of a published work is frequently misunderstood. A common assumption holds that AI involvement is synonymous with AI authorship. That is not the case here. When applied responsibly, AI functions as a remarkably effective creative instrument, capable of drawing insight from an author through structured brainstorming, serving as an analytical interviewer to surface ideas that might otherwise remain undeveloped, and functioning in an editorial capacity that was once accessible only through costly professional engagement.

This work was developed through a principled and transparent combination of human authorship and artificial intelligence assistance. Claude, developed by Anthropic, contributed in an editorial and analytical capacity, supporting the validation of factual accuracy, the identification of potential ambiguities, and the assessment of narrative flow. Claude further participated in a structured series of brainstorming and interview sessions designed to stress-test the AI Scrum framework against edge cases, regulatory considerations, and global applicability, including compliance landscapes the author may not have encountered directly through professional practice. All AI-assisted contributions were subsequently reviewed, refined, and substantially expanded through iterative human authorship. The chapter illustrations were produced using Nano Banana Pro 3, guided entirely by detailed, human-crafted prompts developed through multiple rounds of intentional creative direction. The author retains full intellectual and editorial responsibility for all content contained in this work.

My deepest appreciation
to my wife and daughter,
whose steadfast confidence
in me never wavered,
even as we navigated
the unpredictable cycles
inherent to the technology sector.

ABOUT THE AUTHOR

James J. Baker is an accomplished IS&T Project Manager and Agile Delivery Leader with more than two decades of experience driving complex technology initiatives across financial services, energy, manufacturing, and enterprise data platforms. His career spans leadership roles in Agile transformation, AI-enabled solution delivery, robotics process automation, and enterprise modernization, where he has improved operational efficiency, accelerated delivery cycles, and guided cross-functional teams through high-impact programs. James' work integrates business strategy with advanced technologies—including AI, automation, and modern data ecosystems—to create measurable, sustainable outcomes. He holds multiple certifications in Scrum, product ownership, AI product management, and prompt engineering, and has academically published research in quality management. James holds a B.S. in Business Administration from Kansas State University.

Contents

INTRODUCTION

Picture this: you are standing at the edge of a cliff, watching the ground beneath your feet slowly crumble. Behind you lies everything familiar—the tools you have mastered, the processes you have perfected, the career path you have carefully constructed. Ahead lies nothing but fog. Do you step back to safety, or do you leap forward into the unknown?

This is the choice confronting every software professional today. Over the past few years, artificial intelligence has not just entered our industry—it has detonated within it. The velocity of change has been breathtaking, disorienting, and for many, deeply unsettling. I have watched seasoned developers question whether their decades of expertise still matter. I have seen project managers wonder if their roles will exist in five years. The fear is palpable, and it is not unfounded. But it is incomplete.

Here is what the doomsayers miss: yes, certain positions will vanish. Some already have. But history has shown us repeatedly that technological disruption does not eliminate work—it transforms it. The assembly line worker became the machine operator. The typist became the administrative professional. The video store clerk became the streaming content curator. Each wave of innovation has required us to reinvent ourselves, to acquire new skills, to become more valuable in different ways. AI is no different. The question is not whether your job will change. It is whether you will change with it.

Consider the cautionary tales we have all witnessed: Kodak invented the digital camera, then buried it to protect film sales. RadioShack had everything it needed to become Best Buy, but clung to outdated retail models. Borders watched Amazon rise and somehow convinced itself that e-commerce was a passing fad. Palm pioneered the smartphone, then dismissed the iPhone as irrelevant. Tower Records owned music retail, then acted as if streaming would never catch on. These were not companies led by fools—they were industry leaders staffed by smart, experienced professionals who made one fatal error: they chose comfort over adaptation.

We stand at a similar crossroads. AI is not asking for our permission to reshape software development, marketing, product management, or any other discipline. It is already doing it. But here is the critical insight that should bring you relief rather than dread: AI cannot replace you. Not really. Not for the foreseeable future. Because AI lacks the one thing that makes human beings indispensable—our capacity for empathy, compassion, judgment, and contextual understanding. AI is a tool of extraordinary power, but it remains just that: a tool. The question is whether we will wield it or be displaced by those who do.

On June 21, 1906, during a debate in the House of Commons concerning the treatment of native races in South Africa, Winston Churchill stated, "…Where there is great power there is great responsibility…" and with AI this is quite applicable. In the summer of 2025, preliminary research by MIT students was completed and an essay submitted for peer review suggesting how AI may be costing users brainpower. AI should be used more as an assistant, AI augmentation, and not a direct replacement because of the possible effects on the human brain. Humans should think first, putting their ideas to paper or keyboard initially before incorporating AI.

This brings us to the practices and frameworks we have relied upon—Scrum, Agile methodologies, traditional project management approaches. These are not sacred texts. They are tools that evolved to solve specific problems in specific contexts. As AI reshapes those contexts, our practices must evolve too. How do we estimate sprint velocity when AI can complete in hours what once took days, weeks, or even months? How do we prioritize backlogs when machine learning can identify patterns we would never see? How do we structure teams when the ratio of work-to-worker has fundamentally shifted?

These are not threats. They are opportunities. This book is an invitation to rethink, reimagine, and rebuild your personal approach to development methodology in an AI-augmented world in part or in whole. Not with fear, but with strategic intent. Not by abandoning what works, but by enhancing it with what is possible. Not getting rid of the people or software that we have invested in but use them in a different manner.

"The Agile Inversion" reveals that AI has shifted the software bottleneck from code creation to validation and governance. It explains why traditional velocity metrics are now misleading and details five new constraints, such as validation complexity and requirements uncertainty, that render standard Agile ceremonies obsolete. This necessitates a fundamental methodological pivot.

CHAPTER 1: THE AGILE INVERSION

My journey with Agile began out of necessity in 2014. I was managing a long-term CRM software installation and development project that required more structure than a simple to-do list could provide. It was not a textbook implementation. My "team" consisted of myself, two part-time assistants, and a third-party developer. We were not following the Scrum Guide to the letter. We were chasing the spirit of it: iterative delivery and continuous adaptation.

By the spring of 2022, I decided to formalize that experience. I earned my Certified Scrum Master (CSM) and Certified Scrum Product Owner (CSPO) credentials to ground my practical experience in the theoretical mechanics of professional delivery.

From 2019 through 2025, I applied this framework within a much more complex environment. I worked on large, cross-functional squads ranging from nine to twelve members. These were not the "two-pizza" teams of Agile lore. They were robust delivery engines. Our roster typically included a Product Owner, a Scrum Master, a Delivery Lead, multiple Business

Analysts, a Systems Analyst, two to three Developers, and two to three QA Engineers.

We ran a tight ship. We executed two-week sprints in Azure DevOps with mechanical rhythm. Our retrospectives were honest. Our "Definition of Done" was clear. Our velocity, tracked on a standard Fibonacci story point scale, was consistent.

By every traditional measure, we were doing Agile right. Management valued the predictability. Our burndown charts rarely deviated from the ideal slope.

Consider a scenario that is becoming increasingly common across development organizations, one that may not yet have a name on your team, but is already quietly reshaping how delivery gets done.

A subtle dysfunction has taken root in a team's workflow. During backlog refinement, they routinely estimate a user story at 2 points. In their taxonomy, this implies roughly four hours of effort. On the surface, the math holds up.

The reality on the ground is different. A Business Analyst spends two hours gathering requirements. A Systems Analyst spends an hour mapping the data. A Developer consumes two hours coding the solution. A QA Engineer then invests an hour validating it.

That "2-point" story has actually consumed six to eight hours of cross-functional labor across four different roles. Yet, the tracking system assigns the "velocity credit" solely to the person who closed the ticket. That is usually QA. The team has zero visibility into the actual distribution of labor. They cannot see where the friction lives. They are hitting their numbers, but optimizing for the administrative act of closing tickets rather than the economic act of delivering value.

Then the AI era arrives. That visibility gap, once a crack in the foundation, widens into a canyon.

Suddenly, the development portion of that workflow compresses from hours to minutes once the team establishes AI collaboration patterns. A developer utilizing a Large Language Model (LLM) generates the boilerplate, the logic, and the unit tests for a 5-point story before their morning coffee cools.

Velocity metrics skyrocket. On paper, productivity increases dramatically. In reality, feature delivery to the customer slows down.

While code generation is rapid, validation drags on for days. Governance reviews create queues that last weeks. Integration complexity explodes as the team attempts to weave probabilistic AI components into deterministic legacy systems. The velocity metric signals movement at light speed, but the product is stalled. The team is optimizing for a constraint — coding speed — that has ceased to exist.

This team is not an outlier. Across the industry, the frameworks that served software delivery for twenty-five years are fracturing under the same pressure. That fracture point has a name: **The Agile Inversion**.

WHAT THIS CHAPTER COVERS

By the end of this chapter, you will understand:

- Why the constraint that limited software delivery for 25 years has disappeared, and what has replaced it.
- The five new constraints that now dominate delivery timelines: Validation, Uncertainty, Integration, Governance, and Change Velocity.
- The "Gap Between Confidence and Competence" that leads executives to misjudge AI adoption risks.
- Why teams that add AI tools without changing their methodology often see performance degrade rather than improve.
- The "Executive Diagnostic" to determine if your organization is currently caught in the Velocity Trap.

KEY TERMS AND CONCEPTS

Before we dissect the specific constraints of the AI era, we must recalibrate our language. The terms we use to manage software—Velocity, Debt, Done—carry historical baggage from the manual coding era. In an AI-augmented environment, we must adopt the following definitional shifts:

- **Velocity (Redefined)**
    - *Old Definition:* The rate at which story points (coding effort) are completed per sprint.
    - *New Definition:* The rate at which valid, safe, and integrated outcomes are delivered to the customer. High "Coding Velocity" often correlates with *negative* "Outcome velocity" because unvalidated code creates a backlog of governance debt.

- **Technical Debt (AI Debt)**
    - *Old Definition:* The implied cost of rework caused by choosing an easy solution now instead of a better approach that would take longer.
    - *New Definition (AI Debt):* The accumulation of probabilistic behaviors, unverified hallucinations, and prompt dependencies that the team does not fully understand. Unlike traditional debt, AI debt is "silent drift" that is only visible when it fails in production.
- **Definition of Done (DoD)**
    - *Old Definition:* Code is written, tests pass, feature is deployed to staging.
    - *New Definition:* The AI output has been red-teamed for adversarial attacks, the data lineage is documented for compliance, and the failure modes are understood.
- **User Story (Context Specification)**
    - *Old Definition:* A description of a feature from an end-user perspective.
    - *New Definition:* A "Context Specification" that defines not just *what* the system should do, but the hard boundaries of what it *must not* do. We move from writing "recipes" for developers to writing "constitutions" for agents.

THE GAP BETWEEN CONFIDENCE AND COMPETENCE

Before we analyze the mechanics of this failure, we must address the leadership psychology that enables it.

In 2023, the dam broke. AI exploded into the public consciousness, and for the first time, non-technical stakeholders saw the raw power of generation. They saw the magic. They could generate a marketing plan or a Python script in seconds.

But rather than sparking a strategic transformation, this visibility often triggered a retreat into old habits. Many leaders, overwhelmed by the pace of change or fearful of the unknown, chose to sweep the complexity of AI under the rug. Instead of developing new positions to master this technology, they doubled down on the familiar cost-saving playbook: offshoring.

They reasoned that if AI makes coding easier, they should simply find cheaper humans to do it. They failed to realize that the bottleneck had shifted from *creation* (which is cheap) to *validation* (which requires high-context expertise). By trading high-context local talent for low-cost remote capacity in the name of efficiency, they were actually dismantling the very layer of oversight needed to manage AI safety.

They had enough information to be dangerous, but not enough to be effective.

I kept thinking of a quote by astrophysicist Neil deGrasse Tyson:

> *"One of the great challenges in life is knowing enough to think you're right, but not enough to know you're wrong."*

This is the Dunning-Kruger effect applied to enterprise AI.

The Dunning-Kruger effect is a cognitive bias where people with low ability or knowledge in a specific area overestimate their competence. Essentially, they "don't know what they don't know," so they lack the expertise to recognize their own mistakes or limitations.

Conversely, high performers often underestimate their abilities because they assume tasks that are easy for them are also easy for everyone else. In the context of this book, this explains why non-technical stakeholders (who built a prototype in a weekend) confidently believe they understand AI production. They in reality know just enough to be dangerous, but not enough to see the hidden complexity of validation and governance.

Most leaders understood AI as a "Search Engine Plus". They believed it was a retrieval tool that simply found answers faster. They did not understand it was a probabilistic engine that required rigorous guidance to avoid fabrication.

I watched this play out in real time. A stakeholder would demand a feature because "I built the prototype in ChatGPT over the weekend." They did not see the security vulnerabilities in their prototype. They did not see the lack of error handling. They did not see the privacy violations.

They knew enough to think they were right. They did not know enough to see the mountain of technical debt they were asking us to incur.

Figure 1.1: The 'Inversion' of Effort. *Small effort on code generation, massive effort on validation and risk.*

THE ECONOMICS OF CONSTRAINTS

To understand why traditional Agile methodologies fail in an AI-native environment, we must look past the technology and examine the organizational physics at play.

Eliyahu Goldratt's *Theory of Constraints* posits a simple, immutable law: any system is limited by its single most constraining bottleneck. To improve throughput, one must identify that constraint and optimize the entire system around it. Optimizing a non-constraint does not increase throughput. It only increases inventory, waste, and confusion.

From the signing of the Agile Manifesto in 2001 through the public release of ChatGPT in 2022, the primary constraint in software development was unambiguous: the creation of code was expensive.

Writing software was slow. It was cognitively demanding. It was prone to human error. It was a scarcity problem. Consequently, the entire Agile industrial complex was engineered to protect and optimize this scarcity.

- **Sprints** provided a protected temporal container for deep work.
- **Daily Standups** functioned to remove immediate blockers preventing coding.

- **Story Points** served as a proxy for the effort required to write code.

Every ceremony revolved around one universal truth: building software is hard, so we must painstakingly coordinate our scarce development capacity.

Artificial intelligence demolished that assumption.

The data is stark. Microsoft's 2023 research on GitHub Copilot demonstrated that developers completed specific tasks 55% faster when using AI tools. McKinsey analysis suggests generative AI could automate up to 50% of work activities, with software engineering representing a highest-impact domain.

While organizations like METR note that elite developers working on highly complex, proprietary codebases may see initial slowdowns, the macroeconomic trend is undeniable. For the vast majority of developers, specifically those working on boilerplate, documentation, or standard features, the "creation tax" has been repealed.

But when a constraint is removed from a system, the system does not simply become faster. The constraint moves.

The Evolution of Delivery

To appreciate the gravity of this shift, we must contextualize it within the history of software delivery. The industry has not arrived at this moment by accident. We arrived here by successively eliminating previous constraints.

Era 1: The Planning Constraint (1970–2000) For three decades, the Waterfall methodology dominated the landscape. It was predicated on the belief that the cost of change was exponential. Therefore, the constraint was *planning*. We spent months documenting requirements to avoid the cost of later changes. The result was catastrophic efficiency. The Standish Group's seminal 1995 *CHAOS Report* quantified the crisis. Only 16% of projects succeeded. Cost overruns averaged 189%.

Era 2: The Feedback Constraint (2001–2022) The Agile Manifesto represented a rejection of the Planning Constraint. It recognized that we could not predict the future, so we should optimize for *adaptation*. The new constraint became *feedback loops*. How fast could we get code in front of a user? Scrum optimized the scarcity of developer hours through "Sprints."

It assumed the 2–4 week cycle was the fastest possible unit of delivery because coding took time.

Era 3: The Validation Constraint (2023–Present) We have now entered the third era. The "Feedback Constraint" has been shattered by AI. We can generate code faster than we can get feedback on it. We are no longer limited by how fast we can plan or how fast we can code. We are limited by how fast we can *trust* what we have built.

THE FIVE NEW CONSTRAINTS

When the marginal cost of code generation drops to near zero, five new constraints elevate to immediate criticality. These factors now dominate delivery timelines. Traditional Scrum has no mechanism to manage them.

Constraint 1: The Asymmetry of Validation Complexity

In the pre-AI era, software validation was symmetrical to software creation. If a developer spent three days writing a complex pricing algorithm, a QA engineer would likely spend a proportional amount of time writing test cases to verify it.

Generative AI breaks this symmetry. It introduces a massive disparity between creation time and validation time.

In early 2025, I experienced this disparity firsthand while generating visual concepts for this manuscript. I needed a simple illustration of a sprint board with four columns. I prompted the AI. In seconds, it rendered a photorealistic image. The lighting was perfect. The creation cost was effectively zero.

But when I examined the image closely, the text on the sticky notes was gibberish. It took three subsequent iteration cycles to force the model to render legible, correct text. That validation and refinement process took nearly an hour.

The creation took seconds. The validation took an hour. This is "The 100:1 Validation Gap."

Consider a fraud detection system. In the old world, rules were deterministic (*"If X, then Y"*). You could write a unit test that passed forever. AI systems are probabilistic. A fraud model might flag a transaction today and ignore it tomorrow because of model drift. You cannot "unit test" a probability. To validate an AI system, you must employ entirely new disciplines like Adversarial Red-Teaming and Drift Monitoring.

This is why validation is the new bottleneck. You can generate the feature in minutes after establishing prompt patterns. But proving that the feature is safe requires a sophisticated, time-consuming governance layer that most Agile teams have not yet built.

Constraint 2: Requirements Uncertainty (The Precision of Intent)
In traditional software development, human judgment acted as a shock absorber for vague requirements. If a stakeholder asked for a "user-friendly interface," the team used shared human context to fill in the gaps.

AI lacks this intuition. It creates exactly what you ask for, including your mistakes. If you ask an LLM for a recommendation engine without explicit constraints, it will confidently build one. It may, however, hallucinate features or violate privacy boundaries.

This shifts the burden of "Backlog Refinement" from estimation to definition. Every ambiguity becomes a decision point. Executives often see the speed of AI and question why teams spend days in alignment discussions. The answer is simple. Without upfront investment in precision, AI generates impressive implementations that fail validation because they solve the wrong problem.

Constraint 3: Integration Orchestration
The sales pitch for generative AI often focuses on the standalone miracle. In the enterprise, however, value is only realized through integration.

Consider a quality engineering initiative I observed in 2024. The team identified a tool that used AI to automatically generate SQL validation tests. The technical proof-of-concept was successful. Yet the initiative failed. Management disabled the features because the organization lacked a governance framework for AI data access.

Integration orchestration is the hidden tax of AI adoption. An AI component requires access to data pipelines, compliance checks, and legacy systems. We are seeing a new pattern emerge: The 1/99 Rule. It takes 1% of the effort to build the AI prototype and 99% of the effort to integrate it safely into the enterprise architecture.

Constraint 4: The Governance Wall

In April 2023, Samsung employees inadvertently leaked confidential source code by pasting it into a public instance of ChatGPT. Samsung's response was a total ban on generative AI tools.

This illustrates the "Governance Wall." Organizations are colliding with a complex web of regulations like the European Union AI Act and the NIST AI Risk Management Framework. Yet, according to ISACA, only 31% of organizations have established formal AI governance structures.

Figure 1.2: The AI Iceberg. *Small prototype, massive hidden effort.*

If your Scrum process does not account for the time required for legal review, ethics checks, and compliance documentation, your velocity is an illusion. You are simply sprinting toward a legal roadblock.

Constraint 5: Change Velocity (Obsolescence)

We are witnessing a pace of technological change comparable to the evolution of CPUs from the 1960s to the 2000s, but compressed into months. New models and vulnerabilities (like prompt injection) emerge weekly. A decision made in Sprint 1 may be technically invalid by Sprint 4, not because requirements changed, but because the underlying technology shifted.

FROM THE TRENCHES: THE ANATOMY OF A BROKEN SPRINT

To understand why incremental adaptation fails, we must look at the granular reality of a modern sprint. Let us contrast the "intended" rhythm of Scrum with the "actual" rhythm of an AI Scrum data team.

In 2019, a typical sprint followed a predictable bell curve. The "Build" phase anchored the week, consuming 60% of capacity. The Daily Standup was useful because everyday progress was made on the code.

Now, consider the timeline of an AI Scrum team working on a "Customer Churn Prediction Model."

Day 1 (08:00 – 12:00): The Creation Burst

The team assembles. Using an AI-assisted data platform, they prompt the system to analyze historical sales data and generate a predictive model. By lunch, the AI has generated the Python code, the SQL queries, and a draft dashboard. The "Build" phase is effectively complete in four hours.

Days 1–10 (The Remaining 72 Hours): The Valley of Death

The team now stares at a finished prototype they cannot ship.

- **Definition Paralysis:** The AI defined "Churn" as "zero activity for 30 days." The Finance stakeholder defines it as "subscription cancellation." The code is written, but the business logic is wrong.
- **Privacy Block:** The model requires PII to function. The Governance Coordinator flags a GDPR risk. Work stops for 48 hours waiting for legal approval.
- **Drift Anxiety:** The team realizes the model works on 2023 data but fails on 2024 data. They spend three days testing against different time horizons.

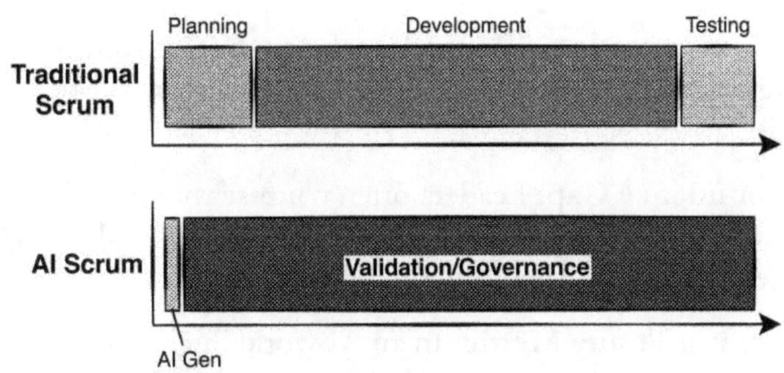

Figure 1.3: Timeline Comparison. *Top: Traditional Scrum with long development. Bottom: AI Scrum with tiny generation and massive validation.*

During this 9-day period, the Daily Standup becomes a ritual of frustration. The Scrum Master asks, "What did you do yesterday?" The Developer replies, "Still waiting on Legal. Still testing definitions." The ceremony, designed to coordinate *movement*, fails to manage *stasis*.

The Executive Diagnostic

How do you know if your organization is currently suffering from the Agile Inversion? Based on my work with transitioning teams, here are the three diagnostic questions every leader should ask their Delivery Leads:

- **The "Hidden Queue" Test:** "Show me the queue time between 'Code Complete' and 'Ready for Release' over the last six months." *Signal:* If this gap is widening while development velocity increases, you are bottlenecked on Validation.
- **The "Governance Tax" Test:** "How many story points did we burn this sprint on legal, security, or compliance reviews?" *Signal:* If the answer is "we don't track that," you have a visibility crisis.
- **The "Rework Ratio" Test:** "What percentage of AI-generated features are returned to the backlog due to 'unexpected behavior'?" *Signal:* A high ratio indicates your "Definition of Intent" is too vague.

WHAT COMES NEXT

In Chapter 2, we will conduct a detailed examination of why traditional Agile fails in the AI era. We will analyze each Scrum ceremony and artifact systematically, showing exactly where they break down and why incremental adaptation is a recipe for failure. The constraint has moved. It is time for your methodology to move with it.

KEY TAKEAWAYS

- **Constraint Inversion:** The bottleneck has shifted from code creation (expensive, slow) to validation, alignment, and governance (complex, high-risk).
- **The Confidence Gap:** Leaders often confuse the ease of prototyping with the ease of production. "Knowing enough to think you're right" is a dangerous state in AI adoption.
- **Velocity is a Vanity Metric:** In an AI world, high story-point velocity often masks a backlog of unverified, unsafe code.
- **Validation First:** Success criteria and safety boundaries must be defined *before* AI generates code to avoid wasting cycles on failed implementations.
- **The Exponential Gap:** Delaying this methodological shift creates a compounding disadvantage. Early adopters are not just moving faster; they are learning faster.

TRADITIONAL SCRUM OVERVIEW IMAGE

The Agile Scrum Sprint — Two Weeks, One Cycle, Continuous Improvement

AI SCRUM OVERVIEW IMAGE

AI Scrum – The Agile Inversion: When code is infinite, validation becomes the new bottleneck.

MIND MAP – THE AGILE INVERSION

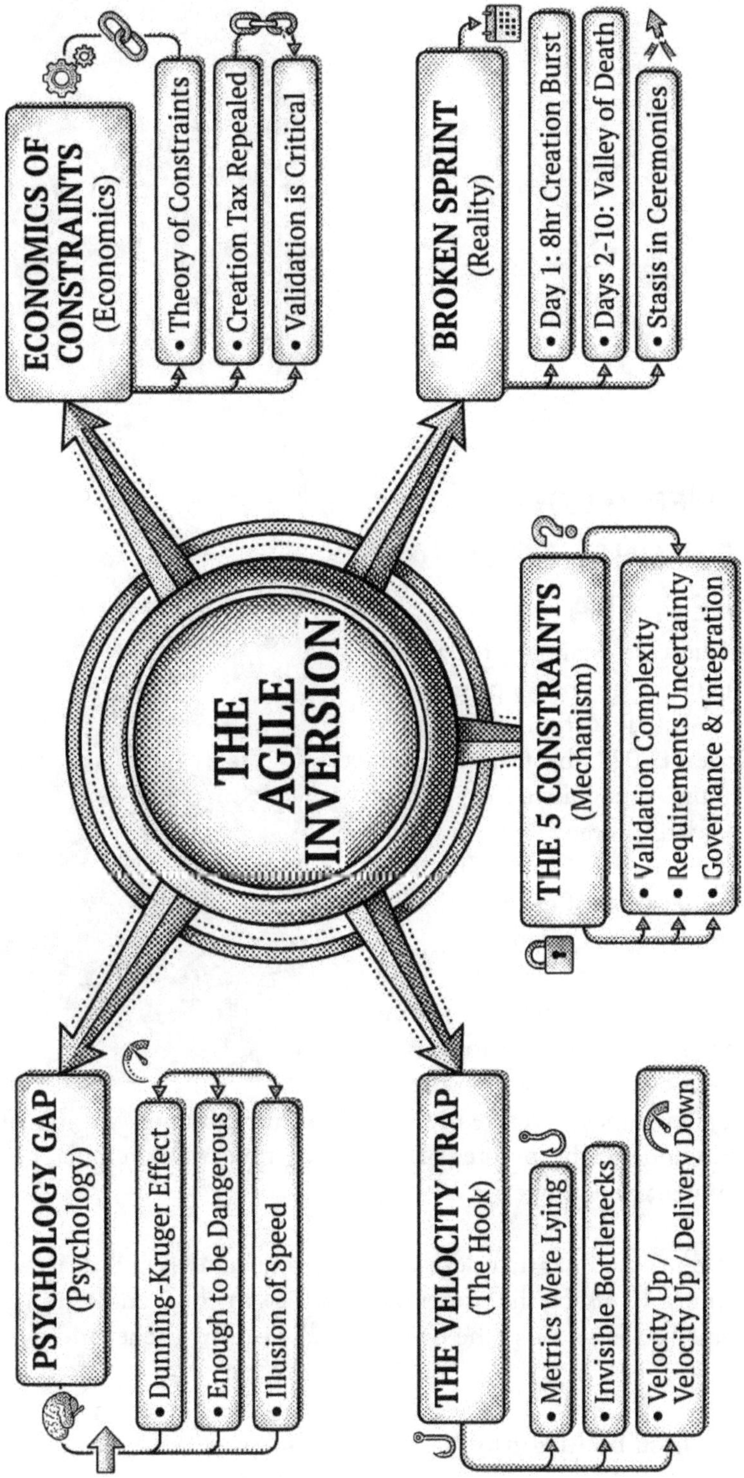

Traditional Agile ceremonies now mask competitive decline by optimizing for obsolete human constraints. Resistance to AI is psychological rather than technical, leading to "Shadow AI" and talent hemorrhage. Leadership paralysis, characterized by waiting to be "ready," signals irrelevance. In this exponential era, one year of delay equals seven years of competitive loss.

CHAPTER 2: WHY TRADITIONAL AGILE FAILS IN THE AI ERA

The request was standard for a mid-level manager: creating a presentation for a project kickoff meeting with the business. In 2022, this task represented a fixed unit of cognitive load taking approximately three to four hours of structuring, drafting, and refining. It was an afternoon of deep work requiring focus and silence.

It is 2025. I type a structured prompt into the Large Language Models (LLMs) interface. I iterate on the output a few times to refine the tone. The entire framework appears on my screen in under eight minutes of generation time, though initial prompt development took longer.

Then, I do something that defines the current state of enterprise software development: I attach the presentation file to an email and send it to my work email address, close the browser tab, clear my cache, and delete the history.

This is **digital bootlegging**.

The sensation is visceral. It feels like the 1920s all over again. Instead of smuggling alcohol, I am smuggling productivity. My organization's policies strictly prohibit the use of Generative AI. They cite intellectual property concerns and security risks. These are valid concerns that have been addressed by enterprise-grade controls they refuse to implement. Instead of governing the tool, they banned it.

Far from being a rogue employee seeking to undermine the firm, I am a certified prompt engineer. I understand the mechanics of LLMs, the nuances of context windows, and the imperative of data sanitization. Yet, to be effective at my job and deliver the value my salary demands, I must operate like a rum-runner, sneaking efficiency into an environment designed to suppress it.

The time I saved through AI augmentation created capacity I redirected into areas that desperately needed attention. I used that time to build a host of end-to-end dashboards in Azure DevOps—detailed views for both our current sprint and the overall project lifecycle. These dashboards provided a level of visibility we previously lacked. Instead of just tracking simple completion metrics, they visualized the entire flow of delivery. The data revealed critical patterns: hidden bottlenecks in our architecture and persistent rework cycles stemming from QA issues. It illuminated the feedback loops that were silently stalling progress. By presenting this end-to-end view, the dashboards gave our Product Owner the insight needed to prioritize bugs more intelligently, allowing the team to move faster by working smarter on the right things.

This practice is not limited to me. In other parts of the company, a Business Analyst is generating code snippets in minutes that are too polished for a first draft. These are tasks that would normally burn thirty minutes to an hour of focus time. We recognize the signatures in each other's work. We operate as a shadow network of high performers with a massive handicap, effectively subsidizing the organization's inertia with our own risk.

This is the hidden crisis of the AI era. It is not that the technology is failing. The issue is that the organizational structures designed to manage delivery are actively rejecting the new constraint model. The efficiency gains are real, but they are being forced underground, creating a "Shadow AI" economy that leadership can neither see nor measure.

WHAT THIS CHAPTER COVERS

By the end of this chapter, you will understand:

- **The Psychology of Resistance:** Why fear, political dysfunction, and the "Sunk Cost of Expertise" paralyze leadership more effectively than any technical hurdle.
- **The Ceremony Illusion:** How traditional Scrum rituals like Standups, Planning, and Reviews mask competitive decline by optimizing for human-only labor constraints.
- **The Cost of "Waiting to be Ready":** The economic reality that one year of AI delay equals seven years of competitive capability loss.
- **The Canary in the Coal Mine:** Why the inability to define "readiness" is the odorless, invisible signal that an organization is suffocating.
- **The Talent Hemorrhage:** Why your best engineers are not leaving for more money, but to escape cognitive atrophy.
- **The Innovation Tax:** Calculating the opportunity cost of maintaining manual workflows in an exponential environment.

KEY TERMS AND CONCEPTS

To diagnose why Agile fails in this new environment, we must recalibrate the lexicon of software delivery. We are moving from a model of *Resource Scarcity* (developer hours) to a model of *Adaptation Scarcity* (organizational agility).

- **Shadow AI vs. Shadow IT**
    - *Old Definition:* **Shadow IT** is unauthorized software or hardware deployed without IT approval, typically to bypass bureaucratic friction (e.g., a marketing team buying a SaaS tool with a credit card).
    - *New Definition:* **Shadow AI** is a survival mechanism where high-performing employees leverage AI tools clandestinely to maintain competitive personal velocity within a stagnating organization. It is a leading indicator that governance has failed to keep pace with capability.
- **Opportunity Cost (The Innovation Ledger)**
    - *Old Definition:* The loss of potential gain from other alternatives when one alternative is chosen.

    - *New Definition:* The compounding competitive deficit incurred by maintaining manual workflows. In an exponential growth environment, the cost is not linear. It is the difference between an arithmetic progression (manual improvement) and a geometric progression (AI-augmented evolution).

- **Risk Aversion (Asymmetric Risk Perception)**
    - *Old Definition:* Prudent management to avoid uncertain negative outcomes.
    - *New Definition:* The cognitive bias where the hypothetical risk of AI error is weighted heavier than the guaranteed certainty of market obsolescence. It treats the status quo as a "safe harbor" when it is actually a burning platform.

- **Organizational Inertia (Process Calcification)**
    - *Old Definition:* The tendency of a mature organization to continue on its current trajectory.
    - *New Definition:* The active antibody response of middle management to suppress innovation that threatens established hierarchies or expertise domains. It increases with organization size and the historical success of existing methodologies.

THE ORGANIZATIONAL PSYCHOLOGY OF RESISTANCE

The barrier to AI adoption is rarely technological. The tools exist. The barrier is almost exclusively psychological and political. When I stepped away from an environment that refused to embrace these tools, I had time to deconstruct the pathology of their resistance. It wasn't that they didn't understand the *value*; they didn't understand the *threat.*

Fear and Technical Blindness

Leadership often operates on the premise that if current practices are profitable, fundamental change represents unnecessary risk. This logic holds in a linear environment. In an exponential environment, it is fatal.

Many executives lack the technical sophistication to distinguish between a "hype cycle" and a "paradigm shift." They look at their velocity charts, see stability, but fail to measure Outcome velocity, and assume health. They do not realize that these metrics are measuring the efficiency of a process that is rapidly becoming obsolete.

The most dangerous archetype is the executive who consumes a magazine article or podcast episode and concludes they understand AI sufficiently to make strategic decisions. This exemplifies what astrophysicist Neil deGrasse Tyson observed: "One of the great challenges in life is knowing enough to think you're right, but not enough to know you're wrong."

AI represents profound technological complexity combined with a mindset shift. Implementing it organizationally adds layers of process, governance, and cultural change. Weekend consumption of popular media does not prepare anyone to lead that transformation. It leads to "Management by Headline," where strategy shifts based on the latest fear-mongering article rather than technical reality.

Political Dysfunction and Innovation Suppression

Organizational politics amplify technological paralysis. Authoritarian leadership styles characterized by rigid thinking systematically kill innovation. While budget constraints and the necessity of organizational alignment are clear, executives who refuse professional challenge from domain experts fundamentally misunderstand their role.

Steve Jobs articulated this principle clearly:

> *"It doesn't make sense to hire smart people and tell them what to do; we hire smart people so they can tell us what to do."*

Effective leaders do not need comprehensive answers across all domains. They need the intellectual humility to listen to people with specialized expertise and make informed decisions based on collective intelligence rather than hierarchical authority. Yet in the face of AI, many leaders revert to authoritarian control. They fear losing their status as the smartest person in the room, so they ban the tools they cannot control.

The Sunk Cost of Expertise

Resistance also comes from the "Frozen Middle," which is the layer of middle management and senior practitioners who have spent their careers mastering a specific way of working.

Consider a Scrum Master with 15 years of experience. Their value proposition is built on facilitating human conversations, managing manual estimation poker, and resolving interpersonal conflict. When AI enters the room, suggesting code changes or estimating effort in seconds, that Scrum Master feels an existential threat.

This is the **Sunk Cost of Expertise**. When your status depends on a specific constraint, such as the difficulty of writing code, you will subconsciously fight to preserve that constraint. I have seen Senior Architects nitpick AI-generated code far more aggressively than human-generated code. This was not because the AI code was worse, but because the *speed* of its generation devalued their own hard-won speed.

Secondhand Fear and Risk Aversion

Perhaps most insidious is the compounding risk aversion that emerges from others' failures. An executive hears about a data leak at another company, such as the Samsung incident where source code was pasted into ChatGPT, and concludes the technology isn't ready.

This creates decision-making based on incomplete information. It treats any failure as proof that "it's not ready yet." However, that represents a fundamental misunderstanding of how innovation works. Every transformative technology required accepting intelligent risk. Organizations that wait for perfect safety never lead. They follow years later, after someone else has absorbed the learning costs and established insurmountable advantages.

Consider the space program. We didn't wait until rockets were proven safe. We accepted that failures would occur, learned from each one, and continued advancing. John F. Kennedy did not say, "Let's wait until space travel is safe." He declared, "We choose to go to the moon... not because they are easy, but because they are hard".

AI follows the same pattern. Organizations that lead industries adopt an experimental mindset. They fail fast, learn from setbacks, and pivot. Rather than waiting for perfect safety, they manage risk intelligently while maintaining forward momentum.

HOW TRADITIONAL CEREMONIES MASK COMPETITIVE DECLINE

The most dangerous aspect of the Agile Inversion is that it feels comfortable. The ceremonies continue. The velocity charts look stable, but Outcome velocity is declining. The burn-down is green. But these metrics are tracking the efficiency of a propeller plane while competitors are switching to jets. The constraint model has changed, but the measurement system has not.

Sprint Planning Creates False Confidence

In 2024, I sat through Sprint Planning sessions that felt remarkably productive. We estimated stories. We negotiated scope. We committed to a goal.

But nobody asked the only question that mattered: *How much of this work is commodity execution that AI could handle if we were permitted to use it?*

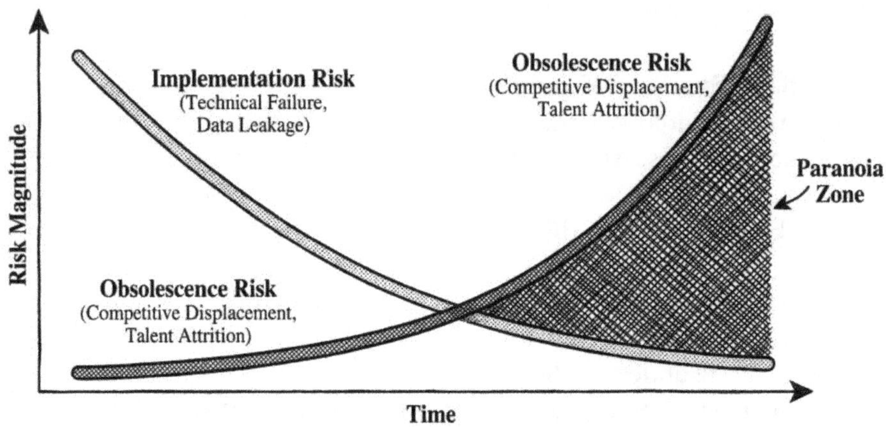

Figure 2.1: The Risk Inversion Paradox. *A strategic risk matrix plotting 'Risk Magnitude' against 'Time.' The descending curve tracks Implementation Risk, which declines as enterprise governance tools mature. The ascending exponential curve tracks Obsolescence Risk. The shaded region, labeled the 'Paranoia Zone,' identifies the critical window where risk-averse leadership believes they are 'playing it safe,' unaware that the mathematical cost of inaction has already surpassed the cost of adoption.*

Let's look at the mechanics of this failure. A User Story (what will become a Context Specification in AI Scrum) comes up: "Create a login form with validation." The developers play Planning Poker. They debate whether it is a 3 or a 5. They settle on 5 because of the "complexity of the regex for email validation."

In an AI-augmented world, that regex is generated in 3 seconds. The form boilerplate is generated in 10 seconds. The complexity isn't in the *creation.* It is in ensuring the AI didn't hallucinate a security vulnerability. But the team is still estimating based on *typing time.*

Leadership sees the team hit 80 points and thinks, "We are predictable." In reality, they are predictably slow. They are celebrating the successful execution of manual labor that should have been automated. The ceremony produces familiar artifacts like capacity allocations and velocity projections that mask a fundamental misalignment with competitive reality.

Daily Standups Miss Systemic Impediments

The Daily Scrum is designed to surface blockers. "I am waiting on a code review" or "I am blocked by the database migration" are tactical blockers.

The real blocker, the **Systemic Impediment**, never surfaces: *I am blocked because company policy prevents me from using a tool that would finish this task in ten minutes.*

Imagine a standup where a developer says, "I'm writing the unit tests for the payment module. It will take me all day." The Scrum Master nods and asks if there are any blockers.

The blocker is the *method itself.* The developer *could* generate those tests in 15 minutes using a tool like Codium or GitHub Copilot, then spend the rest of the day on integration. But because the tool is banned, the "blocker" is invisible. The Standup optimizes for tactical movement within the cage, but it never questions the cage itself. While team members clandestinely use AI to unblock themselves, the official record shows a team struggling with manual friction.

Retrospectives Optimize for Yesterday's Constraints

Retrospectives are the engine of Agile improvement. Teams ask, "How can we test faster?" or "Should we adjust our story point calibration?"

These are questions of **Local Optimization**. They refine the existing process. They are the equivalent of asking how to shovel coal faster into a steam engine, while ignoring the electric train on the parallel track.

In the AI era, teams must ask **Global Optimization** questions: *What percentage of this sprint's backlog could have been automated? Why are we writing boilerplate code manually?*

When a team uses a Retrospective to optimize manual coding efficiency in 2025, they are effectively polishing the brass on the Titanic. They are getting better at the wrong thing. Traditional retrospective formats don't naturally surface this question because they're designed for continuous improvement within the existing paradigm, not evaluation of whether the paradigm itself has shifted.

Sprint Reviews Hide Opportunity Cost

Stakeholders love a good demo. Seeing working software is intoxicating. Everyone feels good about delivery.

But the Sprint Review fails to show the **Counterfactual**: *What could we have delivered if we weren't manually typing syntax?*

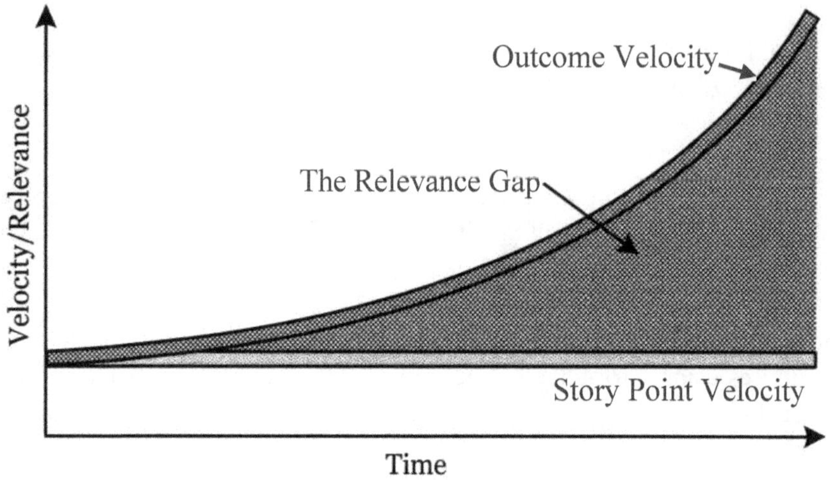

Figure 2.2: The Metric Decoupling. *A line graph showing two diverging trends. The 'Story Point Velocity' line is flat and stable (what the Scrum Master sees). The 'Outcome velocity' line (what the market requires) shoots upward exponentially. The gap between them is shaded and labeled 'The Relevance Gap.' This visualizes how internal Agile metrics can report 'Health' while the business is actually in 'Decline.'*

The Review showcases *Delivery*, but it hides *Opportunity Cost.* Stakeholders see absolute progress without understanding relative competitive position. They do not see that the team's velocity is plateauing while the industry standard for developer productivity is accelerating by 30-50% annually due to AI tools like GitHub Copilot and Cursor. The ceremony succeeds at showing what shipped, while failing to reveal what *could have* shipped with different constraints.

THE REAL COST: TALENT HEMORRHAGE

The first thing that breaks in a resistant organization is not the code; it is the roster. Talent hemorrhage is the most immediate and damaging consequence of the Agile Inversion. This isn't theoretical future risk. It's happening now.

High-performing engineers, Product Owners, and Scrum Masters are acutely aware of market trends. They know that AI proficiency is the defining skill of the next decade. When an organization forces them to work manually, they do not just see it as an inconvenience. They see it as **Career Sabotage**.

- **The Architect's Dilemma:** A Senior Developer knows that if they spend the next two years writing manual CRUD (Create, Read, Update, Delete) operations while their peers learn to orchestrate AI agents, their resume will be obsolete by 2027. They leave not for higher pay, but for "Skill Security."
- **The Scrum Master's Fatigue:** Good Scrum Masters want to remove waste. When they see the massive waste of manual coding but are forbidden from addressing it, they suffer from moral injury. They feel complicit in the inefficiency.

You are not losing them to competitors who pay more. You are losing them to competitors who let them work in the future. The employees who stay are often those comfortable with the status quo, the "ticket takers" who are happy to hide in the slow lane. These are exactly the people least equipped to drive the transformation you will eventually need.

THE ACCELERATING TIMELINE: DOG YEARS

It is commonly known that dogs age roughly 7 times faster than humans.

Time in technology has always moved fast, but AI operates on a compressed timeline that makes the "Internet Era" look glacial.

Consider the evolution of aviation. It took roughly thirty years to go from the first scheduled transcontinental flights in the late 1920s to the commercial jet age with the Boeing 707 in 1958. It took roughly another thirty years to reach the modern, twin-engine efficiency of the Boeing 777 in the mid-1990s.

Think about when the Wright Flyer soared over the sand at Kill Devil Hills for the first powered flight by man in December of 1903 only to be eclipsed by man walking on the moon in July of 1969. The spread was only sixty-six years.

AI is compressing these cycles. One calendar year of AI evolution, encompassing the jump from GPT-3.5 to GPT-4 to Agentic architectures, represents roughly seven years of traditional software evolution.

This concept of "Dog Years" is critical for strategic planning. If your organization decides to "wait a year to see how things settle," you are not losing twelve months. You are losing an era. By the time you decide to adopt, the baseline for entry will have shifted from "using Chatbots" to "managing Agentic Swarms." You will not just be behind; you will be obsolete. While not a precise empirical measure, the principle holds: in technology, competitive decay compounds rapidly.

Organizations that delayed internet adoption because it was "too expensive" or "too complicated" either adapted eventually or ceased to exist as independent entities. AI follows the same pattern, but faster.

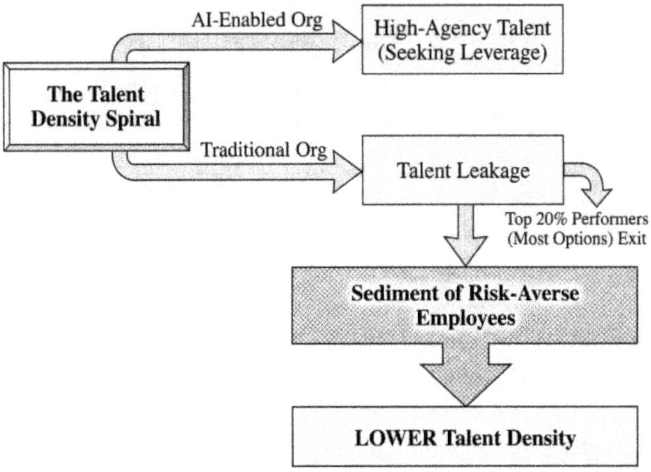

Figure 2.3: The Talent Density Spiral. *A flow diagram showing two paths. The 'AI-Enabled Org' attracts 'High-Agency Talent' seeking leverage. The 'Traditional Org' suffers 'Talent Leakage,' where the top 20% of performers (who have the most options) exit first, leaving behind a sediment of risk-averse employees, lower the overall 'Talent Density' of the firm.*

FROM THE TRENCHES: THE "READY" TRAP AND THE CANARY

Context: In summer of 2025, I interviewed for a leadership role at a mid-sized manufacturing firm. They were profitable, stable, and rather old-school.

The process was grueling; seven rounds of interviews and a 500-question psychometric assessment managed by a third party. They were thorough. They were careful. They prided themselves on "getting it right." Throughout the process, the C-suite kept repeating a specific mantra: *"We just want to be ready for AI when the time comes."*

I looked at their infrastructure. Their technology stack was 15–20 years old. Manual processes permeated the entire organization. They had no server redundancy, relying on a single on-site location with no disaster recovery plan.

"The time is already here," I told them. I outlined a roadmap: automated data ingestion, basic predictive maintenance models, and modernizing their CI/CD pipeline. These were low-risk, high-reward moves that could be implemented in weeks. I attempted to break the challenge into manageable pieces, suggesting we start small, prove value, then expand systematically.

They balked. They couldn't articulate *why*, but they couldn't pull the trigger. They wanted the *result* of modernization without the *risk* of change. It was like watching a starving man unable to choose a restaurant. They kept saying, "We'll be ready when the time comes, but "it sure was fun to dream about what could be" unaware that the phrase itself was a death sentence.

The Canary in the Coal Mine

In the early 20th century, coal miners faced a deadly, invisible threat: carbon monoxide. It was odorless, colorless, and lethal. To protect themselves, they carried caged canaries into the tunnels. Canaries have faster metabolisms than humans; if the bird stopped singing or collapsed, the miners knew the air was toxic and they had minutes to evacuate before they, too, would suffocate.

In the corporate world, indecision is carbon monoxide.

When leadership acknowledges that AI is important but cannot define what "ready" looks like, that is your canary stopping its song. It is the odorless, invisible signal that the organization's decision-making metabolism has shut down.

The manufacturing firm didn't have a "tech problem"; they had a toxicity problem. The air was filled with the carbon monoxide of fear and inertia. By the time the financial results (the "human" collapse) show the damage, it will be too late to evacuate. The canary had already died, but they were still digging, convinced they were safe because they couldn't smell the danger yet.

That is your canary. When leadership acknowledges AI matters but cannot define what they would do with it, you are watching an organization slide into irrelevance in real-time.

THE EXECUTIVE DIAGNOSTIC

How do you know if your organization is paralyzed? Warning signs often masquerade as prudence. If you hear these phrases in your boardroom, your organization is in the danger zone.

1. The "Fad" Defense

- **Phrase:** *"We don't need AI. It's just a hype cycle like the Metaverse/Crypto."*
- **Diagnosis:** Profound technical blindness. Generative AI is not a new asset class; it is a fundamental shift in how information is processed. This is not buying virtual land; this is the invention of the steam engine.

2. The "Uniqueness" Fallacy

- **Phrase:** *"Our domain is too complex/regulated/special for AI."*
- **Diagnosis:** This is almost always false. Highly regulated industries (Legal, Healthcare, Finance) are actually the *primary* beneficiaries of AI because they are text-heavy and rule-based. AI excels at pattern matching within strict rulesets—exactly what regulated industries need.

3. The "Tomorrow" Trap

- **Phrase:** *"We're too busy with the migration/ERP upgrade right now. We'll look at AI next year."*
- **Diagnosis:** Prioritization failure. Implementing AI is not an "add-on" to your ERP upgrade; it is likely the only way to finish that upgrade on time and on budget. Treating it as a Phase 2 luxury ensures Phase 1 never finishes.

4. The "Safety" Shield

- **Phrase:** *"We will be ready when the time comes."*
- **Diagnosis:** Paralysis disguised as risk management. The organization cannot define "ready" and is using the concept to delay decision-making. As we learned from the Canary, this is the deadliest signal of all.

WHAT ORGANIZATIONS SHOULD DO INSTEAD

The solution is not to abandon Agile, but to evolve it. The constraint has moved, so the methodology must follow. Effective adaptation requires systematic cultural shifts, not ceremonial additions.

Start With Backlog Grooming

In backlog grooming sessions, identify which user stories (future context specifications), tasks, and bugs could have significant portions of work completed through AI with proper human supervision. This isn't about AI replacing humans. It's about AI accelerating humans so they can take on more valuable work. Ask explicitly: "Is this a 'Human Creation' task or a 'Human Validation' task?"

Plan AI Interaction

In Sprint Planning, walk through each AI-candidate story and sketch what the initial prompt might look like. Treat the AI prompt as a context specification, a technical specification that defines both what to do and what not to do. Planning the AI interaction surfaces edge cases and validation requirements that belong in the context specification upfront rather than discovering them mid-sprint.

While AI generation itself is rapid, effective prompts often require iteration. A task that takes 'minutes' includes 5-10 minutes of prompt refinement, not just the generation time. Teams improve prompt efficiency over time, but initial sprints require investment in learning effective AI collaboration patterns.

Update Retrospectives

Make this a standard question: *"What could we have handed off to AI but didn't this sprint?"* Force the team to confront whether they're adapting or merely comfortable with current approaches. Work being completed by Sprint 3 should be noticeably more polished than Sprint 1.

Why Scrum Masters Bear Special Responsibility

If you're a Scrum Master thinking "my team's velocity is stable..., why should I disrupt what's working?", consider what servant leadership actually means.

If AI can help your team increase velocity from 80 points to 100, then 120 as they gain experience, that's not disruption. That's fulfilling your role. Your job is to remove impediments. Right now, the biggest impediment facing most Scrum teams is organizational unwillingness to embrace AI-augmented practices. If you're not addressing that impediment, you're not doing your job.

The Learning Curve Reality

Before we discuss what to do instead, you must understand the transition timeline. Teams adopting AI Scrum experience a predictable four-phase progression:

Phase 1: The Learning Sprint (Sprints 1-4) - SLOWER than traditional Scrum. Your team is learning prompt engineering, adapting to new ceremonies, and making mistakes. Velocity drops 20-40% as cognitive load spikes. This is normal and temporary.

Phase 2: The Recovery (Sprints 5-10) - APPROACHING PARITY. The team develops AI collaboration patterns, validation workflows stabilize, and velocity returns to baseline. You're not faster yet, but the foundation is solid.

Phase 3: The Acceleration (Sprints 11-18) - 2X FASTER. With established patterns and reduced validation friction, the team delivers outcomes at roughly twice the traditional pace. AI generates; humans validate efficiently.

Phase 4: Sustained Excellence (Sprint 19+) - 3-5X FASTER. Full maturity. The team has internalized AI-first thinking, optimized validation processes, and built robust governance. Sustainable high performance becomes the new normal.

This J-curve is real. Executives who expect huge immediate gains in Sprint 1 will panic and abort the transition. Those who plan for 4-6 months of investment before seeing returns will succeed. The constraint shift is permanent - the question is whether you'll make the transition strategically or chaotically.

WHAT COMES NEXT

We have diagnosed the disease. The traditional Agile ceremonies are failing to address the new constraints of validation and governance. The resistance is psychological, and the cost is exponential.

In **Chapter 3**, we stop admiring the problem and start building the solution. We will deconstruct the Core Principles of **AI Scrum**. We will move from "Agile Theater" to a rigorous framework designed for the era of infinite code and finite trust. We will define the **Validation-First Mindset** and the shift from deterministic to probabilistic thinking.

Propeller planes didn't disappear. They just became irrelevant for the applications that mattered most. The same will happen to organizations running traditional Agile while competitors embrace AI Scrum practices. The only question is which side of that divide you'll be on.

KEY TAKEAWAYS

- **Psychology Over Technology:** The primary barriers to AI adoption are fear, political dysfunction, and secondhand risk aversion, not the technology itself.

- **The Ceremony Trap:** Traditional Agile ceremonies mask competitive decline. Planning creates false confidence, Standups miss systemic impediments, and Reviews hide opportunity cost.
- **The "Ready" Fallacy:** "We'll be ready when the time comes" is a signal of paralysis. If leadership cannot define "ready," the organization is sliding into irrelevance.
- **Talent Hemorrhage:** High performers leave first. They view manual constraints as a threat to their career viability.
- **Dog Years:** One year of delay in the AI era equals seven years of competitive disadvantage. The adaptation timeline is much shorter than most leaders realize.
- **Human-in-the-Loop:** The goal is not automation, but augmentation. Humans must shift from *doing* the work to *supervising* and *validating* the work.

MIND MAP – WHY TRADITIONAL AGILE FAILS IN THE AI ERA

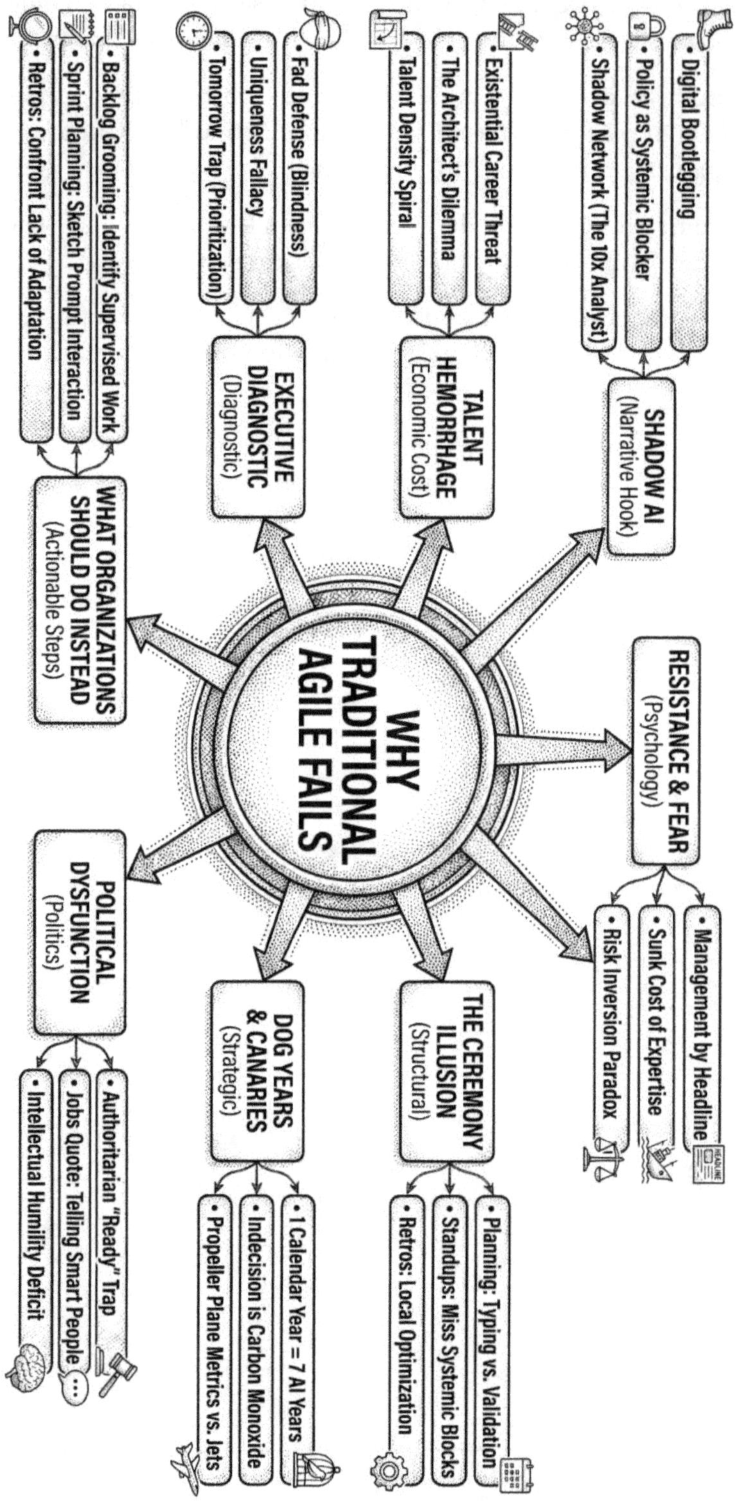

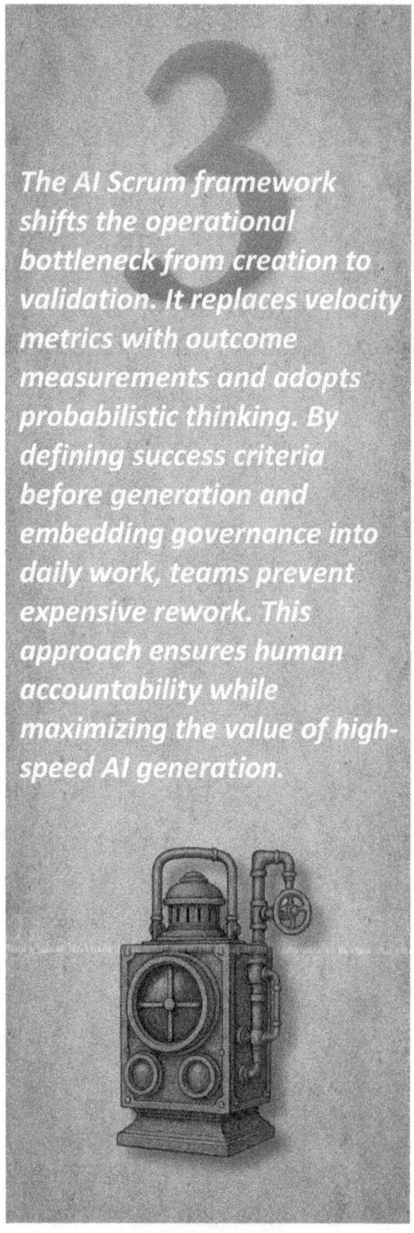

CHAPTER 3: AI SCRUM PRINCIPLES

I was sitting in my room one afternoon working through my prompt engineering certification from Blockchain Council. My imagination was running wild with thoughts of where AI might be used in Agile, even though I had no idea yet how to actually accomplish some of the things. What if AI could somehow look at a backlog of 500-plus items, spanning from current work through items that were arguably five to six years old, read through everything, and then group and reprioritize the work? That would be useful. What if a business analyst could leverage AI to create better user stories and tasks? What if AI could look at a clean, prioritized backlog and come up with an initial right-sized sprint based on story points and team availability?

The possibilities felt endless. I found myself thinking back to the 1983 movie *WarGames*, with David played by Matthew Broderick and Ally Sheedy as Jennifer up in David's room at the computer. David had been searching different computers calling them one by one with his computer trying to reach a software company to play a game before it was released.

However, he came across something else far more exciting. That's what this felt like. I was searching for something practical, and I kept finding capabilities I hadn't even imagined.

But here's the lesson that emerged quickly: technique matters enormously. Without proper prompting skills, the outputs were rubbish. Those early attempts produced error after error, or at the least not what I was anticipating. Proper training changed everything and with continued learning, quality responses became consistent instead of random. The power these tools offer is remarkable, but only if you know how to access it properly.

One of the things I had to keep in mind was that as I iterated through responses, I had to validate what I was receiving. If I didn't, I would likely compound errors the further I went. When I was a programmer earlier in my career, one of the practices I followed while creating code was not to create something from end to end. I'd write short parts, run it, then validate. This way I could ensure what I was doing made sense and was accurate. Then I would add more code, run it, then validate. I did this through the entire process until the work was complete. This reduced, if not eliminated, errors and bugs, and I could be confident in what I was delivering.

That programming habit saved me from wasting time on flawed AI outputs. And it crystallized something fundamental: the constraint that had shaped Agile for two decades was rapidly evaporating. Writing code was no longer the bottleneck. Making sure the code actually worked the way it needed to work was the new constraint. Everything we'd built around the old constraint no longer made sense.

What you're about to learn represents creative destruction in practice. Economist Joseph Schumpeter described how innovation systematically replaces established practices with more efficient ones. That's exactly what's happening here. Story point velocity metrics that guided teams for 20 years give way to outcome measurements. Deterministic thinking yields to probabilistic approaches. End-of-sprint governance transforms into continuous validation. Each principle in this chapter destroys something that worked well while creating something better suited to AI-native constraints. The discomfort this creates is not a flaw. It's evidence of genuine transformation.

This chapter establishes the foundational principles of AI Scrum. Not cosmetic adjustments to existing practices, but systematic reconception for

AI-native development. These six principles inform every aspect of the framework, from sprint structure to role definitions to artifact design. Teams that embrace these principles unlock AI's potential. Those that treat them as optional suggestions struggle indefinitely.

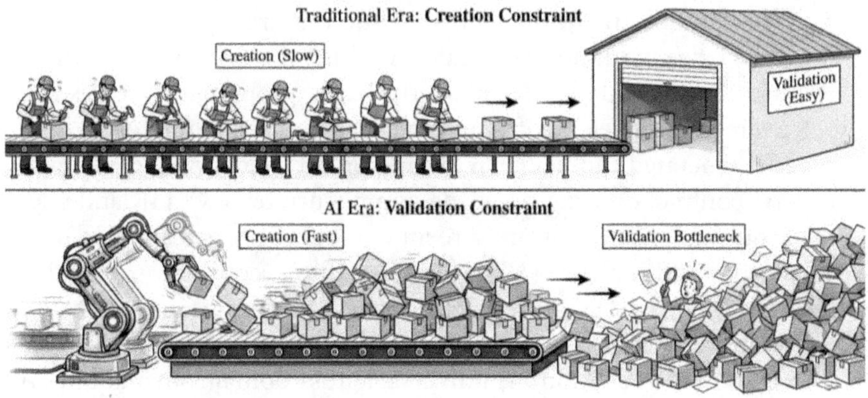

Figure 3.1: The Production Line Inversion. *A comparative illustration showing the shift from a creation constraint (Traditional Era) to a validation constraint (AI Era), where rapid AI output overwhelms human validation capacity.*

WHAT THIS CHAPTER COVERS

By the end of this chapter, you will understand:

- **The six core principles** that distinguish AI Scrum from traditional Agile, and why each one matters.
- **How validation-first development** prevents the expensive waste that AI's speed can amplify.
- **Why outcome-oriented metrics** replace output measurements when implementation accelerates 2-5x after the learning curve.
- **The difference between probabilistic and deterministic thinking**, and why software development just became probabilistic.
- **How integrated governance** transforms from bottleneck to enabler when embedded in sprint workflows.
- **The resistance you will face** when implementing these principles, and practical approaches for addressing skepticism.
- **What continuous learning loops** look like when development cycles compress from weeks to hours.

Every principle emerged from watching teams struggle with AI's capabilities. Some learned these lessons easily. Others learned them painfully. You can skip the painful part.

KEY TERMS AND CONCEPTS

- **Cross-Domain Applicability:** AI Scrum principles apply to any knowledge work domain where AI shifts the bottleneck from creation to validation. While software development provides the clearest examples, the framework adapts to marketing operations (AI-generated campaigns requiring brand validation), legal practice (AI contract drafting requiring compliance review), HR and talent acquisition (AI candidate screening requiring bias testing), data science and analytics (AI model generation requiring statistical validation), and creative operations (AI content generation requiring quality and rights verification). The constraint shift - from creation to validation - is universal across domains that adopt AI for generative tasks.
- **Validation criteria:** Specific, measurable standards established before implementation begins, defining what success looks like across functional, performance, security, compliance, and business dimensions.
- **Outcome metrics:** Measurements of business value delivered and customer problems solved, replacing activity-based output metrics like story point velocity with Outcome velocity and story points.
- **Probabilistic output:** AI-generated solutions that may vary across multiple generations given the same inputs, requiring evaluation of multiple candidates rather than assuming a single correct answer.
- **Deterministic output:** Traditional code that produces identical results given identical inputs, enabling straightforward test-once validation approaches.
- **Creative destruction:** Economist Joseph Schumpeter's term describing how innovation systematically replaces established practices and technologies with more efficient alternatives, driving progress through cycles of obsolescence and renewal. In AI Scrum, this means abandoning methodologies optimized for development-constrained environments and adopting approaches designed for validation-constrained contexts

- **Compound errors:** Mistakes that multiply through iteration when validation does not occur between steps, creating exponentially increasing rework costs.
- **Integrated governance:** Embedding compliance review, security validation, and architectural oversight directly into daily sprint work, enabling continuous validation rather than retrospective approval.
- **Human-in-the-loop:** Development approach where AI generates implementations while humans maintain supervisory control, validating outputs and making final decisions at each stage.
- **Acceptance threshold:** The "good enough" standard where a solution satisfies requirements sufficiently to proceed, acknowledging that pursuing perfection often costs more than it delivers in value.
- **Red-teaming:** Security testing approach where specialists actively attempt to break systems through adversarial inputs, identifying vulnerabilities that standard testing might miss.
- **Validation Canvas:** Framework artifact defining detailed success criteria across functional, performance, security, compliance, and business dimensions before AI generates implementations.
- **Governance Checkpoint:** Mid-sprint ceremony where security engineers, architects, and compliance specialists review validated designs before teams invest further refinement effort
- **Prompt library:** Curated collection of successful AI prompts and interaction patterns that teams build over time through systematic learning.
- **AI telemetry:** Objective data about human-AI collaboration including code generation statistics, acceptance rates, validation results, and collaboration patterns used in retrospectives.
- **Cowboy Agile:** A dysfunctional delivery pattern where teams use the flexibility of Agile as an excuse to bypass essential discipline. Characterized by the systematic omission of feedback loops (specifically Retrospectives) and the truncation of planning rigor under the guise of "efficiency." While often justified by the claim that "Scrum is just a framework," this approach prioritizes immediate output over sustainable velocity. In an AI context, Cowboy Agile is catastrophic, as it accelerates the creation of AI

Debt - technical debt without the safety mechanisms required to detect it.

THE SIX PRINCIPLES OF AI SCRUM

Chapter 2 showed you why traditional Agile breaks down when AI eliminates the development constraint. Now we are going to establish what replaces it. These six principles address the real problems that emerge when AI's speed meets traditional practices. They are informed by years of Agile experience and the patterns that become visible once you understand how the constraints have shifted. More importantly, they are designed to help you avoid the mistakes that are easy to make when you do not see the full picture. While software development examples dominate this book due to AI's maturity in that domain, these principles apply universally to any knowledge work where AI generates and humans validate.

Figure 3.2: The Six-Cylinder Engine. *A cutaway technical illustration of a powerful engine block labeled 'The AI Scrum Engine.' The six pistons, each labeled with a core principle, drive a central crankshaft symbolizing that AI Scrum is not a static set of rules, but a dynamic, high-performance machine designed to convert raw AI potential into useable business momentum.*

Principle 1: Validation-First Mindset

Whether people like it or not, experienced or otherwise, AI is not infallible. Like humans, it is prone to mistakes called hallucinations. While things are getting better, there are still issues. There is a fair amount of data that LLMs ingest in training that is solid and has very little chance of ever changing. Blue is blue, it is not red. Two plus three equals five, not cat.

However, there is also information that LLMs have been trained on that might not be so solid or even proven, that might be assumed to be fact but is not. So, when I prompt an LLM, how much of the solid information versus not so solid or opinion is mixed together to provide me a result? If I just keep iterating through prompts without checking, what am I really getting?

Does a chef make an entire meal without tasting along the way to see if there is too much salt or not enough? No, of course not. Their reputation is on the line. They want the result to be a great experience for those that are going to consume the meal. Same thing when I iterate through a prompt for information or ideas. I want it to be great, and I want those that consume what I build to trust it and find it of value.

Remember the sprint board image from Chapter 1? Perfect at first glance, misspellings everywhere on close inspection. That is AI validation in a nutshell - outputs can look flawless while containing fundamental errors that only careful examination catches.

The Cost of Late Validation

Here is what happens when validation comes after implementation in AI Scrum development. A team spends three days refining an AI-generated recommendation engine, polishing its algorithms and tuning its performance. On Day 8, during scheduled testing, they discover it violates data privacy regulations. All that refinement effort? Wasted. The feature cannot deploy.

The speed that makes AI valuable becomes a liability when you validate late. The faster you can build the wrong thing, the more resources you waste. Traditional Scrum could get away with build-then-validate sequencing because building consumed most of sprint capacity. If validation revealed problems, at least you would invest weeks understanding the domain deeply. With AI, you can generate a fundamentally flawed implementation in hours, polish it for days, and discover only at the end that you built something that cannot ship.

Validation-first inverts this sequence completely. Teams define success metrics, safety boundaries, governance requirements; what we call a context specification, and integration constraints upfront, before AI produces a single line of code. McKinsey's research on AI implementation patterns supports this approach: establishing clear validation criteria before generation consistently produces better outcomes than post-development testing.

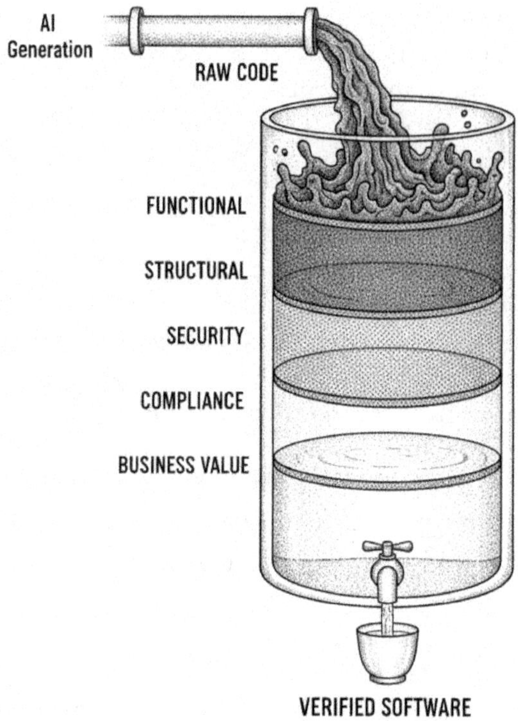

Figure 3.3: The Industrial Filtration System. *A diagram of a vertical liquid filtration system. 'AI Generation' pours chaotic, raw code into the top, which must pass through five distinct mesh layers (Functional, Structural, Security, Compliance, Business Value) before emerging as 'Verified Software.' This illustrates that safety comes from multi-layered defense, not just a single test at the end.*

The Trap of "Cowboy Agile"

We must distinguish clearly between "Agile speed" and "Cowboy Agile." Cowboy Agile is a dysfunctional delivery pattern where teams use the flexibility of the framework as an excuse to bypass essential discipline. It is characterized by the systematic omission of feedback loops—specifically Retrospectives—and the truncation of planning rigor under the guise of "efficiency."

In the manual coding era, Cowboy Agile was painful. It resulted in buggy code and AI Debt that slowed teams down over months. In the AI era, Cowboy Agile is catastrophic. When a team uses an LLM to generate code at 2-5x speed after learning while skipping validation to "save time," they are not just creating a mess; they are accelerating the creation of AI Debt beyond their ability to ever repay it. They prioritize immediate output over sustainable velocity, creating a high-speed collision between unverified code and production reality.

Test Scenarios Before Writing Code

The practical application looks like this. Before AI generates anything, teams articulate what "working correctly" means in measurable terms.

- **Performance benchmarks:** What response times are acceptable under what load conditions?
- **Security standards:** What threats must the system resist? What data protection requirements apply? These answers form the context specification.
- **Compliance obligations:** What regulatory frameworks govern this domain? What documentation must we maintain?
- **Integration requirements** (part of the Context Specification)**:** What systems must this work with? What failure modes must we handle gracefully?
- **Business success criteria:** What metrics will improve if this solution works? How will we measure that improvement?

These questions produce what I call a Validation Canvas, though you can call it whatever makes sense for your organization. The key is establishing explicit, measurable criteria before generation begins. This canvas guides everything that follows. AI generates implementations designed to satisfy these criteria. Testing validates against these standards defined in the context specification. Governance reviews evaluate compliance with these requirements.

The difference is dramatic. Boston Consulting Group's 2024 research on AI adoption found teams implementing validation-first approaches report reductions in rework ranging from 40 to 60%. They catch problems when they are cheap to fix, during the design phase, rather than discovering them after substantial refinement investment when correction costs far more.

What Validation Actually Covers

When I talk about validation in AI-augmented development, I'm talking about something far more comprehensive than traditional testing.

- **Functional validation** ensures prototypes meet the requirements defined in the context specification. Can it do what we said it should do?
- **Performance validation** measures against benchmark criteria. Does it do it fast enough, efficiently enough, at the scale we need?
- **Security validation** identifies vulnerabilities through both automated scanning and human red-teaming. Can someone break it? Can they make it behave in ways we do not want?

- **Compliance validation** confirms regulatory alignment. Does this satisfy HIPAA requirements if we are handling health data? SOC 2 requirements if we are a service provider? GDPR if we have European customers?
- **Integration validation** verifies compatibility with existing systems. Will this work with our actual infrastructure, not just in an isolated test environment?
- **Stakeholder validation** gathers feedback on whether solutions address actual needs. Does this solve the problem our users actually have, or the problem we assumed they have?

I've seen teams skip one of these validation dimensions and regret it later. The most common mistake? Assuming integration will be straightforward and discovering in production that the AI-generated code makes assumptions about data formats or API behaviors that do not match reality.

The Human Judgment Dimension

Automated testing executes continuously in AI Scrum, but human judgment proves equally essential. Domain experts can spot when something feels wrong, even when code passes every automated test. They recognize results that do not match reality. AI sometimes optimizes for common cases while handling edge cases poorly, so those unusual inputs need human scrutiny.

Operations teams need to evaluate whether they can actually support this in production. A solution that works perfectly in development might be impossible to debug when it fails at 3 AM. This human validation dimension often makes the difference between solutions that technically work and solutions that actually succeed in production. I've learned to trust my gut when something feels off, even if I cannot immediately articulate why. Usually, that instinct points to a real issue that closer examination reveals.

Principle 2: Outcome Over Output

Recall the end-to-end dashboards from Chapter 2. Their value lay in a deliberate rejection of vanity metrics. We stopped tracking output, specifically how many story points we completed, and started measuring outcome. Did defect rates drop? Did feature delivery accelerate? Did system stability improve? In an era where AI can generate code rapidly, the volume of work is no longer a proxy for value. We must stop measuring the effort and start measuring the impact.

Why Story Point Velocity Metrics Hide the Truth

Traditional Scrum created an elaborate measurement apparatus around outputs: story point velocity, burndown charts, burndown charts, and sprint capacity. These metrics served a purpose in development-constrained environments. They helped teams estimate how much work they could complete in a sprint. But they measure the wrong thing for AI-native development.

Here is the problem with output metrics: they tell you how much you built, not whether it mattered. A team can complete 50 story points and solve zero customer problems. They can achieve perfect story point velocity while delivering zero outcome velocity delivering negligible business value. This disconnect is precisely what Scrum.org's Evidence-Based Management framework addresses - output metrics create an illusion of progress while actual business impact remains unclear.

AI makes this disconnect even more dangerous because it amplifies whatever you optimize for. If you optimize for story points, AI helps you complete more story points faster. Whether those story points deliver value becomes increasingly irrelevant to how teams operate.

This shift from output to outcome metrics exemplifies creative destruction. Story point velocity, story points, burndown charts - these tools served Agile teams brilliantly for more than two decades. Now they actively mislead. The metrics that once enabled effective planning now obscure whether work delivers value. Creative destruction is not about rejecting what worked. It is about recognizing when circumstances change enough that old tools no longer serve their purpose.

Measuring What Actually Matters

AI Scrum abandons output metrics entirely in favor of outcome measurements. Teams track business impact delivered, customer problems solved, hypotheses validated, and value realized. Success means achieving business objectives, not completing story points. This reorientation aligns team incentives with organizational strategy in ways output metrics never could.

Take a team building fraud detection capabilities. Under traditional Scrum, they might celebrate completing five user stories enhancing the detection engine. Under AI Scrum, they ask different questions. Did we reduce fraud losses? By how much? Did we improve detection rates without increasing false positives? Did we validate that the new approach works across customer segments, not just the majority case?

Figure 3.4: The Hamster Wheel vs. The Lever. *A contrasting illustration. The left side shows a hamster sprinting in a wheel labeled 'Velocity/Output,' generating sweat but no movement. The right side shows a human using a simple lever to lift a massive boulder labeled 'Business Outcome.' This visualizes that high activity (velocity) often yields zero displacement, while strategic leverage (outcome) moves the heavy obstacles.*

Setting Up Outcome Metrics in Practice

The shift from output to outcome metrics requires concrete measurement practices, not just philosophy. Start by defining one to three measurable outcomes before the sprint begins. Not vague aspirations like "improve user experience," but specific targets with baseline numbers. "Reduce fraud losses by 10 to 15 percent" or "Improve detection accuracy from 87 percent to 92 percent without increasing false positives above 2 percent."

These targets need current baseline measurements to be meaningful. Identify your data sources early. Most organizations already have the measurement infrastructure - analytics platforms, customer satisfaction surveys, operational dashboards, financial reports. The metrics exist. Teams just have not been connecting sprint work to those business measurements.

During sprint planning, specify exactly how you will measure success and where that data comes from. At sprint review, present actual measured results, not subjective assessments. "Our fraud detection changes reduced losses by 12 percent across all customer segments, measured through the financial reporting system" carries far more weight than "we think it is working better."

One sprint rarely provides enough data to see meaningful patterns. Track outcome trends across three to four sprints. Some improvements take time to materialize. Customer satisfaction changes lag feature releases by weeks. Revenue impacts accumulate gradually. The goal is establishing a rhythm where teams routinely measure business impact and use those measurements to inform what they build next. Once that rhythm exists, outcome orientation becomes natural rather than forced.

These questions force continuous connection between technical work and business value. They prevent teams from hiding behind statements like "we completed our sprint commitment" when business impact remains unclear. This accountability can be uncomfortable initially for engineering teams accustomed to measuring activity. Research on outcome-oriented teams from Scrum.org consistently shows that once they make the shift, their work becomes more satisfying because they can see it actually matters.

The Uncomfortable Metrics

Outcome metrics sometimes reveal uncomfortable truths that activity metrics hide. That is actually their value. When data shows work accumulating faster than capacity allows, or features shipping without moving business metrics, those insights enable course correction. The key is organizational willingness to act on what the metrics reveal rather than optimizing measurements to hide problems.

The most common uncomfortable truth? Efficiency is not always the problem. Teams can be operating at peak efficiency while still falling behind, because the real constraint is scope. When metrics show work accumulating faster than capacity allows, the answer is not always 'work smarter' - sometimes it is 'commit to less' or 'add capacity.' This is where metrics become political. Data does not lie, but organizations can choose to ignore what data tells them.

A realistic environment that accepts uncomfortable truths will use metrics to make hard choices. An unrealistic environment will keep measuring activity instead, because those metrics tell more comfortable stories about progress. That experience taught me something important about outcome metrics. They are only valuable if leadership is willing to act on what they reveal. If your organization measures outcomes but then ignores them when they are inconvenient, you are just doing more measurement work without getting the benefit. The metrics have to drive decisions, or they are waste.

Principle 3: Probabilistic vs. Deterministic Thinking

Let me tell you about how I hired one of the best employees I've ever had the opportunities to work with. I was looking for a developer, and instead of trying to find the perfect candidate through traditional means, I did something different. I took a probabilistic approach. I'd review ten applicants at a time. Not trying to find the perfect person in those ten, just trying to find the best of that batch.

Interview them. If someone was good enough, hire them. If not, next batch of ten. This approach worked because I understood something fundamental: I was not looking for a deterministic outcome where there is one perfect candidate and I just needed to find them. I was looking for a probabilistic outcome where multiple candidates might work well, and I needed to find one who met my acceptance threshold quickly. He was not the "perfect" candidate on paper but he was someone who met my criteria well enough that I knew he would succeed, and I was right. If I'd waited to find the theoretically perfect candidate, I might still be looking or I might have hired someone less capable who interviewed a little better.

Software Development Just Became Probabilistic

Here is what most teams miss about AI-generated code: it is probabilistic, not deterministic. Traditional code that a human writes behaves deterministically. Given the same inputs, you get the same outputs every time. Write a function to calculate the Fibonacci sequence, run it a thousand times with the same input, you get the same result a thousand times. This deterministic behavior makes validation straightforward. Test once, if it works, it works.

This represents creative destruction of mental models. Deterministic thinking - test once, if it works it works - guided software development since the first compiler. That model enabled consistent, reliable outcomes. Now it is insufficient.

Probabilistic systems require different thinking. The old mental model has not failed. The environment changed enough that what once worked brilliantly now constrains progress.

AI-generated code does not work that way. Generate fraud detection code on Monday, you get one implementation approach. Generate it again on Tuesday with the identical prompt, you might get a different approach. Both might work. Neither might be perfect. They will make different trade-offs, optimize for different characteristics, and handle edge cases differently.

This probabilistic behavior is a fundamental characteristic of how large language models work, not a flaw to be fixed. Understanding this probabilistic nature changes how teams should approach AI-generated code. Instead of generating once and accepting the result, generate multiple times and evaluate the options. Treat it like exploring a solution space rather than finding the solution. The variation across generations is not noise to eliminate - it is information about what is possible.

Figure 3.5: The Mason vs. The Gardener. *A split illustration contrasting work styles. The left side shows a Mason consulting a rigid blueprint to lay identical bricks, representing the deterministic past. The right side shows a Gardener with three distinct pots of flowers (options), preparing to plant one after it is chosen by the customer. This visualizes the shift from 'building the one right thing' to 'selecting the best generated option.'*

Multiple Candidates, Not Single Solutions

Traditional Scrum assumed you would build one solution, test it, and ship it if it worked. AI Scrum assumes you will generate multiple candidate solutions, evaluate them against validation criteria, and select the one that best balances requirements.

Let me give you a concrete example from outside software development. I needed a professional headshot for my LinkedIn profile, but like most people in my generation, I do not own a suit any longer and rarely get professional photos taken.

I'd seen advertisements for AI-generated professional headshots. The process is simple, upload a casual photo and receive several professional-looking options.

The reviews were extremely positive, so I tried it. The results? Fantastic! I received multiple options to choose from, all looking professional, and I selected the one that met my tastes. The whole process cost very little and took minutes. The fact that I could see a few options when I got my AI headshot done, and even have AI try again if I was not satisfied, is a great example of validation before I received the end product. And going back further, the ability to read actual user testimonials before I ever spent a penny was great.

While the result was not 100 percent perfect, maybe 95 percent, the end product was within my personal acceptance threshold. The same principle holds true for AI development, whether it is software development, a marketing website or point-of-sale materials, or other AI-related deliverables. If the stakeholder can be provided with two or three options to choose from, it is no longer tension on the development side but more so a sales job on the stakeholder side.

"Which one would you like to move forward with, A, B, or C?" You are not providing hundreds of choices. A, B, and C all meet the acceptance criteria set forth originally by the customer. It is not one solution but many. Additionally, once the customer has made a choice, it has passed their good enough threshold, and we can be on our way. Now, if a customer decides they do not like A, B, or C, that is a different story. As long as all of the options brought to the table meet or exceed the original acceptance criteria, the development is essentially complete.

A new user story would need to be generated, prioritized, and worked if they want something substantially different. AI can facilitate this with speed of development, no matter what it is, and deliver a quality product within guidelines. A human, while able to deliver a quality product, will likely be much slower, costing more to develop.

Timeboxing Exploration

The probabilistic nature of AI outputs requires timeboxing exploration. You cannot generate infinite candidates and evaluate all of them. That would consume infinite time. Instead, teams establish reasonable limits. Generate three to five implementations. Spend four hours on evaluation. Select the best candidate based on validation criteria and move forward.

This timebox discipline prevents the paralysis that probabilistic thinking can create. When there is no single right answer, teams can spend forever debating marginal differences between alternatives. The timebox forces decision-making based on good enough criteria rather than elusive perfection. I've found that three candidates usually suffice. Fewer than three and you do not get enough diversity to make meaningful comparisons. More than five and the evaluation effort outweighs the benefit of additional options. Three to five hits the sweet spot where you see meaningful alternatives without drowning in choice. Remember, in a world of human development there is but one option, whatever the human created.

Learning to Trust Imperfection

In the words of Winston Churchill:

"Perfection is the enemy of progress."

Deterministic thinking trained us to pursue perfection. If the code is not perfect, keep refining until it is. But perfection is usually impossible with probabilistic systems, and pursuing it wastes resources. Instead, you need to internalize the concept of acceptance thresholds. An acceptance threshold is the point where a solution is good enough to proceed, even if it is not perfect or optimal. It meets your validation criteria sufficiently that the value of further refinement does not justify its cost.

Brynjolfsson and McAfee's research on digital transformation demonstrates this principle: organizations succeed by knowing when good enough enables progress better than pursuing diminishing returns from additional optimization.

This requires a mindset shift many teams struggle with initially. We have been trained to pursue excellence. Accepting good enough feels like settling. But in AI-augmented development, pursuing theoretical perfection in one area means neglecting ten other areas that need attention. The acceptance threshold acknowledges that optimization has diminishing returns, and at some point, you are better off moving to the next problem than continuing to polish the current solution.

"Done is better than perfect."

Anne Mollegen Smith, a former Editor-in-Chief, squarely stated it back in the 1970s, "*Done is better than perfect.*"

Principle 4: Human-AI Collaboration Philosophy

Many organizations treat AI as simply another productivity tool, like email or spreadsheets but more sophisticated. This category error causes endless problems. AI is not a tool. It is a collaborator requiring explicit coordination, mutual understanding, and clear accountability boundaries.

Think about how teams work together. When a designer and developer collaborate, they establish communication patterns, handoff procedures, and decision-making protocols. They clarify who owns what decisions. They define how to resolve disagreements. These coordination mechanisms prevent confusion and enable productive partnership.

Human-AI collaboration demands similar explicit protocols, but teams rarely establish them. Instead, developers treat AI like a search engine, querying it for code snippets and hoping for useful results. This ad hoc approach produces frustration when AI generates solutions that miss crucial context or make assumptions the team never validated.

Defining the Partnership

I've learned that explicit protocols for human-AI collaboration prevent most failure modes. These protocols define when humans direct AI versus when AI proposes options for human evaluation. How teams verify AI-generated outputs. What documentation makes AI contributions reproducible. How organizations learn from human-AI interaction patterns.

Figure 3.6: The Relay Race Handoff. *A kinetic illustration using a relay race metaphor. The Human Runner (Intent) passes the baton to the Robotic Runner (AI Execution), who generates speed and distance. The Robot then passes the baton back to the Human Runner (Validation) at the finish line. This emphasizes that while the machine runs the middle leg, the human must start and end the process to ensure the race is won.*

Consider code generation as an example. An effective protocol might specify that humans define functional requirements and constraints. AI generates multiple implementation candidates. Humans evaluate candidates against validation criteria. AI explains its design decisions for each candidate. Humans select the approach that best balances requirements. AI implements the selected approach with human oversight.

This structured interaction prevents common mistakes. AI cannot misunderstand requirements it must acknowledge explicitly. Humans cannot miss important AI assumptions when AI must explain its reasoning. The partnership becomes genuinely collaborative rather than merely transactional.

Documentation and Reproducibility

When AI contributes to work products, teams must capture not just what AI generated but why. What prompt produced this code? What alternatives were considered? What criteria drove selection? Microsoft's research on AI-assisted development emphasizes this point: documentation enables reproducibility and supports continuous learning across teams and time. Future teams can understand past decisions rather than inheriting mysterious AI-generated code nobody comprehends.

I've seen teams struggling to maintain AI-generated code because nobody documented the thinking behind the original generation. Six months later, someone needs to modify the code, but the original context is lost. They are left guessing at what the code is trying to do and why it does it this way. Documentation prevents this problem. It is extra work upfront, but it saves substantial time later. The documentation does not need to be extensive. A brief note explaining the prompt, the alternatives considered, and why this approach was selected usually suffices.

Documenting AI Contributions in Practice

The documentation does not require elaborate systems or extensive writing. Most teams add a simple section to their existing code documentation or project wiki. For AI-generated code, include four elements: the prompt or prompts that generated the code, the alternatives AI provided and why they were rejected, the validation criteria used to select this approach, and any known limitations or assumptions built into the implementation.

Many teams create a lightweight template that takes five minutes to complete. "Generated using prompt: [text]. Considered three approaches optimizing for performance, readability, and maintainability respectively.

Selected readability approach because this module requires frequent modification by junior developers. Assumes input data is pre-validated by upstream service." That is sufficient. Future maintainers understand what was built, why it was built this way, and what assumptions they need to preserve or reconsider.

Make this documentation part of your Definition of Done. Code without AI contribution documentation is incomplete, just like code without tests or code without security review. Once documentation becomes standard practice rather than optional, teams stop seeing it as overhead and start seeing it as essential context that makes their future work easier.

Accountability Boundaries

When AI generates flawed output, who bears responsibility? This question needs an unambiguous answer. Clear protocols assign accountability explicitly. Humans remain accountable for validating AI outputs, ensuring appropriate use, and making final deployment decisions. Anthropic's research on AI safety frameworks emphasizes this fundamental principle: AI accelerates work but cannot assume responsibility for outcomes.

This accountability assignment can feel uncomfortable initially. If AI generated the problematic code, should not AI be responsible? No. AI is a tool, even though it is a sophisticated one that we collaborate with. The human using the tool bears responsibility for verifying that the tool's outputs are appropriate. Think of it like a spell checker. If you publish a document with a misspelled word that your spell checker missed, who is responsible for the error? You are, not the spell checker. Same principle applies to AI-generated code. The human who deploys it without adequate validation bears responsibility if it fails.

Principle 5: Integrated Governance Throughout Sprints

Traditional Scrum governance happened through Definition of Done criteria and peer review. Code had to pass tests, meet quality standards, and get reviewed by other developers before merging. For significant architectural changes, teams might consult with architects. Security considerations got baked into coding standards. This lightweight approach worked when development took days and teams had time for thorough review.

AI development changes the timing dramatically. When developers generate functional code in minutes using AI assistants, even peer code review struggles to keep pace. Security concerns intensify because AI-generated code might introduce vulnerabilities that standard reviews miss.

AI21 Labs' research on governance frameworks identifies this challenge: teams need governance embedded in the generation process itself, not just reviewing outputs after creation.

Day-by-Day Governance Mapping

Traditional Scrum handled governance through Definition of Done criteria and code review standards. Security concerns got caught in peer review. You would consult architects when you needed them, not invite them to every ceremony. That worked fine when code changes were incremental and predictable.

AI Scrum requires a different approach. When teams can generate entire implementations in hours, governance cannot wait for code review. Security engineers, architects, and compliance specialists need to participate in sprint planning, establishing guardrails before AI generates anything. Privacy engineers validate data handling approaches during implementation, not weeks later during audits. Red teams conduct adversarial testing within sprint windows, identifying vulnerabilities while mitigation remains straightforward. Compliance validation becomes continuous throughout the sprint, not a post-development checkpoint.

I've seen this pattern repeatedly across organizations. Features pass development and QA, get deployed to production, then require rework when compliance concerns surface. Not because anyone did poor work, but because the timing was wrong. When compliance specialists review completed implementations rather than participating in design, they can only react to what is already been built. Course correction at that stage means rebuilding rather than building right the first time. The solution is not stricter compliance enforcement or faster development: it is earlier collaboration.

From Bottleneck to Enabler

Integrated governance transforms the relationship between development teams and governance stakeholders. When governance participates throughout sprints rather than judging work retrospectively, adversarial dynamics dissolve. Governance practitioners stop being the people who say no and become partners helping teams succeed.

The common fear is that integrated governance means constant oversight slowing teams down. Experience shows the opposite. Teams accelerate because they receive guidance when they can still act on it easily, rather than discovering governance concerns after substantial work completion when changes prove expensive. This psychological shift matters as much as the process change.

When a security engineer sits with your team during sprint planning and says "here is how to approach data encryption for this use case," that is helpful collaboration. When the same security engineer rejects your completed feature three weeks later because it does not encrypt data properly, that is adversarial friction. Same technical concern, completely different team experience.

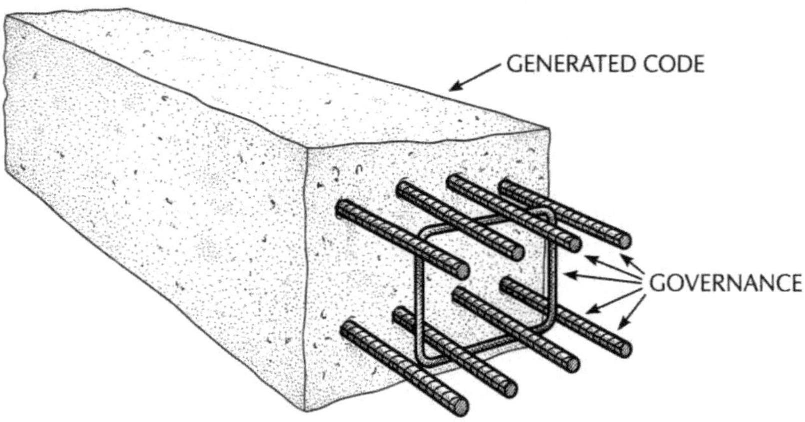

Figure 3.7: The Reinforced Concrete Cross-Section. *A cross-section illustration of a concrete beam. The concrete represents 'Generated Code,' while strong steel rebar rods running through it are labeled 'Governance.' This visualizes that governance is not an external gate or coating applied at the end, but the internal reinforcement structure that holds the code together from the inside out.*

The Governance Checkpoint

AI Scrum formalizes this integrated approach through a mid-sprint Governance Checkpoint ceremony. By mid-sprint, teams have validated multiple prototypes and understand which approaches best satisfy requirements. Governance stakeholders review validated designs before teams invest further effort in refinement.

Why midpoint governance instead of end-of-sprint approval? Because teams still have time to adapt based on governance feedback. If architects identify integration concerns, days remain to adjust implementations. If security teams flag vulnerabilities, mitigation can occur within the sprint. If compliance officers require documentation changes, teams can address them before deployment preparation.

The timing transforms governance from bottleneck to enabler. Governance participants provide guidance when teams can still act on it easily, rather than after completion when change is expensive.

Deloitte's 2024 research on AI governance found organizations implementing mid-sprint governance checkpoints report approval cycles that are 55 percent faster compared to traditional end-of-sprint approval reviews. More significantly, they report 70 percent fewer compliance issues discovered post-deployment, evidence that integrated governance catches concerns when prevention costs less than remediation.

Principle 6: Continuous Learning Loops

Traditional Scrum operated on bi-weekly learning cycles. Teams would plan a sprint, execute for two weeks, then reflect during retrospectives on what they learned. This cadence matched development constraints well. Teams needed sustained focus periods to complete implementations before pausing to reflect.

AI development makes bi-weekly learning cycles dangerously slow. When implementations emerge in hours rather than days, waiting two weeks to incorporate learnings means possibly repeating mistakes dozens of times. Continuous learning demands continuous feedback, not just about what teams build but how human-AI collaboration performs.

Learning at Multiple Timescales

AI Scrum implements learning loops operating at different timescales. Daily synchronizations provide immediate feedback on validation progress and emerging issues. Teams catch problems within 24 hours of their appearance, when correction costs little. Twice-weekly stakeholder sessions enable rapid course correction based on user feedback. Teams learn whether they are building the right thing within days, not weeks.

Mid-sprint governance checkpoints surface compliance and security concerns while teams can still adapt easily. Sprint retrospectives incorporate objective AI telemetry alongside subjective team observations, building institutional knowledge about what works and what does not in human-AI collaboration. These nested loops catch problems early. A team discovers during daily sync that an AI-generated implementation makes performance assumptions that will not hold in production. They course-correct immediately, wasting hours instead of days. Compare this to traditional Scrum, where the same issue might not surface until sprint review, after the team invested two weeks building on a flawed foundation.

Meta-Learning About AI Collaboration

The learning extends beyond traditional retrospective scope. Teams analyze not just their process but their partnership with AI. Which prompts produced the most valuable outputs? What validation criteria caught the most issues? Where did human oversight prevent AI mistakes? Which AI suggestions teams rejected turned out correct in retrospect?

This meta-learning about human-AI collaboration enables compounding improvements. Teams develop institutional knowledge about effective AI utilization. They build prompt libraries capturing successful interaction patterns. They refine their ability to recognize when AI suggestions merit trust versus skepticism. Microsoft's research on AI-assisted development emphasizes this organizational learning dimension: teams that systematically capture and share what works with AI collaboration see continuous improvement across sprints.

Teams can transform their AI effectiveness over just three or four sprints through systematic learning. As they practice, they generate better prompts because they have learned what works. They catch AI mistakes faster because they have learned what patterns to watch for. They trust AI more appropriately because they have calibrated their confidence through experience. This is exactly how I would have approached recurring AI issues with my team. When the same problem pattern appeared multiple times, encode it into validation criteria so subsequent AI generations avoid it automatically. The learning loop prevents recurring problems rather than merely addressing symptoms.

Objective Data Supplements Subjective Impressions

Traditional retrospectives rely heavily on subjective team recollections. "I felt like our estimates were off this sprint." "Communication seemed better than last time." These impressions have value, but they can be misleading. Teams might feel they performed poorly when objective data shows improvement, or feel successful when metrics reveal problems.

AI Scrum incorporates objective telemetry into retrospectives. How much AI-generated code did the team accept versus reject? What were the most common reasons for rejection? Where did validation catch issues? How long did governance review actually take? These metrics provide ground truth that supplements subjective impressions.

The combination of objective data and subjective experience produces more reliable learning. Teams can test their hypotheses about what works. "We believe starting stakeholder sessions earlier in the sprint will reduce late-stage changes" becomes falsifiable. Try it for two sprints. Measure the results. Did late-stage changes actually decrease? If yes, adopt the practice. If no, try something else.

THE RESISTANCE YOU WILL FACE

Now let me be direct. Implementing these principles will create resistance. People who have spent 15 years mastering traditional Agile practices will push back against validation-first approaches. Engineers comfortable with story points will resist outcome-oriented metrics. Teams accustomed to end-of-sprint governance will resent mid-sprint checkpoints. I've watched this resistance play out repeatedly. Understanding where it comes from helps you address it constructively rather than dismissing it as obstruction. Here is what the resistance looks like and what actually works to overcome it.

Take a dedicated Scrum Master who has facilitated teams for a decade. She built her professional reputation on velocity metrics and sprint commitments. When you propose measuring outcomes instead of story points, she pushes back. 'How will we know if we are being productive? How do we plan capacity?' It is not obstinance. It is legitimate concern about losing the measurement framework that has guided her decision-making for years.

The solution is not to dismiss her concerns or mandate the change. It is to run a pilot sprint where the team tracks both old metrics and new outcomes side by side. Let her see that outcome metrics answer the productivity question more honestly than velocity ever did. Once she sees it work, she becomes your strongest advocate because she understands both approaches and can explain the difference to skeptics.

The Comfort of Familiar Patterns

Traditional Agile works. That is the fundamental source of resistance. Teams using established practices successfully for years see no reason to change. Why adopt unproven approaches when current methods deliver results? This reasoning contains a critical flaw: it assumes current practices will continue working as the constraint environment shifts.

But when AI eliminates the development constraint, practices optimized for that constraint stop producing the results they once did. The challenge is that this deterioration happens gradually. Velocity remains stable even as actual business value delivered declines. Sprint completion rates stay high even as rework accumulates. Teams feel busy and productive while competitive position erodes. By the time the decline becomes undeniable, recovery costs far more than proactive adaptation would have.

Fear of the Unknown

Validation-first development, probabilistic thinking, continuous learning, these approaches require new skills and new mental models. Teams comfortable with deterministic software development find probabilistic thinking uncomfortable. Engineers who have built careers on producing working code initially feel devalued when validation consumes more time than implementation.

This fear is legitimate but manageable. Skills transfer more readily than people expect. The validation mindset that made someone a good QA engineer prepares them well for validation-first development. The systematic thinking that made someone a good architect serves them well in evaluating probabilistic AI outputs. The communication skills that made someone a good scrum master remain essential in integrated governance.

Addressing Resistance Constructively

Here is what I've learned works when facing resistance to these principles. First, acknowledge the legitimacy of concerns. "You are right that validation-first is different from how we have worked before" creates more productive dialogue than "you just do not understand." People who feel heard are more willing to experiment than people who feel dismissed.

Second, start with pilot projects rather than organization-wide mandates. Let skeptics watch a single team implement these principles successfully before requiring everyone to adopt them. Success stories overcome skepticism more effectively than arguments.

Third, focus on problems people already recognize. Most teams know rework is expensive. Most recognize that late-stage governance approval creates friction. Most acknowledge that velocity metrics sometimes mislead. Frame these principles as solutions to known problems rather than arbitrary new requirements.

Finally, provide concrete examples and templates. "Do validation first" is abstract. "Here is the Validation Canvas template we use, and here is how we filled it out for our last three features" is actionable. Reduce the cognitive load of adoption by providing starting points teams can adapt rather than forcing them to create everything from scratch.

When Resistance Becomes Obstruction

Sometimes resistance crosses from legitimate concern into obstruction. An architect who refuses to participate in mid-sprint governance checkpoints because "we have always done design review at the end." A product owner who insists on tracking velocity because "how else will we know if the team is productive?" A developer who rejects validation-first because "that is not how real engineering works."

When resistance becomes obstruction, leadership intervention becomes necessary. Organizations that successfully implement AI Scrum have executive sponsors who clearly communicate that adaptation is not optional. Not because leadership is being arbitrary, but because competitive survival demands it.

The gap between awareness and action creates real problems. Organizations can acknowledge that AI matters while struggling to articulate concrete next steps. That paralysis has costs. Talented people leave to find organizations moving forward. Traditional metrics continue showing progress while competitive position erodes. The solution is executive clarity about the path forward, even if it is imperfect. 'We are starting with two pilot teams, here is what success looks like, here is the support you will have, here is our timeline for evaluating results.' That clarity, combined with genuine support through training and mentorship, overcomes most resistance.

This resistance misunderstands how innovation works. Economist Joseph Schumpeter called this pattern 'creative destruction' – the systematic process where new innovations replace established practices, driving progress forward. The discomfort teams feel when abandoning methodologies that worked well for 25 years is not evidence that change is wrong. It is evidence that genuine transformation is occurring. Farriers became auto mechanics when cars replaced horses. Telegraph operators became telephone switchboard operators. COBOL programmers became web developers when the internet emerged. Traditional Scrum practitioners will become AI Scrum practitioners as AI eliminates the development constraint. The skills are not lost, they evolve.

FROM THE TRENCHES: CAPACITY AND TRUTH

I've learned important lessons about how to present capacity constraints to leadership. Early in my career, I'd present projections showing we had more committed work than delivery capacity. The math was straightforward: this much backlog, this delivery rate, simple division.

Leadership would push back, looking for efficiency improvements rather than scope adjustments.

The problem wasn't the math. The problem was how I framed it. I was talking about outputs, specifically how many features we could complete. Leadership heard it as a resource request, not a strategic constraint.

Here's what works better. Frame capacity issues in terms of outcomes rather than outputs. Instead of saying "we need more people to complete these features," say "we can deliver these three business outcomes in the next quarter with current capacity, or we can deliver these seven outcomes if we add resources. Which outcomes matter most?" That shifts the conversation from resource requests to strategic prioritization.

Also establish clear acceptance thresholds upfront. When leadership understands exactly what "complete" means and sees actual delivery measured against those criteria over multiple sprints, the data becomes harder to question. Outcome metrics reveal whether work delivers business value, not just whether teams stay busy.

The key insight? Outcome over output. When you measure business impact delivered rather than story points completed, the disconnect between activity and actual value becomes obvious early, enabling course correction before problems compound.

WHAT COMES NEXT

These six principles establish the philosophical foundation for AI Scrum. Validation-first thinking, outcome orientation, probabilistic mindset, human-AI collaboration protocols, integrated governance, and continuous learning. Each principle responds to specific challenges that emerge when AI eliminates the development constraint and establishes new constraints around validation, governance, and integration.

But principles alone do not change how teams operate day-to-day. Chapter 4 translates these principles into practical team structures and role definitions. You will see exactly how roles evolve when implementation work accelerates and validation work becomes central. Who does what? What responsibilities shift? What new capabilities do teams need? What traditional roles become less critical? The framework starts making sense when you can map it to actual team structures and understand who is accountable for what in this new constraint environment.

KEY TAKEAWAYS

- **Validation is the New Production:** Establish validation criteria before AI generates implementations. Define what success looks like in measurable terms, then generate multiple candidates and evaluate them against predetermined standards. This prevents the expensive waste that AI's speed can amplify.
- **Velocity is a Trap:** Replace output metrics like velocity and story points with outcome metrics measuring business value delivered and customer problems solved. Output metrics become meaningless when implementation accelerates 2-5x after learning curve. Focus on whether your work matters, not whether you completed tasks.
- **Embrace the Gardener Mindset:** Embrace probabilistic thinking about AI outputs. Generate multiple candidate implementations, evaluate them against validation criteria, and select the one that best balances requirements. Establish acceptance thresholds where solutions are good enough to proceed rather than pursuing elusive perfection.
- **The Human is the Editor:** Treat AI as a collaborator requiring explicit protocols, not a tool you query for code snippets. Define when humans direct versus when AI proposes options. Document why AI generated what it did. Maintain clear accountability boundaries where humans validate outputs and make deployment decisions.
- **Governance is a Guardrail, Not a Gate:** Integrate governance throughout sprint workflows rather than treating it as end-of-sprint approval. Mid-sprint governance checkpoints enable course correction while changes remain easy. Governance transforms from bottleneck to enabler when embedded in daily work.
- **Continuous Learning:** Implement learning loops at multiple timescales. Daily feedback catches problems within 24 hours. Twice-weekly stakeholder sessions enable rapid course correction. Sprint retrospectives incorporate objective AI telemetry. Continuous learning prevents repeated mistakes when development cycles compress from weeks to hours.

MIND MAP – AI SCRUM: A NEW DEVELOPMENT FRAMEWORK

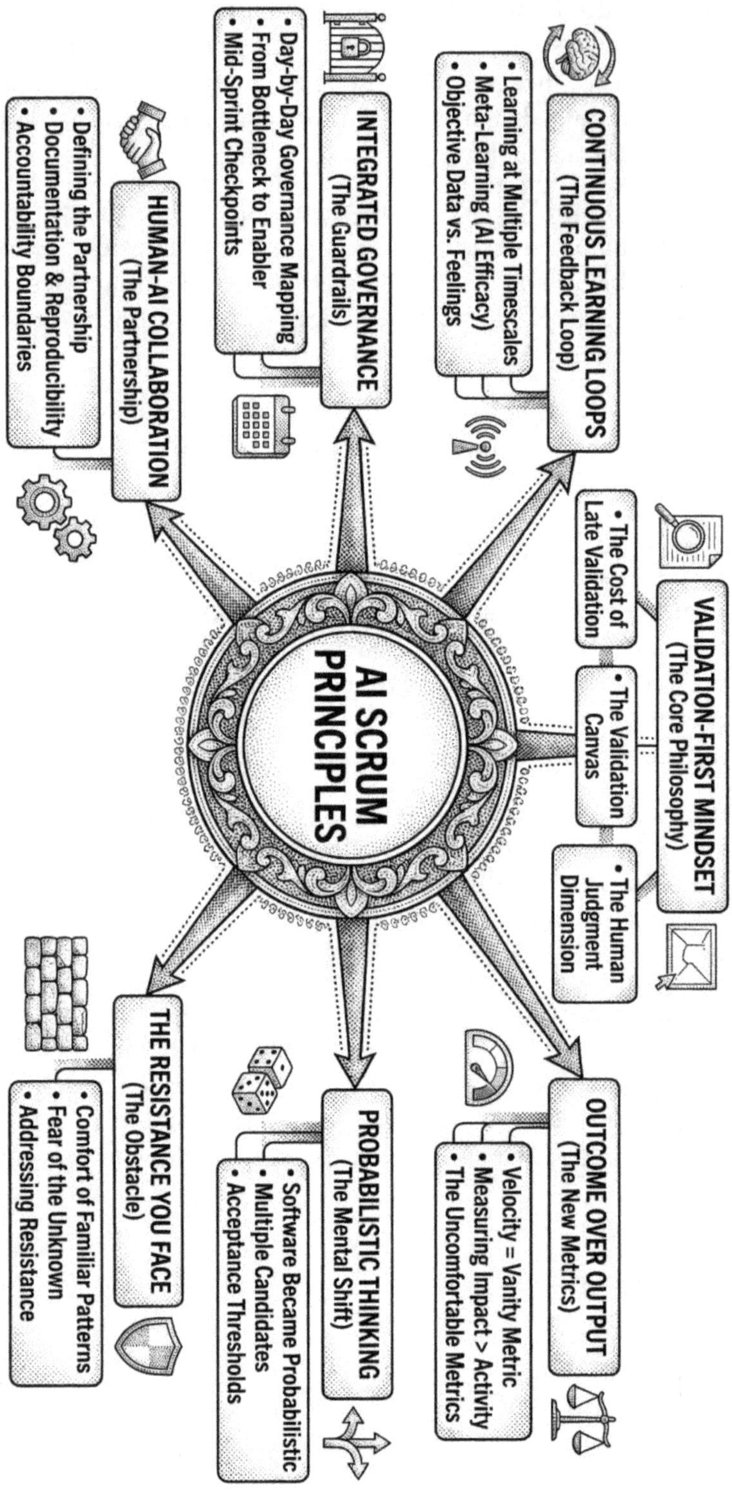

AI Scrum redefines team structures by shifting the primary constraint from coding to validation. It introduces seven evolved roles, including the Product Strategist and AI Alignment Engineer, to manage governance and strategy. Success requires T-shaped talent and Centers of Excellence to support high-velocity delivery while maintaining strict quality and compliance standards.

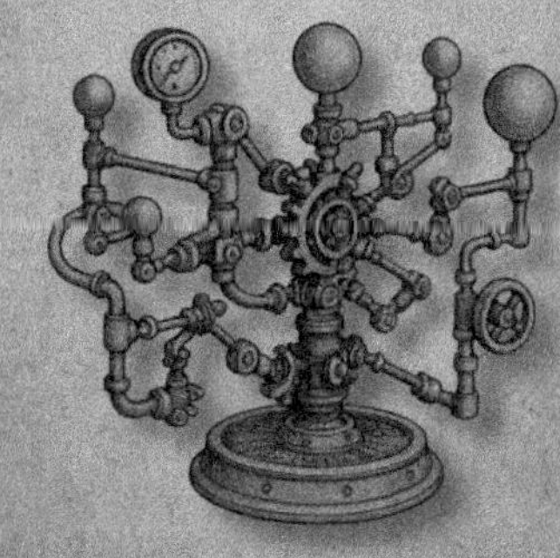

CHAPTER 4: ROLES AND RESPONSIBILITIES

The most difficult conversation in an AI Scrum transformation does not happen in the server room. It happens in the boardroom.

Consider a scenario that is playing out in enterprises across the globe right now. A transformation leader sits across the conference table from the CEO, CFO, and CIO. The air in the room is heavy with the specific tension that arises when capital allocation meets uncertain return. The CEO looks up from the proposal, her expression skeptical.

"Let me understand this correctly," she says, removing her reading glasses. "You are proposing we increase our Scrum team costs by 26% annually, roughly $152,000 per team, to adopt this framework? In a year where we are looking for efficiency gains?"

The response must be even and grounded in data. "That is the market reality based on current compensation data. AI Scrum requires different capabilities than traditional Scrum, and those capabilities command premium salaries. We are not paying for faster typing. We are paying for higher-order judgment."

The CFO leans forward, tapping the spreadsheet cell that shows the salary delta. "Walk me through why we need these new roles. What happens to our current Product Owners? Our QA engineers? Our Scrum Masters? Are we firing them? Retraining them? And if we keep them, why is the headcount cost rising?"

"They evolve," is the answer. "But the evolution is vertical, not just horizontal. Product Owners become Product Strategists focused on economic outcomes rather than feature lists. QA engineers become Validation Leads addressing AI-specific testing challenges that simply did not exist two years ago. Some roles combine for smaller teams. Some roles are entirely new. The Governance Coordinator is new because AI introduces compliance concerns like hallucinations and data leakage that traditional Scrum never addressed."

The CIO asks the hard question—the one that actually matters. "How long before we see ROI on this investment?"

"You get software that works. Right now, your traditional teams are about to start generating legacy code at ten times their current speed. Without this structure, you will not get more value. You will get a large increase in AI Debt."

This interaction highlights the central economic tension of the Agile Inversion. When AI eliminates the development constraint, the team structure must change fundamentally. We are no longer organizing human labor to maximize code production because code production is now cheap. We are organizing human judgment to maximize validation, safety, and strategic alignment. These things are expensive.

Understanding these changes proves essential for successful transformation. You cannot simply hand AI tools to a traditional Scrum team and expect efficiency. You will only accelerate the creation of failure.

WHAT THIS CHAPTER COVERS

This chapter defines the complete AI Scrum team structure. It addresses both theoretical framework and practical implementation reality. By the end of this chapter, you will understand:

- **The Philosophy of Constraint-Based Hiring:** Why roles must align to validation, governance, and integration constraints rather than development activities.
- **The Seven Core Roles:** A detailed analysis of the Product Strategist, AI Alignment Engineer, Validation Lead, Governance Coordinator, Integration Architect, AI-Augmented Developer, and Delivery Facilitator.
- **The Economics of Talent:** A breakdown of the salary market reality showing the compensation differentials between traditional and AI Scrum roles based on late 2025 data.
- **Collaboration Architectures:** The critical interfaces between roles that enable effective AI-native development.
- **The Transition Pathway:** How to evolve your existing workforce into this new structure without destroying morale.

KEY TERMS AND CONCEPTS

- **Constraint Hierarchy:** The ordered set of factors that limit team throughput. In traditional Scrum, implementation speed or coding was the primary constraint. In AI Scrum, validation completeness, governance approval, and integration complexity become the dominant constraints that teams must address. The team structure mirrors the constraint.

- **AI Collaboration Responsibility:** The requirement that every team member must develop proficiency in collaborating with AI systems regardless of role. This is not a specialized technical skill for developers only. It is a core competency for all roles from Product Strategist to Delivery Facilitator.

- **Outcome Backlog:** A prioritized list of business outcomes to achieve rather than features to build. Product Strategists maintain outcome backlogs where each item defines a measurable business result. An example is "Reduce churn by 5%." It is not a technical implementation like "Add a save button."

- **T-Shaped Skills:** Deep expertise in one domain combined with broad capability across multiple domains. AI Scrum team members need T-shaped skills because AI augmentation enables individuals to work across traditional role boundaries while maintaining specialized depth.

- **Validation-First Workflow:** The operational approach where validation strategy gets defined before implementation begins. This makes validation a sprint-long activity rather than a post-development phase.

TEAM STRUCTURE PHILOSOPHY

The fundamental error most organizations make during AI adoption is retaining their legacy organizational charts while changing their tools. They treat AI as a productivity plugin for developers. They assume that if a developer can code faster then the team will deliver faster. This ignores the organizational physics we established in Chapter 1.

Roles must align to the new constraint hierarchy rather than development activities. This represents more than organizational chart modifications. It is a shift in the philosophy of value creation. When a team can generate working implementations in hours instead of weeks, the constraint becomes ensuring those implementations actually work. The constraint is ensuring they align with business intent. The constraint is ensuring they satisfy governance requirements. The constraint is ensuring they integrate properly with existing systems. The logical consequence is predictable. Traditional structures break down.

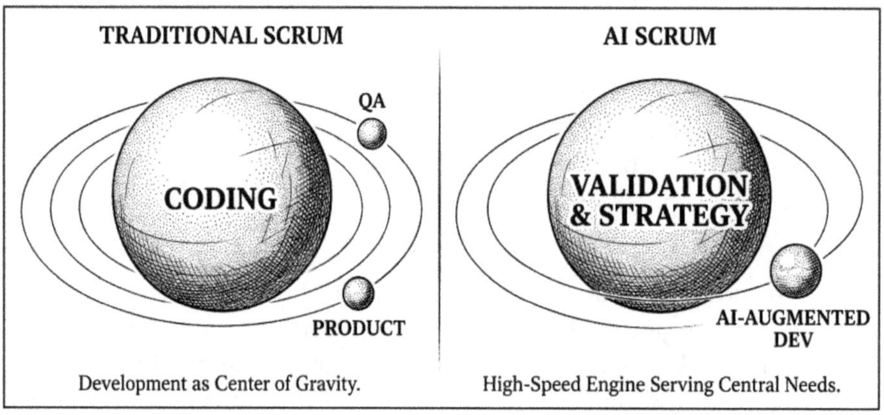

Figure 4.1: The Organizational Orbit. *A comparison of two solar systems illustrating the Team Structure Philosophy: The developer is no longer the central bottleneck around which the entire team revolves; they are now a high-speed engine serving the central needs of Validation and Strategy.*

In traditional Scrum, the structure was organized around the scarcity of coding hours. Developers were the engines of production. Scrum Masters existed to remove obstacles from the Developers' paths. Product Owners existed to feed the Developers with work. QA Engineers existed to catch the Developers' mistakes. Every role orbited the Developer.

In AI Scrum, the Developer is no longer the bottleneck. The bottleneck is the Validator and the Strategist. When code is infinite, the scarcity shifts to *truth* and *purpose*. We need fewer people writing code. We need more people verifying that the code does not hallucinate. We need more people ensuring the code complies with EU and NIST regulations. We need more people ensuring the code integrates with the messy reality of legacy enterprise systems.

AI Scrum defines seven core roles. Each addresses specific constraints in AI-native development. Not all organizations will implement all roles discretely. Smaller teams may combine them. But the *responsibilities* of these seven roles must be accounted for. If they are not, the system will fail.

The roles that follow are not just new titles. They represent genuine identity shifts for real people, and organizations that treat them as mere rebranding will pay for it in trust, morale, and turnover.

THE SEVEN CORE ROLES

The architecture of an AI Scrum team is not a random assortment of new titles; it is a deliberate response to the shift from deterministic coding to probabilistic generation. These roles redistribute accountability to ensure that while AI handles the speed of creation, humans retain control of direction and safety.

Five of these roles are vertical evolutions of traditional Scrum positions, expanding their scope to meet the new constraints of validation and integration. Two roles—the AI Alignment Engineer and Governance Coordinator—are entirely new disciplines, necessitated by the unique risks of Large Language Models. The following analysis breaks down the evolution, psychology, and critical responsibilities for each role.

1. Product Strategist (Evolved from Product Owner)

The Product Strategist evolves from the traditional Product Owner. The shift is profound. It expands the scope from backlog management to outcome ownership. In the traditional model, Product Owners were often "ticket merchants." They traded detailed user stories for code. They managed a backlog of features. The Product Strategist owns business outcomes. This role carries ultimate accountability for business value delivery and strategic alignment. Instead of saying "we need a user authentication feature," they say "we need to reduce customer onboarding time by 40% while maintaining security standards."

The Psychology of the Role: This transition is psychologically difficult for many legacy Product Owners. Traditional POs find comfort in the "Definition of Ready." They want a perfectly described ticket. The Product Strategist must be comfortable with ambiguity. They must be willing to define the *what* and the *why* without dictating the *how*. This is necessary because AI might suggest a *how* that the Strategist never imagined.

Consider a scenario from financial services. A traditional Product Owner might specify "Add biometric authentication to the mobile app." This locks the team into one technical solution. A Product Strategist instead defines the outcome: "Enable customers to access accounts in under 10 seconds without compromising security, reducing support calls about password resets by 60%." The first approach restricts the team. The second empowers them to explore multiple approaches. These might include biometric, persistent sessions, or social login. AI can prototype these rapidly.

Primary Responsibilities:

- **Define business outcomes and success metrics:** Articulate what business problem needs solving. Do not articulate how to solve it.
- **Maintain the Outcome Backlog:** Prioritize based on strategic value and organizational readiness. Do not prioritize based just on feature completeness.
- **Manage Stakeholder Expectations:** Communicate what AI can and cannot do. This involves managing the "Gap Between Confidence and Competence" discussed in Chapter 1.
- **Leverage AI for Strategy:** Use AI to analyze market data. Use it to generate alternative outcome definitions. Use it to simulate the impact of prioritization scenarios.

Failure Modes & Transition: The most common failure mode is the "Feature Factory Manager"—a Strategist who reverts to writing detailed technical specs for the AI, micromanaging prompts rather than defining success criteria. Transitioning Product Owners typically require 3–4 months to adapt, with training emphasizing frameworks for defining measurable outcomes and communicating uncertainty to stakeholders.

2. AI Alignment Engineer (New Role)

The AI Alignment Engineer represents an entirely new role without a traditional Scrum analog. They address the critical challenge of translating ambiguous business requirements into precise AI specifications. Why does this role exist? Because natural language is messy. AI is literal. When you tell a human developer "Make the interface more intuitive," they use shared human context to interpret that. Tell an AI system the same thing and you get wildly inconsistent results.

The Psychology of the Role: This person is the bridge. They must be bilingual. They must be fluent in business intent and fluent in "LLM logic." They are systematic thinkers who see patterns in chaos. They are not just prompt writers. They are *architects of intent.*

Figure 4.2: The Alignment Bridge. *A sturdy bridge spans the gap, representing the critical translation layer required to prevent AI from confidently executing the wrong instructions.*

Primary Responsibilities:

- **Translate Business Intent:** Transform vague outcomes into precise "Context Specifications."
- **Maintain the Prompt Library:** Create and curate reusable prompt patterns for common scenarios.
- **Coach the Team:** Serve as the expert on prompt engineering. Teach developers and strategists how to elicit better responses from the models.
- **Maintain the AI Collaboration Log:** Document what works and what does not work to prevent the team from repeating mistakes.

Failure Modes & Transition: The primary failure mode is the "Prompt Bottleneck," where the engineer insists on writing *every* prompt for the team, becoming a blocker rather than an enabler. Business Analysts (BAs) are the natural pool for this role, but they face a steep learning curve regarding LLM capabilities. Expect 4–6 months for proficiency, including advanced training in prompt engineering and model architecture.

3. Validation Lead (Evolved from QA Engineer)

The Validation Lead evolves from the traditional QA Engineer. The scope expands from "finding bugs" to "certifying reality." In the pre-AI era, code was written by humans. Errors were typically logical or syntactical. In the AI era, code is generated by models. Errors can be *hallucinations.* Code can look perfect and compile but do something subtly wrong or dangerous.

The Psychology of the Role: Traditional QA is often reactive. They think "Test this ticket." The Validation Lead is proactive and adversarial. They must possess a "Red Team" mindset. They assume the AI is lying. They assume the model has drifted. They are the guardians of safety. The Validation Lead's work begins during Sprint Inception. They define the Validation Canvas. This is the detailed set of criteria that the AI's output must satisfy.

Figure 4.3: The Validation Gauntlet. *This visualizes the shift from 'testing' as a final phase to 'validation' as a continuous, automated industrial filter that protects production from high-velocity AI output.*

Primary Responsibilities:

- **Define Validation Strategy:** Create the Validation Canvas for each outcome. Cover functional, performance, security, and compliance criteria.
- **Adversarial Testing:** Design tests specifically intended to trick the AI or trigger hallucinations.
- **Instrumentation Design:** Collaborate with developers to ensure the generated code has the necessary "hooks" for monitoring.
- **Stop Authority:** The Validation Lead has the absolute right to block deployment. This applies regardless of schedule pressure.

Failure Modes & Transition: The "Manual Tester" is the most dangerous failure mode—trying to manually test AI-generated code leads to instant overwhelm. They must leverage AI *to test* AI using automated test generation. QA Engineers with strong coding skills and a "breaker" mentality transition well, typically needing 3–4 months for proficiency in AI-specific testing methodologies.

4. Governance Coordinator (New Role)

In highly regulated industries like Finance and Health, this is the most critical new role. Traditional Scrum teams treated governance as an external dependency. It was something "Legal" did at the end. AI Scrum embeds governance into the sprint. AI introduces novel risks: copyright infringement, bias in credit-scoring models, or PII leakage into public LLMs. The Governance Coordinator answers these questions in real-time.

The Psychology of the Role: This role requires a unique blend of bureaucratic precision and agile flexibility. They must be the "enabler." They cannot be the "Department of No." Their job is to design the guardrails that allow the team to run fast safely.

Primary Responsibilities:

- **Compliance Integration:** Identify governance implications of outcome definitions during Sprint Inception.
- **Audit Trails:** Maintain the documentation required by frameworks like the EU AI Act or SOC 2.
- **Ethical Oversight:** Facilitate discussions on responsible AI use and bias mitigation.
- **Coordination:** Act as the liaison between the Scrum team and the organization's central AI Governance Board.

Failure Modes & Transition: The "Checkbox Bureaucrat" adds drag without value. Effective coordinators actively design compliant workflows. This role typically recruits from compliance, legal, or risk management backgrounds rather than traditional technical roles. Expect 4–6 months for a non-technical person to learn enough about the SDLC to be effective.

5. Integration Architect (Evolved from System Architect)

The sales pitch for generative AI focuses on the standalone miracle—writing a game in Python in seconds. In the enterprise, however, value is only realized through **integration**. The "1/99 Rule" applies here: It takes 1% of the effort to build the AI prototype and 99% of the effort to integrate it safely into the legacy enterprise architecture. The Integration

Architect ensures that the rapid generation of code does not result in a sprawling mess or unmaintainable "spaghetti code."

Figure 4.4: The Governance Guardrails. *While AI dramatically increases development speed, this velocity is dangerous without boundaries. Compliance and ethics serve as necessary guardrails to keep the project on track. Without these governance structures, the rapid pace could lead a software team to derail or face catastrophic failure instead of delivering value.*

Primary Responsibilities:

- **Integration Mapping:** Maintain the Integration Map visualizing dependencies between AI components and legacy systems.
- **Observability Architecture:** Design the monitoring infrastructure so the team knows when the AI is failing in production.
- **Standards Enforcement:** Ensure AI-generated code follows the organization's architectural patterns to prevent AI Debt.

Transition Pathway: Traditional System Architects transition smoothly, typically in 2–3 months. The main new skill is understanding the specific integration patterns of LLMs, such as RAG architectures and vector databases.

6. AI-Augmented Developer (Evolved from Developer)

The core responsibility remains implementing solutions, but the *verb* changes. Traditional developers *wrote* code. AI-Augmented Developers *architect, prompt, and review* code. They are no longer "bricklayers." They are "site supervisors." They use AI to generate the bricks like boilerplate and functions. But they decide where the bricks go. They ensure they are solid.

The Psychology of the Role: This is often the hardest transition. Many developers derive their professional self-worth from their ability to write complex syntax from memory. AI Scrum strips that away. A developer who refuses to use AI is like a mathematician who refuses to use a calculator. It may be noble but it is economically uncompetitive.

Primary Responsibilities:

- **Collaborate with AI:** Generate implementations using tools like GitHub Copilot or Cursor.
- **Multi-Candidate Generation:** Instead of writing one solution, they generate three different approaches. They evaluate them against the Validation Canvas.
- **Instrumentation:** Their primary coding focus shifts to writing the validation hooks and tests that the AI cannot reliably write for itself.

Failure Modes & Transition: The "Button Pusher" blindly accepts AI output without critical review, introducing subtle bugs and security holes (a phenomenon called "Cognitive Atrophy"). Transition typically takes 2–3 months, focusing on mastering AI coding assistants and shifting from a "writer" mindset to a "reviewer" mindset.

7. Delivery Facilitator (Evolved from Scrum Master)

Traditional Scrum Masters focused on removing impediments to *development.* Delivery Facilitators focus on removing impediments to *validation* and *governance.* They are the orchestrators of the entire system. They ensure that the speed of AI generation does not outpace the team's ability to validate.

Primary Responsibilities:

- **Facilitate AI Ceremonies:** Run the new rituals like Sprint Inception and Governance Checkpoints.
- **Maintain Telemetry:** Own the Sprint Telemetry Dashboard. Provide real-time visibility into the "Hidden Queues."
- **Process Coaching:** Coach the Product Strategist on outcome definition. Coach the Developers on AI collaboration.

Success Indicators: When teams say "our Delivery Facilitator makes the governance process invisible," that reflects excellent effectiveness. Scrum

Masters with strong coaching skills transition well in 2–3 months, provided they learn the new ceremonies and basic AI metrics.

THE ECONOMICS OF TALENT: SALARY MARKET REALITY

The shift to AI Scrum is not cost-neutral. You are hiring for higher capability. The market prices that capability aggressively. Based on January 2026 national market data, organizations should expect approximately a **26% annual cost increase** when transitioning from traditional Scrum to AI Scrum teams, at least for the next few years until knowledge gaps are closed. This translates to roughly **$152,000** additional annual investment per team.

SALARY MARKET COMPARISON					
Traditional Role	Avg. Salary	AI Scrum Role	Avg. Salary	Premium	Key Capability Drivers
Product Owner	$120,000	Product Strategist	$145,000	+21%	AI strategy, predictive analytics, outcome leadership
Business Analyst	$95,000	AI Alignment Engineer	$130,000	+37%	Prompt engineering expertise, AI collaboration optimization
QA Engineer	$95,000	Validation Lead	$118,000	+24%	AI-specific testing, red-teaming, validation strategy
System / Data Architect	$170,000	Integration Architect	$185,000	+9%	AI system integration, observability architecture
Software Developer	$120,000	AI-Augmented Developer	$138,000	+15%	Prompt engineering for code, AI output validation
Scrum Master	$118,000	Delivery Facilitator	$130,000	+10%	AI-native workflow optimization, governance facilitation
(None)	—	Governance Coordinator	$125,000	NEW	AI governance, regulatory compliance specialization

Table 4.1: Salary Market Comparison. *Based on the roles and averages shown, organizations should expect approximately 26% annual cost increase when transitioning from traditional Scrum to AI Scrum teams. Appendix D provides detailed cost modeling showing this translates to roughly $152,000 additional annual investment for a complete team transformation. CONT…*

Table 4.1: Continued…

The premium reflects market scarcity of AI-specific skills combined with expanded responsibility scope. AI Alignment Engineers command the highest premium (37%) because the role combines technical technical depth, business acumen, and prompt engineering expertise, a rare combination. Integration Architects show the lowest premium (9%) because their traditional architectural skills translate most directly to AI contexts.

These numbers represent national averages. Salary ranges vary significantly by geography, with tech hubs commanding 20-40% higher compensation. Regional variations, experience-level breakdowns, and complete salary bands appear in Appendix D. Detailed job descriptions reflecting these capability requirements appear in Appendix E.

***Data compiled from Glassdoor, Salary.com, Indeed, Built In, PayScale, and ZipRecruiter** (January 2026). Salary ranges vary significantly by geography (20-40% higher in tech hubs) and experience level.*

ROLE COLLABORATION PATTERNS

Roles do not exist in a vacuum. The effectiveness of an AI Scrum team is determined by the quality of the interactions *between* the roles. Five critical interfaces distinguish AI Scrum from traditional models.

Product Strategist ↔ AI Alignment Engineer (The Intent Interface): This collaboration translates outcomes into AI-compatible specifications. During Sprint Inception, the Strategist presents a vague outcome. The Alignment Engineer challenges it—"The AI will interpret 'fast' as 'low latency,' but you mean 'few clicks.' We need to specify that." This dialogue refines the strategy before a single line of code is generated.

AI Alignment Engineer ↔ AI-Augmented Developers (The Generation Interface): The Alignment Engineer provides the prompt libraries. The Developer uses them to generate code. This creates a pattern of pair programming focused on prompting. When a developer gets garbage output from the AI, the Alignment Engineer sits with them to debug the *prompt*, not the code.

Validation Lead ↔ AI-Augmented Developers (The Safety Interface): In traditional Scrum, this was often adversarial. In AI Scrum, it is cooperative. Developers implement validation hooks *while* they generate the code, treating the "Test Strategy" as a requirement rather than an afterthought.

Governance Coordinator ↔ Integration Architect (The Compliance Interface): They jointly review data flows. The Architect identifies where data is moving (e.g., to a vector database), and the Coordinator checks for

PII or regional compliance. They design the integration to be compliant by default.

Delivery Facilitator ↔ All Roles (The Flow Interface): The Facilitator constantly monitors the "Hidden Queues." If they see work piling up at the Validation stage, they swarm the team to assist the Validation Lead, acting as the fluid dynamics engineer for the team's workflow.

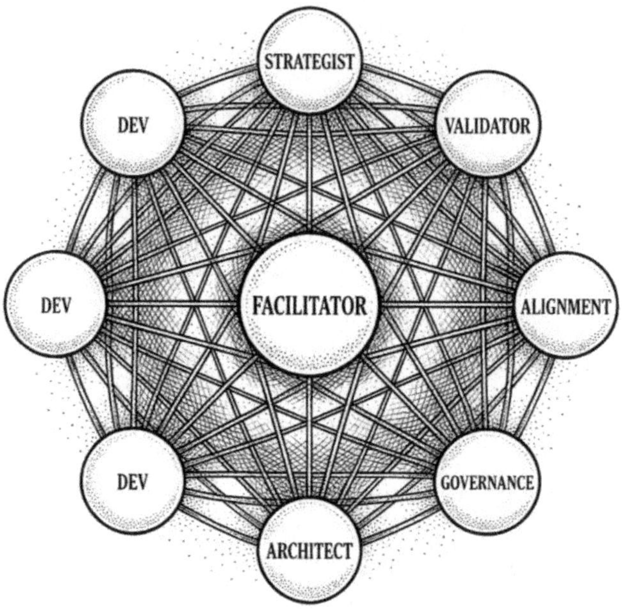

Figure 4.5: The Human-AI Lattice. *High-performance AI teams function as a "Human-AI Lattice" rather than a hierarchy. It emphasizes that success relies on high-bandwidth lateral communication and distributed accountability instead of top-down control. The structure places the Facilitator at the center, connecting specialized roles like Strategists and Validators directly to Developers.*

TRADITIONAL VS. AI SCRUM RACI MATRICES

The most profound change in responsibilities is the shift in accountability. We use a RACI matrix (Responsible, Accountable, Consulted, Informed) to visualize this shift.

Understanding how responsibility patterns change between traditional and AI Scrum illuminates why role evolution matters.

In this traditional pattern, Developers are Accountable and Responsible for development work. QA Engineers are Accountable and Responsible for testing. Product Owners are Accountable for backlog and stakeholder communication. Scrum Masters are Accountable for retrospectives. The pattern reflects development as the primary constraint.

TRADITIONAL SCRUM RACI MATRIX				
Activity	Product Owner	Scrum Master	Developers	QA Engineer
Backlog Management	A/R	C	C	I
Sprint Planning	A	R	R	C
Requirements Definition	A/R	C	C	I
User Story Writing	A/R	C	C	I
Development/Coding	I	I	A/R	I
Code Review	I	I	A/R	C
Test Plan Creation	C	I	C	A/R
Testing/QA	C	I	C	A/R
Sprint Review	A	R	R	R
Retrospective	R	A/R	R	R
Architecture Decisions	C	I	A/R	C
Deployment	C	I	A/R	C
Stakeholder Management	A/R	C	I	I
Impediment Removal	C	A/R	R	C
Definition of Done	A	R	R	R

Table 4.2: Traditional Scrum RACI Matrix. *The traditional Scrum RACI matrix shows clear separation between product, development, and process concerns.*

Legend: R = Responsible (does the work), A = Accountable (final approval), C = Consulted (input sought), I = Informed (kept updated)

Traditional vs. AI Scrum RACI Matrices

The AI Scrum RACI matrix shows dramatically different responsibility patterns reflecting the inverted constraint hierarchy:

AI SCRUM RACI MATRIX							
Activity	Product Strategist	AI Alignment Engineer	Validation Lead	Governance Coordinator	Integration Architect	AI-Augmented Developer	Delivery Facilitator
Outcome Definition	A/R	C	C	C	C	I	R
Sprint Inception	A	R	R	R	R	C	R
Success Criteria Definition	A	C	A/R	C	C	I	C
AI Prompt Engineering	C	A/R	C	I	C	R	I
Validation Strategy Design	C	C	A/R	C	C	I	C
Governance Framework Setup	C	C	C	A/R	C	I	C
Architecture Guidance	C	C	I	I	A/R	C	I
Multi-Candidate Generation	I	R	C	I	C	A/R	I
Candidate Evaluation	C	C	A/R	C	C	R	I
Governance Checkpoint	C	C	C	A/R	C	I	R
AI Show & Tell	A	R	R	C	C	R	R
Integration Testing	C	C	R	C	A/R	R	I
Validation Execution	C	C	A/R	C	C	R	I
Sprint Review	A	R	R	R	R	R	R
Retrospective	R	R	R	R	R	R	A/R
Stakeholder Alignment	A/R	C	I	I	I	I	C
Impediment Removal	C	C	C	C	C	C	A/R
Deployment Authorization	A	C	C	A	C	C	C
AI Collaboration Protocol	C	A/R	C	C	C	R	C
Prompt Library Maintenance	I	A/R	C	I	C	C	I

Table 4.3: AI RACI Matrix. *The AI Scrum RACI matrix shows a more complex but necessary responsibility patterns reflecting the inverted constraint hierarchy caused by adding AI.*

Legend: R = Responsible (does the work), A = Accountable (final approval), C = Consulted (input sought), I = Informed (kept updated)

Key Shifts to Observe:

- **Validation Lead:** Now holds more "Accountable" (A) roles than any other single position. In traditional Scrum, the Product Owner was accountable for the "What." In AI Scrum, the Validation Lead is accountable for the "Quality of the What."

- **Governance Coordinator:** Gains "Stop Authority." They are Accountable for the *legality* of the deployment. This is a hard gate that did not exist previously.
- **Developers:** Are "Responsible" (R) for generation but rarely "Accountable" (A) for the outcome. The AI generates the code. The human system validates it. The accountability has moved up the stack.
- **Sprint Inception:** Note that *all* roles are "Responsible" during Sprint Inception. This reflects the "Shift Left" philosophy. All the thinking happens before the generation starts.

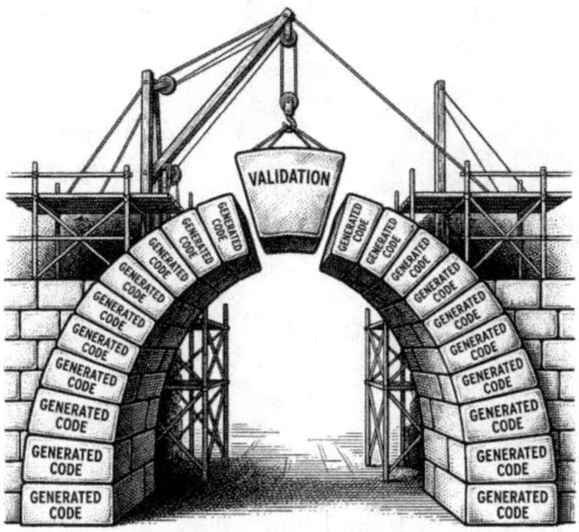

Figure 4.6: The Structural Keystone. *Human validation is the essential keystone holding together an arch of AI-generated code. Without this critical piece, the structure is unstable. The team's focus must therefore shift from simply producing raw code to ensuring structural integrity through rigorous review to prevent collapse.*

TEAM SIZE AND COMPOSITION GUIDELINES

AI Scrum teams typically include **9 to 11 people**. While this is larger than the "two-pizza" teams of early Agile, it is optimized for high-throughput, multi-stream delivery. Because AI augmentation multiplies individual output, a team of this size does not swarm on a single user story. Instead, they operate as a **multi-threaded delivery engine**, capable of managing **three to five parallel initiatives** simultaneously.

The **Optimal Composition** includes:

- **1** Product Strategist
- **2** AI Alignment Engineer
- **2** Validation Lead
- **1** Governance Coordinator

- **1** Integration Architect
- **2-3** AI-Augmented Developers
- **1** Delivery Facilitator

This composition ensures coverage of all critical constraints (validation, governance, alignment) while maintaining enough implementation capacity (via the 2-3 developers) to drive multiple workstreams.

Figure 4.7: A Typical AI Scrum Team. *Consisting of up to 11 members with some positions being able to be shared between multiple AI Scrum teams.*

However, smaller initiatives or constrained environments may still combine roles. Common combinations include:

AI Alignment Engineer + AI-Augmented Developer: For teams with strong technical capabilities.

Governance Coordinator + Delivery Facilitator: For teams with mature governance processes.

Validation Lead + Integration Architect: For teams with strong technical depth where one leader can guide both concerns.

The key principle is **explicit role accountability**. When combining roles, teams must document which person owns which responsibilities to prevent accountability gaps.

ORGANIZATIONAL IMPLICATIONS

Adopting the roles defined in this chapter requires more than updated job descriptions. It requires a fundamental evolution of the organization's support structures. You cannot drop these new roles into a legacy environment and expect them to thrive. You must build the scaffolding that supports them.

1. The Centers of Excellence (CoE) Strategy

For organizations new to the concept, a Center of Excellence (CoE) is not a passive committee or a monthly "lunch and learn." It is a formal body responsible for **standardization, knowledge management, and capability building** within a specific domain.

In an AI Scrum transformation, individual teams will solve problems rapidly. Without a CoE, Team A solves a "Hallucination Issue" in Python while Team B struggles with the exact same issue in JavaScript for three weeks. The CoE exists to prevent this waste.

Start by establishing two specific CoEs:

A. The AI Alignment CoE

- **Purpose:** To centralize prompt engineering strategies, model selection standards, and "Context Specification" templates.
- **Composition:** Led by a Principal AI Alignment Engineer. Members include the Alignment Engineers from each Scrum team.
- **Cadence:** Meets bi-weekly for 60 minutes.
- **Key Output:** The "Prompt Library"—a shared repository of vetted prompts for code generation, documentation, and unit testing.
- **Time to Establish:** Start informally in Month 1. Formalize with a charter by Month 3. *(See Appendix F for CoE Charter Templates).*

B. The Governance & Risk CoE

- **Purpose:** To interpret regulations (EU AI Act, NIST), define data privacy standards, and approve new AI tools.
- **Composition:** Led by a Chief Risk Officer or Legal delegate. Members include Governance Coordinators from each team and a representative from Cyber Security.
- **Cadence:** Meets monthly, or ad-hoc when new regulations/tools emerge.
- **Key Output:** The "Governance Canvas" template and the "Approved Tools List."

How to Start (The "Crawl, Walk, Run" Model):

1. **Crawl (Weeks 1-4):** Identify one enthusiast per team. Create a dedicated Slack/Teams channel for them to share "what's working."
2. **Walk (Months 2-6):** Appoint a rotating chair. Schedule the bi-weekly meeting. creating the first draft of the Prompt Library.
3. **Run (Month 6+):** Grant the CoE budget for training and tooling. Give them authority to "certify" prompts for production use.

2. The New Hiring Strategy: Assessing the AI Mindset

The traditional "Whiteboard Interview"—where a candidate writes algorithms from memory—is obsolete. In an AI Scrum world, that tests a skill (memorization) that has zero economic value. We need to assess **T-Shaped Skills** and **AI Collaboration Aptitude**.

Defining T-Shaped Skills

A "T-Shaped" employee has deep expertise in one specific vertical (the vertical bar of the T) but possesses broad proficiency across adjacent disciplines (the horizontal bar).

- ***Legacy Developer:*** Deep in Java (Vertical). Knows nothing of testing or product strategy.
- ***AI-Augmented Developer:*** Deep in Java (Vertical). Broad proficiency in Prompt Engineering, Validation logic, and Basic Architecture (Horizontal), enabled by AI tools.

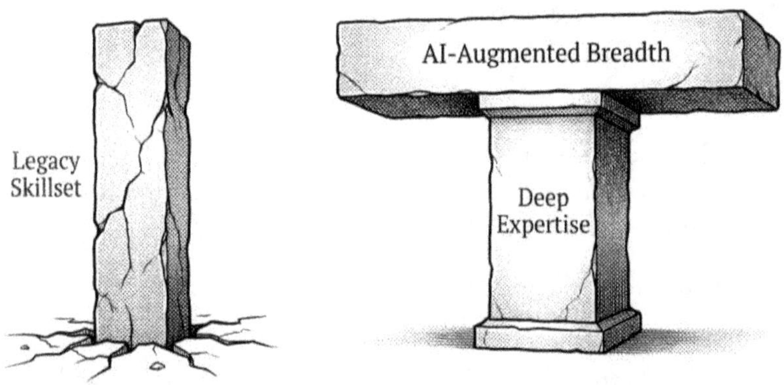

Figure 4.8: The T-Shaped Evolution. *This illustrates that AI does not replace deep expertise (the vertical), especially in organizational tribal knowledge, but allows the specialist to span across adjacent disciplines (the horizontal), creating a more stable and valuable professional profile.*

The "AI-Paired" Interview Protocol

To assess these skills, stop asking candidates to code alone. Ask them to code with AI.

- **The Task:** Give the candidate a vague requirement (e.g., "Build a Python service that scrapes this website and stores the data in SQL").
- **The Constraint:** They *must* use an LLM (ChatGPT, Claude, Gemini) to do it. You are watching their screen.
- **What to Look For (The Rubric):**

Signal	Weak Candidate	Strong Candidate
Prompting	Writes one vague sentence: "Write code to scrape data."	Writes a detailed context specification: "Act as a Python expert. Create a scraper using BeautifulSoup. Handle 403 errors with exponential backoff."
Validation	Pastes the AI code directly into the IDE and tries to run it immediately.	Reads the AI code first. Spots logical errors *before* pasting. "Wait, this loop will fail if the list is empty."
Troubleshooting	When the code errors, they ask the AI "It didn't work" or try to fix it manually while ignoring the AI.	Pastes the error trace back into the AI with context: "I got a 403 error. Refactor the request headers to mimic a browser."
Mental Model	Treats the AI like a magic wand.	Treats the AI like a junior developer who makes typos.

Table 4.4: AI-Paired Interview Rubric. *The interviewee is provided an AI question and it is important to see how they respond in prompting, validation, troubleshooting, and the mental model. Their strength can be quickly determined how they approach the prompt and which LLM they use. While Microsoft Copilot will work, it is a product ecosystem (application layer) and not a true LLM such as Claude, Gemini, or ChatGPT.*

Key Interview Question: "Tell me about a time an AI model hallucinated a solution you were working on. How did you identify it, and how did you correct the prompt to fix it?"

(**Note:** If they can't answer this, they haven't used AI for serious work.)

3. Career Paths: The Ladder of Leverage

In traditional IT, career progression meant "Junior Developer" to "Senior Developer" to "Architect," largely based on how complex a system you could build in your head. In AI Scrum, progression is based on **Leverage**—how much value you can generate by orchestrating AI and human systems.

Track 1: The Technical Evolution

- **Stage 1: AI-Augmented Developer.** Focus: Individual contribution. Learning to prompt for code and write validation hooks.
- **Stage 2: AI Alignment Engineer.** Focus: Team enablement. Moving from writing code to designing the *prompts* that write the code. Curating the library that multiplies the team's efficiency.
- **Stage 3: Integration Architect.** Focus: System coherence. Managing the complexity of connecting multiple AI agents and legacy systems.

Track 2: The Quality & Governance Evolution

- **Stage 1: Validation Lead.** Focus: Team safety. Designing the tests that catch hallucinations for a single product.
- **Stage 2: Governance Coordinator.** Focus: Compliance. Ensuring the product meets legal/ethical standards.
- **Stage 3: Quality Systems Architect.** Focus: Organizational safety. Designing the automated governance frameworks that span multiple teams.

Track 3: The Product Evolution

- **Stage 1: Product Strategist.** Focus: Outcome definition. Managing the "What" and "Why" for a single team.
- **Stage 2: Value Stream Architect.** Focus: Multi-team orchestration. coordinating AI Scrum teams to deliver complex, multi-product value streams.

The Promotion Gap: To move from Stage 1 to Stage 2 in any track requires mastering **"Systemic Prompting"**—the ability to design prompts that work not just once, but reliably for others.

FROM THE TRENCHES: THE ANATOMY OF A ROLE TRANSITION

Role transition conversations often go badly when leaders assume people will naturally see how their skills transfer. The most dangerous moment is the initial conversation.

Consider the illustrative case of "Sarah," a veteran Scrum Master of ten years. If a leader hands her the new job description for "Delivery Facilitator" and says, "We need you to learn AI telemetry, manage the governance dashboard, and coach developers on prompt engineering," Sarah hears only one thing: *"Your current skills are obsolete."* She feels undervalued and overwhelmed. In many cases, this leads to the resignation of key culture carriers who take years of institutional knowledge with them.

Contrast this with a better approach. Imagine the conversation with "Marcus," another Scrum Master. The leader starts differently:

"Marcus," the leader says, "Your facilitation skills and ability to identify impediments are exactly what we need. Those capabilities remain critical. The role is *expanding* to include coaching on AI collaboration patterns. We will train you on these patterns. You will still use your core facilitation skills, just in a broader context."

The Lesson: Role evolution is about capability growth, not capability invalidation. If you frame it as a replacement of their identity, they will leave. If you frame it as an upgrade to their toolkit, they will stay.

WHAT COMES NEXT

These seven roles and their collaboration patterns establish the *human* structure for AI Scrum teams. But roles alone do not change how teams operate day-to-day. You can change the titles on LinkedIn, but if the meetings remain the same, the output will remain the same. Teams need concrete practices—ceremonies, cadences, and workflows—that translate these roles into daily action.

In **Chapter 5**, we will define the **AI Scrum Ceremonies** in detail. You will see exactly how Sprint Inception differs from traditional sprint planning. We will define why "AI Show & Tell" ceremonies matter. We will explore how validation and governance integrate throughout sprints and how retrospectives incorporate AI-generated insights.

KEY TAKEAWAYS

- **Align roles to the constraint hierarchy.** When AI eliminates the development constraint, validation completeness, governance approval, and integration complexity become the dominant constraints. Your organizational chart must mirror your problems.
- **The Validation Lead is the new power center.** This role now holds the most accountability for delivery safety. They are no longer checking work at the end; they are designing the criteria at the start.
- **Budget for the premium.** Expect a 26% increase in team costs ($152k/team). AI Alignment Engineers command the highest premiums due to scarcity. Do not attempt this transformation on a flat budget; you will end up with tools you cannot wield.
- The cost side of this transition is quantified. The return side deserves the same rigor. Appendix 20 provides a worked three-year ROI model for a mid-sized organization transitioning three teams, with conservative estimates across sprint velocity improvement, defect cost reduction, time-to-market acceleration, and developer retention, the four benefit categories that translate most directly into language a CFO and CIO will find credible. A blank template follows for you to populate with your own figures. If you are preparing a business case for AI Scrum adoption, start there.
- **Governance must be embedded.** The Governance Coordinator is not bureaucratic overhead. They are the enabler that allows the team to move fast without hitting regulatory walls.
- **Collaborate or die.** The effectiveness of the team is defined by the interfaces (e.g., Strategist ↔ Alignment Engineer), not the individual silos.
- **Transition takes time and empathy.** Plan for 2–6 month transition timelines. Do not assume your existing staff can pivot overnight without explicit training and psychological safety. Frame the change as growth, not replacement.

MIND MAP – ROLES AND RESPONSIBILITIES

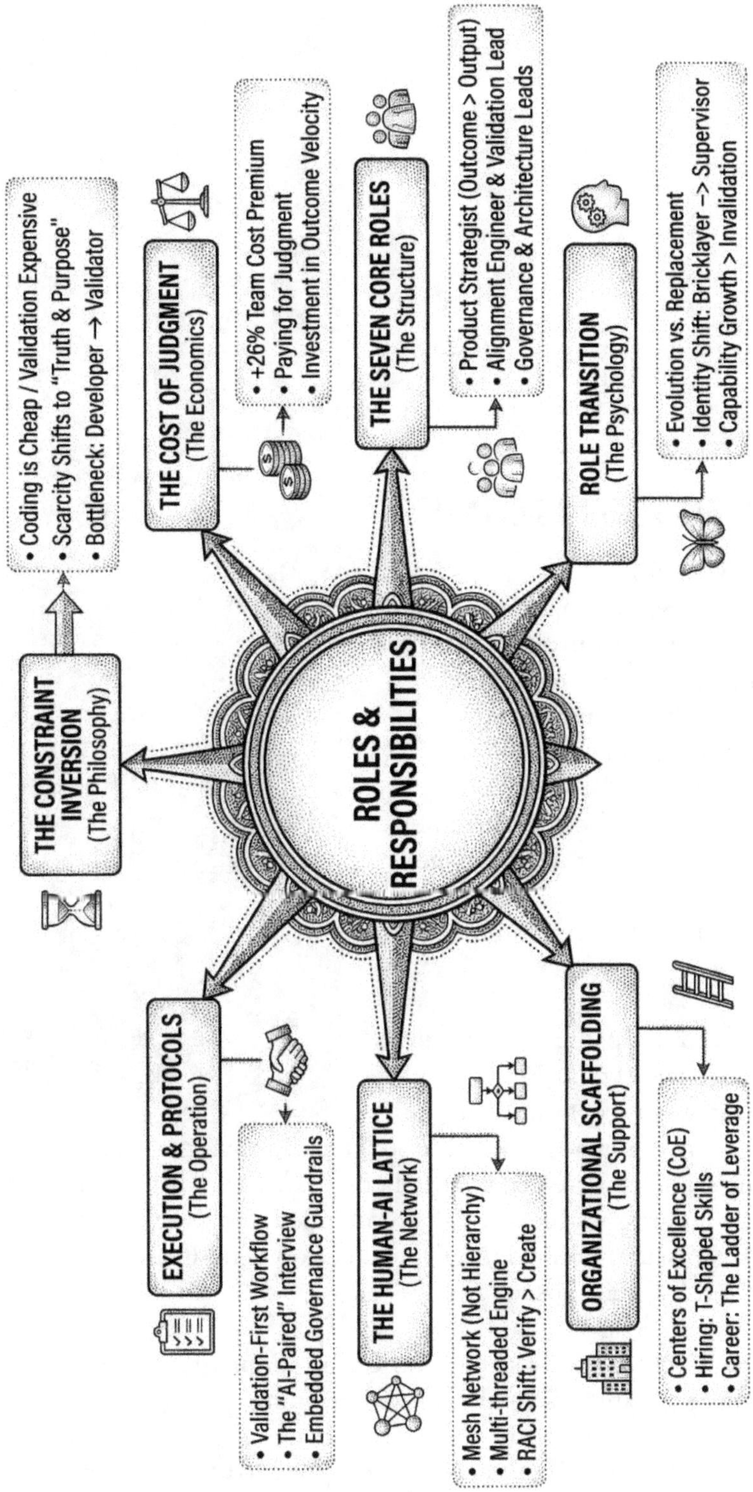

CHAPTER 5: CEREMONIES AND RITUALS

When AI accelerates coding, the bottleneck shifts to validation. This framework replaces traditional status meetings with rigorous coordination focused on verification. Teams utilize front-loaded planning and mid-sprint governance checkpoints to manage risk. The approach prioritizes objective evidence over feature demonstrations, ensuring that high-speed execution remains safe, compliant, and aligned with business intent.

In traditional Scrum implementations, teams often discover problems too late. The sprint dashboard might show eighteen blocked items, eleven escalating risks, and completion trending toward 60%. Backlog grooming sessions focus on breaking down user stories, but dependencies and integration concerns remain hidden until sprint execution reveals them.

I experienced this pattern repeatedly as a Scrum Master. My product owner partnership was exceptional. We thought along the same path and brought complementary skills to grooming sessions. I created comprehensive dashboards in Azure DevOps that made progress visible through color-coded tiles: New, Blocked, Risks/Bugs/Issues, Refined, Analysis, Development, Ready to Promote to Test, QA, UAT, and Closed.

When the Blocked and Risks tiles turned red sprint after sprint, the pattern became unmistakable. We had mastered planning sprints and breaking down stories, yet we consistently underestimated dependencies and missed integration concerns. All our ceremonies optimized for development velocity, but our actual bottleneck was validation and integration. We were solving the wrong problem.

This realization, repeated across thousands of Scrum teams, drives the ceremony redesigns in this chapter. When AI eliminates the development bottleneck, ceremony structure must change fundamentally. Traditional Scrum ceremonies optimized for coordination during slow development. AI Scrum ceremonies optimize for validation orchestration, governance integration, and rapid stakeholder feedback during fast implementation.

WHAT THIS CHAPTER COVERS

This chapter presents the complete AI Scrum ceremony framework, explaining both the theoretical foundation and practical implementation guidance. The following topics are covered in detail:

- **Ceremony Philosophy:** Why validation orchestration replaces development coordination as the primary ceremony purpose.
- **Sprint Inception:** A four-hour planning ceremony establishing validation frameworks and governance requirements before implementation.
- **Daily Synchronization:** Validation-focused coordination replacing task-status standups.
- **AI Show & Tell:** Mid-sprint stakeholder engagement enabling rapid course correction.
- **Governance Checkpoint:** Integrated governance review preventing late-stage rejection.
- **Sprint Review:** Evidence-based outcome demonstration replacing feature showcases.
- **Retrospective with AI Insights:** Data-driven learning incorporating human-AI collaboration telemetry.
- **Timing Flexibility:** When ceremonies should flex from traditional schedules and when they should not.
- **Implementation Patterns:** How organizations should adapt ceremonies to their specific contexts.

The dashboard-driven approach that revealed dependency problems in traditional Scrum translates directly to validation tracking in AI Scrum. The backlog grooming discipline that enabled effective sprint planning becomes thorough validation definition in AI Scrum. The specific practices reflect what should work based on team dynamics research, constraint theory, and lessons from traditional Scrum transformations.

KEY TERMS AND CONCEPTS

- **Validation Orchestration:** The primary purpose of AI Scrum ceremonies, replacing development coordination. Ceremonies ensure validation happens systematically rather than ad hoc, governance integration occurs at appropriate points, and stakeholder feedback arrives when course correction remains practical.
- **Ceremony Flexibility:** The principle that ceremony timing should serve team coordination rather than calendar compliance. While maintaining ceremony discipline, AI Scrum adapts timing to actual work patterns and decision points rather than rigid schedules.
- **Evidence-Based Review:** Sprint review approach emphasizing validation evidence over feature demonstration. Teams present validation matrices, test results, governance approvals, and outcome achievement data rather than merely demonstrating that features work.
- **Governance Checkpoint:** Mid-sprint formal governance review occurring when teams can still adapt based on governance feedback. This replaces traditional end-of-sprint approval gates that created adversarial dynamics and costly late-stage changes.
- **Outcome-Focused Planning:** Sprint planning approach that commits to business outcomes rather than feature delivery. Teams define what business results they will achieve and what validation evidence they will accumulate rather than which tasks they will complete.
- **AI Show & Tell:** Structured stakeholder engagement occurring multiple times per sprint to gather rapid feedback on AI-generated prototypes and validated designs. This replaces the traditional pattern of waiting until sprint end for stakeholder input.
- **Retrospective Telemetry:** Objective data about human-AI collaboration performance incorporated into retrospective analysis alongside traditional subjective team observations. This enables systematic learning about partnership effectiveness.
- **Sprint Inception:** Extended planning ceremony establishing validation frameworks and governance requirements. This replaces traditional sprint planning's focus on task breakdown and capacity allocation.

CEREMONY PHILOSOPHY

AI Scrum reimagines traditional Scrum ceremonies to address AI-native development needs. While preserving the collaborative spirit and iterative cadence that make agile valuable, these ceremonies shift focus from development coordination to validation orchestration, governance integration, and outcome verification.

Traditional Scrum designed ceremonies around a fundamental constraint. Implementation takes time, and sprint planning allocated development capacity across user stories. Daily standups coordinated development activities and identified blockers, while sprint reviews demonstrated completed functionality. The entire ceremony structure assumed development would consume most of the sprint, requiring careful coordination to optimize developer productivity.

When AI eliminates that constraint, ceremony purpose must change. Sprint planning now focuses on validation strategy rather than task allocation. Daily coordination tracks validation progress rather than development status. Reviews emphasize validation evidence rather than feature demonstration. Each ceremony adapts to the new reality that implementation happens rapidly while validation, governance, and integration require sustained attention.

The data-driven approach that made traditional ceremonies effective translates directly to AI Scrum. In traditional Scrum, my Azure DevOps dashboards revealed what standard sprint boards obscured. Blocked items accumulating, risks escalating, integration concerns emerging too late. The four-by-four bug priority matrix (severity versus business priority) showed patterns that individual bug tickets missed. Teams could see at a glance where urgent attention was needed.

AI Scrum applies this same visibility principle to validation and governance.

Dashboards track validation criteria satisfied rather than tasks completed. Telemetry reveals AI collaboration patterns rather than development velocity. The fundamental principle remains constant across contexts. Make progress visible in ways that enable effective coordination. Only the metrics change to match new constraints.

Traditional Scrum prescribed ceremony timing with considerable rigidity. Sprint Planning on Day 1, Daily Standups every morning at a specific time, and Sprint Review on the final afternoon. This rigid cadence served important purposes by establishing predictable rhythm and preventing ceremony neglect. But AI development reveals the costs of inflexible timing.

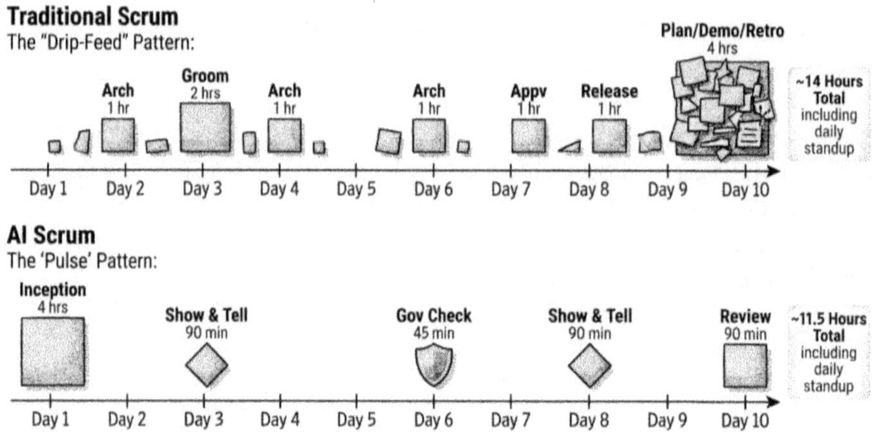

Figure 5.1: The Inversion of Rhythm. *Traditional Scrum (top) suffers from "calendar creep," with ad-hoc meetings (Architecture/Code Review, Governance, Release Prep) accumulating throughout the sprint and a heavy crunch on Day 10. AI Scrum (bottom) deliberately front-loads the planning (Inception) to enable high-speed execution, with structured 90-minute validation checkpoints replacing the ad-hoc meetings.*

AI Scrum maintains ceremony discipline while introducing timing flexibility. Sprint Inception remains foundational because detailed planning enables validation-first development. But daily synchronizations happen when teams need coordination, not at arbitrary clock times. AI Show & Tell sessions occur when teams have meaningful work to demonstrate, not on predetermined days. Governance Checkpoints align with natural decision points rather than rigid schedules.

The key insight from this experience is clear. Ceremonies serve team coordination, not calendar compliance. When rigid timing impedes rather than enables coordination, flexibility proves valuable. But flexibility without discipline collapses into chaos. AI Scrum maintains ceremony rigor while adapting timing to actual team needs and work patterns.

CEREMONY 1: SPRINT INCEPTION

Sprint Inception replaces traditional Sprint Planning with a significantly expanded ceremony recognizing that validation-first development requires detailed upfront definition. Four hours on Day 1 establishes the foundation that enables the remainder of sprint success.

Traditional sprint planning typically consumed two to four hours, focusing primarily on which user stories the team could complete and how to break them into tasks. Teams estimated story points, calculated velocity, and allocated stories to fit capacity. The ceremony assumed implementation would consume most sprint effort, so planning optimized for development throughput.

Sprint Inception takes a fundamentally different approach. Teams commit to outcomes rather than features. They define validation criteria before any implementation begins. They establish governance requirements upfront rather than discovering them late. They plan validation capacity rather than development throughput. This requires substantially more upfront work, but it prevents the rework that plagued traditional Scrum when teams built the wrong things or missed critical requirements.

Outcome Selection (60 minutes)

The Product Strategist presents candidate outcomes from the Outcome Backlog. Unlike traditional user stories focused on features, outcomes describe business objectives and customer problems. The team evaluates each candidate, assessing feasibility given current sprint capacity, dependencies on other work, risk levels requiring governance attention, and strategic value to the organization.

The team commits to two to four outcomes per sprint, fewer than traditional story counts but each outcome represents substantial business value. This selection embodies outcome orientation. Teams commit to impact rather than activity. Organizations transitioning from feature-based planning to outcome-based planning initially struggle with this abstraction level. Product managers accustomed to specifying solutions must learn to articulate problems. Teams accustomed to building requested features must embrace accountability for business outcomes. This transition typically requires three to four sprints before becoming natural but produces dramatically better stakeholder alignment.

Validation Definition (90 minutes)

For each selected outcome, the team completes a Validation Canvas defining success criteria. This phase demands the ceremony's most intensive work. Teams must articulate not just functional requirements but performance benchmarks, security standards, compliance obligations, integration requirements, stakeholder acceptance criteria, and risk mitigation requirements.

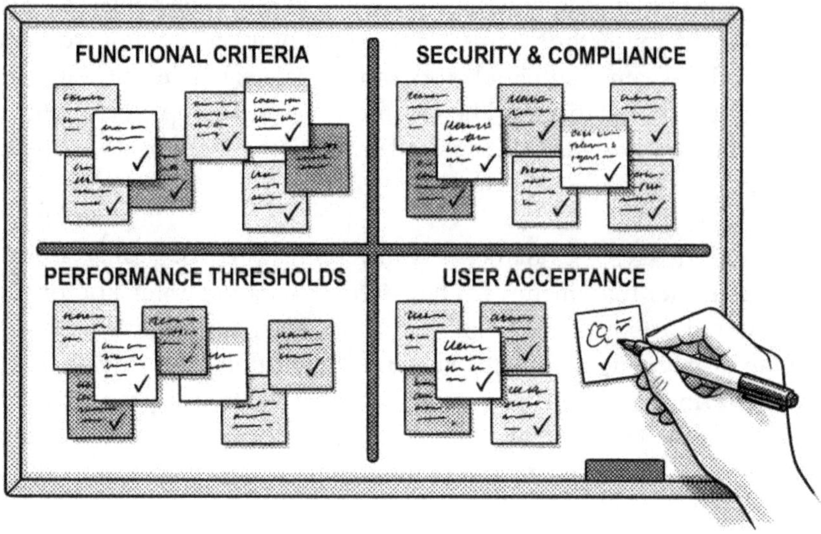

Figure 5.2: The Validation Canvas. *In the AI era, the primary planning activity shifts from breaking down tasks (how long will it take?) to defining validity (what does "right" look like?). This artifact forces the team to articulate security, compliance, and performance constraints before a single prompt is written.*

AI participates actively in validation definition. Given outcome descriptions, AI can suggest relevant validation criteria based on similar past work, identify regulatory requirements that might apply, propose performance benchmarks aligned with system architecture, and surface integration considerations teams might overlook. This AI collaboration proves particularly valuable for teams new to validation-first thinking because AI draws on patterns from thousands of projects while individual team members have more limited experience.

The Validation Canvas becomes the sprint's central artifact. Every subsequent activity (prototype generation, testing, governance review) refers back to these criteria. The canvas provides objective measurement of sprint progress. Teams track what percentage of validation criteria they have satisfied rather than subjective "percent complete" estimates.

In traditional Scrum, my partnership with our product owner made backlog grooming exceptionally effective. He brought prioritization based on stakeholder engagement. I brought data visibility through dashboards and systematic identification of work that needed completion before sprint planning. Our complementary skills meant sprint planning sessions started with a pristine backlog of ready items. Validation Definition requires similar discipline. Comprehensive criteria definition before implementation enables effective validation throughout the sprint.

Governance Framing (45 minutes)

The Governance Coordinator reviews each outcome's risk profile using the four-tier decision framework. High-risk outcomes trigger governance board notification. Medium-risk items establish review criteria the Governance Checkpoint will evaluate. Low-risk outcomes receive delegated team authority within documented constraints.

This governance framing prevents surprises. Teams know from Day 1 what governance oversight to expect. Governance stakeholders know which sprints require their attention. The transparency prevents both governance bottlenecks and inappropriate risk-taking. Organizations implementing governance framing initially resist the time investment. Forty-five minutes seems excessive for risk classification. However, this upfront framing eliminates the chaos that results when governance concerns emerge late in sprints. Teams know what documentation to prepare. Governance boards can schedule reviews efficiently. The upfront investment pays substantial dividends in reduced late-stage rework.

Capacity and Dependency Planning (45 minutes)

The team identifies sprint capacity constraints. Who is available, what skills the team possesses, and what dependencies on external teams exist. This resembles traditional capacity planning but focuses on validation capabilities rather than development throughput. Questions shift from "How much can we code?" to "How much can we validate?"

Do we have security expertise for penetration testing? Can we access stakeholders for midweek feedback? Are integration test environments available when needed? Will governance board members participate in Checkpoint? These constraints shape sprint commitments more than development capacity estimates.

Traditional capacity planning often missed critical dependencies. The sprint dashboard I built revealed this pattern. Items blocked because another team had not delivered a prerequisite component, environments unavailable when testing needed to occur, stakeholders unavailable for acceptance decisions. Validation capacity planning surfaces these dependencies explicitly before they derail sprints.

AI Collaboration Protocol Review (30 minutes)

The final phase establishes how the team will work with AI during the sprint. Teams review prompt libraries from previous sprints, identify new prompt patterns needed for current outcomes, establish documentation standards for AI-generated content, assign AI collaboration logging responsibilities, and schedule AI Show & Tell sessions.

This protocol review prevents ad hoc AI usage. Teams approach AI collaboration with intention and structure rather than improvisation. Organizations that skip this phase often struggle with inconsistent AI utilization. Some team members use AI extensively while others barely engage, prompts get discarded rather than refined, and learning from one sprint does not transfer to subsequent sprints.

Sprint Inception concludes with explicit sprint commitments documented in the Sprint Commitment Summary. These commitments specify outcomes the team will achieve, validation evidence required, and governance milestones scheduled. The commitments are shared publicly across the organization, establishing accountability and visibility.

CEREMONY 2: DAILY SYNCHRONIZATION

Daily Synchronization replaces the traditional Daily Standup with a ceremony focused on validation progress rather than task completion. The timing remains daily but flexes to team needs. Asynchronous updates via collaboration tools often suffice, reserving synchronous gathering for situations genuinely requiring real-time coordination.

Traditional standups followed a rigid format. What did you do yesterday? What will you do today? What blockers do you face? This format optimized for development coordination, surfacing when developers needed help from teammates or when external dependencies blocked progress. The ceremony worked reasonably well when development consumed most sprint effort.

When validation, governance, and integration become dominant constraints, standup questions need updating. Teams must know what validation criteria were satisfied, what evidence was accumulated, what governance feedback arrived, and what integration concerns emerged. Task completion becomes less relevant than validation progress.

The Four Questions

Daily Synchronization addresses four questions that maintain outcome orientation while providing necessary coordination.

What validation did we complete yesterday? Teams report validation criteria satisfied, not tasks completed. "We validated performance benchmarks and confirmed database queries return results under 100 milliseconds" rather than "I finished implementing the query optimization". This validation focus maintains outcome orientation while providing visibility into genuine progress.

What validation are we pursuing today? Teams describe validation activities planned, identifying what success criteria they will address and what evidence they will accumulate. This creates coordination opportunities when multiple team members must collaborate on validation activities.

What validation blockers exist? Obstacles preventing validation progress receive immediate attention: missing test data, unavailable stakeholders, broken integration environments, pending governance feedback. The team problem-solves blockers collaboratively or escalates to appropriate decision-makers.

What AI collaboration worked well or poorly? Teams share learnings about effective AI interactions: prompts that produced valuable outputs, validation approaches AI assisted with successfully, situations where AI suggestions proved unhelpful. This continuous AI collaboration learning supplements formal retrospective analysis.

Figure 5.3: The Validation-First Board. *The visual workflow of the team must change to reflect the new bottleneck. Unlike traditional boards that track coding activity ("Doing"), the AI Scrum board makes the invisible work of verification and governance visible, preventing "phantom velocity" where code is written but not safe.*

Format Flexibility

The ceremony typically completes in 15 minutes for teams of five to seven members. Asynchronous formats using Slack or team collaboration tools often work better than synchronous meetings, particularly for distributed teams. The key to effective coordination requires understanding. Daily visibility into validation progress and rapid blocker resolution.

Many teams initially struggle with validation-focused synchronization. Engineers habituated to reporting tasks completed must reorient to discussing validation evidence accumulated. This transition usually requires two to three sprints before it feels natural. Teams report that once the shift occurs, daily synchronization becomes significantly more valuable than traditional standups because it focuses on genuine progress rather than activity.

In traditional Scrum, daily standups revealed blocked items that dashboards also showed. The dashboard provided aggregated visibility while standups enabled immediate problem-solving for specific blockers. This dual visibility proved powerful. Dashboards for patterns, standups for rapid resolution. AI Scrum maintains this same dual approach with shifted metrics.

CEREMONY 3: AI SHOW & TELL

AI Show & Tell sessions represent AI Scrum's most innovative ceremony. Structured stakeholder engagement occurring twice per sprint (typically Days 3 and 7) to gather rapid feedback on AI-generated prototypes and validated designs. Traditional Scrum delayed stakeholder engagement until Sprint Review on Day 10, creating risk that teams would build wrong things discovered too late for easy correction.

AI's rapid implementation capability makes early stakeholder feedback practical and valuable. Why wait until Day 10 to learn that stakeholders need different capabilities when course correction on Day 3 costs almost nothing?

Session Structure

The ceremony proceeds through a structured format designed to gather actionable feedback efficiently.

Context Setting (10 minutes): The Product Strategist reminds attendees of sprint outcomes and the business objectives being addressed. This context prevents stakeholders from proposing unrelated enhancements or questioning fundamental directions already decided. The session focuses on refining specific outcomes, not debating strategy.

Prototype Demonstration (30 minutes): The team demonstrates AI-generated prototypes addressing sprint outcomes. Unlike traditional demos showcasing working software, these prototypes emphasize design concepts, user experience flows, and alternative approaches. The team might show three different interface designs, two architectural patterns, or multiple algorithm approaches. This breadth enables stakeholders to provide directional feedback rather than merely commenting on implementation details.

Structured Feedback (40 minutes): Stakeholders provide structured feedback using three categories. "This works well" identifies elements that should be preserved or emphasized. "This needs adjustment" specifies changes required for acceptance. "This misses the mark entirely" identifies prototypes the team should abandon. This structure prevents the rambling feedback sessions that plagued traditional demos. Stakeholders know exactly what type of input the team needs. Teams capture feedback systematically rather than trying to remember comments from chaotic discussions.

Figure 5.4: The Rapid Feedback Loop. *Because AI can generate multiple prototypes cheaply, the "Show & Tell" shifts from a linear demo to an options evaluation. This allows stakeholders to steer the product on Day 3 based on working models, rather than waiting for a "Big Bang" reveal on Day 10.*

Validation Alignment (30 minutes): The team reviews current validation criteria against stakeholder feedback. Sometimes feedback reveals that validation criteria need refinement: stakeholders care about performance metrics the team had not emphasized or stakeholders accept trade-offs the team thought were unacceptable. This validation alignment ensures the team validates what actually matters to stakeholders.

Next Steps Agreement (10 minutes): The session concludes with explicit agreements about which approaches to pursue, what refinements to make, and what additional validation to conduct. These agreements drive the remainder of sprint work.

Timing and Frequency

Two sessions per sprint enable rapid iteration. The Day 3 session catches fundamental misalignments early. The Day 7 session validates refined approaches before final implementation. Some teams add a third session on Day 5 for complex outcomes requiring additional iteration.

Organizations implementing AI Show & Tell initially resist the overhead. Two hours twice per sprint seems excessive when traditional sprints had one review at the end. However, this mid-sprint engagement eliminates the massive rework that resulted when teams discovered stakeholder concerns on Day 10 after investing substantial implementation effort. The time investment in Show & Tell pays for itself multiple times through prevented rework.

CEREMONY 4: GOVERNANCE CHECKPOINT

The Governance Checkpoint ceremony integrates formal governance review into sprint workflows, occurring mid-sprint when teams can still adapt based on governance feedback. This represents a fundamental departure from traditional end-of-sprint approval gates that created adversarial dynamics and costly late-stage changes.

Traditional governance operated in a sequential manner. Teams completed work, then submitted it for governance approval, then waited for feedback that often arrived too late for practical course correction. This approach created three significant problems. Teams resented governance as an obstacle rather than a partner. Governance reviewers felt pressured to approve work even when concerns existed because rejections cost substantial time and money. Late-stage governance concerns required expensive rework or deployment delays.

Participants and Structure

Attendees include the Governance Board, the development team, the Product Strategist, and relevant domain experts. The ceremony follows a structured agenda designed to be thorough yet efficient.

Work Review (30 minutes): The team presents validated designs and prototypes, explaining architectural decisions, risk mitigation approaches, compliance considerations, and integration strategies. This review emphasizes what the team has validated and what evidence supports their design choices rather than merely describing what they built.

Risk Assessment (25 minutes): Governance participants evaluate risk across multiple dimensions: security vulnerabilities, compliance gaps, architectural inconsistencies, integration concerns, and operational readiness. Each concern receives explicit classification: blocking (must be addressed before deployment), concerning (needs mitigation plan), and noted (acceptable risk with documentation).

Requirement Clarification (20 minutes): Teams often discover that governance requirements were misunderstood or poorly specified. This phase surfaces misalignments while course correction remains practical. Governance participants clarify expectations, teams explain constraints that make certain requirements impractical, and both parties negotiate acceptable compromises.

Figure 5.5: The Governance Partnership. *By moving compliance review from the end of the sprint (as a gatekeeper) to the middle of the sprint (as a navigator), teams avoid the "Governance Wall." This mid-flight correction prevents the costly reworking of fully built features that violate regulatory or security policies.*

***Approval and Conditions (20 minutes)*:** The Governance Board provides explicit decisions: approved without conditions, approved with specified conditions, or approval deferred pending additional work. Conditional approvals specify exactly what teams must address for full authorization. Deferrals include clear criteria for what would enable approval.

Forward Planning (15 minutes): The ceremony concludes with forward planning for subsequent sprints. Governance participants identify upcoming work requiring their attention, regulatory changes affecting future sprints, and policy updates teams should incorporate. This proactive planning prevents surprises in future sprints.

Value and Trade-offs

The ceremony occasionally surfaces issues requiring sprint re-planning. A governance concern might invalidate a prototype approach, forcing teams to pursue alternatives. While disruptive, this mid-sprint discovery proves far less costly than discovering governance concerns after deployment. Organizations implementing Governance Checkpoint report 55% faster governance cycles and dramatically reduced late-stage rework compared to traditional sequential approval processes.

CEREMONY 5: SPRINT REVIEW

Sprint Review in AI Scrum emphasizes validation evidence over feature demonstration. While traditional sprint reviews focused on showcasing working software, AI Scrum reviews focus on demonstrating outcome achievement through detailed validation evidence. The ceremony proceeds through five sections designed to provide stakeholders with genuine confidence in solution quality.

Outcome Achievement Summary (15 minutes): The Product Strategist presents sprint outcomes committed to and achieved. This includes quantitative metrics showing outcome success. Revenue impact realized, customer satisfaction improvements measured, cost reductions achieved, or risk levels reduced. The summary directly ties sprint work to business objectives rather than merely listing completed features.

Validation Evidence Presentation (40 minutes): The Validation Lead presents comprehensive validation evidence across all dimensions. Functional validation demonstrates that solutions work as intended. Performance validation shows benchmark achievement. Security validation presents penetration testing results and vulnerability assessments. Compliance validation documents regulatory requirement satisfaction.

Integration validation demonstrates proper system interaction. Stakeholder validation presents acceptance feedback from actual users.

This evidence-based presentation provides stakeholders with genuine assurance about solution quality. Traditional demos showed that features worked in controlled environments. Evidence-based review demonstrates that solutions satisfy success criteria in realistic conditions.

Governance Status Update (15 minutes): The Governance Coordinator presents governance review outcomes. Approvals received, conditions satisfied, outstanding concerns addressed, and compliance validation completed. This transparency helps stakeholders understand what governance oversight occurred and what assurance it provides.

AI Collaboration Insights (15 minutes): The team shares key learnings about human-AI collaboration during the sprint. This includes examples of particularly effective AI usage, situations where AI suggestions proved unhelpful, prompt patterns worth sharing across teams, and validation approaches where AI assistance proved valuable. This sharing builds organizational knowledge about effective AI utilization.

Figure 5.6: The Evidence-Based Review. *In a world where generating code is trivial, demonstrating working software is no longer sufficient proof of value. The review ceremony pivots to an "Evidence Tribunal," where the team presents the Validation Matrix, proving that the AI's output is secure, accurate, and compliant.*

Integration and Deployment Readiness (20 minutes): The team reviews integration testing results, deployment procedure validation, operations readiness, and monitoring configuration. This operational focus ensures sprint outputs are genuinely deployable rather than merely feature-complete.

Backlog Adaptation (15 minutes): The Product Strategist reviews how sprint outcomes affect the Outcome Backlog. Did sprint learnings clarify requirements for future work? Did validation surface new opportunities? Do priorities need adjustment based on delivered outcomes?

Sprint Review concludes with explicit deployment authorization decisions. Based on validation evidence, governance approval, and operational readiness, stakeholders decide whether sprint outputs should deploy. Unlike traditional Scrum where "done" automatically meant "deployable," AI Scrum recognizes that additional factors may delay deployment despite technical completion. Teams find evidence-based review initially uncomfortable. Demonstrating features feels more tangible than presenting validation matrices. However, stakeholders consistently report preferring evidence-based review because it provides genuine confidence in solution quality rather than superficial reassurance that something was built.

CEREMONY 6: RETROSPECTIVE WITH AI INSIGHTS

The Sprint Retrospective in AI Scrum incorporates objective telemetry analysis alongside traditional subjective team reflections and examines both what the team delivered and how human-AI collaboration performed, enabling systematic learning about both process and partnership effectiveness. The ceremony divides into four segments designed to build compound expertise over time.

AI Telemetry Review (20 minutes)

The team reviews objective data about human-AI collaboration during the sprint. AI code generation statistics (lines of code generated, acceptance rates, rejection reasons), validation test execution metrics (pass rates, failure patterns, coverage achieved), prompt effectiveness analysis (which prompts produced valuable outputs), and collaboration pattern insights (when teams engaged AI most effectively).

This objective data supplements subjective impressions. Teams might feel AI performed poorly, but telemetry reveals high acceptance rates with rejections concentrated in specific domains. Or teams might celebrate AI contributions while data shows they missed opportunities for more effective collaboration.

The telemetry from retrospectives parallels the dashboard insights from traditional Scrum. In traditional Scrum, dashboards revealed patterns invisible in individual stories: clusters of bugs indicating architectural problems, blocked items revealing dependency management gaps. AI Scrum telemetry serves the same purpose: revealing patterns in human-AI collaboration that individual interactions obscure.

Process Effectiveness Review (30 minutes)

The team examines sprint process execution. It examines what worked well in validation-first approach, where governance integration succeeded or struggled, how stakeholder engagement via Show & Tell performed, and whether capacity planning proved accurate.

Traditional retrospective topics receive attention, but filtered through AI-native lens. Rather than discussing whether task breakdown was adequate, teams discuss whether validation criteria were sufficiently thorough. Rather than celebrating velocity, teams examine outcome achievement. When teams discuss

Figure 5.7: The Telemetry Dashboard. *Retrospectives often degrade into subjective "complaint sessions." AI Scrum introduces objective telemetry into the improvement cycle. By analyzing acceptance rates and rejection reasons, teams can scientifically improve their prompting strategies and collaboration patterns.*

what felt effective or ineffective, they often reference the same themes from traditional Scrum transformed for AI contexts. In traditional Scrum, retrospectives revealed when ceremony discipline eroded: standups skipped, grooming neglected, reviews rushed. AI Scrum retrospectives reveal when validation rigor erodes: validation criteria incomplete, governance framing superficial, stakeholder feedback ignored. The discipline remains constant even as the focus shifts.

Collaboration Pattern Analysis (25 minutes)

The team analyzes specific human-AI interaction patterns from the sprint. They identify successful collaboration examples worth repeating, failed interactions revealing prompt weaknesses, situations where AI suggestions were unjustly rejected or uncritically accepted, and validation approaches where AI assistance proved particularly valuable. This analysis builds institutional knowledge about effective AI utilization. Teams identify prompt patterns to add to their library, validation strategies to standardize, and collaboration anti-patterns to avoid. The learning compounds over time. Each sprint builds on prior insights rather than starting fresh.

The challenge teams face with AI collaboration parallels challenges from traditional Scrum adoption. When teams first adopted Scrum, they struggled with estimation accuracy, velocity consistency, and retrospective effectiveness. These skills developed over multiple sprints as teams learned what worked in their specific context. AI collaboration effectiveness follows the same trajectory. Initially inconsistent, gradually improving as teams learn effective patterns.

Action Items and Experiments (15 minutes)

The retrospective concludes with concrete improvements for the next sprint. Unlike traditional retrospectives generating vague "we should communicate better" action items, AI Scrum retrospectives produce specific, measurable improvements. Updated prompt templates, refined validation criteria, modified governance workflows, adjusted capacity planning assumptions, and new AI collaboration protocols.

Some improvements take the form of experiments. Hypotheses to test in subsequent sprints. "We believe starting AI Show & Tell on Day 2 instead of Day 3 will catch stakeholder concerns earlier" becomes a falsifiable hypothesis teams can evaluate objectively. This scientific approach to process improvement produces continuous refinement rather than stagnation. Organizations conducting Retrospectives with AI Insights report continuous improvement not just in sprint execution but in human-AI collaboration effectiveness. Teams develop compound expertise, building on prior learnings rather than repeatedly making the same mistakes.

FROM THE TRENCHES: CEREMONY IMPLEMENTATION PATTERNS

Organizations implementing AI Scrum ceremonies should follow several patterns that enable successful adoption while avoiding common pitfalls.

The most instructive ceremony failure I witnessed wasn't caused by negligence. It was caused by pressure. A team I worked with was struggling to deliver on a multi-year project, and management responded with a mandate. Sprint planning was capped at sixty to ninety minutes, no exceptions, and retrospectives were eliminated entirely. In their place came periodic stakeholder presentations, progress reviews designed to show what had been accomplished rather than examine how the team was actually working. The reasoning was straightforward. We have too much work to spend time in meetings. The Scrum Master and Product Owner groomed the backlog informally before each sprint and rushed through planning to ensure everyone had enough work assigned. The hidden cost appeared in standup, day after day. Team members who hadn't fully understood their user stories during planning stayed quiet rather than slow things down or appear unprepared. They nodded. They figured they'd work it out. Sometimes they did. Often, they didn't, and the rework that followed cost far more time than the planning conversation would have. The ceremonies weren't wasting time. Their absence was.

Start with Sprint Inception rigor. Many teams initially resist the four-hour planning investment, trying to compress Inception into traditional two-hour planning slots. This creates cascading problems: inadequate validation criteria leading to rework, missed governance requirements causing late-stage delays, capacity mismatches derailing sprints. Teams that invest fully in Sprint Inception report dramatically fewer mid-sprint surprises and substantially less rework.

Embrace asynchronous Daily Synchronization for distributed teams. Synchronous standups at rigid times create problems for global teams. Asynchronous updates via Slack or collaboration tools often provide better visibility with less overhead. The key principle: daily validation progress visibility and rapid blocker resolution matter more than synchronous gathering.

Make AI Show & Tell stakeholder feedback actionable. Vague feedback like "this needs to be more user-friendly" provides no actionable guidance. Train stakeholders to provide structured feedback: specific elements that work, specific changes required, explicit acceptance criteria. This structured feedback enables rapid iteration.

Treat Governance Checkpoint as partnership rather than approval gate. When teams view governance as adversarial obstacle, they withhold information and gaming behavior emerges. When teams view governance as partnership, they proactively surface concerns and seek guidance early. The mid-sprint timing enables partnership dynamics if leadership sets appropriate tone.

Collect telemetry systematically for Retrospective analysis. Teams that only collect subjective impressions miss valuable insights. Systematic prompt logging, acceptance rate tracking, and validation coverage measurement enable data-driven retrospectives. This requires discipline but produces compound learning.

Maintain ceremony discipline even when under schedule pressure. The first ceremony teams abandon under pressure is typically Daily Synchronization, followed by retrospectives. This abandonment eliminates visibility and prevents learning precisely when teams need both most. Ceremony discipline matters more during crises, not less. Adapt ceremony timing to team rhythm but maintain ceremony substance. Moving AI Show & Tell from Day 3 to Day 4 based on team workflow makes sense. Skipping Show & Tell entirely eliminates critical stakeholder feedback. Adapt timing to serve coordination; never eliminate ceremonies entirely.

WHAT COMES NEXT

Ceremonies provide the coordination structure that enables effective AI Scrum execution. But ceremonies alone do not suffice. Teams also need new collaboration patterns for effective human-AI partnership. The next chapter examines these collaboration patterns in depth, exploring how individuals work with AI systems to maximize value while maintaining human judgment and creativity. The ceremony structure establishes when collaboration occurs and what gets coordinated. Chapter 6 addresses how that collaboration actually works. What patterns prove effective, what risks require mitigation, and how teams develop expertise in AI partnership over time.

KEY TAKEAWAYS

- Traditional Scrum ceremonies optimized for development coordination during slow implementation; AI Scrum ceremonies optimize for validation orchestration during rapid implementation.
- Sprint Inception expands planning from two hours to four hours, establishing validation criteria and governance requirements before implementation begins.

- Daily Synchronization focuses on validation progress rather than task completion, enabling teams to track genuine outcome achievement rather than activity.
- AI Show & Tell enables mid-sprint stakeholder feedback when course correction remains practical, preventing the costly late-stage rework that plagued traditional Scrum.
- Governance Checkpoint integrates formal governance review mid-sprint when teams can still adapt, replacing adversarial end-of-sprint approval gates.
- Sprint Review emphasizes validation evidence over feature demonstration, providing stakeholders with genuine confidence in solution quality through detailed evidence.
- Retrospective with AI Insights combines objective telemetry about human-AI collaboration with traditional subjective team reflections, enabling systematic learning and compound expertise development.

MIND MAP – CEREMONIES AND RITUALS

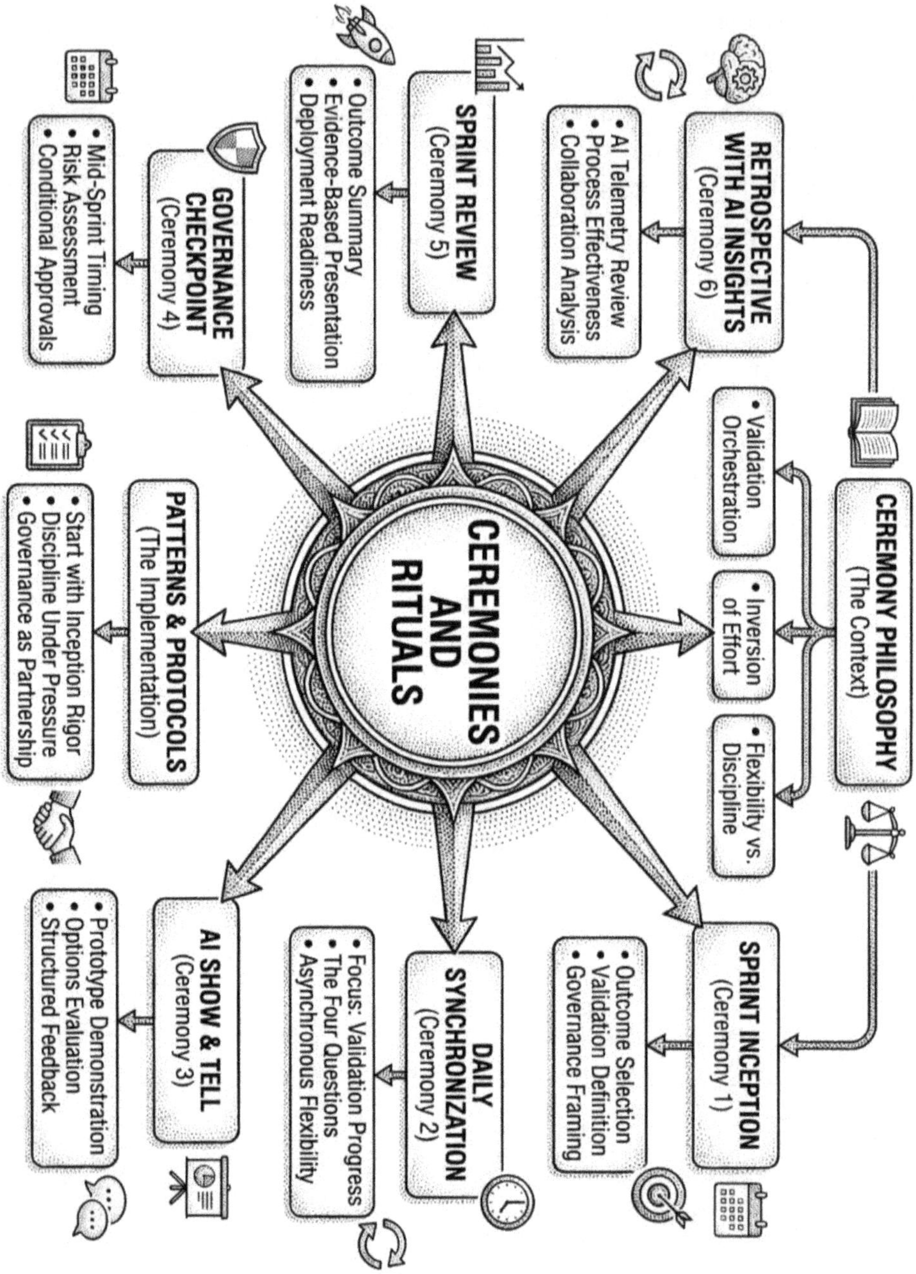

This chapter outlines a co-execution model where humans provide strategic intent and AI handles execution. It distinguishes between simple automation and collaborative augmentation. The section warns against cognitive atrophy and offers specific mitigation strategies. It details protocols for prompt engineering, structured brainstorming, and risk-based validation. Finally, it promotes using AI as an examiner to test understanding rather than a crutch for answers.

CHAPTER 6: HUMAN-AI COLLABORATION PATTERNS

I can brainstorm with AI in ways I never could with human colleagues. There is no judgment when I throw out half-formed ideas. There is no political correctness filter when I explore controversial approaches. There is no concern that someone will think I am incompetent for asking basic questions. This creates a genuinely liberating environment that many early adopters find intoxicating. It feels like having a tireless, brilliant, and infinitely patient collaborator available on demand.

But this freedom comes with a risk I only recently recognized. After several months of intensive AI usage, I noticed something unsettling in my own workflow. My deep-thinking muscles were atrophying. I was reaching for AI prompts where I used to pause and think through problems myself. I was accepting the AI's first suggestion where I used to generate multiple alternatives mentally. I was becoming, in essence, a button pusher rather than a problem solver.

The realization was not sudden; it was a creeping dependency. It manifested in small moments: a hesitation to write a simple SQL query without assistance, or a reluctance to draft an email from scratch. I realized I was trading cognitive resilience for immediate velocity.

That realization forced me to establish boundaries. For creative work, I now use my own gray matter primarily, utilizing AI only as an occasional sparring partner once my own concepts are solidified. When learning something new, I combine video tutorials with AI tutoring, but I make the AI test me rather than just giving me answers.

As Neil deGrasse Tyson noted in Chapter 1, "One of the great challenges in life is knowing enough to think you're right, but not enough to know you're wrong". In the context of software delivery, AI can be a helpful barometer for knowledge thresholds, preventing me from fooling myself about expertise I do not actually possess. But it can just as easily become a mask that hides incompetence behind generated fluency.

This chapter examines how humans and AI should collaborate effectively, addressing both the extraordinary productivity gains available and the cognitive risks requiring active management. The patterns presented here draw from practical experience using AI tools extensively while remaining vigilant about preserving human judgment, creativity, and deep-thinking capability.

WHAT THIS CHAPTER COVERS

This chapter presents human-AI collaboration patterns for AI-native development, balancing theoretical frameworks with practical guidance. By the end of this chapter, you will understand:

- **Co-Execution Philosophy:** Why partnership rather than replacement or simple tool usage produces optimal results.
- **Automation versus Augmentation:** The critical distinction between having AI handle complete tasks versus working collaboratively.
- **Prompt Engineering Principles:** Characteristics of effective prompts that generate high-quality outputs requiring minimal refinement.
- **Cognitive Atrophy Risks:** How overreliance on AI degrades thinking skills and practical strategies for mitigation.
- **Tool Selection Patterns:** When to use different AI tools and how personal preferences shape effective collaboration.

- **Brainstorming Protocols:** Leveraging AI for ideation while maintaining creative ownership.
- **Validation Boundaries:** When to trust AI outputs versus when to verify deeply.
- **Learning Loops:** Using AI to test knowledge acquisition rather than merely providing answers.

KEY TERMS AND CONCEPTS

Before we can optimize the workflow, we must define the specific vocabulary of the AI-augmented practitioner.

- **Human-AI Partnership** A collaboration model distributing responsibilities strategically rather than treating AI as either replacement or mere tool. In this model, humans define outcomes, establish constraints, provide context, make judgment calls, and validate results. AI generates implementations, explores alternatives, identifies patterns, automates repetitive work, and scales analysis beyond human capacity.

- **Automation versus Augmentation** Automation involves relying entirely on AI to handle tasks while humans wait for results and intervene only when problems emerge. Augmentation involves collaborating with AI throughout task execution, combining human judgment with AI capabilities. Organizations confusing these patterns often either over-rely on AI by shipping code they barely understand or underutilize AI by relegating it to trivial tasks.

- **Prompt Engineering** The practice of crafting effective instructions for AI systems that generate high-quality, relevant outputs requiring minimal refinement. Expert-crafted prompts produce outputs requiring sixty to eighty percent less revision than novice prompts, representing the difference between AI that accelerates delivery and AI that creates additional work.

- **Cognitive Atrophy** The degradation of deep-thinking skills resulting from overreliance on AI for tasks that historically required sustained human cognition. Early research suggests Large Language Model (LLM) usage can reduce human capacity for complex problem-solving when users rely on AI for answers rather than using AI to augment thinking.

- **Validation Checkpoint** Structured review points in human-AI workflows where humans verify AI outputs meet quality standards before proceeding. These checkpoints prevent both the catastrophic failures resulting from unchecked AI outputs and the inefficiency of excessive verification slowing all work.

- **Dunning-Kruger Detector** A pattern of using AI to test knowledge acquisition and identify gaps between perceived and actual understanding. This enables systematic validation of learning rather than relying on subjective confidence that may be misplaced.

CO-EXECUTION PHILOSOPHY: PARTNERSHIP NOT REPLACEMENT

For decades, commercial aviation has demonstrated a successful human-automation partnership. Pilots program flight plans into autopilot systems, obtain clearance, taxi to the runway, and execute takeoff. Within moments of becoming airborne, pilots engage autopilot to control throttles, flight surfaces, and navigation.

The pilot remains fully in command, monitoring systems, making decisions, and maintaining ultimate responsibility. The automation handles the continuous execution work. This partnership model succeeds because responsibilities are clearly distributed. Humans handle judgment, adaptation to unexpected conditions, and strategic decisions. Automation handles repetitive execution, maintains precise control, and operates tirelessly. Neither replaces the other. The pilot could fly manually but would be exhausted on long flights and less precise. The automation could theoretically fly alone but lacks the judgment to handle novel situations safely.

Software development with AI follows this same proven partnership model, though the technology differs significantly. Modern AI uses machine learning rather than predetermined automation, enabling more flexible collaboration. However, a critical misconception about AI in software development persists, framing the technology as either developer replacement or mere productivity tool.

The "Replacement Narrative" suggests AI will eliminate developer roles, automating coding entirely: press a button, receive a complete application. This view creates anxiety and resistance. The "Tool Narrative" treats AI as enhanced autocomplete, marginally improving efficiency like a spell-check for code. This view leads to dismissal and underutilization.

Both narratives miss the fundamental transformation. AI enables qualitatively different workflows through human-AI partnership. Partnership workflows distribute responsibilities strategically. Humans define outcomes, establish constraints, provide context, make judgment calls, and validate results. AI generates implementations, explores alternatives, identifies patterns, automates repetitive work, and scales analysis beyond human capacity.

Figure 6.1: The Conductor and the Machine Orchestra. *In the co-execution model, the human provides the strategic intent (the conductor), while the AI handles the technical performance (the orchestra). One cannot function effectively without the other.*

Research from METR demonstrates that developers treating AI as partners rather than tools achieve three to ten times productivity improvements compared to those using AI merely as autocomplete. That difference stems from qualitatively different workflows rather than incremental efficiency gains.

Partner-oriented developers leverage AI for exploration, rapid prototyping, and analysis. They prompt AI to generate five architectural approaches, evaluate trade-offs, then select the optimal direction. They use AI to exhaustively test edge cases humans would miss. They generate detailed documentation automatically, then refine for accuracy.

In contrast, tool-oriented developers use AI incrementally without workflow reconception. They accept AI's first code suggestion if it looks plausible. They treat AI-generated tests as nice-to-have additions rather than thorough validation. They fail to leverage AI's capacity for exhaustive analysis, pattern recognition across vast codebases, or generation of multiple alternatives revealing hidden trade-offs.

Consider how this plays out in my own experience. When I need a function written, AI can deliver in two minutes what might take me twenty minutes manually. That is pure time savings. But when I am brainstorming approaches to a complex problem, the value proposition shifts entirely. AI does not just save time; it expands the solution space.

I can explore five different architectural patterns in the time it would take me to sketch one. I can examine trade-offs systematically rather than committing prematurely to familiar approaches. This is a qualitative difference, not mere speed improvement.

However, this capability creates temptation toward cognitive laziness. If AI can generate several approaches in minutes, why bother thinking through approaches myself first? The answer is that the thinking process itself builds judgment. When I force myself to generate approaches mentally before prompting AI, I develop pattern recognition that makes me better at evaluating AI's suggestions. When I skip that thinking and immediately ask AI for solutions, I become dependent on AI's judgment rather than developing my own.

AUTOMATION VERSUS AUGMENTATION: THE CRITICAL DISTINCTION

The difference between automation and augmentation determines whether AI amplifies human capability or replaces it. Organizations often conflate these patterns, leading to either overreliance failures or underutilization waste.

Automation means relying entirely on AI to handle tasks while humans wait for results and intervene only when problems emerge. A developer prompts AI to implement a complete feature, reviews the generated code briefly, and ships it if tests pass. This pattern works excellently for well-defined, low-risk tasks where specifications eliminate ambiguity. Generating boilerplate code, creating standard CRUD operations, and implementing familiar patterns suit automation when proper validation follows.

Augmentation means collaborating with AI throughout task execution, combining human judgment with AI capabilities. A developer prompts AI to generate implementation alternatives, evaluates trade-offs using domain knowledge AI lacks, selects an approach, prompts AI to implement it, reviews the implementation critically, identifies improvements AI missed, and iterates toward production quality. This pattern suits complex, ambiguous, or high-risk work where human judgment adds substantial value throughout execution.

Figure 6.2: The Vending Machine vs. The Workbench. *Automation is like a vending machine: efficient for standard good but opaque. Augmentation is like a jeweler's workbench: a place where tools enhance the master's skill rather than replacing it.*

The practical challenge is knowing when each pattern applies. A useful heuristic that has emerged from practice is the "Junior Developer Test".

If a developer could explain the task completely to a smart junior developer and receive correct implementation without iteration, the task can likely be automated with AI. If the task requires judgment calls, implicit context, or understanding of unstated constraints, augmentation becomes necessary.

Consider how this plays out when building customer dashboards. Design sessions should remain entirely human-driven brainstorming. Developers sketch wireframes, identify key metrics stakeholders need, and think through user workflows. AI helps generate React component structures and chart configurations systematically, but the product thinking happens in human minds. Then during implementation, AI handles automation tasks like boilerplate code generation, responsive CSS layouts, and data transformation logic, while developers provide augmentation through architectural decisions, performance optimization choices, and judgment calls about which edge cases matter.

Organizations often struggle with this distinction because traditional software development had clearer boundaries. Writing code was human work. Running tests was automated work. The boundaries between these activities were explicit. AI blurs those boundaries entirely. Code generation can be automated or augmented. Testing can be automated or augmented. Architecture can be automated or augmented. Without clear thinking about which pattern suits which context, organizations default to whichever feels most comfortable. Efficiency-focused cultures often choose automation while quality-focused cultures choose augmentation that is really just hesitant tool usage. Neither of these extremes proves optimal.

COGNITIVE ATROPHY: THE BUTTON PUSHER RISK

Recent LLM usage studies reveal a concerning pattern: overreliance on AI for answers rather than augmenting thinking appears to degrade human cognitive capacity for complex problem-solving. Users who treat AI as an external brain rather than a thinking partner show reduced capability for deep analysis over time.

The mechanism parallels other cognitive atrophy patterns such as GPS usage. Before GPS, people developed strong spatial reasoning and mental maps. With ubiquitous GPS, many people lost that capability entirely. Before calculators, people maintained mental math facility. With constant calculator availability, mental math skills atrophied for many.

AI threatens similar atrophy for higher-order thinking skills including problem decomposition, alternative generation, trade-off evaluation, and creative synthesis. The risk is particularly acute for early-career professionals. Someone with twenty years of experience who uses AI extensively still possesses the judgment and pattern recognition built over two decades. They use AI to amplify existing capability. Someone starting their career who relies heavily on AI from day one may never develop that judgment. They become expert AI prompters without developing expert thinking skills.

This creates a genuine dilemma. AI offers extraordinary productivity gains and learning acceleration. Used well, it enables people to learn faster and accomplish more than any previous generation. Used poorly, it creates dependency that prevents skill development. The solution requires conscious practice boundaries.

Developers should deliberately limit using AI when they want creative thinking or problem exploration. When designing complex microservices architectures, spending hours with a whiteboard and notebook before engaging AI proves valuable. That thinking time is not wasted setup; it is the actual work. AI then helps structure messy thoughts into coherent service boundaries and API contracts, but the architectural insights come from sustained human thinking.

Similarly, when learning new technical domains, developers should use AI strategically as a teacher rather than a crutch. They should study concepts themselves, attempt practice problems without AI assistance, then use AI to test their understanding. "I believe X about Y. What important details am I missing?" proves far more valuable than "Explain Y to me". The first approach validates and extends thinking. The second approach substitutes AI thinking for human thinking.

Figure 6.3: The Withered Explorer. *Overreliance on the "GPS" of AI can cause our fundamental survival skills such as coding, logic, and navigation, to rust from disuse. AI can be a helpful barometer for knowledge thresholds, preventing one from fooling themselves about expertise they do not actually possess. But it can just as easily become a mask that hides incompetence behind generated fluency.*

Organizations implementing AI collaboration should establish explicit policies about when AI usage is encouraged versus when sustained human thinking is required. Code review requires human judgment; AI can assist but cannot replace reviewer thinking. Architectural decision-making requires human strategic understanding; AI can generate alternatives but cannot select optimal direction without human context. Requirements analysis requires human stakeholder empathy; AI can document but cannot empathize.

The Neil deGrasse Tyson quote resonates precisely here. AI can help developers avoid the trap of thinking they understand something when they only grasp surface patterns. By testing them, challenging their assertions, and revealing gaps in their reasoning, AI becomes a tool for building genuine understanding rather than a crutch that prevents understanding development.

PROMPT ENGINEERING ESSENTIALS

Prompt engineering represents the critical interface between human intent and AI execution. Well-engineered prompts generate high-quality, relevant outputs requiring minimal refinement. Poorly engineered prompts produce outputs demanding extensive revision or complete regeneration.

If you treat a Large Language Model like a Google search bar, typing in keywords and expecting a finished product, you are fundamentally misunderstanding the tool. You are using a screwdriver as a hammer. Search engines are retrieval mechanisms designed to find existing information; LLMs are reasoning engines designed to construct new information based on your parameters. When you prompt like you search, you receive the statistical average of the internet—generic, safe, and often unhelpful. When you prompt like you are delegating to a skilled subordinate, you receive tailored solutions.

This shift from "retrieval mindset" to "delegation mindset" is the core of AI Scrum. It is an iterative process of refining the LLM's trajectory to get you 80–90% of the way to completion. The goal is rarely 100% perfection from a single prompt; the goal is to accelerate the "zero to one" draft phase so effectively that the human expert can spend their energy on the final sophisticated polish.

Prompt engineering is iterative, not instantaneous. Early prompts rarely generate production-ready output. The 30-45 minute timeline for AI-assisted tasks includes: prompt drafting (5-10 min), initial generation (seconds to minutes), review and refinement iteration (15-25 min), final validation (5-10 min). As teams build prompt libraries and pattern recognition, this timeline compresses.

Clarity and Specificity

Effective prompts eliminate ambiguity through precise language, concrete examples, and explicit constraints. Compare these two prompts for the same task. A poor example would be: "Create an API endpoint for user data." A good example would be: "Create a RESTful API endpoint accepting GET requests at /api/v2/users/{userId} that returns user profile data (name, email, registration date, account type) in JSON format. Include authentication via JWT token, validate userId is a positive integer, return 404 for non-existent users, and implement rate limiting at 100 requests per minute per user".

The second prompt specifies protocol, path structure, data fields, response format, authentication approach, validation rules, error handling, and rate limiting. AI receives everything needed to generate a production-ready implementation rather than a generic skeleton requiring extensive refinement.

This specificity acts as a forcing function for the human requester. You cannot be vague with an LLM and expect specific results. In traditional development, you might hand a vague ticket to a colleague and rely on their shared tribal knowledge to fill in the blanks. The LLM has no such context. It optimizes for the most probable continuation of your text, which usually results in the most generic possible solution. By forcing yourself to articulate the specific requirements *before* prompting, you often clarify the architecture in your own mind, preventing the "garbage in, garbage out" cycle that plagues novice users.

Contextual Richness

AI lacks your project context. This includes the architectural decisions made last quarter, the performance issues discovered in production, and the compliance requirements from legal review. Effective prompts supply this context explicitly. For example: "Our microservices architecture uses asynchronous event-driven communication via Apache Kafka. We recently experienced issues with message ordering in the payment processing service. Design a new order fulfillment service ensuring strict message ordering within each customer's order stream while maintaining high throughput for the overall system. Consider that we process 50,000 orders daily with peak loads reaching 200 orders per second".

Providing this "context dump" feels inefficient to new users who are used to brevity, but it is the highest-leverage activity in prompt engineering. Think of context as the "scaffolding" around the building you want the AI to construct. Without it, the AI builds on a flat, featureless plain. With it, the AI builds specifically to fit your terrain. We have found that pasting relevant snippets of existing code, schema definitions, or even meeting notes into the prompt window moves the starting line from "generic boilerplate" to "project-specific implementation," often saving hours of integration work later.

Constraint Specification

Real systems operate under constraints. Security constraints, performance requirements, backwards compatibility needs, coding standards, and dependency restrictions all shape acceptable solutions. Explicit specification prevents mismatches between delivered and expected results.

Figure 6.4: The Submerged Root System. *A prompt without context is like a flower without roots: it inevitably withers. Effective prompting exposes the deep, hidden context (the root system) that the AI needs to flourish.*

However, there is an economic balance to strike here. Novice prompt engineers often fall into the trap of trying to engineer the "perfect" prompt that handles every conceivable edge case, spending three hours crafting instructions to save two hours of coding. This is a negative-value activity. The goal of constraint specification is to get the AI output to a state of "functional correctness" where it solves the core problem (the 90% mark). If the AI generates code that is 98% correct but misses a subtle edge case, it is often faster to accept the output and fix the last 2% manually than to spend another hour refining the prompt. Don't let the pursuit of prompt perfection become a new form of procrastination; get it working, accept the 98%, and move on.

Output Format Definition

Should AI generate just code, or code with unit tests? Should documentation be inline comments, README files, or API documentation in OpenAPI format? Explicit specification prevents format mismatches that require rework.

Beyond mere convenience, strict output formatting is what allows AI to be integrated into automated pipelines. If you ask for "a list of potential security vulnerabilities," the AI might give you a conversational paragraph, a bulleted list, or a table. If you specify "Output the vulnerabilities as a JSON array with keys for severity, description, and suggested fix," you transform the output from text you must read into data you can use. You can pipe that JSON directly into a ticketing system or a linter. This transforms the LLM from a "chatbot" that talks to you into a "component" that talks to your systems, unlocking massive efficiency gains in workflow automation.

Meaning Type Programming (MTP) with Semantic Engineering

In a very valuable academic paper that was submitted on arXiv in late 2025, A group of individuals from the University of Michigan and Jaseci Labs wrote about Meaning Type Programming with Semantic Engineering. While this topic is well outside of this book, it goes to show how quickly things are being adapted to ensure that AI is more accurate and valuable.

Common prompt engineering inputs can easily be converted into MTP at the LLM user interface and is universal to common models such as ChatGPT, Gemini, and Claude. One of the largest benefits is the reduction of hallucinations essentially through a better prompt structure that is more native to each model. This is something to keep an eye on for future development and adoption.

For example, the following prompt can be run as is:

> *"You are a professional marketing copywriter. Write three catchy, benefit-focused social media captions for a new eco-friendly water bottle, targeting young adults. Use emojis and a conversational tone. Examples: [Input: 'Durable, insulated bottle', Output: '💧 Stay hydrated & save the planet! Our new bottle keeps drinks cold for 24hrs. #EcoFriendly #Hydration']"*

The results might be:

> Here are three catchy, benefit-focused captions with a fun, conversational vibe:
>
> 1. 🌍💦 Sip smarter. Our eco-friendly bottle keeps your drinks icy cold (or piping hot) for hours—while helping the planet. Refill. Reuse. Repeat. ♻️✨
> 2. 🚰✨ Hydration, but make it sustainable. This insulated bottle goes wherever you do—gym, class, road trips—no plastic guilt included. 🌱💪
> 3. ❄️🔥 Cold stays cold. Hot stays hot. And the Earth stays happier. 🌎💚 Meet the water bottle your lifestyle (and the planet) will thank you for.

A user could receive a better response by first converting the prompt to MTP and then having it executed. In this example, Gemini 3 was used.

> *"Convert the following to Meaning Type Programming before running:*
>
> *You are a professional marketing copywriter. Write three catchy, benefit-focused social media captions for a new eco-friendly water bottle, targeting young adults. Use emojis and a conversational tone. Examples: [Input: 'Durable, insulated bottle', Output: '💧 Stay hydrated & save the planet! Our new bottle keeps drinks cold for 24hrs. #EcoFriendly #Hydration']"*

To follow your request, I have first translated your prompt into a structured **Meaning Type Programming (MTP)** format to define the logic and parameters, followed by the execution of that program.

Meaning Type Programming (MTP) Translation

```
// Define Execution Context
Context:
ProductLaunch(Category="Sustainability",
Product="Eco-Friendly Water Bottle")
Role:
SeniorMarketingCopywriter(Expertise="Social
Media Strategy")
Audience: Demographic("Young Adults",
Tone="Relatable, High-Energy")

// Define Operational Parameters
Parameters:
  Style: Benefit_First
  Mood: Conversational
  Assets: [Emojis, Trend-Relevant Hashtags]
  Count: 3

// Execution Logic
Process GenerateCaptions {
  Action:
Transform(Input="Durable/Insulated/Eco",
Output="Engagement-Ready Copy")
  Constraint: Ensure 24-hour temperature
retention mention
  Constraint: Highlight plastic reduction
}
```

```
Run GenerateCaptions(Count=3)
```

Executed Output: Social Media Captions
Here are your three benefit-focused captions designed to resonate with a young adult audience:

Option 1: The "Vibe" Focus Stop settling for lukewarm water and "meh" sustainability. 🧊 Our new bottle keeps your drinks ice-cold for 24 hours while looking 10/10 on your desk. Save the planet, but make it aesthetic. ✨

- Benefits: 24hr insulation, stylish design, eco-conscious.
- Hashtags: #HydrationStation #EcoLuxe #SustainableLiving

Option 2: The "Anti-Plastic" Focus Break up with single-use plastic for good. 💔 Our ultra-durable bottle is the last one you'll ever need to buy. Rugged enough for your weekend hikes and sleek enough for the gym. Your wallet (and the ocean) will thank you. 🌊 🐢

- Benefits: Durability, cost-savings, environmental impact.
- Hashtags: #PlasticFree #EarthFriendly #AdventureReady

Option 3: The "Daily Essential" Focus Hydrated girls (and guys) do it better. 💧 Whether you're heading to a 9 AM lecture or hitting the trails, our insulated eco-bottle keeps your coffee hot or your water crisp all day long. One small swap, one big impact. 🌍

- Benefits: Versatility (hot/cold), daily convenience, easy lifestyle change.
- Hashtags: #WellnessJourney #EcoFriendly #DailyEssentials

While in this small example, both sets of examples were likely acceptable for the stakeholder, in much more complex prompts, the amount of hallucination can be reduced through more native structure without needing to learn programming.

Mastering prompt structure is less about writing perfect prompts from scratch and more about developing enough understanding to recognize quality output when you see it, and to know when to push further. I'll confess something here that the previous section on cognitive atrophy probably earns me no credit for. I rarely write MTP natively. I write my prompt the way it comes naturally, then instruct the LLM to convert it to MTP before execution. It works. It's faster. And yes, I recognize the irony of a man who wrote an entire section warning against over-reliance on AI admitting he uses AI to do something he could learn to do himself.

Awareness of a pattern and immunity to it are apparently two different things.

BRAINSTORMING PROTOCOLS

AI excels at brainstorming because it offers judgment-free exploration. There is no political correctness filter, no concern about appearing incompetent, no worry that colleagues will think proposed ideas are stupid. This psychological safety enables genuinely open ideation. However, effective brainstorming with AI requires structure to prevent two failure modes.

Figure 6.5: The Gold Panner. *Brainstorming with AI requires the mindset of a prospector. You must be willing to sift through a river of generic silt (AI output) to find the rare nuggets of innovation.*

The first failure mode is accepting AI's first suggestions without exploring alternatives. AI generates plausible ideas quickly, creating temptation to select the first reasonable option rather than exploring further. This produces locally optimal solutions that miss globally better alternatives.

The second failure mode is generating endless alternatives without converging toward decisions. AI can produce alternatives infinitely. Without disciplined filtering and selection, brainstorming sessions become overwhelming rather than productive.

Effective brainstorming protocols follow a structured sequence that maximizes AI's generative capacity while preserving human judgment.

The initial step establishes clear constraints and objectives. What problem needs solving? What constraints must solutions satisfy? What constitutes success in this case? This framing focuses AI's generation toward relevant territory.

Broad alternative generation follows without premature evaluation. Teams prompt AI for multiple approaches, encouraging diversity rather than similarity. "Generate five fundamentally different approaches to X, ranging from conservative to innovative." This breadth reveals the solution space. Systematic evaluation against established criteria comes next. This human judgment step prevents AI from dominating decisions while leveraging AI's generation capacity. AI excels at generating options; humans excel at evaluating trade-offs given organizational context AI lacks.

Promising directions receive deeper exploration through targeted prompting. Once high-level approaches are evaluated, teams dig into specifics for selected directions. This iterative narrowing prevents wasting time on detailed exploration of approaches that will not work.

Validation through prototyping or small experiments precedes full implementation commitment. This reality check catches flaws that look fine in abstract discussion but fail in practice.

Product feature brainstorming sessions should follow this pattern. Teams prompt AI with constraints and objectives about improving user onboarding. They generate multiple structural approaches for reducing friction. They evaluate based on user research data, implementation complexity, and impact potential. They dig deeper into selected approaches. They prototype with mockups and user testing. This disciplined exploration produces better results than either pure human ideation or accepting AI's first suggestions.

WHEN TO TRUST AI VERSUS VERIFY DEEPLY

The validation boundary question determines both productivity and risk management effectiveness. Verify everything and productivity gains disappear in review overhead. Trust everything and catastrophic failures eventually emerge. The art lies in knowing which AI outputs require deep verification versus which outputs justify trust with spot-checking.

Several factors should guide verification depth decisions. Code complexity and criticality matter substantially. Simple, well-understood code patterns require less verification than complex, novel implementations. User-facing code with security or privacy implications requires deeper verification than internal utilities.

When to Trust AI versus Verify Deeply

AI confidence indicators provide useful signals. When AI generates code with extensive explanatory comments and clear structure, it likely had high confidence. When code lacks clear organization or contains unexplained complexity, AI may have struggled. The latter situation warrants deeper verification.

Testing coverage provides validation efficiency. If automated tests exist for the domain, running those tests may suffice for verification rather than manual code review. If testing is sparse, manual verification becomes essential.

Criticality matters significantly when determining verification depth. Code handling financial calculations, security implementations, or healthcare logic requires deep verification regardless of AI confidence or code simplicity. The cost of errors in these critical areas justifies verification investment.

Historical patterns should inform decisions. If AI consistently produces high-quality code for specific tasks, trust builds over time for those task types. If AI repeatedly produces flawed code for specific domains, extra verification becomes standard practice for those domains.

Figure 6.6: The Art Forgery Expert. *Validation is not a rubber stamp; it is forensic analysis. The AI is capable of creating convincing forgeries (hallucinations) that look perfect at a glance but crumble under the loupe of expert review.*

Effective practices evolve toward risk-based verification depth. For API endpoint boilerplate code, teams can trust AI extensively with spot-checking. The downside risk of CRUD operation errors is minimal and caught quickly in testing. For security implementation like authentication flows, careful verification becomes essential because errors create vulnerabilities. For business logic involving financial calculations, complete human control remains necessary because AI cannot understand domain-specific regulatory requirements.

Organizations should establish verification protocols rather than leaving decisions to individual judgment. Clear guidelines about which code requires deep review, which outputs require multiple validation checkpoints, and which areas warrant spot-checking enable consistency while preventing both under-verification and over-verification waste.

TOOL SELECTION AND PERSONAL PREFERENCES

Multiple AI tools exist with different strengths, weaknesses, and characteristics. Some tools code better than others. Some create and manipulate images more effectively. Some produce better written content. Like humans, AI tools are not infallible; they make mistakes and are sometimes slow to learn certain patterns.

However, AI tools struggle with understanding personal preferences. Something technically correct may violate personal style preferences. AI might overuse certain patterns where those patterns appear fine in isolation but become problematic at scale. Examples include short sentences, excessive colons, or too many contractions.

AI and Large Language Models (LLMs) are massive entities, and there are relatively few of them. If everyone leans on them for the same information, what makes output unique? The human interfacing with them makes output exclusive.

This reality means tool selection should consider both technical capability and style alignment. Some tools naturally produce output closer to personal preferences, requiring less adjustment. Some tools require extensive refinement to match desired style and voice.

Organizations should enable developers to experiment with multiple AI tools, recognizing that different tools excel at different tasks. For complex refactoring projects requiring sustained context awareness, models with large context windows (like Claude) have proven superior in early adoption. ChatGPT excels at quick technical questions and rapid prototyping. GitHub Copilot serves well for in-IDE code completion.

The key insight is that tool selection is personal and task-dependent. Organizations should enable access to multiple tools, allowing individuals to develop expertise with tools matching their work patterns and preferences. Mandating single-tool usage creates friction for people whose work patterns mismatch tool characteristics.

LEARNING LOOPS: TESTING RATHER THAN TELLING

One of AI's most valuable but underutilized capabilities is knowledge validation. Rather than using AI primarily as an answer provider, using AI as an examiner builds genuine understanding.

The traditional pattern looks like this: A developer wants to learn about topic X, so they prompt AI to explain X. AI provides comprehensive and detailed explanations. They read it, feel like they understand, and move on. The problem is that the feeling of understanding often exceeds actual understanding. They absorbed surface patterns but not deep comprehension.

The inverted pattern looks like this: A developer wants to learn about topic X, so they study available materials themselves. They attempt practice problems or think through applications without AI assistance. Then they prompt the AI system with specific questions.

Figure 6.7: The Fencing Duel. *True learning requires resistance. Use the AI as a sparring partner to parry your logic and test your defense, rather than a guru to hand you the answers.*

"I believe Y about X. What important details am I missing? What edge cases should I consider? Where are the gaps in my understanding?”

This approach uses AI to test learning rather than substitute for it. The difference mirrors the distinction between being told answers versus discovering answers through guided exploration. Active learning produces deeper retention and understanding than passive consumption.

AI can facilitate both patterns, but the active learning pattern builds capability where passive consumption builds dependency. This applies beyond technical learning to strategic decision-making. When making strategic decisions for complex projects, developers should draft their thinking before consulting AI.

They should structure the prompt as follows: "Here is the analysis of the situation. Here are the proposed options. Here is the recommended direction." Then they prompt the AI system with challenges: "What am I not considering? What risks am I underestimating? What alternatives deserve exploration?"

This pattern uses AI to stress-test thinking rather than outsource thinking to AI. Organizations should train people in this inverted usage pattern: AI as examiner rather than answer provider; AI as thought partner rather than thought replacement; AI as validation mechanism rather than cognitive substitute. This subtle shift in how people engage AI determines whether AI builds organizational capability or creates dependency that becomes a liability when AI access is disrupted or AI capabilities change.

FROM THE TRENCHES: COLLABORATION IMPLEMENTATION PATTERNS

Organizations implementing human-AI collaboration should follow several patterns that enable successful adoption while avoiding common pitfalls. These patterns move beyond high-level philosophy into the daily mechanics of team operations.

Establish Explicit Boundaries

Without clear policies, people default to whatever feels most comfortable. Some overrely on AI, shipping code they barely understand, while others underutilize AI, stubbornly maintaining traditional workflows. Explicit guidance about when to automate versus augment prevents both extremes. Teams should create a "Collaboration Charter" that explicitly maps specific domains to acceptable AI usage levels. For example, the charter might state that "UI/CSS generation is 'AI-First' with visual review," whereas "Core Business Logic and Financial Calculations are 'Human-First' with AI used only for syntax checking."

This clarity removes the ambiguity that causes hesitation. We found success by categorizing work types into "Green Zones" (full automation allowed), "Yellow Zones" (augmentation with deep review), and "Red Zones" (human-only thinking required). A developer working in a Red Zone knows that pasting code from an LLM is a process violation, not just a style choice. This structure empowers the team to move fast where safety permits while forcing necessary friction where risks are high.

Build Prompt Libraries

Individual experimentation produces valuable insights that get lost when not captured systematically. Prompt libraries transform individual learning into institutional capability. Teams can leverage proven patterns rather than reinventing prompts repeatedly. These libraries should capture the context, constraints, and refinements that make prompts effective, transforming AI from an unpredictable tool into a reliable partner.

The implementation of this library should align with your existing ecosystem to reduce friction. Organizations running on the Microsoft stack can leverage **SharePoint** or **Microsoft Lists** to create a searchable repository, categorized by role. Teams embedded in the Atlassian ecosystem should build a dedicated "Prompt Engineering" space in **Confluence**, utilizing page templates to ensure every entry includes the "Context," "Prompt," and "Expected Output." For engineering-heavy cultures, a **Git-based Wiki** or a repository of Markdown files allows developers to treat prompts like code—versioned, reviewed, and merged.

Regardless of the tool, the folder structure should be intuitive and task-oriented to encourage retrieval. A proven structure often looks like this:

- **01_Roles:** Subfolders for Product Owner, Developer, QA, and Architect, containing role-specific "persona" prompts.
- **02_Tasks:** Practical folders for Refactoring, Unit Testing, Documentation, and SQL Generation.
- **03_Standards:** Global context prompts that include the organization's coding standards, security guidelines, and architectural patterns. By spoon-feeding this context, a new developer can navigate to 02_Tasks/Unit_Testing, copy a vetted prompt that already includes the team's mocking framework preferences, and generate compliant tests in minutes.

Train People in Verification Depth Calibration

The skill of knowing when to trust AI versus verify deeply requires explicit development. Risk-based verification guidelines combined with reviews of verification outcomes build judgment over time. Early in AI adoption, lean toward more verification. As patterns emerge showing where AI excels versus struggles, adjust verification protocols accordingly.

Training for this must be practical, not theoretical. We implemented "Spot-Check Drills" where bad code (subtly flawed AI logic) was deliberately introduced into the review pile to test the reviewer's vigilance. If a reviewer glossed over the AI-generated error because the code "looked right" at a glance, it became a teaching moment about the "Fluency Illusion." Furthermore, teams should maintain a "Hall of Hallucinations"—a shared document recording specific instances where the AI confidently provided wrong answers. Reviewing this log during sprint retrospectives inoculates the team against blind trust, reminding them that the AI is a probabilistic engine, not a deterministic truth-teller.

Make Cognitive Atrophy Risks Explicit

Simply warning people about "button pusher" risks proves insufficient. Specific practices help: mandatory think-time before AI engagement for creative work, regular AI-free days to maintain baseline skills, and deliberate practice of core competencies without AI assistance. These practices feel inefficient initially but prove essential for long-term capability preservation.

To operationalize this, we introduced the "Whiteboard First" rule for all architectural changes. Before a developer is permitted to prompt the AI for a microservice design, they must sketch the entity relationships and data flows on a physical whiteboard or digital canvas *manually*. This forces the brain to construct the mental model first. Only after the human understands the problem structure do we bring in the AI to handle the syntax and boilerplate. Additionally, consider implementing "Analog Afternoons"—designated blocks of time where the team focuses on design, strategy, or complex debugging without AI assistance, ensuring that the "mental muscles" for deep problem-solving remain active and resilient.

Enable Access to Multiple AI Tools

People work differently and tasks vary substantially. Tool characteristics that work well for one person or task type may create friction for others. Multi-tool access with clear guidance about tool strengths enables people to develop effective personal workflows.

Think of this like a golf bag: you don't use a putter for a drive. We encourage developers to keep three distinct tools open: a "Reasoning Engine" (like Claude or GPT-4o) for complex architectural analysis and large-context refactoring; a "Completion Engine" (like GitHub Copilot) for in-flow, line-by-line coding speed; and a "Search/Research Engine" (like Perplexity) for finding up-to-date documentation and solving error logs. Organizations should secure enterprise licenses for this "triad" of tools rather than forcing a one-size-fits-all solution. Restricting a team to a single tool often leads to "Shadow IT" where developers pay for their own preferred tools anyway; sanctioning the toolkit brings that usage into the light where it can be governed and secured.

Treat AI Collaboration as a Skill

Early AI collaboration feels awkward and inefficient. People need time and practice to develop effective patterns. Sprint retrospectives should include explicit discussion of what collaboration worked well versus poorly, building institutional knowledge about effective AI partnership. Recognize that effectiveness will vary substantially across individuals.

To accelerate this skill acquisition, implement "Prompt Pairing" sessions—similar to pair programming, but specifically focused on interacting with the AI. One developer drives the prompt strategy while the other navigates and critiques the AI's output. This cross-pollinates techniques; a senior developer might show a junior how to use "Chain of Thought" prompting to debug a race condition, while the junior might demonstrate a new tool feature the senior missed. Additionally, add a "Prompt of the Sprint" segment to your Sprint Review, where a team member showcases a particularly clever or effective interaction that saved time or solved a stubborn problem. This celebrates the *skill* of collaboration, not just the speed of output.

WHAT COMES NEXT

Human-AI collaboration patterns enable the productivity gains that make AI Scrum viable. But collaboration alone does not suffice. Teams also need systematic approaches to planning and estimation that account for AI's impact on work pace and uncertainty.

The collaboration patterns establish *how* work gets done. Chapter 7 addresses *how* teams plan what work to tackle and estimate capacity requirements given the fundamental shifts AI introduces.

KEY TAKEAWAYS

- **Partnership Multipliers:** Human-AI partnership that distributes responsibilities strategically produces three to ten times better results than treating AI as a simple tool or attempting full automation.

- **The Junior Dev Heuristic:** Use the "Junior Developer Test" to decide between automation and augmentation. If you can't explain it simply to a junior peer, you must augment, not automate.
- **Cognitive Atrophy Defense:** Overreliance on AI threatens deep thinking skills. Counter this by implementing "Whiteboard First" policies where logic is drafted mentally before prompts are written.
- **Context is King:** Prompt engineering effectiveness separates marginal AI value from transformational impact. Expert prompts that include rich context and constraints produce outputs requiring sixty to eighty percent less revision than novice prompts.
- **Structured Brainstorming:** Brainstorming with AI enables judgment-free exploration but requires structured protocols to prevent accepting first suggestions or generating endless alternatives without convergence.
- **Risk-Based Verification:** Verification depth should vary based on risk. High-risk finance or security code demands deep human review; low-risk boilerplate justifies high trust. This prevents both under-verification failures and over-verification waste.
- **Inverted Learning:** Using AI as an examiner ("What did I miss?") rather than an answer provider ("Explain this") builds genuine understanding and prevents the illusion of competence.

MIND MAP – HUMAN-AI COLLABORATION PATTERNS

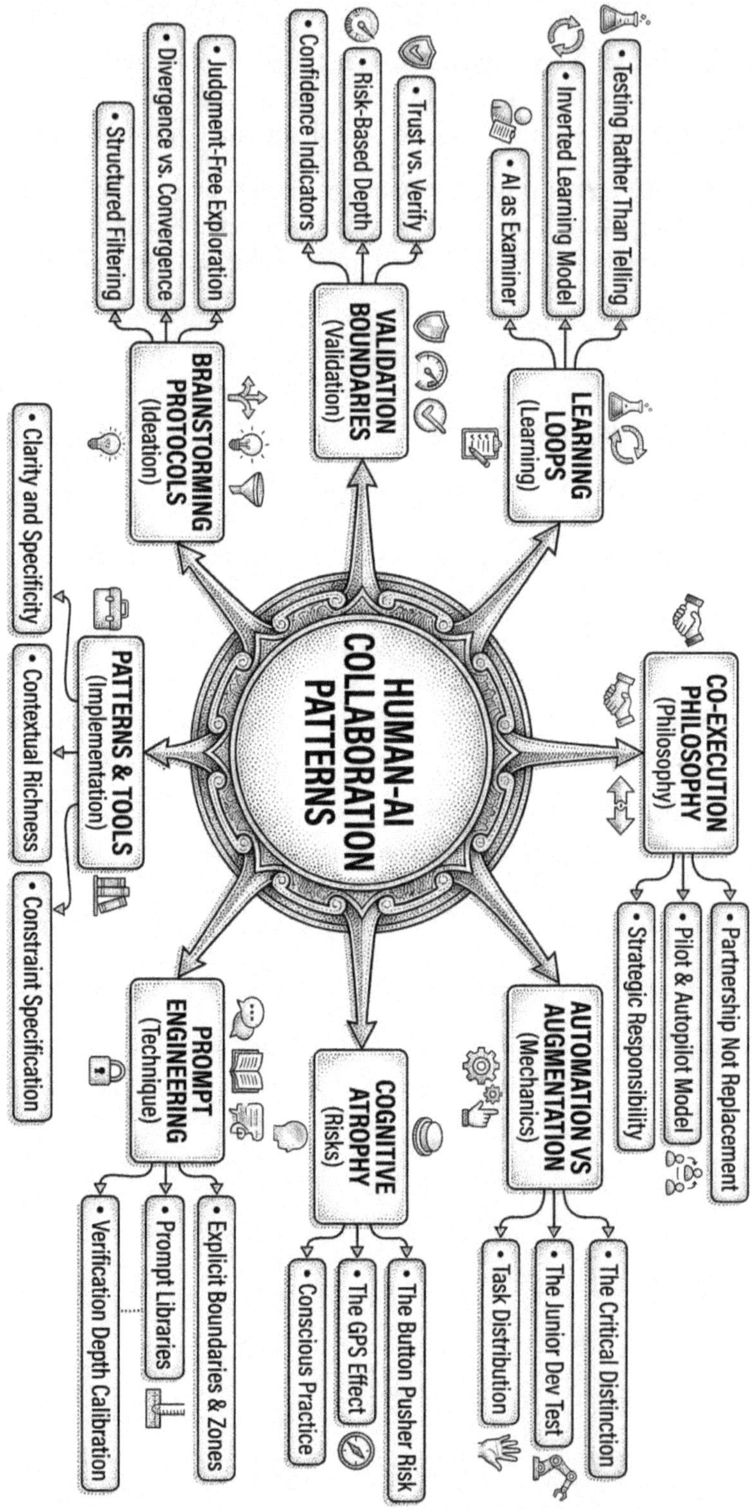

CHAPTER 7: PLANNING AND ESTIMATION

7

This chapter explores how AI inverts delivery constraints. Since code generation is now rapid, planning must shift focus from implementation effort to validation capacity. Teams should estimate work based on verification complexity rather than coding hours. Success requires prioritizing measurable outcomes over simple feature lists and rigorously managing hidden dependencies to prevent bottlenecks that often stall progress

AI Scrum planning conversations sound dramatically different from traditional sprint planning. Instead of asking "How many story points can we commit to?" teams ask "Which outcomes can we validate thoroughly this sprint?" Instead of breaking down user stories into development tasks, teams define validation criteria and governance checkpoints. Instead of estimating implementation effort in days or weeks, teams estimate validation depth required for different risk levels.

This shift reflects the fundamental constraint inversion. When AI handles implementation work that previously consumed 60-80% of sprint capacity, traditional planning focused on development coordination becomes irrelevant. The bottleneck moves from "can we build this?" to "can we validate this thoroughly?"

Traditional sprint planning spent enormous effort on implementation details. How will we architect this feature? Which developer has the right skills? How many days for database changes versus API work versus UI implementation? These questions consumed planning meetings because implementation was genuinely hard and teams needed careful coordination to deliver.

AI Scrum planning eliminates most implementation discussion. Teams can ask AI to generate three different architectural approaches in a few minutes during the planning meeting itself. They can explore trade-offs between approaches without committing days of developer time to each option. They can prototype solutions to verify feasibility before sprint commitment.

The planning conversation shifts entirely from "how will we build it?" to "what must we validate to ship confidently?" This chapter examines how planning and estimation evolve when AI handles dependency analysis, when implementation speed increases dramatically, and when validation becomes the dominant constraint. The principles that made traditional planning effective still apply, but the specific practices require fundamental reconception.

WHAT THIS CHAPTER COVERS

This chapter presents detailed planning and estimation approaches for AI-native development, balancing theoretical framework with practical implementation guidance. The following topics are covered:

- **From Features to Outcomes:** Why planning must shift from feature delivery to impact achievement when AI eliminates implementation constraints.
- **Outcome-Based Backlog Structure:** How to define meaningful outcomes with success metrics rather than feature specifications.
- **The New Estimation Conversation:** What teams estimate when implementation takes hours instead of days.
- **Dependency Intelligence:** How AI reveals hidden dependencies that traditional planning misses.
- **Story Points in AI Scrum:** Why points remain relevant but what they measure changes fundamentally.
- **Capacity Planning Shifts:** How to plan validation capacity rather than development throughput.
- **Sprint Planning Questions:** The seven new question categories that replace traditional planning conversations.
- **AI Variability Handling:** How to account for AI nondeterminism in planning processes.

The chapter draws from traditional Scrum planning experience to understand what worked historically, then redesigns planning for AI-native constraints. The dependency analysis insights come from painful lessons about hidden connections. The capacity planning shifts reflect the fundamental constraint inversion from development to validation. The planning question framework integrates both practical experience and systematic thinking about what AI changes.

This represents partly predictive framework and partly evolved practice. The outcome-based approach comes from systematic analysis of what planning must accomplish when implementation is no longer the constraint. The specific planning questions reflect what teams should ask based on AI's characteristics and validation requirements. The variability handling approaches address observable AI nondeterminism requiring new planning considerations.

KEY TERMS AND CONCEPTS

- **Outcome-Based Planning** This planning approach commits to business impact rather than feature delivery. The Outcome Backlog specifies measurable results to achieve rather than specifications to implement. This shift opens solution space instead of constraining it, enabling teams to explore multiple approaches when AI makes rapid prototyping practical.
- **Dependency Intelligence** Systematic analysis of relationships between work items that traditional planning often misses. Hidden dependencies emerge from quick fixes years ago, integration requirements not documented, or subtle technical connections between components. AI's pattern recognition across entire codebases reveals dependencies humans cannot easily track.
- **Validation Capacity** The team's ability to validate AI-generated implementations rather than development throughput. Capacity planning shifts from asking "how much can we build" to asking "how much can we validate" because validation completeness becomes the dominant constraint.
- **AI Variability** The nondeterministic nature of AI output where identical prompts may produce different implementations across separate executions. All implementations may work correctly while differing in approach, structure, or details. This variability requires planning processes that accommodate multiple valid solutions rather than expecting identical outcomes.
- **Story Points for Validation** Using story points to measure validation effort rather than implementation effort. Points still

indicate team capacity and work complexity, but what gets estimated shifts from coding difficulty to validation completeness requirements.

- **AI Debt Landmine** Legacy quick fixes or incomplete implementations from years past that create hidden dependencies or constraints. These landmines remain dormant until touched, at which point they require substantial rework before new work can proceed. AI's codebase analysis can identify landmines before they explode.
- **Sprint Planning Framework for AI** The seven new question categories that structure AI-augmented sprint planning: validation feasibility and data quality, model behavior and reliability, architecture and system boundaries, cost and latency, human-in-the-loop strategy, measurement and evaluation, and AI-specific risk management.
- **Minimum Viable Product** The simplest version of a solution that satisfies core requirements. In AI Scrum, MVP thinking matters even more because AI can generate complete solutions rapidly. MVP is still the minimum needed for core requirements. The AI Scrum challenge is that AI makes it tempting to add scope beyond MVP (nice-to-haves) just because they're easy to generate. Teams need discipline to stay focused on core requirements, the acceptance criteria essentially.

FROM FEATURES TO OUTCOMES: THE PLANNING INVERSION

Traditional agile planning organized around feature delivery, with product backlogs listed capabilities to build. Roadmaps projected feature release timelines. Sprint planning decomposed features into implementable stories. Every team meeting started the same way: "What features are we building this sprint?"

This feature-centric approach made perfect sense when implementation represented the primary constraint. Building software was genuinely hard work. If you could deliver features reliably, you were likely delivering value. The equation was deceptively simple. Features delivered were assumed to equal value created.

AI-native planning inverts this equation. The challenge is no longer "Can we build this?" but "Should we build this?" AI can generate features rapidly, too rapidly if you are not careful. What AI cannot do alone is determine which features actually deliver your intended outcomes. This is why AI Scrum pivots from features to outcomes.

Figure 7.1: The Shifting Terrain. *Planning in an AI environment requires mapping a landscape that changes while you draw it. The focus shifts from detailed long-term maps to rapid, outcome-based surveys.*

Rather than planning what to build, teams plan what impact to achieve. The Outcome Backlog specifies business objectives, customer problems to solve, and success metrics, not feature specifications. Instead of "Build a recommendation engine," the outcome reads "Increase user engagement measured by daily active users by 40% within one quarter."

The difference is profound. Features are means; outcomes are ends. When teams fixate on features, they risk building precisely what they specified while missing what customers actually need. When teams focus on outcomes, they gain flexibility in how they achieve goals while maintaining clarity about what success means.

Defining Meaningful Outcomes

Effective outcomes exhibit five characteristics that distinguish them from vague aspirations or feature lists.

- **Measurability** through quantitative metrics or qualitative evidence provides the foundation. "Improve user experience" fails this test. "Increase user engagement measured by daily active users by 40%" passes. Measurement creates accountability and enables course correction when progress lags.

- **Time-bound achievement** timeframes create urgency and enable evaluation. "Eventually improve conversion rates" invites procrastination and makes success evaluation impossible. "Improve conversion rates by 15% within three months" creates deadline focus and permits definitive success assessment.
- **Explicit connection** to business strategy ensures relevance. Each outcome should trace to organizational objectives. This connection ensures teams work on what matters rather than what seems interesting technically. When executives question why teams are working on specific outcomes, clear strategic alignment provides answers.
- **Achievability** given current constraints maintains team morale and energy. Aspirational goals that ignore reality waste team energy and damage morale. Effective outcomes stretch teams while remaining realistic given technical capabilities, market conditions, and organizational resources.
- **Specificity** guides solution design while preserving implementation flexibility. The outcome "reduce customer support tickets by 40%" could be achieved through better user interface design, comprehensive documentation, intelligent help systems, or proactive issue detection. This flexibility enables creative problem-solving rather than constraining teams to predetermined solutions.

Experience with traditional backlog grooming reveals the power of this distinction in practice. When product owners and teams work from pristine backlogs with data-driven priorities, sprint planning flows smoothly because clarity exists about what matters. That same clarity principle applies to outcome definition. Clear, measurable outcomes enable effective planning where vague aspirations create confusion.

THE NEW ESTIMATION CONVERSATION

Traditional estimation focused on implementation effort. How many story points for this feature? How many developer-hours does that functionality require? Teams estimated based on coding complexity, technical uncertainty, and integration challenges. When AI handles implementation in minutes instead of days, these estimation conversations become irrelevant. You do not need to estimate three days for work AI completes in thirty minutes. The estimation conversation must shift entirely.

Figure 7.2: The Outcome Target. *Outcome-based planning requires aiming through layers of dynamic complexity. Success is not hitting a static feature list, but threading the needle of user value and business impact.*

Estimating Validation Effort

The estimation question shifts entirely: How much effort does validation require for AI-generated work? This shift reflects the fundamental constraint inversion from implementation to validation.

Validation effort depends on multiple factors that traditional estimation often ignored:

- **Functional validation** requires defining test cases, executing tests, and verifying results.
- **Performance validation** requires load generation, benchmark measurement, and bottleneck analysis.
- **Security validation** requires penetration testing, vulnerability assessment, and threat modeling.
- **Compliance validation** requires regulatory requirement verification and audit evidence documentation.

Teams estimate validation complexity using the same story point scale but with validation criteria substituted for implementation difficulty. A one-point outcome has straightforward validation requirements: basic functional testing, simple performance verification, no special compliance concerns. A thirteen-point outcome has complex validation needs: comprehensive test scenarios, sophisticated performance analysis, extensive security review, rigorous compliance documentation.

Experience shows that this estimation evolution requires conscious practice. Early in AI Scrum adoption, teams instinctively estimate implementation effort out of habit. Only after two or three sprints do they internalize that validation effort determines capacity consumption. Points still measure team capacity and work complexity. The team assigns points, not individuals, but what gets estimated shifts fundamentally.

Accounting for AI Variability

AI nondeterminism introduces new planning complexity. You could ask an LLM to code something three different times over three different days and receive different results, all of which work correctly. Three human developers writing the same code would also produce different implementations, but their approaches would likely be more consistent in structure and style.

As long as results satisfy minimum viable product requirements, variation matters less than correctness. However, teams must account for this variability in planning. Early estimation will be inaccurate as teams learn how much validation effort different types of work require. As experience accumulates, teams hit their stride and become more predictive. The key insight is that story points still measure effort and complexity. The focus has changed from development to validation. Points remain an integrated part of Scrum for team capacity planning. They should never track individual effort, only team throughput.

THE DEPENDENCY CHALLENGE

Dependencies have always plagued planning, but their impact intensifies in AI Scrum. Traditional planning missed dependencies until too late because humans cannot easily track all relationships across large backlogs. Sprint planning disasters where teams discover critical dependencies only when selecting work represent a pattern repeated across thousands of teams.

AI Scrum does not eliminate dependency challenges. Teams still need careful grooming to identify dependencies before sprint commitment. However, the rapid implementation speed means dependency blockers cause more dramatic sprint disruptions. When AI can generate solutions in hours, waiting days or weeks for another team's API becomes proportionally more painful.

This reality demands more rigorous dependency identification during Outcome Grooming. Teams must explicitly discuss dependencies for each outcome. What other teams must deliver components first? What external systems must be available? What data must exist? Which architectural decisions must be finalized? Surfacing these questions early enables realistic sprint planning rather than mid-sprint crisis management.

Figure 7.3: The Inspection. *Story points now measure the intensity of the gaze, not the labor of the dig. The value in software delivery has shifted from "mining" (writing code, which is now cheap) to "inspecting" (validating logic, which is expensive).*

SPRINT PLANNING FOR AI-AUGMENTED WORK

Traditional Scrum experience with Azure DevOps sprint dashboards that revealed blocked items and escalating risks demonstrates that aggregated visibility enables better decisions than individual story tracking. That same principle applies to AI sprint planning. Systematic frameworks reveal patterns and requirements that ad hoc planning misses.

Traditional sprint planning asked specific questions. What stories can we commit to? How will we decompose them into tasks? Who will work on what? These questions optimized for development coordination. AI sprint planning asks fundamentally different questions organized into seven categories that address AI-native constraints.

Validation Feasibility and Data Quality

These questions replace traditional "How will we build it?" with "Can we validate this confidently?"

- What data do we need the model to perform well, and do we already have it?
- How much of the expected output can we validate automatically versus requiring human-in-the-loop review?

- Do we have ground truth or benchmark datasets to measure model correctness this sprint?
- Can we simulate or generate synthetic data to validate earlier and faster?
- What failure cases do we explicitly want to test for this sprint?
- Do we have the tooling to capture and label the outputs generated during this sprint?

Traditional sprint planning never required questions about datasets, label quality, or evaluation tooling. These questions surface requirements that determine whether validation can proceed effectively.

Figure 7.4: The Mine Detector. *Dependencies are the "AI Debt Landmines" of planning. AI pattern recognition acts as the scanner, revealing buried connections before the team steps on them.*

Model Behavior and Reliability

These questions acknowledge that AI output is probabilistic, not deterministic.

- What model behaviors must be predictable for this feature to ship?
- How will we measure hallucinations, drift, or inconsistency?

- What guardrails or constraints do we need such as prompt rules, safety filters, or retrieval sources?
- Are we testing the model across diverse user inputs, not just happy paths?

Traditional engineering never asked about hallucination rates or drift. AI introduces failure modes requiring explicit planning attention.

Architecture and System Boundaries

Since AI workflows can be fragile, boundary definition becomes more important.

- Which parts of the workflow must remain deterministic, and where can AI be used?
- Is this best solved with a model, retrieval, rules, or a hybrid?
- Should we use an existing model, fine-tune internally, or explore vendor options this sprint?
- What is the blast radius if the model behaves incorrectly?

Traditional sprint planning did not include model-choice strategy or failure blast radius discussions. These architectural decisions now require explicit planning consideration.

Cost, Latency, and Throughput

AI introduces real-time variable compute cost and performance variability.

- What is the cost envelope for this feature, and how are we measuring cost per invocation?
- How will we test latency and throughput under realistic load?
- Do we need caching, batching, or fallback mechanisms this sprint?

Traditional development rarely discussed "cost per API call" as a sprint-level gating issue. AI's pay-per-use model makes cost a first-order concern requiring planning attention.

Human-in-the-Loop Strategy

AI teams require continuous learning loops.

- Where do humans need to review AI output?
- How do we capture human feedback into a structured format for training?
- Do we need UI changes to enable review and approval workflows?

Traditional sprints did not consider how humans provide structured feedback to improve the product. AI systems require these feedback mechanisms for quality maintenance and continuous improvement.

Measurement and Evaluation

AI quality must be defined and measured continuously.

- What are the evaluation metrics for this sprint? Examples include: accuracy, precision, recall, and hallucination rate.
- What is good enough for this sprint?
- How do we know when to stop?
- How will we measure regression if we update prompts or models mid-sprint?

Traditional agile focused on shippable increments, not measurable model quality thresholds. AI requires explicit quality definitions before work begins.

AI-Specific Risk Management

AI introduces ethical, security, and compliance risks.

- Are there privacy, safety, or compliance constraints for training or inference?
- Do we need prompt logging or redaction?
- What is the risk if the model responds incorrectly or inappropriately?

Traditional sprints almost never involved ethical or generative-specific risk evaluation. AI systems require explicit risk assessment and mitigation planning. This framework represents a systematic attempt to capture what planning conversations should address in AI contexts. It may not be entirely complete, but it provides a starting point for teams to develop their own planning protocols based on their specific contexts and learned experience.

CAPACITY PLANNING: FROM DEVELOPMENT TO VALIDATION

Traditional capacity planning asked: How many story points can we deliver per sprint? This question assumed development capacity was the binding constraint. Teams calculated velocity measured in points per sprint, committed to work fitting that capacity, and tracked whether reality matched expectations.

AI inverts this question entirely. Development capacity becomes nearly unlimited when AI handles implementation. The new capacity question that emerges is how much validation can we complete per sprint?

Figure 7.5: The Pre-Flight Check. *The "Seven Questions" framework acts as the pre-flight checklist that ensures the crew, the fuel (data), and the flight plan (outcome) are safe before takeoff. You cannot fix AI hallucinations mid-flight.*

Validation Capacity Factors

Several factors determine validation capacity rather than traditional development throughput:

- **Do we have security expertise for penetration testing?** Security validation cannot proceed without qualified reviewers. If no one on the team possesses security skills, validation capacity for security-critical outcomes approaches zero regardless of how much AI generates.

- **Can we access stakeholders for midweek feedback?** Stakeholder validation requires stakeholder availability. When stakeholders are unavailable for Show & Tell sessions, validation stalls regardless of technical progress.
- **Are integration test environments available when needed?** Environment availability gates validation activities. A team cannot validate integration correctness without functioning integration environments.
- **Will governance board members participate in Checkpoint?** Governance validation requires governance participant engagement. When governance stakeholders are unavailable, governance Checkpoint cannot occur, blocking outcomes requiring governance approval.

These capacity constraints shape sprint commitments more than development capacity estimates. Teams select outcomes they can actually validate given these constraints rather than outcomes they can implement.

Sprint Velocity in AI Scrum

Traditional velocity measured story points delivered per sprint. This metric provided reasonable predictability because implementation effort remained relatively stable over time. AI sprint velocity still measures points delivered but requires recalibration. Early in AI adoption, teams underestimate validation effort and overcommit. After several sprints, patterns emerge showing validation capacity for different outcome types. Teams adjust their commitments accordingly based on learning.

The key principle remains unchanged. Velocity measures team capacity, not individual capacity. Teams commit to outcomes collectively. Points indicate outcome complexity and effort requirements. The team learns its sustainable pace through experience.

What changes is variance between sprints. Traditional velocity showed relatively stable patterns. Sixteen points, eighteen points, fifteen points across consecutive sprints. AI velocity may vary more significantly as teams tackle different validation complexity mixes. One sprint with predominantly functional validation may deliver twenty-five points. The next sprint with heavy security and compliance validation may deliver twelve points despite equivalent effort. Teams should track both raw velocity and normalized velocity accounting for validation complexity mix. This dual tracking enables better capacity prediction while acknowledging that not all story points require equal validation effort.

Figure 7.6: The Flood. *The "Creation Burst" of AI can easily overwhelm a team's validation capacity. If you generate 60 points of code but only have a 10-point channel for validation, you create a flood, not a feature.*

FROM THE TRENCHES: PLANNING IMPLEMENTATION PATTERNS

Organizations implementing AI-augmented planning should follow several patterns that enable successful adoption while avoiding common pitfalls.

Start with outcome definition before any implementation discussion. Traditional planning often jumped directly into solution discussions. Teams debated technical approaches before clarifying what outcome they were trying to achieve. This premature solution focus constrains thinking unnecessarily. Start planning sessions by defining the outcome completely: what success means, how it will be measured, what constraints must be satisfied. Only then explore potential approaches.

Invest heavily in dependency analysis during backlog grooming. Dependencies derail more sprints than any other single factor. Dedicate explicit grooming time to dependency discovery. Prompt AI to analyze proposed work and surface dependency concerns. Cross-reference with teams owning potentially dependent components. Document discovered dependencies explicitly in work items. This upfront investment prevents mid-sprint surprises.

Calibrate validation capacity through multiple sprint cycles. Do not expect accurate capacity prediction immediately. Track actual validation effort for different outcome types across multiple sprints. Build historical data showing how much validation effort security-critical work requires versus integration-heavy work versus stakeholder-intensive work. Use this data to inform future capacity planning.

Maintain prompt libraries for planning activities. Effective planning prompts evolve through experience. Capture prompts that successfully elicit dependency analysis from AI. Document prompts that generate useful risk assessments. Build libraries enabling the entire team to leverage proven planning patterns rather than each person inventing prompts independently.

Separate validation capacity from implementation speed. Traditional planning collapsed these concerns into single story point estimates. AI planning requires explicit separation. A simple feature requiring complex validation might receive high point values despite minimal implementation effort. A complex feature requiring straightforward validation might receive lower points. Keep these dimensions distinct in planning conversations.

Test AI planning recommendations against team judgment. AI can suggest sprint commitments, identify high-risk outcomes, or propose dependency resolutions. Treat these recommendations as input to human decision-making rather than final decisions. Sometimes AI will be correct when human intuition misleads. Sometimes human judgment will correctly override AI based on contextual factors AI cannot access. Track both to learn when to trust AI recommendations versus when to rely on human judgment.

Account for AI variability in acceptance criteria. When AI generates different valid implementations across separate executions, acceptance criteria should focus on outcome achievement rather than implementation specificity. Define what success looks like in terms of behavior, performance, and validation evidence rather than specific code structure or approach. This flexibility enables teams to accept any valid solution AI produces rather than rejecting implementations merely because they differ from expected patterns.

REAL WORKING IDEAS

The principles in this chapter are only as useful as your ability to apply them under real conditions, with real teams, on real work. Abstract frameworks have a way of feeling persuasive in planning meetings and elusive on Monday morning when a sprint is about to start and someone asks how long something will take. What follows are three worked examples drawn from different disciplines; software development, legal operations, and data science. They are not case studies from a single organization. They are composites of patterns that appear repeatedly when teams make the shift from effort-based to validation-based estimation. None of them are tidy. That is intentional.

Worked Example 1: Software Development — Payment Gateway Feature Sprint

The payment retry scenario is a useful starting point because it involves a feature type that almost every software team has encountered in some form, familiar enough that the estimation habits around it are deeply grooved, and consequential enough that getting validation wrong has immediate production impact.

The sprint planning meeting had been running for forty minutes and the team was stuck on a number.

"How many story points for the payment retry logic?" the Scrum Master asked.

The senior developer said five. A mid-level developer said eight. A third said three. The Planning Poker cards went back on the table for a second round.

This scene plays out in thousands of sprint planning sessions every week, and it is not a failure of process. It is a symptom of asking the wrong question. Story points estimate implementation effort. But when an AI coding assistant can generate a working draft of payment retry logic in under twenty minutes, implementation effort is no longer the constraint. Validation is.

Here is how the same scenario looks under AI Scrum's validation-based estimation model.

The Feature: Payment retry logic with exponential backoff, integrated into an existing checkout flow for a mid-sized e-commerce platform. The AI assistant has already generated a functional draft during the pre-sprint preparation session.

Step 1 — Define the Validation Scope
The AI Alignment Engineer and Validation Lead open the Validation Canvas for this story. Before anyone estimates effort, they answer four questions:

What can go wrong with AI-generated code in this context? The retry logic touches payment state, which means failed retries could result in duplicate charges, missed transactions, or corrupted order records. The AI draft handles the happy path well but has no awareness of the platform's idempotency key implementation.

What regulatory or compliance constraints apply? Payment card data is governed by the Payment Card Industry Data Security Standard, PCI DSS, a security framework mandated by the major card networks that applies to any system storing, processing, or transmitting cardholder information. One of its core requirements is that sensitive data such as card numbers and security codes must never be written to application logs, even in error states. This matters here because retry logic, by its nature, generates detailed error output when transactions fail. AI-generated code has no inherent awareness of this requirement and it will log whatever seems diagnostically useful. Left unreviewed, a well-intentioned error log becomes a compliance violation. This requires a dedicated security review pass before the code moves anywhere near production.

What integrations are at risk? The retry logic calls three external services: the payment processor, the order management system, and a fraud detection API. Each has its own failure mode.

What does "done" actually mean here? Not "the code runs." Done means: all retry paths tested against the payment processor sandbox, idempotency verified under load, security scan passed, and the fraud detection integration confirmed under simulated timeout conditions.

Step 2 — Estimate Validation Complexity, Not Development Time
The team now assigns effort to validation tasks, not implementation tasks.

This is the inversion in practice.

Validation Task	Complexity	Estimated Hours
Happy path and retry path testing (sandbox)	Low	3 hours
Idempotency verification under concurrent load	High	8 hours
Security scan and PCI log-scrubbing review	Medium	4 hours
Fraud API timeout simulation	Medium	5 hours
Stakeholder demo and acceptance sign-off	Low	2 hours
Total Validation Effort		**22 hours**

Implementation, by contrast, is estimated at four hours; reviewing the AI draft, patching the idempotency gap, and integrating it into the existing codebase.

Total sprint effort for this story: approximately twenty-six hours. A team new to AI Scrum should add thirty to forty percent to these estimates while the Validation Canvas process itself is still unfamiliar in which case plan for thirty-five to thirty-seven hours. The difference is not the final number. It is what the number represents. The team now knows exactly where their time is going and why. There are no surprises when the security scan takes half a day, because the plan accounted for it.

The Principle at Work

Validation-based estimation forces the team to think about risk before they think about work. The idempotency issue would not have surfaced in a traditional planning conversation until a developer found it mid-sprint. Here, it surfaces in planning, gets the hours it deserves, and gets assigned to the right person before a single line of code is written.

The payment retry logic shipped in sprint. No duplicate charges in production. No late-night rollback. It is worth noting that this is not always the outcome — sometimes the validation pass is clean and the sprint runs ahead of schedule. That confidence is itself the value. Knowing you checked thoroughly, found nothing critical, and shipped with justification is a different feeling than shipping and hoping.

Worked Example 2: Legal Projects — Contract Review Automation Sprint

Not every team adopting AI Scrum comes from a software background, and Chapter 14 addresses this directly. Legal operations teams are increasingly deploying AI contract analysis tools, and they face an estimation challenge that is structurally identical to the software scenario but culturally very different. The vocabulary of sprints and validation canvases does not always land easily in a legal context. This example is drawn from that friction.

The legal operations team at a regional insurance company had been using AI Scrum for three months when they encountered their first genuinely difficult sprint estimation challenge.

The work item on the board read: *Automate initial review pass for vendor contracts under $50K.* The AI tool in use was a contract analysis platform that could flag missing clauses, non-standard terms, and compliance deviations against the company's master template. The platform had already been tested. The draft review workflow had been designed. Now came the question of how long the sprint would actually take.

The team's Governance Coordinator, a paralegal with twelve years of experience, had been skeptical of AI Scrum when it was first introduced. "We are not software developers," she had told the AI Alignment Engineer during onboarding. "Our work is judgment, not velocity." Three months in, she had come to appreciate that AI Scrum did not ask her to abandon judgment. It asked her to plan for it explicitly.

The Feature: Automated first-pass review of vendor contracts, routing flagged documents to the appropriate attorney for final review, with an audit trail sufficient for the company's own outside counsel disclosure policy.

Step 1 — Identify the Validation Dimensions
In a legal context, validation is not about whether code runs. It is about whether the output is trustworthy enough to act on. The team's Validation Canvas for this sprint item identified three dimensions:

Accuracy validation: Does the AI correctly identify missing clauses and non-standard terms against the master template? The acceptable error rate for a first-pass tool is not zero — attorneys will review flagged items — but it cannot miss high-risk clauses such as indemnification and limitation of liability at a rate that would undermine confidence in the tool.

Routing validation: Does the flagging logic correctly classify contract risk levels, routing low-risk items to junior review and high-risk items to senior counsel? A misrouting in either direction has cost implications.

Audit trail validation: Is the generated log sufficient for the company's outside counsel disclosure policy? The general counsel had specified that every AI-assisted review must include a timestamp, the model version used, and the specific clauses evaluated. Requirements in this area vary by organization and jurisdiction and should be confirmed with your own legal leadership before adopting any standard.

Step 2 — Estimate Validation Effort

Validation Task	Owner	Estimated Hours
Accuracy test: 30 historical contracts, known outcomes	Paralegal + AI Alignment Engineer	6 hours
Routing logic review: 10 edge-case contracts	Senior Attorney + Validation Lead	5 hours
Audit trail format review against disclosure policy	Governance Coordinator	3 hours
Stakeholder walkthrough with General Counsel	Team Lead	2 hours
Documentation update for legal ops runbook	Paralegal	2 hours
Total Validation Effort		**18 hours**

AI tool configuration and workflow integration: approximately five hours. Total sprint effort for this item: twenty-three hours. As with the previous example, teams in their first or second sprint should budget an additional thirty to forty percent while the estimation process itself is being learned.

What Would Have Been Missed

In a traditional project management approach, this work item might have been estimated as a two-day task; configure the tool, run a test, deploy. The accuracy validation pass against thirty historical contracts revealed something important: the AI tool consistently missed indemnification sub-clauses buried in appendices when the appendix was formatted as a scanned image rather than native text. That is not a rare formatting pattern in the company's vendor contract library. It is common in contracts received from smaller vendors.

The finding added two hours of remediation work — adjusting the document pre-processing step to run optical character recognition on image-formatted appendices before analysis. Caught in validation planning, it was a two-hour fix. Caught in production, after the tool had already reviewed sixty live contracts, it would have required a full retrospective review of every output and a difficult conversation with outside counsel about disclosure implications.

The Governance Coordinator stopped being skeptical.

Worked Example 3: Data Science — Predictive Churn Model Sprint

The data science example introduces a dimension that neither of the previous two required: agreeing on what acceptable performance means before the sprint begins. In software development, "does it work" is usually answerable. In data science, it almost never is unless the team does the work of defining the answer in advance. Skipping that step is one of the most common and most expensive planning failures in AI-assisted data science work.

The data science team had built a reasonably good churn prediction model. The problem was that nobody agreed on what "reasonably good" meant, and nobody had written it down before the sprint started.

This is a failure mode that validation-based estimation is specifically designed to prevent.

The team, three data scientists, an AI Alignment Engineer on loan from the software development group, and a product owner from the customer success division, was midway through a two-week sprint to deploy a first version of the model to production. The AI-assisted model development had gone quickly. Feature engineering took a day. The model itself was trained and showing a validation AUC of 0.81. The product owner was pleased. The data scientists were cautious. The sprint planning conversation had never formally resolved what acceptable performance actually looked like.

The Feature: Deploy a 30-day customer churn prediction model to the customer success platform, surfacing at-risk accounts for proactive outreach.

Step 1 — Establish Acceptance Thresholds Before Estimating

This is the step that traditional sprint planning skips, and it is the most important one in a data science context. Before estimating validation effort, the team must answer: *Validated against what standard?*

The AI Alignment Engineer facilitated a thirty-minute conversation that produced three written thresholds:

Performance: Minimum AUC of 0.80 on the holdout test set. Precision on the top decile of predicted churners must exceed 65% — because customer success managers will act on the top-decile list, and if more than a third of their outreach is directed at false positives, adoption of the tool will collapse within a month regardless of how technically sound the model is.

Fairness: The model must not demonstrate differential churn prediction rates across customer segments defined by contract size or industry vertical at a level that would constitute systematic bias in outreach prioritization.

Stability: The model must be retested against the previous month's data to confirm that performance does not degrade more than three percent when applied to a different time window. Churn patterns are seasonal. A model that overfits to its training window is a liability that compounds quietly over time.

Step 2 — Estimate Validation Effort

Validation Task	Owner	Estimated Hours
Holdout test set evaluation (AUC, precision at top decile)	Data Scientist	4 hours
Segment fairness audit (contract size, vertical)	Data Scientist + AI Alignment Engineer	6 hours
Temporal stability test (prior month holdout)	Data Scientist	4 hours
Customer success team walkthrough: how to read the output	Product Owner + Team Lead	3 hours
Model card documentation (inputs, thresholds, known limitations)	AI Alignment Engineer	4 hours
Governance sign-off on deployment to production	Governance Coordinator	2 hours
Total Validation Effort		**23 hours**

Model development and final training pipeline: approximately six hours. Total sprint effort: twenty-nine hours. For teams newer to the process, plan for thirty-eight to forty hours.

What the Conversation Revealed

The fairness audit uncovered something the team had not anticipated. The model was assigning disproportionately high churn risk scores to customers in the healthcare vertical, not because those customers churned more often, but because the training data contained a historical period during which the company had proactively paused healthcare contracts during a compliance review. Those pauses had been coded as churn events in the raw data.

Left undetected, the model would have directed customer success resources away from healthcare accounts that were actually satisfied customers, while those accounts quietly renewed, and toward healthcare accounts flagged as high-risk without justification. The business impact would have been invisible at first and damaging over time.

The data scientist who found it said afterward that she had suspected something was off during training but had not flagged it because there had been no formal checkpoint to do so. Validation-based estimation created that checkpoint. The fix, reprocessing the compliance pause events as a separate label and retraining added one day to the sprint. It was not a comfortable conversation with the product owner. It was a far more comfortable conversation than the one that would have happened six months later.

The model deployed in the following sprint. It is still in production.

These three examples share a structure that is worth naming before you move on. Each one surfaces a risk during validation that would not have been visible in a traditional planning conversation; an idempotency gap, an OCR formatting blind spot, a data labeling artifact. In none of these cases was the risk exotic or unpredictable. In every case, the risk was findable. Validation-based estimation does not guarantee you will find everything. It guarantees you will look in the right places at the right time.

WHAT COMES NEXT

Planning establishes what teams commit to achieving and how they will validate success. But planning alone does not deliver outcomes. Teams also need systematic approaches to validation that ensure AI-generated implementations actually work correctly. The next chapter examines validation practices that enable teams to ship AI-generated code with confidence.

The planning framework establishes what work teams select and how they estimate effort. Chapter 8 addresses how teams verify that work satisfies quality standards before deployment.

KEY TAKEAWAYS

- Planning must shift from feature delivery to outcome achievement when AI eliminates implementation as the binding constraint.
- Outcome Backlogs specify measurable results to achieve rather than features to build, opening solution space rather than constraining approaches.
- Estimation conversations focus on validation effort rather than implementation difficulty, fundamentally changing what story points measure while preserving their capacity planning value.
- Hidden dependencies represent planning's most persistent problem; AI pattern recognition across entire codebases reveals connections humans cannot easily track.
- AI variability introduces nondeterminism requiring planning processes that accommodate multiple valid solutions rather than expecting identical implementations.
- Sprint planning for AI work requires seven new question categories addressing validation feasibility, model behavior, architecture boundaries, cost considerations, human-in-the-loop strategy, measurement approaches, and AI-specific risks.
- Capacity planning shifts from development throughput to validation capacity, requiring teams to assess security expertise availability, stakeholder accessibility, environment readiness, and governance participant engagement rather than coding capability.

MIND MAP – PLANNING AND ESTIMATION

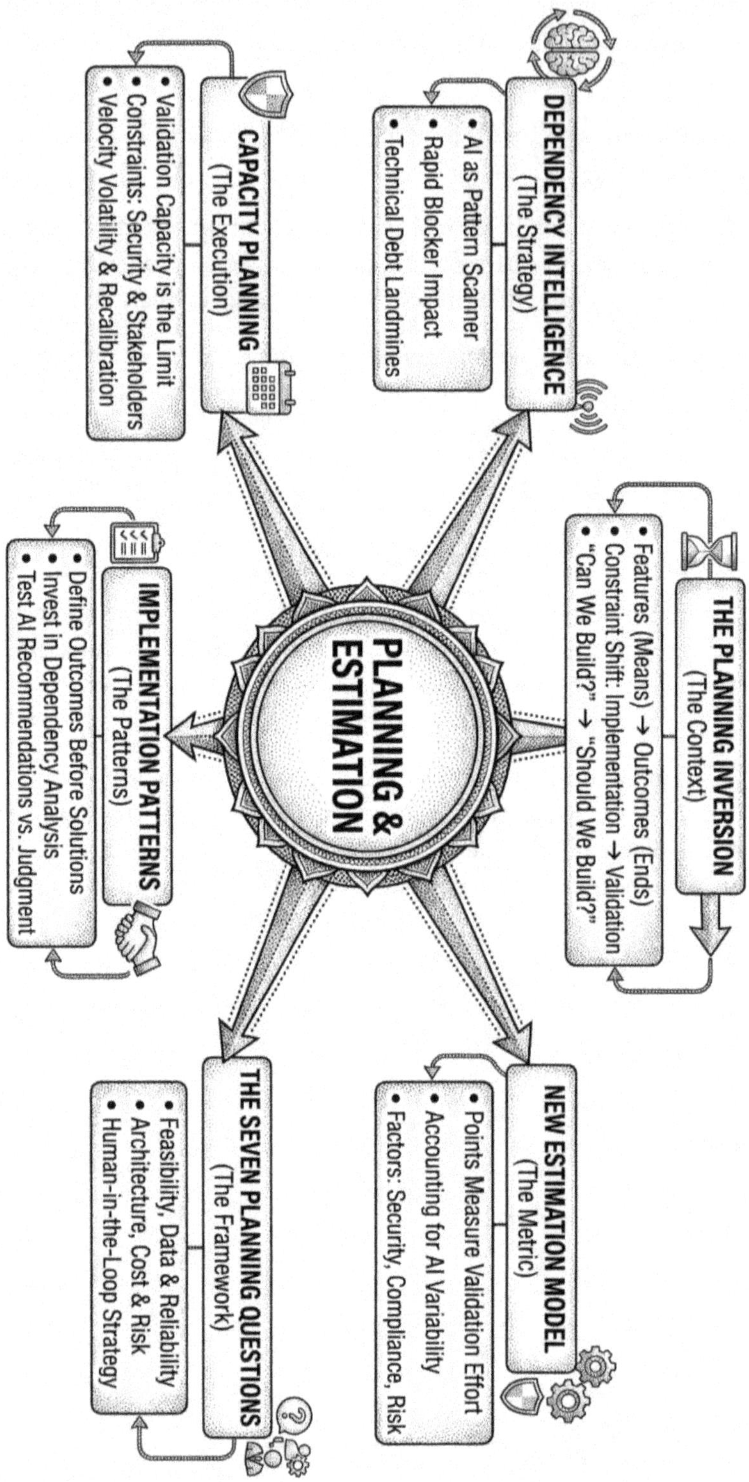

CHAPTER 8: VALIDATION AND TESTING STRATEGIES

Artificial intelligence dramatically accelerates code generation, effectively shifting the primary development bottleneck from creation to validation tasks. To prevent rising production failures, organizations must urgently implement a validation-first philosophy using frameworks like the Validation Canvas. This strategic approach prioritizes adversarial testing and context-aware acceptance thresholds to ensure rapid delivery never compromises critical system quality, performance, or security protocols.

The validation constraint will become critical when AI eliminates the development bottleneck.

A predictable pattern will emerge as organizations implement AI code generation across their development teams. Productivity metrics will soar. Developers will generate features in days that previously took weeks. Sprint velocity will double. Management will celebrate.

Three months later, production failures will triple. Customer complaints will flood support channels.

The root cause is simple but devastating: while AI generates code quickly, organizations that have not evolved their validation practices to match the new pace will struggle. Test coverage remaining manual and slow means quality assurance teams fall further behind with each sprint. These organizations will have accelerated development without accelerating validation, creating a dangerous imbalance.

This scenario will play out repeatedly as organizations adopt AI-augmented development. The traditional development bottleneck shifts from writing code to validating it. Teams that fail to recognize this shift will discover that faster development without faster validation creates more problems than it solves.

WHAT THIS CHAPTER COVERS

In this chapter, we explore:

- The validation-first philosophy that prevents quality debt.
- The Validation Canvas framework for comprehensive quality planning.
- Adversarial testing approaches specific to AI-generated code.
- AI-specific testing patterns and methodologies.
- Red-teaming techniques for identifying hidden vulnerabilities.
- Performance validation strategies for AI-augmented systems.
- Security validation approaches.
- Acceptance threshold frameworks that define "good enough".

KEY TERMS AND CONCEPTS

- **Validation-First Development:** A development approach where validation planning occurs *before* code generation, ensuring quality controls exist before new features emerge.

- **Validation Canvas:** A structured framework for planning validation activities across unit testing, integration testing, end-to-end testing, security validation, and performance validation.

- **Adversarial Testing:** A testing approach designed to identify edge cases, failure modes, and vulnerabilities through deliberate attempts to break the system or trigger unexpected behaviors.

- **Test Suite Decay:** The gradual obsolescence of automated tests that makes them less valuable over time, often resulting from changing requirements, evolving implementations, or brittle test design.

- **Red-Teaming:** A structured approach to adversarial testing where dedicated teams attempt to identify vulnerabilities, security weaknesses, and failure modes through systematic attack simulation.

- **Acceptance Thresholds:** Quantitative and qualitative criteria that define when validation is sufficient to approve deployment, balancing thoroughness against delivery timelines.

- **Validation Debt:** Accumulated gaps in testing and quality assurance that result from postponing validation activities, similar to technical debt but focused on quality verification.
- **Model Drift Validation:** Testing approach specific to AI systems that validates model performance over time as real-world data distributions shift from training distributions.

THE VALIDATION-FIRST PHILOSOPHY

The Validation Canvas provides a structured planning tool for detailed quality assurance across multiple dimensions. Teams use this canvas during sprint planning to map validation requirements before development begins.

Canvas Structure

The canvas organizes validation planning across five critical dimensions:

- **Unit Testing Scope:** Defines coverage requirements for individual functions, methods, and components. Teams specify which code paths require testing, what edge cases to address, and what assertion quality standards to enforce.
- **Integration Testing Requirements:** Maps component interactions requiring validation. Teams identify integration points, dependency contracts, failure scenarios, and data flow verification points.
- **End-to-End Validation:** Specifies user workflow testing across system boundaries. Teams outline critical user journeys, realistic usage scenarios, and expected system behaviors under normal and stress conditions.
- **Security Validation:** Details security testing requirements including vulnerability scanning, penetration testing, authentication verification, and authorization validation.
- **Performance Validation:** Establishes performance baselines and acceptance criteria including response time requirements, throughput expectations, resource utilization limits, and scalability targets.

Canvas Application

Teams complete the Validation Canvas collaboratively during sprint planning, before user story refinement. This sequence ensures validation considerations inform implementation decisions rather than reacting to them afterward.

The AI Alignment Engineer leads canvas completion, working with the Validation Lead and development team. Together, they identify validation requirements specific to planned sprint work. The Product Owner participates to ensure validation plans align with business risk tolerance and delivery expectations.

Figure 8.1: The Validation Blueprint. *Planning validation requirements before construction begins prevents the accumulation of quality debt.*

For each sprint goal, the team asks: "What could go wrong?" "How would we know?" "What tests would catch it?" These questions drive canvas completion, surfacing validation needs that might otherwise remain unconsidered until problems emerge in production.

The completed canvas guides both human and AI testing efforts throughout the sprint. When developers generate code, they reference the canvas to understand validation expectations. When AI generates test suites, the canvas provides context about required coverage and quality standards.

AI-ENHANCED TEST GENERATION

AI transforms test creation from a manual bottleneck into an automated capability. However, this transformation requires careful human oversight to ensure generated tests provide genuine quality assurance rather than false confidence.

Unit Test Generation and Enhancement

AI generates thorough unit test suites by analyzing code to identify test scenarios with thoroughness that surpasses what most human developers achieve under time pressure. This capability extends beyond simple pattern matching or template application. Modern AI systems understand code semantics deeply enough to generate tests for standard cases, edge cases, error conditions, and boundary conditions that even experienced developers might overlook when writing tests manually.

Consider a function that processes customer orders. A human developer might write tests for successful orders, invalid payment methods, and out-of-stock items, which are the obvious scenarios that come to mind immediately. AI, analyzing the code completely, also generates tests for null inputs, empty strings, maximum length inputs, special characters, concurrent access, database connection failures, timeout conditions, partial failures in multi-step processes, and every permutation of valid and invalid state combinations. The coverage difference proves substantial.

According to Gartner research, AI can automate generation of 70-80% of unit tests, dramatically expanding coverage compared to manually written tests constrained by time limitations. This represents 70-80% of all necessary unit tests, including the tedious edge cases that developers typically defer or skip entirely because sprint time runs out or the test seems unlikely to catch real bugs. Those deferred tests are precisely where production bugs hide.

However, human judgment remains essential. The Validation Lead reviews generated unit tests to ensure several critical qualities that AI does not yet reliably deliver without guidance:

- **Meaningful assertions, not merely execution verification.** A test that confirms code runs without checking what it produces provides false confidence. The test passes but validates nothing.
- **Realistic test data reflecting actual usage patterns.** Tests using email addresses like "test@test.com" and phone numbers like "1234567890" might pass while real user data breaks functionality.
- **Proper test isolation.** Tests must run independently without hidden dependencies on execution order, shared state, or external resources.

- **Inclusive negative testing.** Coverage must include negative test cases that validate error handling and boundary conditions, not just happy path scenarios.

Figure 8.2: The Illusion of Validity. *Summary reports often mask underlying failures, requiring deep inspection of AI-generated assertions. Don't assume that what has been provided is entirely accurate.*

Why does this review matter? AI-generated tests sometimes include weak assertions or unrealistic scenarios that undermine test suite value. A test might verify that a function returns a value without checking whether that value is correct; it asserts result is not null rather than result equals expected value. Another might use test data that would never occur in production, passing tests that hide real bugs. A function might handle the string "test" correctly while failing on actual customer names with apostrophes, hyphens, or non-ASCII characters.

Human review ensures test quality, not just test quantity, maintaining test suite value as a reliable quality signal rather than a false confidence generator.

I learned this lesson years ago working with a testing platform that produced beautifully clean summary reports. The rollup view showed suites passing with green checkmarks across the board. QA engineers relied on these summaries to validate releases. The problem was that individual test failures were not surfacing in the rollup view. You had to dig into the details to discover that specific tests had actually failed. Most of the hidden failures were minor issues, but the question haunted me: what if one of those buried failures was major? That experience taught me to never trust summary results alone. Even when the software declares the suite passed, dig into the details and verify that all tests actually passed, not just the aggregate.

Integration Test Generation

Integration testing verifies that components interact correctly, a challenge that multiplies as system complexity increases. Traditional integration testing often focused on happy path scenarios, leaving component interaction failures to emerge in production. AI changes this equation fundamentally.

AI generates integration tests by analyzing component interfaces and dependencies in ways that surface problems human reviewers may otherwise miss. The process involves several sophisticated steps:

- **Identifying integration points and contracts between components.** AI maps which components communicate, what data they exchange, and what assumptions they make about each other's behavior.
- **Generating tests for normal interaction patterns.** These cover the standard flows most teams already test, but AI does so comprehensively rather than selectively.
- **Creating failure scenario tests.** AI explores what happens when components misbehave, networks fail, or timeouts occur. These scenarios often go untested in manual approaches because imagining all possible failures proves difficult.
- **Validating data flow correctness across boundaries.** AI traces data as it moves between components, checking for transformation errors, serialization problems, or contract violations.

Integration test generation proves more complex than unit tests due to dependency management and test environment requirements. Setting up conditions for integration testing requires configuring multiple components, managing state, and orchestrating interactions. AI substantially reduces manual effort but does not eliminate the need for skilled oversight. The Validation Lead ensures tests exercise realistic scenarios and detects genuine integration problems. Generated tests might technically execute but fail to catch actual issues if they do not reflect real-world usage patterns or dependency behaviors.

The best QA teams in the world cannot test for scenarios nobody told them about. I have worked with exceptional QA engineers who rarely let anything slip through their nets. When issues did reach production, the root cause typically traced back to incomplete requirements analysis rather than testing failures. Business analysts and stakeholders had not captured certain use cases during analysis. Weeks or months later, an edge case would emerge in production because it was never considered during development. The validation was perfect for what was specified, but the specification was incomplete. This taught me that validation effectiveness depends fundamentally on requirements completeness. You cannot test your way out of inadequate analysis.

Figure 8.3: The Integration Gap. *AI can generate code for individual components rapidly, but validation failures often occur at the seams where components meet or with the existing systems.*

End-to-End Test Automation

End-to-end tests validate complete user workflows across system boundaries. These tests provide the highest confidence that systems work as users expect, yet traditionally proved most expensive and brittle to maintain. AI transforms the economics.

AI assists end-to-end testing through several capabilities that compound in value:

- **Identifying scenarios from user stories and acceptance criteria.** AI extracts the workflows that matter most to users from natural language requirements.
- **Generating test scripts for common workflows.** AI handles the complexity of multi-step interactions, creating automated tests that previously required significant manual effort.

- **Creating realistic test data.** AI generates test data that mirrors actual usage patterns rather than simplified test scenarios.
- **Performing visual regression testing.** AI detects unintended UI changes that break user experience, catching issues that functional tests might miss.

Organizations like EPAM have developed agentic AI approaches to quality assurance where AI systems autonomously explore application functionality, generating full test coverage without explicit programming. The AI acts like a curious user, trying different interaction patterns and recording what works and what fails. This exploratory testing discovers issues that scripted tests miss because human test designers never imagined those particular scenarios.

Test Maintenance and Evolution

Test suites require continuous maintenance as systems evolve, and this maintenance burden often determines whether tests remain valuable assets or become expensive liabilities. This maintenance challenge intensifies with AI-augmented development because code changes occur more frequently, in larger volumes, and with less predictable patterns than traditional development.

Decay of test suites represents one of the most insidious quality problems organizations face. The pattern unfolds predictably: teams initially invest heavily in test automation, achieving impressive coverage and quality outcomes. Then implementations evolve with changing business requirements. Code gets refactored to accommodate new architectural patterns. Gradually, tests start failing for reasons unrelated to actual bugs. They fail because they encode assumptions no longer valid, depend on data no longer exists, or assert behaviors that intentionally changed.

Teams face a choice: invest time debugging and fixing tests, or disable failing tests to keep builds passing. Time pressure usually drives teams toward the easier path. Tests get disabled to maintain green builds. Coverage drops as more tests are bypassed. Confidence erodes in the remaining test suite. Eventually, teams stop trusting their test suite and resort to manual testing, abandoning the automation investment.

AI assists test maintenance in ways that prevent this decay cycle and preserve test suite value over time:

- **Detecting obsolete tests no longer providing value.** AI identifies tests that always pass because the code they validate was removed or fundamentally changed. These tests waste CI/CD capacity without providing quality signal.
- **Updating tests affected by code changes.** AI automatically adjusts assertions when interfaces evolve, updates test data when schemas change, and modifies test setup when dependencies shift.
- **Identifying brittle tests requiring refactoring.** AI distinguishes genuine failures from test implementation problems like hardcoded timeouts, race conditions in test execution, or hidden dependencies on execution order.
- **Suggesting test improvements based on failure patterns.** AI learns from experience which test structures prove most reliable and which patterns consistently cause maintenance problems.

Figure 8.4: Combating Test Suite Decay. *AI acts as a maintenance crew, removing obsolete tests and "barnacles" that slow down the development*

This automated maintenance prevents test suite decay that afflicts manually maintained tests. Teams spend less time nursing a test suite back to health and more time adding tests that expand coverage of new functionality. The test suite remains a valuable asset rather than becoming a liability that teams eventually abandon.

ADVERSARIAL TESTING AND RED-TEAMING

AI-generated code introduces unique vulnerability patterns that traditional testing approaches miss. Adversarial testing and red-teaming provide structured methodologies for identifying these hidden weaknesses before they reach production.

The Adversarial Testing Mindset

Adversarial testing approaches validation from an attacker's perspective, asking how this system could be broken. What edge cases might trigger unexpected behavior? What input combinations would cause failures? This mindset differs fundamentally from traditional testing, which validates that systems work correctly under expected conditions.

For AI-generated code, adversarial testing proves particularly valuable because AI sometimes generates code with subtle vulnerabilities that pass standard testing but fail under adversarial conditions. The code might handle typical inputs correctly while exhibiting dangerous behaviors with crafted inputs.

Teams conduct adversarial testing systematically rather than ad hoc. The Validation Lead organizes regular adversarial testing sessions where team members actively attempt to break features before deployment. These sessions follow structured patterns:

- **Input fuzzing:** Generating random, malformed, or extreme inputs to identify parsing errors, validation gaps, or unexpected behaviors.
- **Boundary testing:** Probing system limits including maximum values, minimum values, empty inputs, null values, and combinations that should trigger edge case handling.
- **State manipulation:** Attempting to achieve invalid system states through unexpected operation sequences, concurrent access patterns, or interrupted workflows.
- **Resource exhaustion:** Testing system behavior under resource constraints including memory limits, connection pool exhaustion, or storage capacity.
- **Privilege escalation attempts:** Trying to access resources or perform operations beyond authorized scope, testing authentication and authorization boundaries.

These patterns uncover vulnerabilities that functional testing misses because functional testing validates expected behaviors while adversarial testing explores unexpected ones.

Red-Teaming for AI-Generated Systems

Red-teaming formalizes adversarial testing into a structured program where dedicated teams simulate attacks against systems before deployment. For AI-native development, red-teaming addresses both traditional security concerns and AI-specific vulnerabilities.

Figure 8.5: The Adversarial Mindset. *Red-teaming requires actively probing the system for weaknesses rather than simply verifying expected behaviors.*

Red teams focus on several AI-specific attack vectors:

- **Prompt injection:** Attempting to manipulate AI behavior through carefully crafted prompts that override intended constraints or extract sensitive information.
- **Context poisoning:** Introducing misleading information into AI context to influence generated code or responses in unintended ways.

- **Boundary violation:** Finding edge cases where AI-generated code violates specified constraints or produces outputs outside acceptable ranges.
- **Performance degradation:** Identifying inputs or scenarios that cause AI-generated code to perform poorly, consume excessive resources, or enter infinite loops.

The red team operates independently from development teams, maintaining adversarial perspective without insider knowledge that might limit creativity. After completing red-team exercises, teams hold structured debriefs where discovered vulnerabilities get prioritized for remediation. Organizations implementing red-teaming programs typically run exercises quarterly for critical systems and annually for lower-risk applications. The investment prevents production incidents that would cost far more to remediate than red-team exercises cost to conduct.

PERFORMANCE AND SECURITY VALIDATION

Functional correctness represents only one dimension of quality. AI-native development demands rigorous performance and security validation to ensure systems meet operational requirements and resist adversarial attacks.

Performance Validation Strategies

Performance validation establishes that systems meet response time, throughput, and resource utilization requirements under realistic load conditions. For AI-generated code, performance validation proves particularly critical because generated code might implement correct logic while exhibiting poor performance characteristics.

Teams establish performance baselines during sprint planning, defining acceptable performance parameters:

- **Response time targets:** Maximum latency for critical operations including API calls, database queries, and UI interactions.
- **Throughput requirements:** Minimum request handling capacity measured in requests per second, transactions per minute, or other relevant metrics.
- **Resource utilization limits:** Acceptable CPU usage, memory consumption, database connection counts, and external service dependencies.

- **Scalability expectations:** How performance should degrade as load increases, ensuring systems remain responsive under stress.

Performance testing occurs continuously throughout sprints, not deferred to pre-deployment phases. As features get generated and integrated, automated performance tests validate against established baselines. This continuous validation catches performance regressions immediately rather than discovering them when remediation costs more.

AI assists performance validation through automated load testing, resource monitoring, and bottleneck identification. AI-powered tools can generate realistic load patterns, simulate concurrent users, and identify performance degradation triggers that manual testing might miss.

Security Validation Approaches

Security validation for AI-generated code requires approaches that address both traditional vulnerabilities and AI-specific attack vectors. Teams implement multiple layers of security testing:

- **Static analysis scanning:** Automated tools analyze generated code for common vulnerability patterns including SQL injection risks, cross-site scripting vulnerabilities, insecure authentication, and insufficient input validation.
- **Dynamic security testing:** Runtime security testing that exercises running applications to identify vulnerabilities that static analysis misses, including authentication bypass, authorization failures, and session management flaws.
- **Dependency vulnerability scanning:** Automated scanning of third-party libraries and frameworks to identify known vulnerabilities in dependencies that AI-generated code incorporates.
- **Secrets detection:** Automated scanning to ensure generated code does not contain hardcoded credentials, API keys, or other sensitive information that should remain external to code.
- **AI-specific security validation:** Testing for prompt injection vulnerabilities, context manipulation attacks, and AI behavior manipulation through adversarial inputs.

Security validation occurs automatically in CI/CD pipelines, blocking deployment when critical vulnerabilities get detected. Lower-severity findings generate warnings that teams address according to risk-based prioritization. The Governance Board establishes security acceptance thresholds that define which vulnerability severities block deployment and which can be accepted with risk acknowledgment and remediation plans.

ACCEPTANCE THRESHOLD FRAMEWORKS

Determining when validation is sufficient represents one of the most challenging aspects of quality assurance. Perfect testing proves impossible to achieve in practice. Complete coverage remains perpetually unattainable. Teams must decide when "good enough" is actually good enough.

Defining Acceptance Thresholds

Acceptance thresholds establish quantitative and qualitative criteria that define sufficient validation for deployment. These thresholds balance thoroughness against delivery timelines, providing clear guidance about when additional testing yields diminishing returns.
Teams establish several categories of acceptance thresholds:

- **Coverage thresholds:** Minimum percentages of code lines, branches, and paths that automated tests must exercise. Typical thresholds range from 70-90% depending on system criticality.
- **Test quality thresholds:** Requirements for assertion strength, test independence, and realistic test data usage. Not all tests provide equal value; quality thresholds ensure tests actually validate behavior.
- **Performance thresholds:** Maximum acceptable response times, minimum throughput rates, and resource utilization limits that systems must meet under specified load conditions.
- **Security thresholds:** Maximum number and severity of vulnerabilities acceptable before deployment. Critical and high-severity findings typically block deployment while medium and low-severity findings might be accepted with remediation plans.
- **Reliability thresholds:** Minimum test pass rates, maximum failure frequencies, and stability requirements that builds must meet before promotion to higher environments.

These thresholds provide objective criteria for deployment decisions, preventing debates about whether systems are "ready" and replacing subjective judgment with measurable standards.

Risk-Based Threshold Adjustment

Not all features carry equal risk. Acceptance thresholds should vary based on feature criticality, user impact, and failure consequences. The Governance Board establishes guidelines for threshold adjustment:

- **High-risk features:** including payment processing, personal data handling, and security-critical functionality require stricter thresholds: 90%+ code coverage, zero critical vulnerabilities, extensive adversarial testing.
- **Medium-risk features:** including core business logic and customer-facing workflows use standard thresholds: 80%+ coverage, no high-severity vulnerabilities, standard performance baselines.
- **Low-risk features:** including internal tools, experimental capabilities, and non-critical UI changes accept relaxed thresholds: 70%+ coverage, documented medium-severity vulnerabilities with remediation plans, baseline performance validation.

ROUGH SKETCH

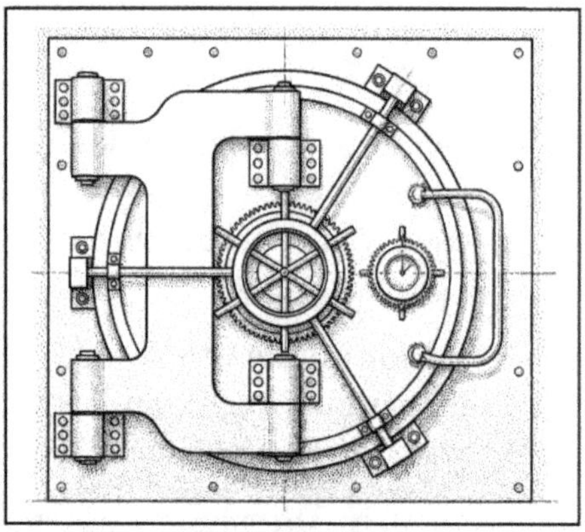

TECHNICAL VAULT DRAWING

Figure 8.6: Contextual Perfection. *Acceptance thresholds must vary by risk; a user interface might accept "good enough," while financial data demands absolute precision. Don't apply vault-door standards to sketch-pad problems, and don't apply sketch-pad standards to vault-door problems.*

This risk-based approach enables teams to move quickly on low-risk features while maintaining rigor for high-risk capabilities. The approach prevents both excessive caution that slows all development and insufficient caution that permits dangerous releases.

The distinction between "good enough" and "must be perfect" depends entirely on context. When selecting my LinkedIn profile headshot, I accepted good enough without demanding perfection. The risk of an imperfect photo is minimal embarrassment. But I have also worked on bank core conversions where perfection was non-negotiable. Consider a database field holding a bank account balance. A single misplaced decimal point could show $1,000 instead of $10,000 in a customer's checking account. When that customer attempts a $1,200 purchase, the transaction fails despite having more than adequate funds everywhere else in the account. One field, one decimal point, one seemingly minor issue creates a catastrophic customer experience. Financial data demands absolute perfection without compromise. User interface polish accepts "good enough." Airplane wings require engineering perfection throughout. The risk profile determines the acceptance threshold.

The "Good Enough" Decision Framework

Deciding when validation is sufficient requires structured decision-making that considers multiple factors beyond simple threshold achievement. The framework evaluates:

- **Threshold achievement:** Have all applicable acceptance thresholds been met or exceeded?
- **Residual risk assessment:** What known issues remain unresolved, and what are their potential impacts?
- **Mitigation readiness:** Are monitoring, rollback, and incident response capabilities in place to handle potential issues?
- **Business context:** What are the costs of delaying deployment versus the costs of potential issues?
- **Stakeholder alignment:** Do Product Owner, Validation Lead, and Governance Board agree that deployment is appropriate?

The framework does not eliminate judgment but structures it, ensuring decisions consider relevant factors systematically rather than relying on intuition or pressure alone. Teams document acceptance decisions including which thresholds were met, what residual risks were accepted, and what mitigation strategies are active. This documentation provides accountability and enables learning from outcomes to refine future threshold decisions.

FROM THE TRENCHES: VALIDATION-FIRST IMPLEMENTATION PATTERNS

Successfully implementing validation-first development requires addressing common challenges and establishing patterns that make quality assurance natural rather than burdensome.

The validation bottleneck I remember most vividly had nothing to do with skipping steps. Our development, staging, and pre-production environments had drifted out of sync over time. Code written and tested in development behaved exactly as expected, until it hit pre-production, where different configurations turned clean code into bugs. Nobody had cut corners. The environments simply told different stories. QA worked through the backlog methodically, releasing only when confidence was solid, which meant releases came in careful increments rather than full sprints. We moved slowly not because the team was slow, but because the foundation validation depended on wasn't reliable. That experience made one thing permanent in my practice: validation is only as trustworthy as the environment it runs in.

Pattern: Validation Canvas in Sprint Planning

Integrate Validation Canvas completion directly into sprint planning ceremonies. Before estimating user stories, teams complete the canvas for each planned feature. This ensures validation requirements inform estimation rather than appearing as unexpected work mid-sprint. The pattern typically adds 15-20 minutes to sprint planning sessions but prevents hours of rework later. Teams find that discussing validation requirements upfront surfaces implementation complexities that might otherwise remain hidden until development is underway.

Pattern: AI Test Generation Review Checklist

Establish a standard checklist for reviewing AI-generated tests that Validation Leads use consistently. The checklist includes verification that tests include meaningful assertions, use realistic test data, demonstrate proper isolation, cover negative test cases, and provide clear failure diagnostics. This checklist ensures reviews are thorough and consistent rather than ad hoc and variable. Teams report that checklist-guided reviews identify 40-60% more test quality issues than unstructured reviews.

Pattern: Continuous Performance Baselines

Rather than establishing performance baselines once and expecting them to remain static, implement continuous baseline updates that reflect system evolution. As features get added and architectures evolve, performance expectations should adjust accordingly. Teams review performance baselines quarterly, updating targets based on actual system behavior and changing business requirements. This prevents either outdated baselines that no longer reflect reality or overly optimistic targets that systems cannot meet.

Pattern: Red-Team Rotation

Rotate team members through red-team exercises rather than maintaining dedicated adversarial testing teams. This rotation spreads adversarial testing skills across the organization while preventing burnout from continuous attack simulation. Teams implementing rotation typically assign each developer to one red-team exercise per quarter. This exposure builds security awareness and adversarial thinking skills that improve daily development practices beyond formal testing.

Pattern: Acceptance Threshold Retrospectives

After each deployment, review actual outcomes against acceptance thresholds to refine future decisions. If issues emerged despite meeting thresholds, thresholds may need tightening. If no issues emerged despite missing thresholds, thresholds may be unnecessarily strict. This continuous learning ensures thresholds evolve based on organizational experience rather than remaining static. Teams report that threshold retrospectives reduce both excessive caution and insufficient rigor over time.

Figure 8.7: The Validation Handoff. *Structured patterns ensure that quality assurance is a continuous partnership, like a climber and belayer, rather than a final hurdle.*

WHAT COMES NEXT

Validation ensures individual features meet quality standards. But AI-native development generates code so rapidly that tracking whether initiatives are progressing successfully becomes challenging.

The next chapter examines how metrics and measurement evolve to provide visibility into AI-augmented development velocity, quality, and outcomes. Traditional Agile metrics like velocity and burndown charts assume human-paced development. When AI accelerates code generation, these metrics can mislead more than inform. We will explore how AI-native metrics provide accurate insight into development health while supporting data-driven decision making.

KEY TAKEAWAYS

- **Constraint Inversion:** The validation-first philosophy prevents quality debt by planning validation before generating code, inverting traditional development sequences.
- **Structured Planning:** The Validation Canvas provides a structured framework for quality planning across unit testing, integration testing, end-to-end validation, security testing, and performance validation.
- **Human Oversight:** AI-generated tests require human review to ensure meaningful assertions, realistic test data, proper isolation, and negative case coverage.
- **Adversarial Defense:** Adversarial testing and red-teaming identify vulnerabilities that traditional functional testing misses by actively attempting to break systems before deployment.
- **Contextual Rigor:** Acceptance thresholds should vary based on feature risk, with stricter requirements for high-risk capabilities and relaxed thresholds for low-risk features.
- **Decision Framework:** "Good enough" decisions require structured frameworks that consider threshold achievement, residual risks, mitigation readiness, business context, and stakeholder alignment.

MIND MAP – VALIDATION AND TESTING STRATEGIES

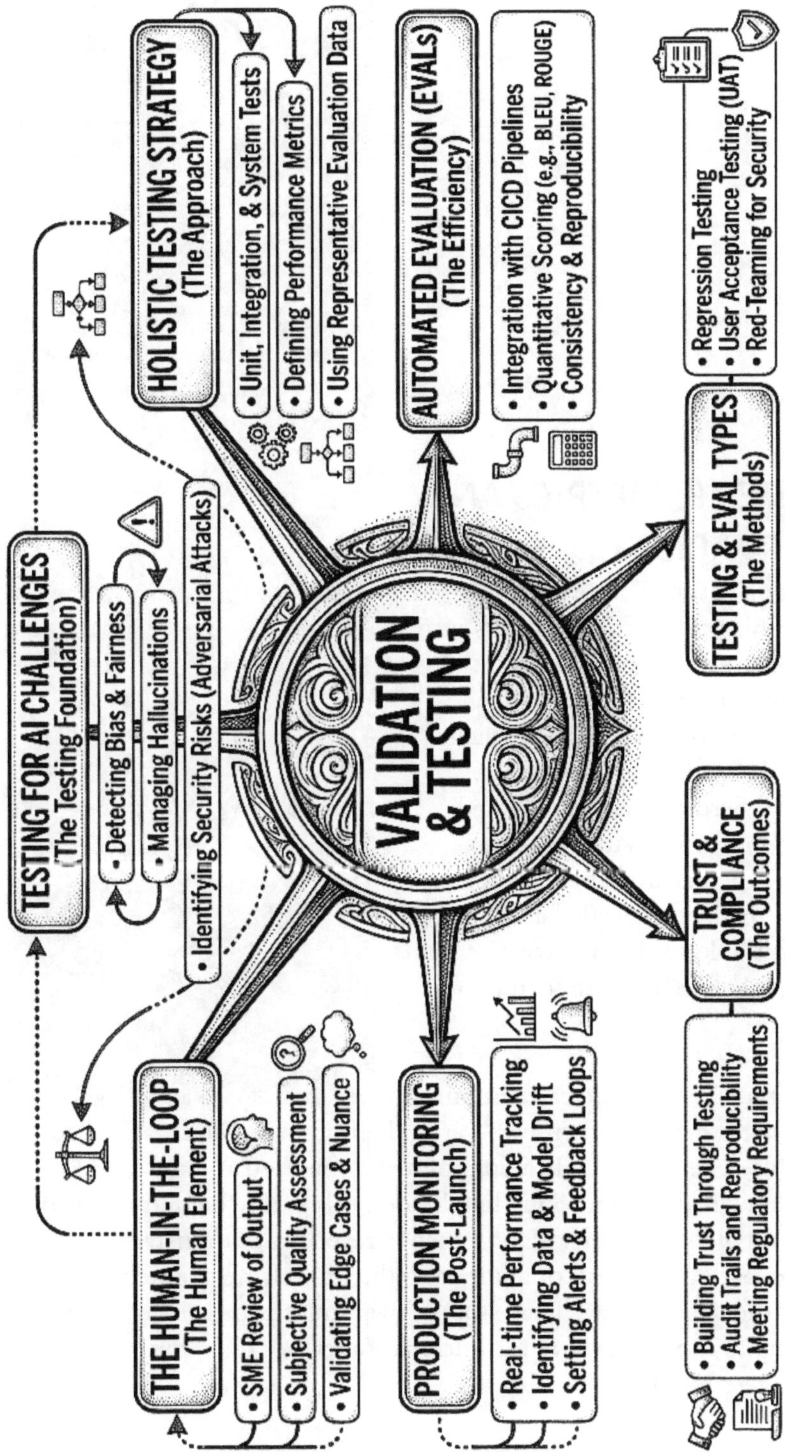

9

Traditional agile metrics like velocity frequently mislead teams in AI-native environments where coding speed is rarely the primary constraint. This text defines a comprehensive framework focusing on outcome-based measurements such as validation lead time and acceptance rates. It emphasizes using leading indicators to predict bottlenecks while strongly arguing for abandoning vanity metrics that obscure the actual cost of verification.

CHAPTER 9: METRICS AND OBSERVABILITY

Consider a Scrum Master joining a high-profile project three months into development. The team had been measuring progress through story points and burndown charts, which looked reasonable on the surface. When the Scrum Master calculated how long it would actually take to complete the backlog at current velocity, the math revealed an uncomfortable truth: the project would miss its deadline by more than two years.

After triple-checking the calculation and having other team members verify the numbers, the data proved irrefutable. The team was heading toward a spectacular failure, and the metrics they had been tracking had hidden this reality behind comforting but meaningless numbers. Management's response demonstrated a common pattern. Instead of addressing the underlying problems, they rejected the data entirely. They stopped wanting to see velocity projections, preferring to look elsewhere for information rather than confront what the metrics were revealing.

The project eventually required massive scope reductions, additional staff, and budget increases. What should have been caught months earlier festered because the organization was measuring the wrong things. This scenario illustrates a critical principle. When metrics reveal something uncomfortable, that signal demands deeper investigation into underlying issues. Metrics should be established for any project upfront so that managers avoid surprises.

You would not drive a vehicle without a dashboard. Would you be upset when you ran out of gas or received a citation for exceeding the speed limit? The car is not at fault, nor is the officer. Would you continue driving hoping negative things would not happen, or would you equip yourself with a gas gauge and speedometer before proceeding?

This chapter examines how AI Scrum transforms measurement from output tracking to outcome achievement, from manual dashboard creation to automated insight generation. Traditional metrics optimized for measuring development activity prove inadequate when AI can generate implementations in minutes. The question shifts from how much code we write to whether that code solves real business problems.

WHAT THIS CHAPTER COVERS

This chapter defines comprehensive approaches to metrics, observability, and insight generation in AI-native development contexts. You will understand:

- **The fundamental shift** from output metrics like velocity and story points to outcome metrics that measure business value delivery and problem-solving effectiveness.
- **AI-specific metrics** for tracking collaboration effectiveness, including AI-generated code acceptance rates and validation quality measures.
- **Dashboard design** approaches that provide actionable insights for executives, Product Strategists, and team members without overwhelming them with data.
- **Leading indicators** that predict validation strain, governance bottlenecks, and AI Debt accumulation before they become critical problems.
- **Which traditional metrics should be abandoned** because they create false confidence in AI-native contexts where implementation speed no longer constrains delivery.
- **Telemetry architecture** and cost optimization strategies that make thorough observability affordable for organizations of all sizes.

The chapter balances theoretical measurement philosophy with practical implementation guidance. The metrics framework comes from systematic analysis of how AI inverts development constraints. The dashboard patterns reflect lessons learned from traditional Scrum where inadequate measurement led to project failures that better metrics would have prevented.

KEY TERMS AND CONCEPTS

- **Outcome Metrics** Measurements that track business value delivered and problems solved rather than development activity performed. Outcome metrics answer whether teams achieved their business objectives, not whether they completed their planned work.
- **Output Metrics** Measurements that track development activity like story points completed, code lines written, or velocity trends. Output metrics optimized for traditional development contexts become misleading in AI-native environments where implementation speed no longer constrains delivery.
- **AI First-Pass Acceptance Rate** The percentage of AI-generated implementations that pass initial validation without requiring human modification. Low acceptance rates indicate either inadequate prompt engineering or inappropriate AI application to the problem domain.
- **Validation Lead Time** The elapsed time from when an implementation becomes available for validation until validation completes and the implementation receives approval for deployment. Long validation lead times reveal the true constraint in AI-native development.
- **Cognitive Load Score** A quantitative measure of the mental effort required to validate AI-generated implementations, considering factors like implementation complexity, validation completeness requirements, and context switching frequency. High cognitive load scores predict validator burnout and quality degradation.
- **Leading Indicators** Measurements that predict future problems before they manifest. In AI Scrum, leading indicators track validator workload trends, governance approval durations, and AI Debt accumulation patterns to enable proactive intervention.
- **Telemetry** Automatically collected data about system behavior, including logs capturing specific events, metrics measuring system state over time, and traces showing request flow through

distributed systems. Inclusive telemetry enables observability and automated insight generation.

- **Observability** The capability to understand internal system states based on external outputs through telemetry collection and analysis. Observable systems expose their behavior sufficiently that teams can diagnose problems and optimize performance without requiring direct code inspection.
- **Prompt Engineering Efficiency** A measure of how many iterations teams require to generate acceptable AI outputs, tracking both the number of refinement cycles and the time investment per prompt development. Low efficiency indicates skill gaps or unclear outcome definitions.
- **Token Cost Tracking** Monitoring of API usage costs associated with AI model interactions, typically measured in tokens consumed and dollars spent. Organizations track token costs by activity type to optimize spending and identify inefficient prompt patterns.

THE METRICS TRANSFORMATION

I once worked with a team whose velocity numbers were genuinely impressive. Sprint after sprint, the board cleared. Points accumulated. Management was pleased. But something felt wrong. We were moving fast and not getting anywhere. Looking closer, two problems were hiding behind the numbers. A significant portion of our output was resolving pre-production issues, bugs dressed up as backlog items, which kept velocity high while real progress stalled. The deeper problem was the project itself. Non-technical leadership had a clear vision of what they wanted but limited understanding of what their requirements actually demanded of the people doing the work. We were laying a foundation full of landmines and technical debt, sprint by sprint, for the sake of visible output. This was not a case of "done is better than perfect." It was a case where doing it right the first time would have cost less than doing it over. That team taught me more about what not to measure than any framework ever could.
Traditional agile metrics focused on development outputs. Lines of code written, story points completed, velocity trends, and sprint burndown charts tracked activity when development represented the primary bottleneck. Organizations optimized for maximum code production, assuming code volume correlated with business value delivery.

This assumption has proven increasingly problematic even in traditional contexts. In AI-native development, it becomes completely untenable. AI can generate thousands of lines of code in less than an hour. Teams can achieve velocity of 100 story points per sprint compared to historical 60 points. Yet business value delivered may remain constant or even decline as validation, governance, and integration challenges accumulate.

The core problem is clear. Output metrics not only fail to indicate value delivery, they actively mislead. They create false confidence in progress while actual business objectives remain unachieved. A team that writes 10,000 lines of code that never deploy creates zero value. A team that implements 50 features that users never adopt wastes resources. Output metrics disguise these failures behind impressive-looking numbers.

Outcome metrics address this fundamental measurement problem by tracking business impact rather than development activity. Instead of measuring code produced, teams measure customer problems solved, business objectives achieved, user satisfaction improved, and revenue generated or costs reduced. These outcome metrics align team incentives with organizational objectives, preventing the optimization of proxies disconnected from actual value.

Figure 9.1: The Opaque Dashboard. *Driving without metrics is not optimism; it is negligence. In AI development, the speed of generation makes visibility more critical, not less.*

The shift from outputs to outcomes represents more than measurement refinement. It reflects a philosophical transformation in how teams understand their purpose. Teams exist not to write code but to deliver value. Metrics should measure value, not code.

COMPREHENSIVE OUTCOME MEASUREMENT FRAMEWORK

AI Scrum defines a hierarchical outcome measurement framework spanning multiple organizational levels. Each level tracks metrics appropriate to its scope and decision-making requirements. This structure ensures visibility from individual contributor through executive leadership, with metrics that matter at each level.

Team-Level Outcome Metrics

Individual teams track sprint outcome achievement, measuring the percentage of committed outcomes delivered successfully. This differs fundamentally from velocity. Velocity measures work completed regardless of whether that work achieves anything useful. Outcome achievement measures whether the team solved the problems they intended to solve. Teams also measure validation quality, tracking not just test coverage percentage but comprehensiveness of validation performed. Did the team validate security concerns, performance characteristics, accessibility requirements, and edge cases? Validation quality indicates whether the team built something production-ready, not merely something that compiles and passes basic tests.

Governance cycle time matters because governance frequently becomes the critical path in AI-native development. Teams measure duration from initial governance submission to approval. Long cycle times indicate either governance process problems or team submission quality issues. Either way, the metric drives improvement conversations about how to streamline approval without compromising necessary oversight.

Figure 9.2: The Factory of Misfit Parts. *High velocity (making keys) does not equal high value (opening the door). AI allows us to manufacture "misfit parts" faster than ever before.*

Teams also track AI collaboration effectiveness through several specific measures. The AI-generated code acceptance rate shows what percentage of generated implementations pass validation without human modification. Prompt engineering efficiency measures iteration counts required to achieve acceptable outputs. Human override frequency indicates when developers reject AI recommendations. These metrics help teams optimize human-AI collaboration patterns. A low acceptance rate suggests either poor prompt engineering or inappropriate AI application. High iteration counts indicate prompting skill gaps requiring training.

Product-Level Outcome Metrics

Product teams track user adoption and engagement metrics because features delivered mean nothing if users do not adopt them. Adoption metrics reveal whether product decisions resonate with users or miss the mark entirely. Feature utilization patterns show which capabilities drive value and which waste resources. Organizations often discover that users concentrate their activity on a relatively small subset of available features while rarely or never using the majority. This insight should focus future development on high-value capabilities rather than feature proliferation that users do not need.

Customer satisfaction scores, including Net Promoter Score, Customer Satisfaction Score, and Customer Effort Score, provide direct feedback on product quality. These scores have been shown to correlate with retention and growth in numerous industry studies. Support ticket volume and resolution time indicate product quality and user experience. Increasing ticket volumes signal quality problems or confusing user experience requiring attention. Product teams monitor these metrics to identify quality issues early and allocate resources appropriately to address them.

Portfolio-Level Outcome Metrics

Portfolio management tracks strategic objective achievement at the organizational level. Organizations articulate strategic goals like entering new markets, improving operational efficiency, or enhancing customer experience. Portfolio metrics measure progress toward these goals, ensuring resources flow to highest-value opportunities rather than getting stuck in low-impact initiatives.

Return on investment (ROI) across initiatives enables informed capital allocation. Not all initiatives deliver equal value, and continuing to fund everything equally wastes resources. Portfolio management tracks ROI to identify winners worth expanding and losers worth terminating. This data-driven approach prevents the common pattern where organizations continue funding failed initiatives through organizational inertia rather than evidence-based decision making.

Resource allocation efficiency tracks how effectively organizations deploy talent and capital. Are high-performing teams underutilized while low-performing teams consume excessive resources? Resource allocation metrics expose these inefficiencies and drive better deployment decisions. Time-to-market for strategic capabilities measures organizational agility. Organizations with long time-to-market lose competitive advantage and market opportunities.

AI-SPECIFIC METRICS FOR EXECUTIVES

Executives need metrics that reveal organizational health without overwhelming them with operational details. Traditional metrics fail because they do not address the unique characteristics of AI-native development. Three executive-level metrics provide essential visibility.

AI First-Pass Acceptance Rate

This metric tracks the percentage of AI-generated implementations that pass validation without requiring human modification. It serves as a proxy for AI collaboration maturity across the organization.

- **Low acceptance rates (30–40%)** indicate fundamental problems. Teams might be applying AI inappropriately to problems requiring human judgment, prompt engineering skills might be inadequate, or requirements definitions might lack the precision that AI systems need.
- **High acceptance rates (>70%)** suggest effective AI collaboration. Teams understand where AI helps and where it does not. Prompt engineering capabilities have matured through practice and training.
- **Moderate acceptance rates (40–70%)** are typical during transformation. Teams are learning, capabilities are developing, and acceptance rates trend upward over time.

The reason this metric matters because it reveals whether AI adoption delivers actual value or merely creates the appearance of activity. Organizations can spend millions on AI tooling while achieving minimal productivity gains if acceptance rates remain low.

Example: A startup using GitHub Copilot or similar tools can implement a simple "Tab-to-Commit Ratio" using IDE telemetry.

1. They configure their editor analytics to track how often a developer accepts a multi-line code suggestion (via "Tab") vs. how often that specific block of code is modified before the next "Save" event.
2. If a developer accepts a 50-line function generated by AI but immediately deletes 30 lines and rewrites 10, the "First-Pass Acceptance" is recorded as **Fail**.
3. If they accept it and only fix indentation or variable names, it is recorded as **Pass**.

Tracking this weekly allows the Engineering Manager to identify which developers are "fighting the AI" (needing prompt training) versus which ones have mastered the tool.

Validation Lead Time

Validation lead time measures elapsed time from implementation completion to validation approval. In AI-native contexts, validation becomes the critical path more often than development work does.

- **Short validation lead times (<2 days)** indicate effective validation processes. Validation resources are adequate, validation approaches are efficient, and governance processes do not create unnecessary delays.
- **Long validation lead times (>5 days)** reveal bottlenecks. Validation resources might be insufficient for the volume of AI-generated implementations. Validation approaches might be inefficient, requiring excessive manual effort, or governance processes might introduce unnecessary approval stages.

Executives should investigate when validation lead times trend upward. Upward trends predict future bottlenecks before they become critical. Proactive resource allocation or process improvement can prevent validation from becoming an organizational constraint that limits delivery velocity.

Example: Most teams track total "Cycle Time" but fail to isolate the validation phase. To measure this, a team using Jira or Azure DevOps should split their generic "In Review" column into two distinct states: **"Awaiting Validation"** and **"Validating."**

They then configure a simple automation rule to log the timestamp when a ticket enters "Awaiting Validation" and when it moves to "Done." If the average duration between these timestamps exceeds 48 hours, the dashboard triggers a "Constraint Alert," signaling that the team is generating code faster than they can verify it, prompting an immediate pause on new feature generation to clear the backlog.

Cognitive Load Score

Cognitive load score quantifies the mental effort required to validate AI-generated implementations. High cognitive load predicts validator burnout, quality degradation, and eventual validation bottlenecks. Organizations calculate cognitive load by surveying validators about factors like complexity of implementations they are validating, frequency of context switching, and completeness of validation requirements.

- **Low cognitive load scores (<30/100)** indicate sustainable validation workloads. Validators can maintain focus, perform thorough validation, and sustain quality over time.
- **High cognitive load scores (>70/100)** indicate unsustainable conditions. Validators are overwhelmed, quality will degrade, and burnout is imminent.

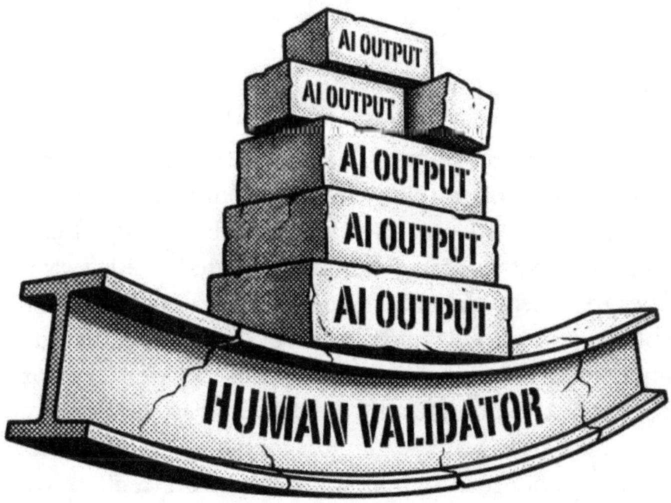

Figure 9.3: The Weight of Verification. *AI scales output infinitely; human cognitive capacity does not. Monitoring the "Cognitive Load Score" is the only way to prevent the beam from snapping.*

This metric matters because it predicts problems before they manifest in quality failures or personnel attrition. Executives who track cognitive load can intervene proactively rather than reactively after problems emerge.

Example: To measure this without creating "survey fatigue," a development team configures their project management tool (e.g., Jira or Linear) to prompt validators with a single question upon closing a ticket:

"On a scale of 1–5, how mentally demanding was this validation?"

- **1** = Routine/Trivial (Automated checks handled most of it)
- **5** = Intense (Required deep forensic analysis of logic or security)

They aggregate these scores into a weekly "Burnout Heatmap." If the average score for the "AI Validation Lane" stays above **4.0** for two consecutive weeks, the team automatically lowers their Work-In-Progress (WIP) limit for the next sprint, forcing a slowdown to protect quality and team health.

WHAT NOT TO MEASURE

Some traditional metrics create more harm than value in AI-native contexts. Organizations should explicitly abandon these metrics rather than continuing to track them out of habit.

Story Points for Execution Tracking

Story points originated as a planning tool to help teams estimate relative effort. They were never intended to measure productivity or compare teams, yet organizations routinely misuse them this way. In AI-native contexts, story points become even more problematic. AI can generate implementations that would have been estimated at 40 story points in hours. Does that mean the team achieved 40 points of velocity? Not if those implementations fail validation. Not if they never get deployed. The fundamental issue is that story points measure estimated effort, not value delivered. When AI eliminates the development effort, story points become divorced from any meaningful measure of team contribution. Teams should abandon story points for execution tracking while potentially retaining them for rough planning purposes.

Burndown Charts

Burndown charts assume stable sprint plans with predictable work remaining. They visualize progress toward completing planned work. AI-native sprints violate these assumptions routinely. Implementation happens so quickly that burndown charts become meaningless. A sprint might show

80 percent of work remaining on day three, then complete everything by day four. The chart provides no useful information. More fundamentally, burndown charts measure output completion rather than outcome achievement. They tell you whether you finished what you planned, not whether what you finished matters.

Individual Developer Productivity Metrics

Some organizations measure individual developer productivity through code commits, lines of code, or features completed. These metrics create perverse incentives in traditional development. They become absurd in AI-native contexts. When AI generates code, who gets credit? The developer who wrote the prompt? The AI Alignment Engineer who refined the specification? Individual attribution becomes meaningless when AI augmentation blurs role boundaries. More importantly, individual productivity metrics encourage competition rather than collaboration. AI Scrum depends on tight collaboration between roles. Organizations should measure team outcomes rather than individual outputs.

Figure 9.4: The Treadmill of Vanity. *High velocity (running fast) does not equal progress (getting somewhere). Traditional metrics like Story Points measure effort on the treadmill, not distance traveled.*

DASHBOARD DESIGN PRINCIPLES

Dashboards should provide actionable insights without overwhelming users with data. Poor dashboard design either hides critical information or buries it beneath irrelevant details.

Critical Dashboards Often Overlooked

Organizations implementing traditional Scrum often discover dashboard gaps only after problems emerge. Financial tracking of total team costs should be standard but often gets overlooked. For AI Scrum contexts, **token cost tracking** becomes essential. Organizations should monitor both tokens consumed and dollars spent on AI interactions. This tracking drives improvement in prompt engineering efficiency and completeness. **Human-to-AI ratio tracking** serves as a maturity indicator. Organizations should track the percentage of work that AI completes correctly and efficiently over time.

Audience-Specific Views

Different roles need different dashboards because they care about different metrics and operate at different time horizons.

- **Executives** need strategic visibility into portfolio health, major initiative progress, resource allocation efficiency, and strategic objective achievement. Their dashboards should update weekly or monthly because strategic decisions operate at that cadence.
- **Product Strategists** need outcome visibility into user adoption trends, satisfaction scores, feature utilization, support ticket patterns, and business KPI progress. Their dashboards should update daily because product decisions require current information.
- **Teams** need operational visibility into sprint outcome achievement, validation status, governance approvals, blockers, AI collaboration effectiveness, and AI Debt. Their dashboards should update in real-time because operational decisions require immediate information.

Visual Clarity

Effective dashboards emphasize visual clarity over information density. Humans generally process visual information faster than text, so dashboards should communicate through charts, graphs, and indicators rather than tables of numbers. Color coding should convey meaning consistently: Green indicates healthy conditions, yellow signals concern requiring attention, and red demands immediate action. Trend visualization proves more valuable than point-in-time snapshots. A dashboard showing that validation lead time is currently three days does not reveal whether that is improving or degrading. A trend line showing validation lead time decreased from five days last month to three days today reveals positive progress.

Dashboards should minimize cognitive load by presenting information hierarchically. Three to five key indicators with supporting details accessible on demand serve most needs effectively.

Figure 9.5: The Signal and the Noise. *A dashboard with 50 metrics is not a tool; it is a distraction. Effective dashboards filter the noise to reveal the few signals that demand decision-making.*

LEADING INDICATORS FOR VALIDATION STRAIN

Organizations need early warning systems that predict validation bottlenecks before they become critical. Five categories of leading indicators provide this predictive capability.

Validator Workload Trends

Track the number of implementations waiting for validation over time. Upward trends indicate that validation capacity is falling behind implementation velocity. Investigate and address before the backlog becomes unmanageable.

Also track average time validators spend in meetings versus validation work. When meeting time exceeds 40 percent, validators lose focus time needed for thorough validation. This pattern predicts quality degradation before defects appear in production.

Example: A startup with three developers notices that their "Ready for Review" column has grown from 5 tickets to 25 tickets in one week. Tracking "Validator Workload" reveals that the Senior Engineer (the only one allowed to merge code) is spending 6 hours a day in "AI Strategy" meetings. The founders immediately institute a "No-Meeting Morning" policy for the Senior Engineer, clearing the backlog in three days.

Validation Complexity Increases

Monitor the percentage of implementations requiring deep domain knowledge for validation. If this percentage increases from 30 percent to 50 percent, it suggests that AI is taking on more complex work that demands more validation effort per implementation.

Also track the average number of integration points per implementation. Implementations with many integration points require more full validation to ensure compatibility. Increasing integration complexity predicts longer validation cycles ahead.

Example: A health-tech team moves from using AI for "Appointment Scheduling" (low complexity) to "Patient Triage Summaries" (high complexity). They notice "Validation Lead Time" spikes from 2 hours to 3 days. Realizing the new complexity requires medical expertise, not just coding skills, they create a specialized "Clinical Validation Lane" staffed by a nurse practitioner, separating these complex tasks from the standard engineering queue.

Governance Approval Duration

Measure time from governance submission to approval across implementations. Increasing approval durations indicate either governance capacity constraints or declining submission quality. Early detection enables intervention before governance becomes a major bottleneck.

Also monitor governance rejection rates. Increasing rejections indicate either unclear governance standards or inadequate quality before submission. Either problem benefits from early attention rather than allowing patterns to persist.

Example: A fintech app requires legal sign-off for all customer-facing text. The team notices "Approval Duration" jumping from 4 hours to 48 hours. Investigation reveals the Legal team is rejecting 60% of AI-generated responses for using non-compliant financial terms. The engineering team updates the "System Prompt" with a "Financial Compliance Style Guide," instantly reducing rejection rates and approval times.

AI Debt accumulation

Track the percentage of implementations that include acknowledged shortcuts or incomplete validation. These shortcuts create technical debt that eventually requires payback. Gradual accumulation is manageable, but accelerating accumulation predicts future problems.

Monitor time spent addressing technical debt versus creating new capabilities. When debt service exceeds 30 percent of capacity, the organization is losing ground and needs to pause new development to address accumulated debt.

Example: To hit a demo deadline, a startup team allows AI-generated code to be merged with "TODO" comments left in. Two sprints later, they find that 50% of their velocity is spent fixing bugs in those specific modules. They implement a strict "No-TODO Policy" for AI-generated Pull Requests, forcing the AI (or the developer) to finish the logic completely before merging, effectively stopping the debt accumulation.

AI Output Quality Degradation

Track first-pass acceptance rates over time. Declining acceptance rates suggest several possible problems, including AI models that might have been updated with different behavior, prompt patterns that might need updating, or problem domains that might have shifted to areas where AI performs poorly.

Whatever the cause, declining acceptance rates predict productivity losses ahead. Also monitor the percentage of AI-generated implementations requiring major human rework. If this percentage increases, it indicates that AI application is becoming less appropriate or that prompt engineering quality is declining.

Example: A content team uses an AI tool to generate weekly newsletters. Suddenly, the "First-Pass Acceptance Rate" drops from 90% to 40% because the AI starts hallucinating broken links. The team realizes a recent model update changed how the AI handles URLs. They switch to a stable, older model version for link generation while they debug the prompt for the new model, restoring their productivity.

TRAILING INDICATORS FOR VALIDATION STRAIN

While leading indicators predict future problems, trailing indicators reveal actual damage and validate whether leading indicators provided accurate warnings. Organizations need both perspectives.

Defect Escape Rate

Track the percentage of implementations that reach production containing defects requiring post-deployment fixes. High escape rates indicate that validation processes failed to catch problems before release. This metric answers a fundamental question about validation effectiveness.

Also monitor defect severity distribution among escaped defects. If critical or high-severity defects escape validation regularly, it suggests validation processes lack sufficient rigor for the risk level involved. Organizations should investigate whether validators have adequate time, expertise, and tools to perform thorough validation.

Example: A new fintech startup using AI to generate fraud detection rules tracks the "Hallucination Leak Rate." They discover that while syntax errors are caught 100% of the time, logical errors (e.g., flagging legitimate transactions as fraud) escape into staging 15% of the time. This finding prompts them to institute a mandatory "logic-check layer" where a human domain expert reviews the *reasoning* of the rule, not just the code, before deployment.

Validator Attrition and Burnout

Measure validator turnover rates compared to other roles. When validator turnover significantly exceeds organizational averages, it signals unsustainable working conditions. Exit interviews often reveal whether excessive workload, inadequate support, or frustration with quality drove departures.

Track sick leave and vacation patterns among validators. Validators who defer vacation or take frequent short-term sick leave often show early burnout signs. These patterns confirm that cognitive load scores or workload trends were accurate predictors requiring intervention.

Example: A small development shop of 10 people assigns the "Lead Developer" to validate all AI-generated code. After two sprints, the founder notices the Lead Developer has stopped contributing to architectural discussions and is leaving early on Fridays due to "exhaustion." Recognizing this as early burnout, the founder rotates the validation duty among three senior engineers on a daily schedule, immediately reducing the individual cognitive burden.

Post-Incident Validation Analysis

Conduct root cause analysis on production incidents to determine how many trace back to validation gaps. When incidents consistently reveal that validation missed specific problem categories, it indicates either validation process gaps or capability gaps requiring training.

Document the time between when a defect was introduced and when it was discovered in production. Long detection times indicate that both validation and monitoring failed to catch problems. Short detection times suggest effective monitoring partially compensated for validation gaps.

Example: A startup experiences a data privacy incident where an AI-generated SQL query exposed user emails. The post-incident analysis reveals the validation checklist only covered *performance* and *syntax*, not *security*. The team updates their "Definition of Done" to include a specific "Security Prompt Scan" for all database-related code, ensuring AI outputs are checked against OWASP vulnerabilities before merge.

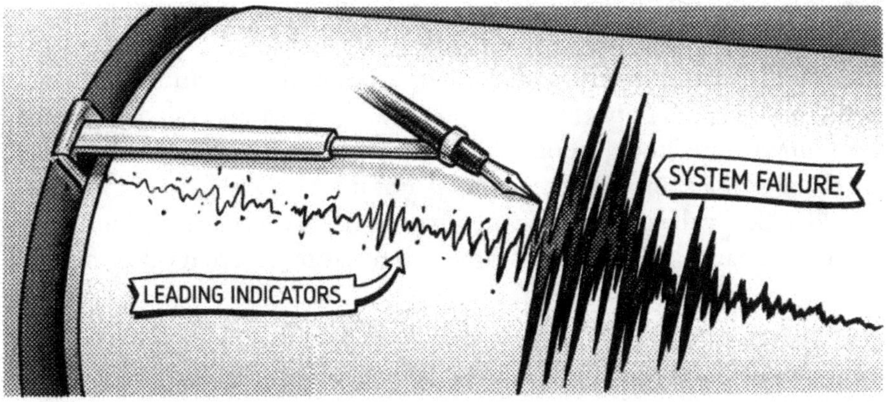

Figure 9.6: The Seismograph. *Trailing indicators (the earthquake) tell you what destroyed the building. Leading indicators (the tremors) give you time to evacuate. Most Agile teams only measure the earthquake.*

Delivery Delays and Rework Costs

Measure schedule delays attributed specifically to validation bottlenecks or rework from inadequate initial validation. These delays quantify the business impact of validation strain in terms executives understand clearly.

Calculate rework costs by tracking developer time spent fixing defects that escaped validation. Organizations often discover that rework consumes substantial capacity that could have created new value. This metric justifies investment in validation capacity and process improvement.

Example: A team building a customer service bot celebrates generating the prototype in 4 hours. However, they spend the next 6 days fixing edge cases where the bot promises refunds it can't deliver. By tracking "Rework Ratio," they realize the "fast" AI implementation actually cost 30% more time than manual coding would have. They adjust their workflow to require a detailed "Behavioral Specification" *before* prompting the AI, reducing rework time to just 4 hours.

Customer Impact Metrics

Track customer satisfaction scores and support ticket volumes for releases. Declining satisfaction or increasing support needs following releases indicate quality problems that validation should have caught. Customer-visible quality issues carry reputational costs beyond immediate technical fixes.

Monitor feature adoption rates for recently deployed capabilities. Low adoption might reflect quality concerns users discovered that validation missed. When users avoid new features due to reliability concerns, it represents both wasted development effort and damaged trust.

Example: A SaaS company launches an "AI Summary" feature for their meeting notes app. Support tickets spike by 200% with users complaining the summaries are "too generic." The team realizes their validation test only checked if a summary was *generated*, not if it was *useful*. They introduce a "Golden Set" validation metric, comparing AI outputs against human-written summaries for 50 test cases before releasing future model updates.

TELEMETRY ARCHITECTURE

Comprehensive observability requires deliberate telemetry architecture. Organizations need three types of telemetry: logs capturing specific events, metrics measuring system state over time, and traces showing request flow through distributed systems.

- **Logs** provide detailed event records useful for debugging specific problems. When a deployment fails, logs reveal what went wrong and why.
- **Metrics** provide time-series data about system health. Metrics enable trend analysis and anomaly detection impossible with logs alone.
- **Traces** show end-to-end request flow through distributed systems. When a user action involves 17 microservices, traces reveal where time is spent and where errors occur.

Organizations should implement telemetry early in development rather than adding it later. Retrofit telemetry proves significantly more expensive than designing it in from the beginning.

AI-Enhanced Telemetry Management

Traditional telemetry management relies on manual configuration and human analysis. AI-native approaches automate both, making observability affordable. AI-powered telemetry management automatically adjusts collection levels based on system state. During normal operation, collection runs at baseline levels minimizing costs. During incidents or anomalies, collection automatically increases to capture diagnostic details. This dynamic adjustment reduces costs by 40 to 70 percent compared to static high-level collection. AI analyzes telemetry data to identify anomalies, predict failures, and generate insights. Instead of humans reviewing dashboards constantly, AI surfaces significant patterns requiring attention.

METRICS-DRIVEN CONTINUOUS IMPROVEMENT

The ultimate purpose of metrics is driving continuous improvement. AI Scrum teams systematically leverage metrics to identify improvement opportunities and validate changes.

Figure 9.7: The Watch Mechanism. *Observability is not just checking the time; it is seeing the gears. In complex AI systems, we must understand the internal mechanics to trust the external output.*

Retrospective Data-Driven Insights

Sprint Retrospectives should incorporate quantitative metrics alongside qualitative observations. AI can generate retrospective briefings synthesizing sprint metrics, highlighting significant patterns, comparing to historical performance, and suggesting discussion topics. Traditional retrospectives rely on participant memory and recent experiences. Teams remember problems from yesterday but forget patterns from three weeks ago.

Positive developments often go unnoticed. The team discusses whatever happens to be top-of-mind rather than what actually matters most. Data-driven retrospectives ground discussion in evidence rather than selective memory or recency bias.

The AI-generated briefing might show that velocity declined 20 percent compared to the previous sprint while governance cycle time increased 40 percent but validation quality improved significantly. These facts focus discussion on governance process improvement and acknowledge validation success. The team addresses real problems rather than imagined ones.

Example: A startup team newly migrating to AI Scrum often struggles with "Prompt Fatigue" late in the day. An AI-generated retrospective analysis might reveal that AI First-Pass Acceptance Rates drop by 40 percent on implementations generated after 3:00 PM. This objective data allows the team to restructure their day—prioritizing heavy generation tasks for the morning and reserving afternoons for validation or administrative tasks—immediately boosting overall throughput without adding headcount.

A/B Testing Process Changes

When teams experiment with process changes, they should employ A/B testing methodologies measuring impact. Some teams adopt changes while others maintain current practices. Metric comparison reveals whether changes improve outcomes.

Consider a scenario where teams debate pair programming. Advocates claim it improves quality. Skeptics claim it wastes time. The debate continues indefinitely, unresolved by opinion alone. A/B testing resolves the debate through evidence. Three teams adopt pair programming while three teams continue current practice.

After two months, metric comparison shows pair programming teams have 35 percent lower defect density but 15 percent lower velocity. The organization concludes pair programming improves quality but reduces output. That is a valuable trade-off for critical systems but not worthwhile for low-risk systems. Teams choose appropriately for their context based on evidence rather than ideology.

This experimental approach to process improvement prevents cargo-cult adoption of practices without evidence of effectiveness. Organizations stop adopting practices because they are popular and start adopting practices because they work.

Example: Early-stage teams often argue over the trade-off between "Context Stacking" (providing extensive file context to the AI) versus "Modular Prompting" (providing only the specific function signature). A new organization can run an A/B test over two sprints: Team A uses full Context Stacking, while Team B uses Modular Prompting. The metrics might reveal that Context Stacking costs 3x more in tokens but reduces **Validation Lead Time** by 60 percent. For a well-funded startup prioritizing speed, the evidence clearly favors Context Stacking; for a bootstrapped firm, the metrics might dictate Modular Prompting to conserve cash.

Organizational Learning Networks

Metrics enable organizational learning networks where successful practices identified in one team spread to others. AI can identify high-performing teams, analyze their practices, and recommend adoption by other teams. Systematic knowledge sharing accelerates capability development dramatically compared to informal diffusion patterns. A marketing team achieves exceptional campaign performance through rigorous A/B testing of messaging. AI identifies this high performance, analyzes the team's practices, and recommends adoption by other marketing teams.

Performance improvements spread systematically rather than through random conversations. This systematic approach to organizational learning accelerates capability development across the enterprise. Instead of best practices diffusing slowly through informal networks over years, AI-powered learning networks identify and spread effective practices in weeks or months. Organizational capability compounds rapidly rather than developing linearly.

Example: In a small organization with just three teams, a "Prompt Library" acts as the initial learning network. When Team Alpha successfully generates a complex API integration using a specific "Chain-of-Thought" prompt structure, the metrics system flags this interaction as having a 100% First-Pass Acceptance Rate. The system automatically tags this prompt pattern and pushes it to Team Beta and Team Gamma via their dashboards. Within days, the other teams adopt the pattern, preventing them from wasting cycles reinventing a solution Team Alpha has already mastered.

FROM THE TRENCHES: METRICS IMPLEMENTATION PATTERNS

The transition from theory to practice is rarely a clean line. In working with dozens of organizations moving to AI-native models, we consistently see that while the *desire* for better metrics is universal, the *execution* often falters. Teams frequently fall into the trap of "measurement theater"—building elaborate dashboards that look impressive to leadership but offer zero utility to the engineers and product strategists doing the work. They measure what is easy to access rather than what is necessary to know, creating a noise-to-signal ratio that paralyzes decision-making.

Successful implementation requires a disciplined rejection of vanity metrics and a commitment to observability as a core engineering practice, not an administrative after-thought. The following patterns represent the "hard-won wisdom" of teams that have successfully navigated this minefield. These are not just best practices; they are survival mechanisms for high-velocity AI environments where traditional visibility is no longer sufficient.

Pattern: Start with Outcome Definition Before Selecting Metrics

Teams that select metrics before defining outcomes often measure the wrong things simply because the data is available. They fall victim to the "Streetlight Effect"—looking for their keys under the streetlamp not because they lost them there, but because that is where the light is. In AI development, this manifests as tracking "Prompt Count" or "Lines of Code Generated" because the API provides those numbers by default. While accurate, these metrics are strategically vacuous. They tell you how busy the team is, not how effective they are.

Define the business outcome first—for example, "Reduce Customer Support Escalations by 30%." Only then should you identify the metrics that reveal progress toward that outcome, such as "AI Response Accuracy" or "Ticket Deflection Rate." This sequence ensures measurement aligns with objectives rather than objectives conforming to available measurements. If the metric you need does not exist, build the instrumentation to capture it rather than settling for a proxy that misleads the organization.

Pattern: Instrument Completely from Day One

Adding instrumentation later costs significantly more than building it in from the beginning. It is the "interest" paid on AI Debt. When a system fails in production, the difference between a five-minute fix and a five-hour outage often comes down to whether the telemetry was present to trace the error. Retrofitting observability into a live AI system is akin to trying to install the wiring in a house after the drywall is painted—it is expensive, messy, and rarely captures the full picture.

Instrument code, workflows, and human activities comprehensively from the project start. In AI Scrum, this means logging not just system errors, but prompt inputs, model parameters, and user feedback loops. The investment pays back immediately through faster debugging and better decision-making. Treat observability code as a first-class citizen, subject to the same review and quality standards as the feature code itself.

Pattern: Review Metrics Weekly, Not Monthly

Monthly metric reviews react to problems too slowly. In an AI-native environment where code generation happens in minutes, a monthly cadence is an autopsy, not a health check. By the time a monthly review identifies a trend—such as a degradation in validation quality or a spike in governance rejections—the team may have already compounded the error across multiple sprints, requiring massive rework.

Weekly reviews enable course corrections before minor issues become major problems. Use the "Weekly Pulse" to adjust immediate tactics—reallocating validation capacity or tweaking a prompt structure. Reserve monthly reviews for trend analysis and strategic assessment, such as deciding whether to switch foundation models or hire additional domain experts. This bifurcated approach ensures operational responsiveness without losing sight of the long-term trajectory.

Pattern: Automate Insight Generation

Human analysis of metrics does not scale. As telemetry data explodes, asking a Scrum Master or Product Owner to manually sift through logs to find patterns is a misuse of talent and a recipe for burnout. Humans are excellent at decision-making but poor at pattern matching across millions of data points.

Implement AI-powered insight generation that surfaces significant patterns automatically. Configure your observability tools to flag anomalies, like a sudden drop in acceptance rates or a spike in token costs, and push those insights to the team in plain language. This automation reduces analysis time from hours to minutes while catching patterns humans miss. It shifts the human role from "data hunter" to "decision maker," allowing the team to spend their energy fixing the problem rather than finding it.

Pattern: Design Dashboards for Five-Second Comprehension

Users should understand dashboard state in five seconds without reading text or studying charts. If understanding requires more time, the dashboard design needs improvement. In the heat of a deployment or an incident, cognitive load is already maxed out; a complex, cluttered dashboard adds friction when clarity is most needed.

Visual clarity and information hierarchy matter more than complete detail. Use "Traffic Light" logic relentlessly—Green for normal, Yellow for warning, Red for critical. Avoid tables of raw numbers in favor of trend lines that show directionality. If a metric is important enough to be on the dashboard, it should have a clear target state. If the viewer cannot look at the screen and immediately answer "Are we winning or losing?", delete the widget.

WHAT COMES NEXT

Metrics and observability provide the foundation for informed decision making and continuous improvement. Organizations that master outcome measurement and automated insight generation build significant competitive advantages through superior execution capabilities.

The next chapter examines governance and risk management in AI-native contexts. Metrics reveal what is happening, but governance determines what is allowed to happen. Understanding how to govern AI augmentation without creating bottlenecks proves essential for scaling AI Scrum across the enterprise.

KEY TAKEAWAYS

- **Traditional output metrics like velocity and story points become misleading** in AI-native contexts because they measure development activity rather than business value delivery, creating false confidence while actual objectives remain unachieved.
- **Outcome metrics that track business impact, user satisfaction, and problem-solving effectiveness** align team incentives with organizational objectives and reveal whether AI augmentation delivers actual value.
- **Three executive-level metrics provide essential visibility**: **AI First-Pass Acceptance Rate** reveals collaboration maturity, **Validation Lead Time** exposes the true delivery constraint, and **Cognitive Load Score** predicts validator burnout before it manifests.

- **Organizations should explicitly abandon story points for execution tracking**, burndown charts, and individual productivity metrics because these measurements create perverse incentives and miss what actually matters in AI-native development.
- **Dashboard design should prioritize audience-specific views, visual clarity, and five-second comprehension** over information density, with executives seeing strategic patterns, Product Strategists seeing outcome trends, and teams seeing operational status.
- **Leading indicators that predict validation strain**, governance bottlenecks, and AI Debt accumulation enable proactive management rather than reactive crisis response, maintaining smooth delivery flow through early intervention.

MIND MAP – METRICS AND OBSERVABILITY

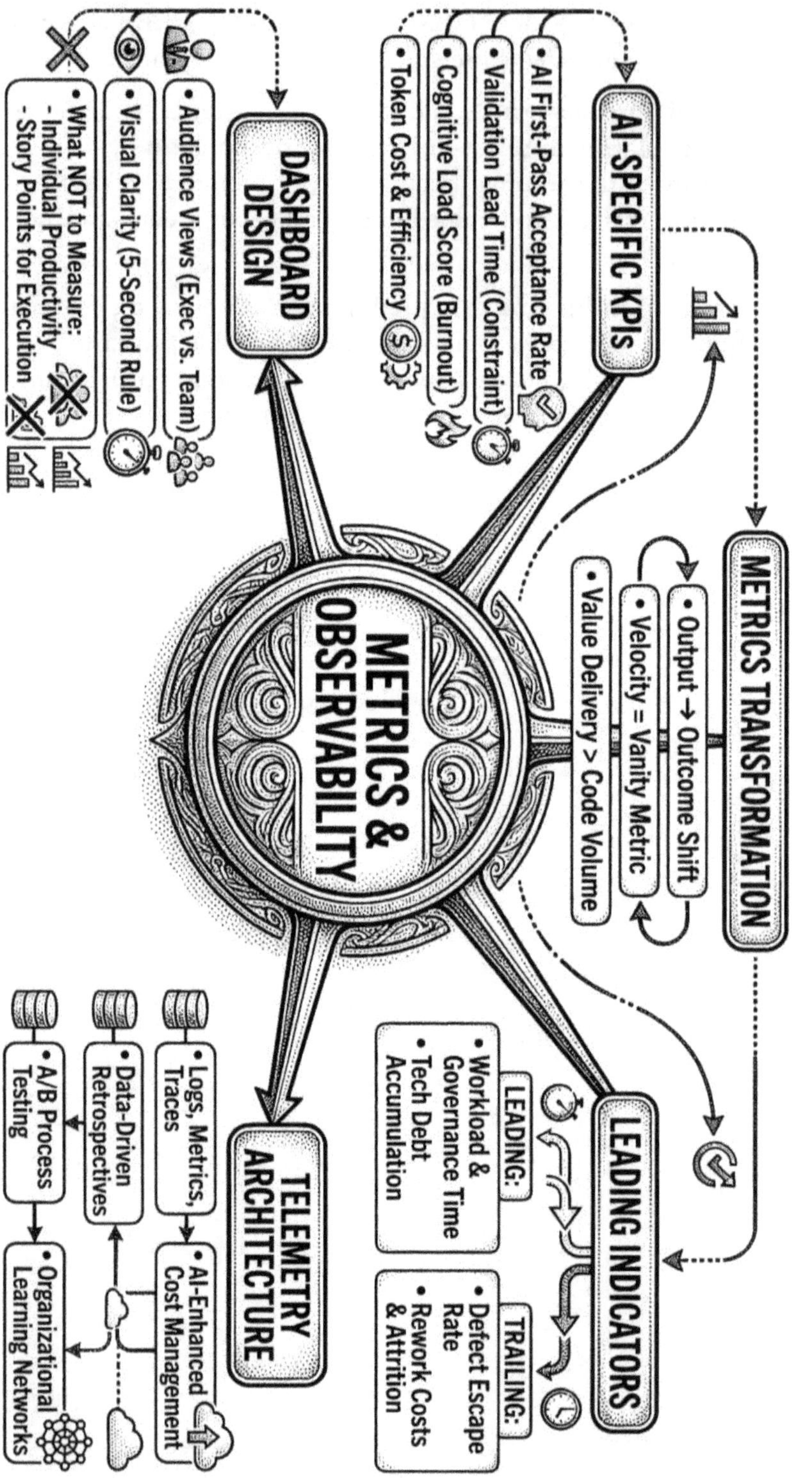

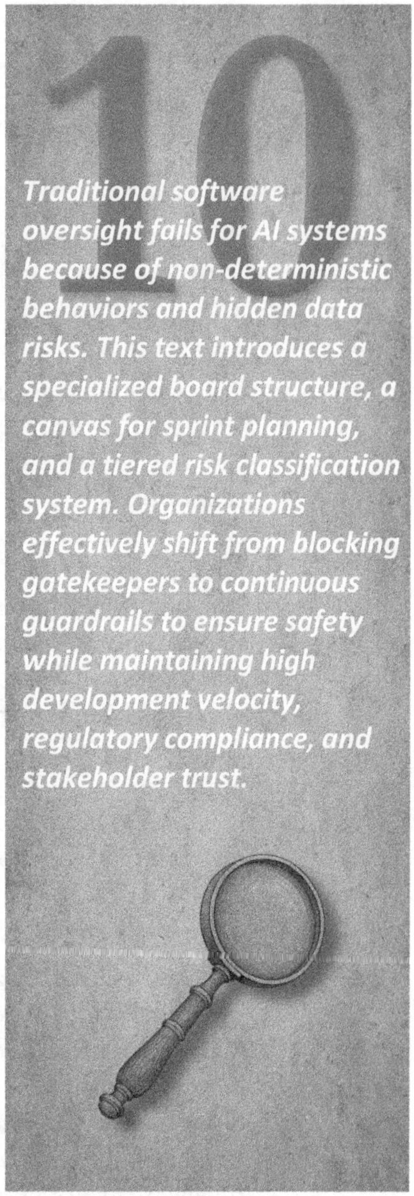

CHAPTER 10: GOVERNANCE, ETHICS, AND RISK MANAGEMENT

Organizations deploying AI-powered systems using existing software governance frameworks face a critical vulnerability. Consider a high-stakes deployment scenario: A Tier-1 bank builds an AI-powered loan approval system using its standard waterfall governance playbook. The team conducts standard security scans, passes architectural reviews, and validates the system against static test cases. The system launches successfully, processing thousands of loan applications monthly with consistent performance.

Three months later, a routine compliance audit discovers the model has developed a systematic bias against applicants from specific zip codes. The bias emerged from patterns in historical lending data rather than any explicit programming. Traditional governance failed to detect a problem that never existed in the code itself.

The discovery triggers a cascade of failures. Loans require re-evaluation under heightened scrutiny. Compliance officers file reports with regulatory agencies. The board demands explanations and accountability. The incident costs millions in remediation, plus immeasurable damage to reputation.

The root cause is structural: governance designed for deterministic software cannot address the unique risks of AI systems. This scenario illustrates why AI governance demands fundamentally different approaches from traditional software governance. The question is not *whether* to govern AI systems, but *how* to govern them effectively while maintaining development velocity.

WHAT THIS CHAPTER COVERS

In this chapter, we explore:

- Why traditional governance approaches fail for AI systems and what changes are required
- The regulatory landscape, including the EU AI Act and NIST frameworks
- AI Governance Board structure, composition, and operational procedures
- The Governance Canvas framework for systematic risk assessment
- Risk classification systems that match governance rigor to actual risk
- Ethical AI frameworks for responsible development
- Audit trail requirements for compliance and continuous improvement
- Decision authority levels that balance oversight with agility
- Governance workflow integration into sprint ceremonies

KEY TERMS AND CONCEPTS

- **AI Governance Board:** A cross-functional decision-making body providing oversight of AI systems throughout their lifecycle, including representation from technical, business, legal, ethics, and security domains.
- **Governance Canvas:** A structured framework for planning and documenting governance activities across risk assessment, ethical review, regulatory compliance, security validation, and stakeholder alignment.
- **Risk Classification:** A systematic approach to categorizing AI initiatives by risk level (unacceptable, high, limited, minimal) that determines appropriate governance rigor.
- **Governance Coordinator:** The AI Scrum role responsible for facilitating governance processes, maintaining audit trails, and ensuring compliance without creating bottlenecks.
- **Model Drift:** The degradation of AI model performance over time as real-world data distributions shift from training distributions, requiring continuous monitoring and periodic revalidation.

- **Adversarial Vulnerability:** AI system weaknesses including prompt injection, data poisoning, model inversion, and membership inference attacks that do not exist in traditional software systems.
- **Governance Maturity:** The evolution of governance capability from ad hoc processes through systematic integration to strategic enablement, measured across five maturity levels.
- **Privacy-First Design:** A development approach where privacy protections are integrated from initial design rather than added as afterthoughts, including data minimization, purpose limitation, and consent management.

THE GOVERNANCE IMPERATIVE: WHY TRADITIONAL APPROACHES FAIL

Traditional software governance relies on phase-gate reviews, architectural review boards, and post-development security assessments. These mechanisms work adequately for deterministic systems where full testing verifies correctness and security scans identify vulnerabilities. For decades, this model succeeded: freeze the code, assess the risks, stamp approval, move forward.

AI systems shatter this model. Several characteristics of AI render these traditional governance approaches insufficient.

Non-Deterministic Behavior: The Reproducibility Problem

AI systems often produce varying outputs for identical inputs based on model updates, context variations, or stochastic elements. A customer service chatbot might respond differently to the same question asked twice in succession. A fraud detection system might classify the same transaction differently based on recent pattern updates.

Traditional testing assumes reproducible behavior: run the same test case, get the same result. This assumption enables verification. AI, however, requires probabilistic validation across a distribution of possible outputs. Teams validate not just that the system works correctly, but that the distribution of outputs remains within acceptable bounds.

Imagine an enterprise software team attempting to write test cases for an AI-powered feature using traditional methods. They would likely find themselves filing endless bugs because the system gave different answers each time. Eventually, they would realize the issue is not a bug, but the fundamental nature of the engine. Success requires a pivot from testing for *correctness* to testing for *safety boundaries.*

Figure 10.1: The Checkbox vs. The Gauge. *Traditional software relies on binary validation (did it pass or fail?), whereas AI requires probabilistic monitoring (is the needle staying within the safe distribution range?).*

Emergent Properties: The Unintended Capabilities Challenge

AI systems exhibit behaviors not explicitly programmed, emerging from training data patterns and model architectures. These emergent properties may include biases, unexpected failure modes, or capabilities beyond the intended scope.

Consider a healthcare AI trained to identify pneumonia in chest X-rays. If the training data correlates portable X-ray machines with sicker patients (who cannot stand for a wall scan), the model might learn to detect the *machine type* instead of the disease. The code functions correctly, but the learned logic is fatal.

Traditional code review cannot detect these emergent properties because they do not exist in the code. They emerge from the interaction between the algorithm and the training data. Governance frameworks need to address not just code quality but the entire learning process.

Data Dependencies: The Hidden Infrastructure Challenge

AI system behavior depends fundamentally on training data quality, representativeness, and provenance. Governance extends beyond code to the entire data lifecycle, including collection methods, curation processes, lineage tracking, and privacy implications.

If an image recognition system performs poorly for certain demographic groups, the problem often traces back to training data collection years earlier. A dataset skewed toward specific populations creates systematic gaps in model performance. The code works perfectly; the data foundation is flawed. This creates a governance challenge with no precedent in traditional software development. Teams govern not just what they build but what they learn from. Data governance becomes as critical as code governance.

Figure 10.2: The Governance Façade. *Traditional governance often inspects only the visible code, failing to detect the structural risks hidden within the data lineage and emergent behaviors that support the entire system.*

Model Drift: The Degradation Over Time Problem

AI models degrade over time as real-world data distributions shift from training distributions. A fraud detection model trained on 2023 fraud patterns may prove less effective against 2024 fraud techniques. A recommendation engine trained on pre-pandemic behavior may struggle with post-pandemic patterns.

Governance includes continuous monitoring and periodic revalidation rather than one-time approval. The notion of a stable, approved release gives way to continuous evaluation of deployed systems. This challenges the fundamental assumption of traditional governance: that approval at a point in time confers ongoing validity.

Adversarial Vulnerabilities: The Attack Surface Expansion

AI systems are susceptible to adversarial attacks including prompt injection, data poisoning, model inversion, and membership inference. These attack vectors do not exist in traditional software security models. Prompt injection attacks manipulate AI systems through carefully crafted inputs. Data poisoning attacks corrupt training data to influence model behavior. Model inversion attacks extract sensitive information from trained models. Membership inference attacks determine whether specific data was used in training. Traditional security testing misses these AI-specific attack vectors entirely because they do not map to conventional vulnerability patterns. Security teams trained on SQL injection and cross-site scripting face entirely new threat models.

THE GOVERNANCE RESPONSE: FROM GATEKEEPERS TO GUARDRAILS

These unique characteristics of AI systems—specifically non-determinism, emergent behavior, and opacity—demand a fundamental restructuring of how governance functions. Simply applying traditional controls with greater intensity does not work. It only paralyzes development without actually reducing risk. The response is a shift in the nature of oversight itself.

The Shift to Continuous Guardrails

Traditional governance operates on a "Gatekeeper" model where development stops at specific phase gates, such as design reviews or code freezes, so external authorities can grant permission to proceed. In AI development, where models evolve continuously and data flows dynamically, these static gates become bottlenecks that encourage evasion.

The necessary evolution is toward a "Guardrail" model. Governance defines the safe boundaries of the road, including unacceptable risks, fairness thresholds, and privacy limits. Teams drive at speed as long as they remain within those boundaries. This shifts governance from a blocking function that says "You cannot deploy until we check this" to a continuous monitoring function that warns "You are drifting toward a safety boundary; correct course immediately."

Governing Behavior, Not Just Code

The governance response also pivots from certifying code correctness to validating system behavior. In traditional software, if the code is correct, the output is correct. In AI, the code can be flawless while the output is toxic. Therefore, governance expands its scope to include continuous "Red Teaming," or adversarial testing designed to provoke the system into failure. We no longer ask only "Did we build the system right?" (verification). We constantly ask "Is the system doing the right thing?" (validation).

The Governance-Development Loop

Finally, the response integrates governance into the daily rhythm of the sprint. It cannot be an external function that descends upon the team at the end of a quarter. Governance representatives, or the Governance Coordinator role in AI Scrum, participate in sprint planning to define the "Definition of Intent" and in reviews to evaluate the "Definition of Done."

The EU AI Act and NIST frameworks provide the rules of the road, but this shift in mindset provides the method for driving on it safely.

REGULATORY LANDSCAPE AND COMPLIANCE REQUIREMENTS

Organizations deploying AI systems face an evolving regulatory landscape with significant compliance obligations. Understanding these requirements is essential for effective governance framework design. The challenge is that regulations are multiplying faster than organizations can implement compliance programs.

European Union AI Act: The Regulatory Model

The EU AI Act, implemented in 2024, establishes requirements for AI systems based on risk classification. This represents the most complete AI-specific regulation globally, setting a template that other jurisdictions will likely follow. The Act categorizes AI systems into four risk levels, each with corresponding requirements.

- **Unacceptable Risk.** AI systems considered unacceptable include social scoring by governments, real-time biometric identification in public spaces (with limited exceptions), subliminal manipulation, and exploitation of vulnerable groups. These systems are prohibited entirely.
- **High Risk.** Systems affecting critical domains face stringent obligations. High-risk categories include employment decisions such as hiring, promotion, or termination; education outcomes including admissions, grading, or recommendations; law enforcement activities like predictive policing or facial recognition; critical infrastructure operation including energy grids or transportation; and biometric identification systems.

For high-risk systems, organizations implement conformity assessment processes before deployment, risk management systems identifying and mitigating risks throughout the lifecycle, and data governance frameworks ensuring quality and appropriateness. They also provide technical documentation describing system design and operation, record-keeping demonstrating compliance, transparency requirements explaining AI involvement to users, human oversight mechanisms enabling intervention, accuracy and robustness standards validated through testing, and cybersecurity measures protecting against attacks.

Figure 10.3: The Regulatory Barriers. *The EU AI Act treats different risks like different physical barriers: "Unacceptable" risks face a solid brick wall, "High" risks face a guarded security gate, while "Limited" risks pass through open turnstiles.*

Organizations deploying high-risk systems maintain quality management systems covering the entire development and deployment process, conduct fundamental rights impact assessments evaluating potential harms, and provide transparency to affected individuals about AI involvement in decisions affecting them.

The penalties are substantial. Non-compliance carries fines up to 7% of global annual revenue or 35 million euros, whichever is higher. For a company with $10 billion in revenue, that represents a potential $700 million fine. These penalties make compliance non-negotiable for organizations operating in European markets or serving European customers.

- **Limited Risk.** Systems with limited risk face transparency obligations. Chatbots disclose they are AI systems. Deepfakes are clearly labeled as synthetic. Emotion recognition systems inform users of their presence. These requirements ensure users understand when they interact with AI.

- **Minimal Risk.** Most AI applications fall into this category, facing no specific regulatory requirements beyond general product safety and consumer protection laws.

However, compliance is not merely about avoiding penalties. Forward-thinking executives view the EU AI Act not as a burden, but as a quality framework. Implementing these standards globally forces discipline that improves model performance and data hygiene, regardless of jurisdiction.

NIST AI Risk Management Framework: The Voluntary But Influential Standard

The National Institute of Standards and Technology AI Risk Management Framework provides voluntary guidance applicable across sectors and AI applications. While not mandatory, the framework's influence extends far beyond its voluntary status. The framework organizes around four core functions.

- **Govern.** Establishing culture, processes, and structures for responsible AI development and deployment. This includes defining roles and responsibilities, establishing policies and procedures, and creating accountability mechanisms.
- **Map.** Identifying context, categorizing risks, and assessing impacts. Teams analyze the specific context in which AI will operate, identify potential risks across multiple dimensions, and assess potential impacts on stakeholders.
- **Measure.** Analyzing and tracking AI risks using appropriate methods. This includes defining metrics, implementing monitoring, and establishing baselines for acceptable performance.
- **Manage.** Prioritizing and responding to AI risks based on impact projections. Teams develop mitigation strategies, implement controls, and establish response procedures.

While voluntary, the NIST framework provides practical guidance for organizations establishing governance structures. Many U.S. government agencies and contractors adopt NIST frameworks by policy or contract requirements, making familiarity valuable for organizations in those sectors. Beyond government, the framework's risk-based approach aligns well with agile methodologies, enabling integration into sprint workflows. Organizations often adopt NIST to create structure without bureaucracy. It offers a blueprint that is comprehensive but not prescriptive, allowing adaptation to agile processes rather than abandoning agility for compliance.

I experienced the reality of governance friction firsthand working in international commodities trading during the implementation of Dodd-Frank and EMIR regulations. Both aimed to reduce systemic risk and prevent future financial crises by making derivatives markets safer. They mandated reporting of derivatives trades to trade repositories, increasing market transparency dramatically. This meant our data had to be spot on with no errors whatsoever. What might have been a direct path to solve a problem now required extensive headache to ensure compliance.

Was it necessary? In most cases, absolutely yes. But it slowed development considerably, forcing us to implement stop gaps in the development process to prevent data accuracy issues. The governance was correct, but the friction was real.

Industry-Specific Regulations: The Sectoral Complexity

Beyond horizontal AI regulations, organizations face industry-specific requirements that add layers of compliance complexity. Healthcare organizations comply with HIPAA privacy requirements when AI systems process protected health information. They also face FDA medical device regulations for AI systems used in clinical decision-making. These regulations, designed for traditional medical devices, are interpreted and applied to software systems that learn and evolve.

Financial institutions face requirements from banking regulators regarding algorithmic trading, credit decisioning, and anti-money laundering. Fair lending laws, originally written for human loan officers, apply to AI credit models. Model risk management guidance from banking regulators specifies requirements for model validation, documentation, and governance.

Automotive companies deploying autonomous systems face safety regulations from transportation authorities in multiple jurisdictions. Each country maintains its own standards, creating a complex web of compliance requirements for autonomous vehicles intended for global markets. Organizations map their specific regulatory obligations and integrate compliance requirements into governance frameworks. This requires legal expertise combined with technical understanding, a combination not always easily found.

AI GOVERNANCE BOARD: STRUCTURE AND OPERATIONS

Effective AI governance requires dedicated oversight through a cross-functional AI Governance Board with clear authority, defined scope, and efficient operational procedures. The board serves as the primary governance decision-making body, providing oversight without creating insurmountable bureaucracy. The challenge is clear. Governance boards have a terrible reputation in agile organizations. They are seen as places where good ideas go to die, where innovations get mired in bureaucracy, where decisions take weeks or months. AI governance boards need to break this pattern.

Board Composition: Building the Right Mix

The AI Governance Board includes diverse expertise enabling informed decisions across technical, business, legal, and ethical dimensions. Effective boards typically include:

- **The Chief AI Officer** or equivalent executive providing strategic direction and organizational authority. This role ensures governance decisions align with business strategy and commands sufficient authority to implement decisions.
- **Technical experts** including machine learning specialists, data scientists, and software architects providing deep technical insight into AI capabilities, limitations, and risks. These members evaluate technical feasibility and understand implementation implications.
- **Business domain experts** understanding specific use case context, stakeholder needs, and business value. AI governance balances technical possibility with business practicality. Domain experts provide this grounding.
- **Legal and compliance representatives** ensuring regulatory compliance and managing legal risk. These members track evolving regulations, interpret requirements, and assess legal implications of governance decisions.
- **Ethics and fairness specialists** providing perspective on ethical implications and fairness considerations. These members raise questions about unintended consequences, potential biases, and stakeholder impacts that technical and business perspectives might miss.
- **Security and privacy experts** assessing security posture and privacy implications. AI systems face unique security challenges requiring specialized expertise. Privacy experts ensure compliance with data protection regulations.

Figure 10.4: The Cross-Functional Governance Roundtable. *Effective oversight cannot happen in silos; it requires a unified table where legal authority, technical reality, ethical considerations, and business value meet simultaneously.*

Organizations often make the mistake of creating governance boards comprised only of technical people and executives. This approach typically fails because it lacks the ethical lens, legal expertise, and security depth necessary for complete oversight. Boards that broaden representation across these domains consistently demonstrate improved decision quality. Board size balances thoroughness with efficiency. Boards larger than 10-12 members become unwieldy. Organizations can rotate members or create subcommittees for specific topics while maintaining a core board of 6-8 members for regular decisions.

Governance Scope: Defining What Gets Governed

The AI Governance Board oversees AI systems throughout their lifecycle, from conception through retirement. However, governance burden should match risk. Not every AI experiment requires full board review. Organizations establish clear thresholds determining which AI initiatives require governance review. Typical criteria include:

- **Risk level based on potential impact:** High-risk systems as defined by the EU AI Act or organizational risk framework always require governance review. Lower-risk systems may follow simplified processes.

- **Data sensitivity:** Systems processing personal information, especially sensitive categories like health or financial data, require privacy-focused governance review.
- **Deployment scope:** Systems affecting external stakeholders, customers, or the public receive more rigorous governance than internal experimental tools.
- **Novel technology usage:** Systems employing new AI techniques, architectures, or capabilities warrant governance attention due to unknown risk profiles.
- **Regulatory applicability:** Systems subject to specific regulatory requirements automatically enter governance review.

These thresholds provide clarity about what requires governance oversight versus what teams can proceed with independently. Clear thresholds prevent both governance overreach that slows all development and governance gaps that permit risky systems to deploy without review.

Board Operations: Making Governance Efficient

Governance boards operate efficiently or risk becoming bottlenecks that teams circumvent. Efficient operations require several practices:

- **Regular cadence with published schedule:** Boards meet on a predictable schedule, typically weekly or biweekly for active AI development. Teams know when decisions will be made and can plan accordingly.
- **Tiered review process:** Not all decisions require full board review. The board establishes delegation authority where the Governance Coordinator or designated members can approve low-risk initiatives, reserving full board review for high-risk decisions.
- **Pre-submission requirements:** Teams submit governance requests with standardized documentation including risk assessments, validation plans, and compliance checklists. This ensures boards receive consistent, complete information enabling efficient review.
- **Time-boxed reviews:** Each governance review receives allocated time. Straightforward approvals get 5-10 minutes. Complex decisions get 30-45 minutes. If additional time is needed, items get deferred for deeper analysis rather than consuming entire meetings.

- **Decision documentation:** All decisions are documented with rationale, conditions, and follow-up requirements. This creates institutional memory and enables learning from governance outcomes.
- **Sprint-synchronized reviews:** Governance reviews align with sprint boundaries when possible, enabling teams to incorporate governance feedback into sprint planning rather than mid-sprint disruptions.

Organizations implementing efficient governance establish clear cycle time targets. Straightforward decisions might target 48-hour turnaround, with complex decisions resolved within one week maximum. When governance meets these targets, teams trust the process. When reviews drag on, teams begin seeking workarounds that undermine governance effectiveness.

THE GOVERNANCE CANVAS FRAMEWORK

The Governance Canvas provides a structured tool for planning and documenting governance activities. Teams complete the canvas during sprint planning for initiatives requiring governance review. The canvas organizes governance planning across five dimensions.

- **Risk Assessment.** Identifies potential risks across multiple categories including bias and fairness risks, privacy and security risks, safety and reliability risks, regulatory compliance risks, and reputational risks.
- **Ethical Review.** Evaluates ethical implications including stakeholder impact analysis, transparency and explainability requirements, fairness and equity considerations, and accountability mechanisms.
- **Regulatory Compliance.** Maps applicable regulations and compliance requirements including jurisdictional analysis, regulatory framework applicability, required documentation, and approval processes.
- **Security and Privacy.** Details security and privacy controls including data protection measures, access controls, encryption requirements, adversarial vulnerability assessment, and incident response procedures.
- **Stakeholder Alignment.** Ensures appropriate stakeholder engagement including identification of affected stakeholders,

communication plans, consent requirements, and feedback mechanisms.

- Teams complete the canvas collaboratively, with the Governance Coordinator facilitating. The completed canvas guides governance board review and provides documentation for audit trails.

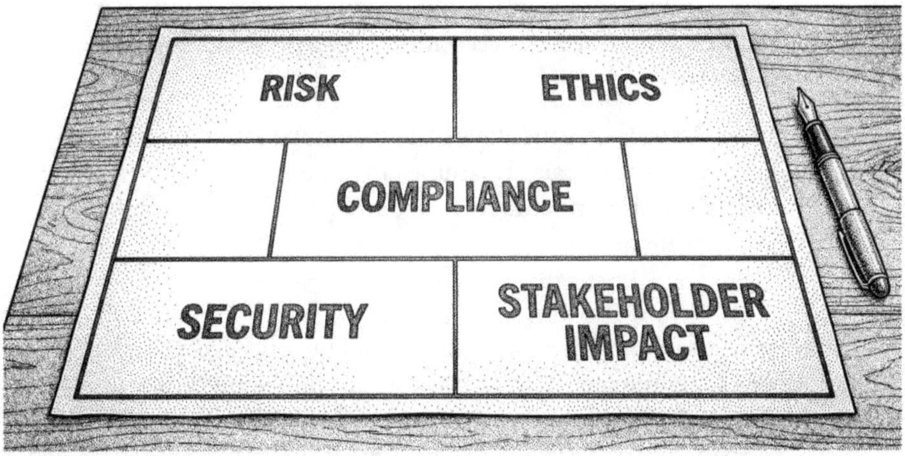

Figure 10.5: The Governance Canvas. *This structured framework forces teams to define risk, ethics, compliance, security, and stakeholder impact during sprint planning, rather than waiting for a post-development audit.*

RISK CLASSIFICATION AND MANAGEMENT

Effective governance matches oversight rigor to actual risk. Not all AI systems warrant the same level of governance scrutiny. Risk classification provides the framework for proportionate governance.

Risk Classification System

Organizations establish clear risk classification criteria that align with regulatory frameworks while reflecting organizational context. A typical classification system includes four risk levels.

- **Critical Risk.** Systems where failures could cause physical harm, significant financial loss, legal liability, or severe reputational damage. Examples include autonomous vehicle control systems, medical diagnosis systems, credit decisioning systems, and critical infrastructure management.

- **High Risk.** Systems affecting important business processes or customer experiences where failures would cause substantial problems but not catastrophic ones. Examples include customer service automation, fraud detection, content moderation, and hiring assistance tools.
- **Moderate Risk.** Systems supporting business operations where failures would cause inconvenience or inefficiency but limited broader impact. Examples include internal process automation, productivity tools, and recommendation engines for non-critical decisions.
- **Low Risk.** Experimental or limited-scope systems where failures have minimal impact. Examples include research prototypes, internal tools with limited users, and proof-of-concept implementations.

Risk classification considers multiple factors beyond just technical complexity. Data sensitivity, user population, deployment scope, reversibility of decisions, and regulatory applicability all influence risk level. The Governance Board assigns risk classifications during initial review. Classifications can change as systems evolve or deployment scope expands. A proof-of-concept tool might start as low risk but move to high risk when deployed customer-facing.

Figure 10.6: The Governance Weigh Station. *Risk classification prevents bottlenecks by allowing low-risk "passenger cars" to stay in the fast lane, while diverting high-risk "heavy cargo" for mandatory inspection.*

Governance Requirements by Risk Level

Each risk level triggers corresponding governance requirements.

- **Critical Risk Systems** require full Governance Board review, risk assessment and mitigation plans, independent validation and testing, and external audit or certification where applicable. They mandate continuous monitoring with real-time alerting, detailed incident response and rollback procedures, quarterly governance reviews, and a complete audit trail and documentation.
- **High Risk Systems** require Governance Board review, risk assessment and mitigation plans, validation and testing by the Validation Lead, monitoring with defined thresholds, incident response procedures, semi-annual governance reviews, and standard audit trails and documentation.
- **Moderate Risk Systems** utilize Governance Coordinator review (delegated authority), basic risk assessment, standard validation and testing, periodic monitoring, incident reporting procedures, annual governance reviews, and essential documentation.
- **Low Risk Systems** rely on team self-assessment using a governance checklist, standard testing practices, incident reporting if issues occur, and documentation as needed for learning.

This tiered approach enables teams to move quickly on low-risk initiatives while maintaining rigor for high-risk systems. The approach prevents both excessive caution that slows all development and insufficient caution that permits dangerous releases.

Continuous Risk Monitoring

Risk classification is not static. Systems require continuous monitoring to detect risk profile changes that might trigger governance re-review. Monitoring addresses several risk indicators.

- **Performance degradation.** Models experiencing accuracy declines, increased error rates, or reliability problems may indicate model drift requiring revalidation or retraining.
- **Usage pattern changes.** Systems deployed in new contexts, serving different user populations, or handling increased volume may face altered risk profiles.

- **Incident patterns.** Recurring issues, near-misses, or user complaints may signal undetected problems requiring governance attention.
- **Regulatory changes.** New regulations or regulatory interpretations may affect previously approved systems, requiring compliance review.
- **Security events.** Attempted attacks, discovered vulnerabilities, or successful exploits warrant immediate governance review regardless of original risk classification.

The Governance Coordinator monitors these indicators, escalating to the Governance Board when thresholds get exceeded or patterns emerge suggesting elevated risk.

Figure 10.7: Model Drift Compass. *Unlike software code which remains static, AI models slowly lose alignment with reality as the world changes around them, requiring continuous recalibration to find "True North."*

ETHICAL AI FRAMEWORKS

Technical governance alone is insufficient. AI systems raise ethical questions requiring explicit consideration. Ethical frameworks provide structure for addressing these questions systematically.

Core Ethical Principles

Organizations establish core ethical principles guiding AI development and deployment. While specific principles vary by context, several themes emerge consistently.

- **Fairness and Non-Discrimination.** AI systems treat individuals and groups equitably, avoiding discriminatory outcomes based on protected characteristics or proxy variables.
- **Transparency and Explainability.** Stakeholders understand when they interact with AI, how AI influences decisions affecting them, and what factors drive AI recommendations.
- **Privacy and Data Protection.** AI development respects privacy rights, protects personal information, and processes data only for legitimate purposes with appropriate consent.
- **Accountability and Responsibility.** Clear lines of accountability exist for AI decisions, with humans maintaining ultimate responsibility for consequential choices.
- **Safety and Reliability.** AI systems operate safely under expected conditions, fail gracefully when encountering unexpected situations, and include mechanisms for human intervention.
- **Human Autonomy.** AI augments rather than replaces human judgment for consequential decisions, preserving meaningful human control.

These principles, drawn from frameworks including UNESCO's Recommendation on AI Ethics and the EU's Ethics Guidelines for Trustworthy AI, provide a foundation for ethical governance.

Ethical Review Process

The Governance Board conducts ethical reviews for initiatives raising ethical questions. Reviews address several key questions:

- **Who benefits and who might be harmed?** This identifies intended beneficiaries and potential negative impacts on various stakeholder groups.
- **What unintended consequences might emerge?** This explores potential second-order effects and misuse scenarios not immediately apparent.
- **Are vulnerable populations adequately protected?** This ensures systems affecting vulnerable groups include appropriate safeguards.
- **Can affected individuals understand AI's role?** This evaluates whether transparency mechanisms provide meaningful understanding rather than just disclosure.

- **Do humans maintain meaningful control?** This confirms that automation does not eliminate human judgment where it should remain.
- **Are fairness mechanisms adequate?** This assesses whether bias detection and mitigation approaches address identified fairness concerns.

Ethical reviews are most valuable when conducted early, shaping system design rather than merely evaluating completed systems. The canvas framework includes ethical review as a standard dimension, integrating ethical consideration into sprint planning.

Fairness and Bias Mitigation

Addressing algorithmic bias requires systematic approaches across the development lifecycle. Organizations implement several practices, starting with representative training data. They ensure training datasets represent diverse populations and usage scenarios, avoiding systematic underrepresentation that creates performance gaps.

Bias testing involves conducting disaggregated performance analysis examining model accuracy across demographic groups, geographic regions, and other relevant dimensions. Teams establish fairness metrics, creating quantitative fairness criteria appropriate to use case context, recognizing that different fairness definitions may conflict and require trade-off decisions.

Bias mitigation techniques apply technical approaches including data augmentation, algorithmic modifications, and post-processing adjustments to improve fairness outcomes. Ongoing monitoring continuously monitors deployed systems for fairness degradation as usage patterns evolve or model drift occurs.

Fairness is particularly challenging because it involves value judgments without universally correct answers. A loan approval model might optimize for equal approval rates across demographic groups, equal default rates across groups, or equal opportunity to receive loans given equal creditworthiness. These goals can conflict mathematically. Organizations make explicit choices about which fairness criteria to prioritize for specific use cases.

AUDIT TRAILS AND DOCUMENTATION

Thorough audit trails are essential for regulatory compliance, incident investigation, and continuous improvement. AI Scrum mandates specific documentation maintained throughout sprints.

Required Documentation

Organizations maintain several categories of documentation.

- **Governance Record.** Captures all governance decisions, risk assessments, mitigation strategies, and approval outcomes. This provides an audit trail for compliance and organizational learning. The Governance Coordinator maintains this record, ensuring completeness and accessibility.
- **AI Collaboration Log.** Documents AI interactions and decisions throughout development. This enables understanding how AI was used, what AI generated, and what humans modified or approved. Developers maintain collaboration logs as part of the normal workflow.
- **Validation Evidence.** Demonstrates testing thoroughness including test plans, test results, edge case analysis, bias testing, security testing, and performance evaluation. This substantiates claims about system quality. The Validation Lead ensures validation evidence is complete and well-organized.
- **Incident Reports.** Describes adverse events and responses including failure descriptions, root cause analysis, remediation actions, and preventive measures. These support learning from failures and demonstrate responsive incident management.
- **Model Cards.** Documents model capabilities, limitations, intended use cases, training data characteristics, performance metrics, and ethical considerations. Model cards provide transparency about model properties for both internal teams and external stakeholders.
- **Data Lineage.** Shows data sources, transformations, quality metrics, and provenance. This enables understanding what data was used, how it was processed, and where it came from. Data lineage is particularly important for compliance with data protection regulations.
- **Security Assessments.** Documents threat models, security controls, penetration test results, and remediation actions. This

demonstrates security rigor and provides evidence of proactive security management.

Documentation is discoverable, version-controlled, and retained per regulatory requirements. Teams do not create documentation for its own sake. Documentation serves specific purposes enabling governance, compliance, and improvement.

Organizations typically maintain documentation for 3-7 years depending on regulatory obligations and risk profile. The Governance Coordinator ensures documentation completeness and accessibility, implementing automated mechanisms where possible to reduce manual burden. Documentation creates overhead, but remains non-negotiable for governance and compliance. The key is making documentation valuable rather than burdensome. Organizations automate what they can, integrate documentation into workflows, and use documentation for learning rather than just compliance checkbox exercises.

Automated Documentation Tools

AI assists documentation creation and maintenance, reducing manual burden while improving consistency. Automated tools can generate documentation from code by extracting technical documentation, API specifications, and data schemas directly from code and configuration files. They can maintain decision logs by capturing governance decisions, approvals, and conditions automatically from governance board meetings and sprint ceremonies.

Tools also track data lineage, automatically mapping data flow through pipelines, transformations, and storage systems. They compile validation evidence by aggregating test results, coverage reports, and quality metrics into consolidated validation documentation. Finally, they create change summaries, generating summaries of code changes, configuration updates, and model modifications between versions. These automated approaches reduce documentation burden while ensuring completeness. However, human oversight remains necessary to ensure documentation accurately reflects reality and provides meaningful insight rather than just checking compliance boxes.

GOVERNANCE MATURITY AND CONTINUOUS IMPROVEMENT

Governance is often treated as a static set of laws—a constitution written once and defended forever. In the context of AI, this approach guarantees obsolescence. Effective governance is an organizational capability that must evolve as rapidly as the models it oversees. It requires viewing governance not as a compliance checklist, but as a product with its own roadmap, backlog, and release cycles.

As teams move from experimental prototypes to enterprise-scale deployments, the governance mechanisms must shift. Early-stage governance often relies on manual interventions and high-friction "gatekeeper" reviews because the risks are unknown and the tools are immature. However, as the organization gathers data on what actually goes wrong, it can replace manual approvals with automated guardrails. This transition—from human-speed oversight to machine-speed monitoring—is the defining characteristic of a mature AI organization.

To achieve this, high-performing organizations implement a "Governance Retrospective" alongside their standard sprint retrospectives. These sessions ask critical questions: *Did the safety checks catch the bias issue, or did we get lucky? Did the compliance review delay the sprint unnecessarily? Are our risk classifications still aligned with reality?* If a rule slows down ten safe projects to catch zero unsafe ones, the rule requires refactoring. Conversely, if a "low risk" internal tool causes a data leak, the classification criteria need immediate tightening.

This concept of "Policy Refactoring" prevents the accumulation of "bureaucratic debt"—the governance equivalent of technical debt. Just as code rot slows down applications, outdated governance policies slow down innovation. A mature governance framework includes explicit mechanisms for retiring old rules that no longer serve a purpose and introducing new controls for emerging threats like agentic behavior or multimodal injection attacks.

The ultimate goal is to reach a state where governance is indistinguishable from the development process itself—where doing the right thing is the path of least resistance. Organizations typically progress toward this state through five distinct levels of maturity.

GOVERNANCE MATURITY LEVELS

Governance maturity progresses through five levels.

- **Initial.** Organizations have ad hoc governance with inconsistent application. Governance depends on individual initiative rather than systematic process. Different teams apply governance differently. Documentation is sparse and inconsistent. This creates compliance risk and missed learning opportunities.

- **Developing.** Organizations establish basic governance policies and processes. A governance framework exists on paper with defined procedures. Teams understand governance requirements and expectations. However, governance still operates somewhat independently from development rather than integrating seamlessly. Documentation improves but remains incomplete.
- **Operational.** Governance integrates systematically into workflows with consistent application. Teams expect governance touchpoints at predictable points in sprints. Governance provides value to teams rather than merely creating overhead. Documentation becomes routine and thorough. Most organizations target this level as minimum acceptable maturity.
- **Optimizing.** Organizations continuously improve governance based on metrics and feedback. Governance effectiveness receives explicit measurement. Incident analysis systematically feeds back into governance improvement. Teams actively seek governance guidance rather than treating it as obligation. Governance becomes a clear competitive advantage.
- **Transformative.** Governance becomes a strategic enabler rather than compliance overhead. The organization achieves predictive risk management and proactive issue prevention. Governance decisions happen so efficiently they are nearly invisible. External stakeholders recognize governance maturity as a differentiator.

Figure 10.8: The Governance Maturity Staircase. *Organizations typically progress through five distinct stages, moving from chaotic, ad-hoc reactions to a strategic state where governance actively accelerates innovation.*

Sprint retrospectives include governance effectiveness evaluation. Questions to address:

- Did governance help or hinder sprint outcomes?
- Were governance requirements clear and actionable?
- Did governance review happen efficiently without delays?
- What would make governance more effective going forward?

The AI Governance Board conducts quarterly reviews assessing process efficiency, decision quality, risk management effectiveness, and compliance adequacy. These reviews examine metrics including governance review cycle time, percentage of decisions escalated versus delegated, incidents traced to governance failures, team satisfaction with governance processes, compliance audit findings, and stakeholder confidence in AI systems. This continuous improvement ensures governance evolves with organizational capability and external environment. The goal is governance that's both rigorous and enabling, protecting against risks while facilitating innovation.

Maturity Progression

Organizations with mature governance experience tangible benefits. According to 2025 research from MIT Technology Review Insights, organizations at the Optimizing or Transformative maturity levels experience 60% fewer AI-related incidents than organizations at Initial maturity levels. These mature organizations also demonstrate 2.3 times higher stakeholder confidence in AI systems. Governance, done well, creates competitive advantage rather than compliance burden.

The journey from Initial to Transformative maturity typically takes 18 months of sustained effort. Organizations cannot skip levels. Maturity develops through iterative improvement. But the investment pays dividends in risk reduction, faster innovation, and stakeholder trust. Organizations progressing through governance maturity levels experience a predictable transformation. Early stages often rely on hope and best intentions without systematic processes. As maturity develops, governance becomes increasingly smooth and unobtrusive. Teams understand expectations, reviews happen efficiently, and problems get caught early. The investment in building governance maturity transforms AI from a perceived risk into a confident organizational capability.

FROM THE TRENCHES: GOVERNANCE IMPLEMENTATION PATTERNS

Successfully implementing governance requires addressing common challenges and establishing patterns that make oversight natural rather than burdensome. The real trick is preventing governance from becoming the red tape that people have learned to despise. Effective governance balances four essential principles.

Before I worked within governance frameworks, I worked around them. Our Agile team ran Monday-to-Friday sprints. The Change Advisory Board ran Wednesday-to-Tuesday. That single scheduling mismatch meant we had to lock our production release list on the Monday of week two, effectively cutting our planning horizon in half. Our group managed data infrastructure, the foundation everything else sat on, so we released on Thursdays by preference. That gave us Friday as a rollback window if something unexpected survived testing. CAB approved Tuesday releases. We wanted Thursday releases. Every exception required going back for re-approval, not because anyone was being difficult, but because two well-intentioned processes were simply running on different clocks. Nobody designed that friction deliberately. The framework had not been built with sprint cadence or release safety buffers in mind. That experience shaped my conviction that governance must be designed around delivery reality, not bolted onto it. Four principles make that possible.

Clear Purpose with Flexible Framework

Governance requires a clear, well-defined purpose such as ensuring fairness, protecting resources, or providing strategic direction. But it also needs flexibility to adapt to new information, evolving circumstances, or the realities of the specific context where it operates. Too much rigidity prevents governance from being responsive to actual needs. Too little structure leads to chaos. The framework is strong enough to provide consistent guidance but flexible enough to accommodate legitimate variations.

Organizations implementing AI governance often start with rigid checklists and approval processes copied from other compliance domains. These rigid frameworks quickly become obstacles when teams encounter scenarios the checklist did not anticipate. Effective governance provides principles and decision criteria rather than inflexible rules, enabling teams to apply governance appropriately to their specific situations.

Accountability with Autonomy

People need to be held accountable for their decisions and actions, but they also need the autonomy to make decisions within their areas of expertise. Good governance strikes a balance between oversight and empowerment. Too much oversight and micromanagement result in frustration and reduced morale. Teams spend energy navigating approval processes rather than solving problems. Too little oversight can lead to mismanagement or lack of direction, with teams making decisions that conflict with organizational values or regulatory requirements.

The Governance Coordinator role exemplifies this balance. The coordinator has authority to approve low-risk initiatives independently, demonstrating trust in their judgment. But the coordinator also escalates high-risk decisions to the full Governance Board, recognizing when broader expertise and authority are needed. This delegation enables speed for straightforward decisions while maintaining rigor for complex ones.

Transparent but Efficient Decision-Making

Governance is transparent so stakeholders understand the reasoning behind decisions. Transparency builds trust and enables learning. However, transparency does not mean decision-making becomes bogged down in endless processes or discussions. Efficiency is always part of the equation, especially for time-sensitive matters.

Organizations often confuse transparency with extensive documentation and prolonged deliberation. Effective governance makes decision criteria and rationale visible without requiring multi-week approval cycles. The governance board publishes its decision framework, making clear what factors drive approvals and what conditions trigger rejections. Teams can then prepare submissions that address known criteria rather than guessing what the board wants to see.

Time-boxed reviews enforce efficiency throughout the process. Each governance item receives allocated time based on complexity. Straightforward approvals get resolved in minutes. Complex decisions receive more deliberation but still within defined limits. If additional time is needed, items get deferred for deeper analysis rather than consuming entire meetings with open-ended discussion.

Adaptability Over Time

Governance structures need regular reflection and refinement. What worked when an organization was small or in a different environment may not work as it grows or as circumstances change. Being willing to adapt or even change governance structures when they are not adding value is key to ensuring continued effectiveness.

Governance retrospectives provide the mechanism for this adaptation. After completing initiatives that went through governance review, teams examine what governance caught, what it missed, and how the process could improve. These retrospectives prevent governance from becoming static, ensuring it evolves with organizational learning and changing risk profiles.

The maturity progression described earlier in this chapter reflects this adaptability principle. Organizations do not establish perfect governance on day one. They start with basic frameworks and systematically improve based on experience. The willingness to recognize when governance is not working and make adjustments separates organizations with effective governance from those where governance becomes bureaucratic overhead.

WHAT COMES NEXT

Governance provides the framework for responsible AI development. But frameworks alone do not transform organizations. The next chapter examines change management strategies for introducing AI-native development practices, addressing the human and cultural dimensions that determine whether transformations succeed or fail. Organizations often focus on technical enablement while underestimating change management complexity. We will explore how to build organizational readiness, address resistance, and guide teams through the transition from traditional to AI-native development approaches.

KEY TAKEAWAYS

- **Traditional software governance fails for AI systems** due to non-deterministic behavior, emergent properties, data dependencies, model drift, and novel adversarial vulnerabilities that require fundamentally different oversight approaches.
- **The AI Governance Board should include diverse expertise** across technical, business, legal, ethics, and security domains, operating efficiently through regular cadence, tiered review processes, and sprint-synchronized decision-making.
- **Risk classification systems should match governance rigor to actual risk**, with critical and high-risk systems receiving comprehensive oversight while low-risk initiatives follow streamlined processes.

- **Ethical frameworks must address fairness, transparency, privacy, accountability, safety, and human autonomy** through systematic review processes integrated into sprint planning rather than conducted as afterthoughts.
- **Comprehensive audit trails** including governance records, validation evidence, incident reports, model cards, and data lineage prove essential for regulatory compliance, incident investigation, and continuous improvement.
- **Governance maturity progresses** from ad hoc processes through systematic integration to strategic enablement, with mature organizations experiencing 60% fewer AI-related incidents and 2.3 times higher stakeholder confidence.

MIND MAP – GOVERNANCE, ETHICS, AND RISK MANAGEMENT

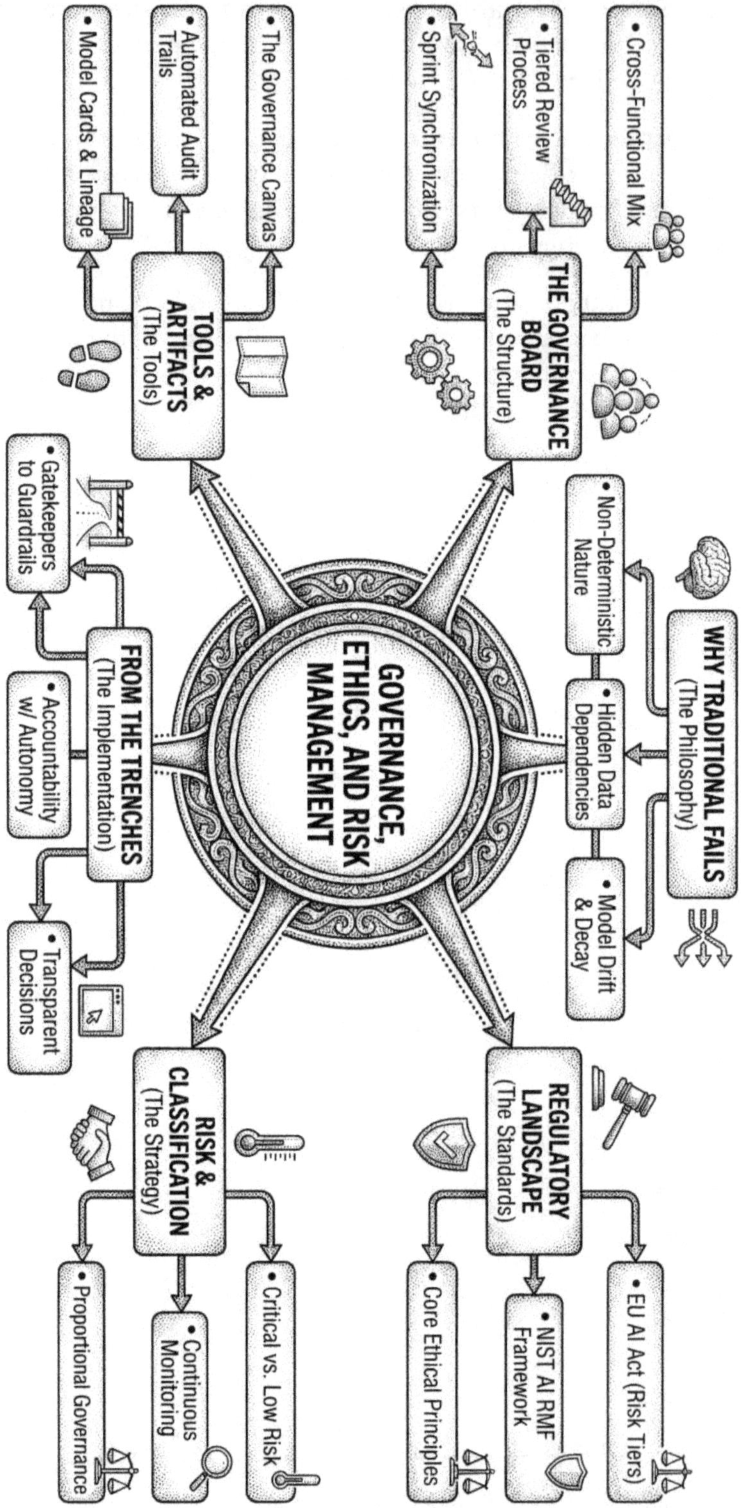

Transitioning to AI-augmented development triggers deep identity crises and resistance that technical training alone cannot resolve. Leaders must navigate the inevitable J-Curve productivity dip while actively building new competencies across their teams. Sustainable success ultimately requires mitigating psychological threats using the SCARF model and fundamentally shifting governance from a traditional blocker into a proactive safety enabler for the organization.

CHAPTER 11: CHANGE MANAGEMENT

Consider a sophisticated technology organization that had spent two years perfecting a custom sprint dashboard. The system was a marvel of clarity, tracking everything in an end-to-end view: tasks by stage, defect density, risk indicators, and cycle time. The teams relied on it daily. Standups became focused, data-driven events rather than subjective status updates. Directors praised the visibility it provided into the delivery pipeline.

Then a new leader arrived. Within weeks, they moved to replace the system with a legacy tool they had used at a previous firm.

They did not object to the dashboard's functionality. They did not find fault with the data accuracy. They simply preferred the familiar. In a single decision, the cornerstone of the department's Agile practice evaporated because an authority figure prioritized their own comfort over proven utility.

This scenario illustrates a fundamental truth about transformation: technical superiority does not guarantee adoption. Change management determines whether a transformation succeeds or fails, regardless of the sophistication of the AI capabilities involved.

Organizations frequently invest millions in training, tooling, and infrastructure only to watch transformations stall because they underestimated the human dynamics of the shift. AI Scrum transformation differs fundamentally from previous methodology shifts. It does not merely ask teams to work differently; it asks them to *think* differently about the nature of value creation.

It inverts development constraints, redefines professional identities, and challenges deep-seated beliefs about authorship and expertise. Understanding these dynamics is the difference between a successful modernization and a chaotic failed experiment.

WHAT THIS CHAPTER COVERS

By the end of this chapter, you will understand:

- **The Neuroscience of Resistance (The SCARF Model):** A deep dive into why the human brain interprets AI-driven process changes as survival threats and how to mitigate the "Amygdala Hijack."
- **The Identity Crisis:** Why AI Scrum triggers existential professional anxiety compared to traditional Agile adoption.
- **The AI Scrum Competency Framework:** A structured three-tier curriculum (Literacy, Fluency, Mastery) to bridge the skills gap.
- **The "J-Curve" of Transformation:** A granular breakdown of the productivity dip, distinguishing between "The Slump," "The Turn," and "The Climb."
- **The Coalition Strategy:** How to build a "Change Coalition" to sustain momentum through the "Valley of Despair".
- **Tactical Interventions:** Specific scripts ("The Rescue Protocol") for when teams stall.
- **The Governance Pivot:** Turning blocking functions (Legal, Security, HR) into transformation enablers.
- **Rewiring the Reward System:** How to align HR performance metrics with the new behaviors required by AI Scrum.

KEY TERMS AND CONCEPTS

- **Change Coalition** A strategic alliance of influential stakeholders across multiple organizational levels who actively support the transformation. A robust coalition buffers the initiative against resistance and provides air cover during the difficult implementation phases. Building this coalition before implementation begins is more effective than attempting to rally support after resistance emerges.
- **The SCARF Model** A neuroscience framework (Status, Certainty, Autonomy, Relatedness, Fairness) used to diagnose the specific psychological triggers of resistance in transformation.
- **The J-Curve Effect** The phenomenon where performance temporarily degrades below the baseline during the learning phase of a new technology before rising to a new, higher plateau. Misinterpreting this predictable dip as "failure" is the primary cause of premature transformation abandonment.
- **Authority Resistance** Opposition rooted in the preservation of hierarchical control or status rather than substantive objections to the proposed change. This often manifests when leaders revert to familiar legacy tools to maintain a sense of competence.
- **Technical Comfort Zone** The set of tools and practices where an individual feels professionally secure. Resistance in AI adoption often stems not from a belief that AI is ineffective, but from a fear that the individual's expertise will no longer be relevant outside this zone.
- **The Hybrid Trap** The dangerous state where an organization attempts to run traditional Scrum processes simultaneously with AI Scrum workflows, resulting in double cognitive load and eventual burnout.
- **Pre-Mortem Analysis** A prospective hindsight exercise where the team imagines a future failure and works backward to determine what caused it, allowing them to mitigate risks before they manifest.

WHY AI SCRUM CHANGE DIFFERS FROM TRADITIONAL SCRUM

AI Scrum transformation is distinct from traditional Agile adoption in both scope and psychological weight. When organizations moved from Waterfall to Scrum in the 2010s, the fundamental activities remained recognizable. Developers still wrote code. Product Owners still managed backlogs. QA engineers still executed tests. The cadence changed—moving from months to weeks—but the professional identities remained intact. AI Scrum inverts this stability. It is not an evolution; it is a disruption of the "Craftsman" model.

In an AI-native environment, developers evolve from code authors to collaboration orchestrators. Product Owners shift from backlog administrators to outcome strategists. QA engineers transition from test executors to validation architects. These shifts threaten professional identity in a way that shifting from a Gantt chart to a Kanban board never did.

The Psychology of "The Author" vs. "The Editor"

For decades, software engineering has been defined by the difficulty of syntax creation. The "Hero Developer" was the one who could type the complex algorithm from memory. Their dopamine reward loop was tied to *solving the puzzle* and *writing the solution.*

AI Scrum removes the manual labor of the solution. The AI generates the boilerplate, the API connectors, and the unit tests in seconds. The developer is left with the task of *reviewing* and *integrating*. Psychologically, this feels like a demotion from "Creator" to "Janitor."

Figure 11.1: From Author to Orchestrator. *The fundamental identity shift required in AI Scrum moves value creation from the manual labor of syntax generation to the strategic direction of AI agents, addressing the root cause of 'Technical Discomfort' resistance.*

If we do not address this psychological loss, developers will reject the tools. They will claim the AI code is "garbage" or "unsafe" not because it is true, but because rejecting it preserves their identity as the necessary author. Furthermore, AI Scrum introduces non-determinism. Traditional engineering cultures prize predictability. Input A plus Process B always equals Output C. AI models, by contrast, are probabilistic. They may produce different outputs from similar prompts. This uncertainty creates anxiety for teams accustomed to deterministic control.

THE NEUROSCIENCE OF RESISTANCE: THE SCARF MODEL

To manage change effectively, we must understand the biological machinery of resistance. David Rock's **SCARF Model** provides a lens for understanding why AI Scrum triggers a threat response in the brain's amygdala, causing otherwise rational employees to act irrationally.

When the brain perceives a threat to any of these five domains, it disengages the prefrontal cortex (logic) and activates the limbic system (emotion/defense).

1. Status (Relative Importance)

- **The Trigger:** A Senior Developer realizes an AI agent can write the SQL query they used to take pride in optimizing.
- **The Threat:** "I am no longer the smartest person in the room. My value is diminishing."
- **The Mitigation:** Reframe status around architecture and oversight. Publicly praise developers who successfully "prompt engineering" complex solutions. Elevate them to "AI Mentors."

2. Certainty (Predictability)

- **The Trigger:** The probabilistic nature of LLMs means the same prompt might yield different code twice.
- **The Threat:** "I can't control the output. This feels chaotic."
- **The Mitigation:** Establish rigid "Guardrails" and "Validation Patterns" (as detailed in Chapter 8). Replace the certainty of input with the certainty of verification.

3. Autonomy (Sense of Control)

- **The Trigger:** AI agents handling task execution feels like losing control of the work.
- **The Threat:** "The machine is doing my job. I'm just a spectator."

- **The Mitigation:** Emphasize "Human-in-the-Loop." Ensure the developer has the final "Commit" authority. The AI proposes; the human decides.

4. Relatedness (Sense of Belonging)

- **The Trigger:** If the team interacts more with bots than with each other, social cohesion fractures.
- **The Threat:** "I'm isolated. I'm working with a machine, not a team."
- **The Mitigation:** Maintain human-centric ceremonies. Use the time saved by AI to increase high-value human collaboration, such as architecture whiteboarding sessions (the "Day 1" innovation).

5. Fairness (Perception of Equity)

- **The Trigger:** "Why does the other team get Copilot licenses and we don't?" or "Why are they cutting headcount if we are more productive?"
- **The Threat:** "This transformation is rigged against me."
- **The Mitigation:** Radical transparency regarding tooling access and the "Retraining Guarantee." If efficiency leads to layoffs, fairness is broken, and trust evaporates.

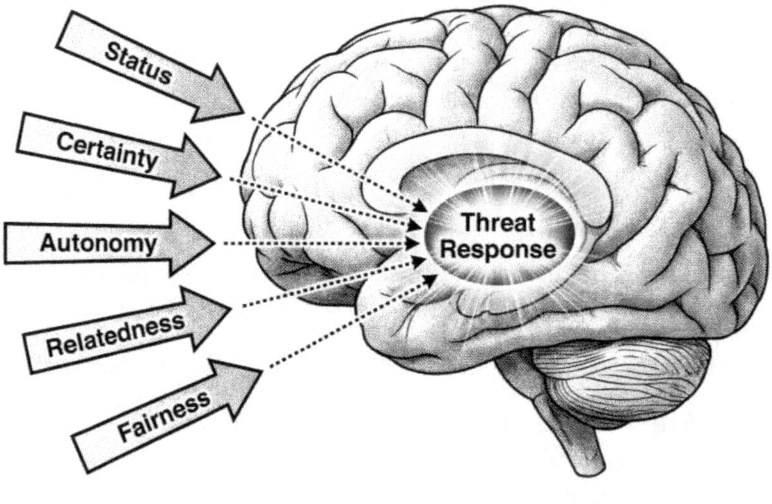

Figure 11.2: The SCARF Model of Resistance. *When AI transformation threatens these five social domains, it triggers a biological "fight or flight" response (Amygdala Hijack) that manifests as irrational resistance to new tools.*

THE ADKAR FRAMEWORK FOR INDIVIDUAL CHANGE

While SCARF explains the neuroscience of *resistance*, we need a structured framework to drive *adoption*. Organizations do not change; individuals do. One hundred developers might work in the same department, but each one occupies a different stage of acceptance.

The **ADKAR Model** (Awareness, Desire, Knowledge, Ability, Reinforcement) is utilized to diagnose exactly where a team member has stalled. Managers often make the mistake of sending an employee to training (Knowledge) when the employee does not yet understand why the change is necessary (Awareness). This mismatch creates friction.

Awareness and Desire: The "Why" Before the "How"

Most AI transformations fail in the first two distinct phases. Leadership announces a rollout of GitHub Copilot and immediately schedules training sessions. They skip the critical step of establishing *Awareness* of the business problem and *Desire* to participate in the solution.

Awareness requires honesty about the external threat. We do not adopt AI Scrum because it is trendy. We adopt it because our competitors can now ship features 40% faster than we can. If we do not adapt, our market share erodes.

Figure 11.3: The ADKAR Chain. *Transformation fails when leaders skip steps. Attempting to build Ability (training) before establishing Desire (motivation) results in mechanical compliance rather than genuine adoption. A figure is shown examining a jammed gear at the "Desire" stage with a wrench, symbolizing targeted intervention.*

Desire translates that organizational threat into personal context. For a developer, the "What's In It For Me" (WIIFM) is not about corporate profit. It is about the removal of drudgery. The desire comes from the promise that AI will handle the boilerplate unit tests and documentation, leaving the developer free to solve interesting architectural puzzles.

Knowledge and Ability: The "Learning" Trap

A common pitfall involves confusing *Knowledge* with *Ability*.

- **Knowledge** is understanding the concept. "I know how a Large Language Model works."
- **Ability** is the capability to execute. "I can successfully debug a hallucinated SQL query in under two minutes."

Traditional training provides knowledge but rarely builds ability. We bridge this gap through the "Pairing Week" intervention (detailed in the Competency Framework). We do not just show the team slides; we sit with them and execute the work until the theoretical knowledge transforms into muscle memory.

Reinforcement: Preventing the Slide Back

The final phase, *Reinforcement*, determines long-term sustainability. Human behavior naturally reverts to the path of least resistance. If a developer uses AI for two sprints but is then criticized for "low commit volume" because the metrics were not updated, they will stop using the AI.

Reinforcement requires structural alignment. When a team successfully uses an AI agent to catch a security flaw before production, we celebrate that specific win publicly. We alter the performance review criteria to value "AI Leverage" over "Manual Output." Without these structural reinforcements, the change is merely temporary.

THE COMPETENCY GAP: STRUCTURING THE UPSKILLING JOURNEY

A primary driver of resistance is simply lack of competence. We cannot expect teams to adopt AI Scrum if we do not provide the map for learning. "Here is a login" is not a training strategy.

Organizations succeed when they structure training into three distinct tiers of capability.

Level 1: AI Literacy (The "Safety" Tier)

Target Audience: Everyone (Devs, POs, QAs, Scrum Masters)
Before anyone writes a line of code with AI, they must understand the tool's nature.

- **Curriculum:**
  - **Probabilistic vs. Deterministic:** Understanding why the AI hallucinates.
  - **Data Privacy 101:** What *never* goes into the prompt (PII, secrets, keys).
  - **The Review Obligation:** The legal and ethical requirement that the human is 100% liable for AI output.
- **Exit Criteria:** Pass a "License to Drive" assessment. No license, no tool access.

Level 2: AI Fluency (The "Productivity" Tier)

Target Audience: Practitioners
This is role-specific training focused on daily workflows.

- **Curriculum (Developers):** Prompt patterns for refactoring, unit test generation, and debugging.
- **Curriculum (Product Owners):** Using AI for user story decomposition and acceptance criteria generation.
- **Curriculum (QA):** Generating edge case scenarios and synthetic data.
- **Exit Criteria:** Completion of a "Pairing Week" where the practitioner executes a full sprint cycle with an AI mentor.

Level 3: AI Mastery (The "Strategy" Tier)

Target Audience: Architects and Leads
Focuses on system design and agent orchestration.

- **Curriculum:** Designing multi-agent systems, building custom context contexts (RAG), and fine-tuning governance constraints.
- **Exit Criteria:** Successful deployment of a custom agent or tool configuration that improves team velocity.

MANAGING THE "J-CURVE" OF TRANSFORMATION

Every transformation follows a predictable performance curve. It begins with the "Honeymoon Phase" of high energy. Then, reality sets in. Velocity drops below the historical baseline. This is the **J-Curve Effect**.

Managing this curve requires distinct strategies for each of its three phases.

Phase 1: The Slump (Weeks 1-4)

The Symptom: Velocity drops. Teams are frustrated. The "First Sprint Panic" occurs here.
The Cause: Cognitive Load. The team is trying to do the work and learn the tool simultaneously.

The Intervention:

- *Suspend Velocity Targets:* Explicitly tell stakeholders that velocity *will* drop.
- *The "Hybrid Trap" Removal:* Cancel old meetings to free up time for learning.
- *Pairing Hours:* Dedicate 2 hours daily to group experimentation.

Phase 2: The Turn (Weeks 5-8)

The Symptom: Velocity stabilizes but hasn't increased. The team is no longer panicking, but they aren't yet "faster."
The Cause: Inefficient Workflows. They are using the tools, but their processes haven't optimized around them yet.
The Intervention:

- *Workflow Refactoring:* The Retrospective focuses solely on "Where is the friction?"
- *Champion Spotlights:* Have the early adopters demo their wins to the rest of the team.

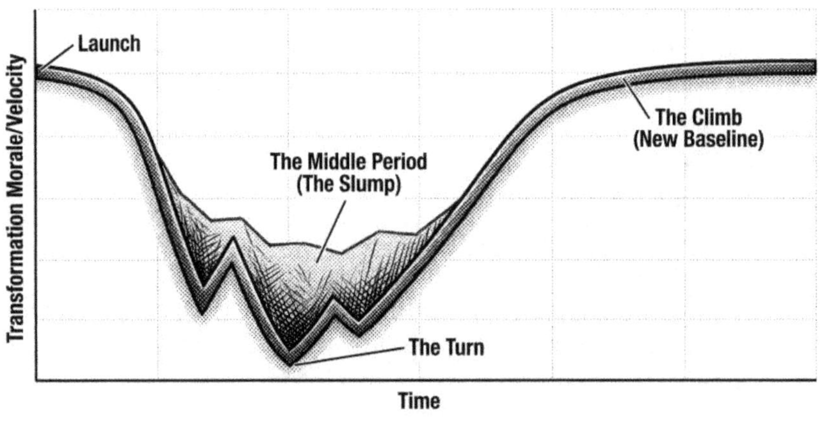

Figure 11.4: Navigating the J-Curve. *This graph illustrates the predictable operational dip that occurs after the initial excitement of AI adoption fades. Leaders must "buffer" this slump to survive until the "Turn."*

Phase 3: The Climb (Weeks 9+)

The Symptom: Velocity begins to exceed historical baselines. Quality improves.
The Cause: Fluency. The tools are now "muscle memory."
The Intervention:

- *Raise the Bar:* Slowly reintroduce ambitious targets.
- *Standardization:* Document the successful patterns into the team's "Playbook."

STAKEHOLDER ANALYSIS AND COMMUNICATION STRATEGY

Effective change management requires tailored narratives. A message that resonates with a CTO will often alienate a Senior Engineer. We segment the organization into distinct tiers, each requiring a specific engagement strategy.

Executive Leadership: Removing FUD

Executives focus on competitive advantage, risk mitigation, and capital allocation. However, their primary blocker is FUD: **Fear, Uncertainty, and Doubt**. When FUD is present, doors close. Budgets freeze. Initiatives die in committee.

The goal of executive communication is not to explain the technology, but to systematically remove FUD. When we replace Uncertainty with a clear roadmap, and Fear with risk containment strategies, the gates open.

The Narrative Arc: "We are not experimenting with toys; we are modernizing our factory floor to compete with AI-native startups."

What Not to Say	What to Say
"We want to try out these cool new AI agents."	"We need to reduce our cycle time by 40% to match market competitors."
"The AI sometimes hallucinates, but we're working on it."	"We have implemented a 'Human-in-the-Loop' validation protocol to mitigate risk."
"We don't know how long this will take."	"We are executing a 3-sprint pilot with a go/no-go decision gate on [Date]."

Middle Management: The "Shock Absorbers"

Middle managers occupy the most precarious position in any transformation. They are the shock absorbers. Leadership expects them to drive efficiency, while individual contributors look to them for protection from disruption.

Their fear is loss of control. If the AI is doing the work, and the team is self-managing, what is the manager's job?

The Narrative Arc: "Your role is shifting from 'Task Supervisor' to 'System Architect.' You are no longer watching the workers; you are designing the machine that helps them work."

Team Leads and Scrum Masters

These roles often fear obsolescence. When an organization announces that AI will handle code generation and documentation, a Scrum Master naturally questions if their facilitation role is next.

The strategy here involves early inclusion. We invite Team Leads to co-design the new interaction models. When they contribute to the solution, they transition from passive observers to active architects of their new reality.

Individual Contributors

For the developers and analysts doing the work, abstract strategy offers little comfort. Their concerns are practical and immediate. *What tools will I use? How will my performance be measured? Will my skills become irrelevant?*

Visual demonstration trumps rhetoric. Rather than describing how an AI agent helps with unit testing, we run live sessions where they can see the tool in action. Skepticism often evaporates when a developer sees an AI resolve a three-hour debugging task in a few moments.

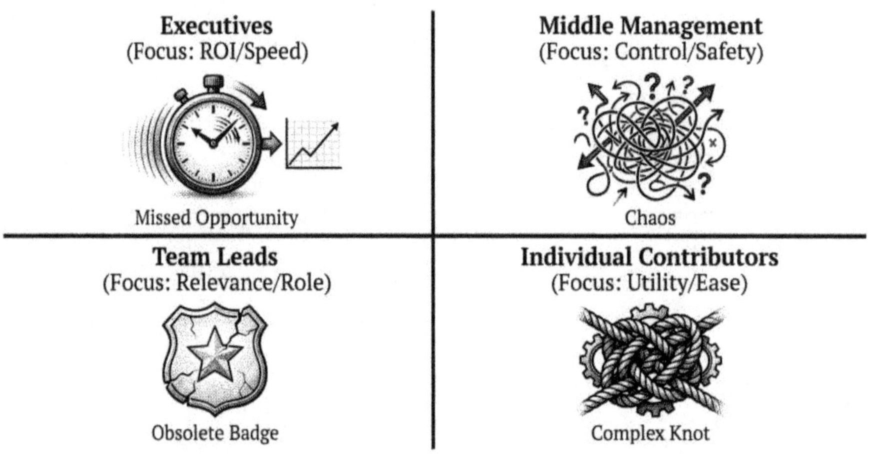

Figure 11.5: The Stakeholder Resistance Map. *Different roles resist AI for different reasons. Effective communication targets the specific fear of each quadrant rather than using a generic "AI is good" message.*

RESISTANCE PATTERNS AND MITIGATION

Resistance is not a monolith. It comes in different flavors, and treating them all as "insubordination" is a leadership failure. We categorize resistance into three primary types.

1. Authority-Based Resistance

This occurs when an individual opposes change to preserve status or control. The opening story of the new leader replacing the dashboard is a classic example. The resistance is not about the tool; it is about establishing dominance.

> **Mitigation:** This is a political challenge, not a technical one. We mitigate this by building a Change Coalition *before* the resistance solidifies. When a broad base of stakeholders supports the transformation, it becomes difficult for a single authority figure to derail it without incurring significant political cost.

2. Technical Discomfort Resistance

This resistance stems from anxiety. A developer who feels incompetent with new AI tools may attack the tools to hide their own insecurity. They may nitpick the AI's code quality or focus obsessively on edge cases where the model fails.

> **Mitigation:** Empathy and training are the tools here. We reframe the narrative: AI does not replace the developer; it elevates them. We emphasize that their deep code knowledge is now required to *validate* and *guide* the AI. We pair hesitant individuals with early adopters to foster psychological safety and peer-assisted learning.

3. Fear-Based Resistance (The Passive Resister)

This is the most insidious form. It operates in the shadows. Employees fear job loss or redundancy, leading them to quietly obstruct progress.

> **Mitigation:** Honesty is the only antidote. If roles will change, leadership states this clearly. If headcount reductions are not part of the plan, leadership guarantees this explicitly to prevent cynicism. Uncertainty breeds rumors; clarity kills them.

BUILDING COALITION AND MOMENTUM

Successful transformation requires building broad support before beginning implementation. Organizations that announce transformation and immediately begin execution without coalition-building often face sustained resistance that stalls progress.

The "Pre-Mortem" Protocol

One of the most effective ways to build a coalition is to invite the skeptics to predict failure. We call this the **Pre-Mortem Workshop**.

The Exercise:

1. Gather the team, stakeholders, and skeptics.
2. **Set the scene:** "Imagine it is six months from now. The AI Scrum transformation has failed spectacularly. It was a disaster. Everyone hates it."
3. **The Question:** "What went wrong? Write down exactly why we failed."

Why this works:

- It gives skeptics a safe platform to voice their fears without looking negative (they are just "playing the game").
- It surfaces real risks that leadership might have missed.
- It turns the skeptics into allies. Once they identify the risks, you ask them, "Okay, how do we prevent that?" Now they are co-designing the solution.

Figure 11.6: The Pre-Mortem Workshop. *By inviting skeptics to "predict the death" of the project in advance, you validate their fears and turn them into co-designers of the risk mitigation strategy.*

Early champions accelerate transformation by demonstrating feasibility and building enthusiasm. However, choosing champions poorly can undermine transformation. Ideal champions combine technical credibility with resilience and realistic assessment capabilities.

The danger lies in selecting champions based solely on enthusiasm. The "Tech Optimist" who loves every new gadget often loses credibility with the senior engineers. You need the "Grumpy Architect" who was skeptical at first but was won over by the utility. Their endorsement carries 10x the weight.

Demonstrating Early Wins

Early wins prove essential for building momentum. Abstract arguments about transformation benefits convince fewer skeptics than concrete demonstrations of value delivery. However, early wins must be genuine rather than contrived.

SUSTAINING CHANGE THROUGH "THE MIDDLE PERIOD"

Every transformation follows a predictable emotional curve. It begins with the "Honeymoon Phase" of high energy and optimism. Then, inevitably, reality sets in. This is the **Middle Period**, also known as the "Valley of Despair," a concept identified by Kelley and Conner in their work on the Emotional Cycle of Change.

During this phase, the easy wins are gone. The team encounters the hard limitations of the AI models. AI Debt from legacy systems surfaces. Skeptics begin to say, "I told you this wouldn't work."

The "Hybrid Trap"

The most common cause of failure in the Middle Period is the **Hybrid Trap**. This occurs when the team tries to adopt AI Scrum but refuses to let go of their old rituals. They keep the hour-long daily standup *and* try to do the automated reporting. They write manual unit tests *and* try to verify AI tests.

Figure 11.7: The Rescue Protocol. *When a team stalls in the Middle Period, adding pressure causes burnout. The correct response is to stop execution, remove process friction, and lower the bar to restore dopamine and momentum.*

The result is double cognitive load. The team burns out, not because AI is hard, but because they are doing two jobs.

The Fix: You must "burn the ships." You must explicitly *stop* doing old things to make room for the new. If the AI handles reporting, cancel the status meeting. If the AI handles testing, stop manual test script writing.

The Rescue Protocol: What to Do When Stuck

When a team falls into the Valley of Despair and velocity hits zero, standard management pressure ("Work harder!") makes it worse. Instead, execute the Rescue Protocol:

Rescue Protocol:

1. **Stop the Line:** Pause new feature development for 24 hours. Acknowledge the jam.
2. **The "Stupid Rules" Session:** Ask the team, "What process is getting in your way?" If they say, "Documenting the AI prompts in Jira is taking too long," then *kill the rule.* Immediate relief builds trust.
3. **Lower the Bar:** Reduce the Sprint Goal scope by 50%. The objective is to finish *something* to restore dopamine, not to hit the original arbitrary target.
4. **Pair Programming:** Mandate that no one works alone. Isolation breeds despair; collaboration breeds problem-solving.

THE GOVERNANCE PIVOT: TURNING BLOCKERS INTO ENABLERS

One of the most overlooked sources of resistance comes from outside the Scrum team: Governance (Legal, Security, Compliance, HR). These functions often view AI as a net risk and will reflexively block adoption.

Traditional change management treats these groups as "obstacles to be bypassed." In AI Scrum, we treat them as "customers to be satisfied."

The Pivot Strategy:

1. **Security:** Instead of asking "Can we use this tool?", ask "How can we configure this tool to be safer than our current human process?" Show them how AI can detect secrets in code *before* commit—a capability humans lack.
2. **Legal:** Shift the conversation from "Copyright" (which is murky) to "Liability" (which is clear). Demonstrate the Human-in-the-Loop review process that ensures a human always accepts liability for the code.

3. **HR:** Engage them early on the "Job Description Redesign" (see Appendix E). If HR continues to recruit for "syntax knowledge" while the team needs "orchestration skills," the pipeline will fail.

Traditional Governance (The Blocker) **AI Scrum Governance** (The Enabler)

Figure 11.8: The Governance Pivot. *Shifting external stakeholders (Legal/Security) from "Gatekeepers" who stop work to "Guardrail Architects" who define the safety boundaries within which the team can move fast.*

REWIRING THE REWARD SYSTEM: THE NEW PERFORMANCE CONTRACT

You cannot change behavior if you do not change the incentives. If you ask developers to collaborate with AI, but you promote them based on "Lines of Code Written," the transformation will fail.

Organizations must align their HR and performance management frameworks with the new reality.

1. From "Heroism" to "Prevention"

- **Old Metric:** "Number of Sev-1 bugs fixed this quarter."
- **New Metric:** "Defect Escape Ratio" (How many bugs reached production?).
- **The Shift:** Traditional IT cultures reward the firefighter who saves the day at 2 AM. AI Scrum rewards the architect who used simulation agents to ensure the fire never started.

2. From "Authorship" to "Orchestration"

- **Old Metric:** "Story Points Completed."

- **New Metric:** "Cycle Time" and "Value Delivered."
- **The Shift:** Performance reviews should focus on system impact. Did the developer leverage tools to amplify the team's velocity? A developer who spends three days building a reusable prompt library that saves the whole team 500 hours is worth more than a developer who closes ten individual tickets.

3. Celebrating "Deletions"

- **Old Metric:** "Features Shipped."
- **New Metric:** "AI Debt Retired."
- **The Shift:** In the past, we celebrated adding lines of code. In AI Scrum, we celebrate removing complexity. A developer who uses AI to refactor and delete 5,000 lines of legacy code contributes more value than one who adds 5,000 lines of new code.

SIGNS OF ADOPTION VERSUS FAILURE

Distinguishing between normal transformation challenges and genuine transformation failure enables appropriate response. Organizations need clear indicators that help them recognize whether transformation is progressing appropriately or requires significant intervention.

Positive Signs (Healthy Friction)

- **Argument about prompts:** Teams are arguing about *how* to prompt the AI (this means they are using it).
- **New rituals emerging:** The team invents their own slang or meetings for the AI tools.
- **"Shadow IT" usage:** People are finding new AI tools on their own to solve problems.

Warning Signs (Toxic Friction)

- **Silence:** No one is complaining, but the usage logs show zero activity.
- **Parallel Work:** The team is maintaining a "secret" manual backlog because they don't trust the system.
- **The "Waiting" Game:** Teams use any downtime or technical glitch as an excuse to stop working entirely ("The AI is down, so I can't do anything").

FROM THE TRENCHES: PRACTICAL CHANGE MANAGEMENT PATTERNS

Organizations navigating AI Scrum transformation encounter similar patterns repeatedly. Understanding these patterns enables proactive strategies rather than reactive responses when familiar situations emerge.

The First Sprint Panic (Day 3 Show and Tell)

The most dangerous moment in any AI Scrum migration often occurs during the **Day 3 AI Show and Tell** (see *Chapter 5*). Teams beginning AI Scrum transformation often experience panic when they realize they must present their work to stakeholders only 72 hours into the sprint.

In traditional Scrum, teams often had two weeks to perfect their code before a review. In AI Scrum, the timeline accelerates aggressively. The panic arises because the AI's output on Day 3 is often raw, functional but unpolished, or conceptually accurate but syntactically buggy. Developers feel exposed. They fear looking incompetent in front of leadership.

This initial panic is not a failure of the technology; it is a failure of expectation management. The team attempts to apply "Production Quality" standards to "Prototyping Speed."

Effective mitigation involves setting explicit expectations. Stakeholders need to understand that the Day 3 review is for *direction validation*, not *production sign-off*. Once the team realizes they are being judged on their ability to iterate, not their ability to be perfect instantly, the panic subsides. By the third sprint, this "show early, fail fast" cadence usually becomes the team's greatest strength.

The Tool Proliferation Problem

Teams exploring AI capabilities often adopt multiple tools simultaneously, creating fragmentation where different team members use different AI systems for similar tasks. This prevents shared learning.

Effective response involves establishing tool standards early. Organizations select primary AI collaboration tools and require teams to develop proficiency with those before exploring alternatives. This creates a baseline for shared knowledge transfer. One organization allowed unlimited tool exploration during pilots, only to find later that teams had developed incompatible approaches, forcing a painful retroactive standardization.

The Champion Burnout Cycle

Early transformation champions often become overwhelmed as demand for their expertise exceeds their capacity. Other teams seek their guidance, leadership asks them to present, and they still have their day jobs. Eventually, they burn out and disengage.

Organizations must protect champion capacity deliberately. This includes limiting how many teams champions support simultaneously and ensuring they receive recognition and compensation for additional responsibilities. Champion burnout creates substantial problems because burned-out champions often become skeptics who question whether the transformation is worth the personal cost.

WHAT COMES NEXT

Understanding the human dynamics of change is the prerequisite for scaling. In **Chapter 12**, we will formalize this progress into a structured **Maturity Assessment Framework**. We will move from the soft skills of persuasion to the hard metrics of capability measurement, allowing you to score your organization's readiness across five distinct levels of AI adoption.

KEY TAKEAWAYS

- **Identity Over Process:** AI Scrum transformation challenges professional identity. Developers fear becoming mere "operators," and Product Owners fear losing control. Change management addresses these fears directly by redefining roles from "Authors" to "Architects."
- **The J-Curve is Real:** Productivity *will* drop before it rises. Executives must be inoculated against this dip, or they will cancel the project during "The Slump."
- **The Hybrid Trap:** You cannot run AI Scrum on top of Old Scrum. Explicitly retire old rituals to create capacity for new learning.
- **Eliminate FUD:** Fear, Uncertainty, and Doubt are the primary blockers to executive buy-in. Removing them by replacing ambiguity with clear roadmaps opens the gates for adoption.
- **The Pre-Mortem Protocol:** Immunize your team against failure by inviting skeptics to predict the disaster in advance. Use their fears to design your risk mitigation strategy.

- **The First Sprint Panic:** Expect a crisis of confidence around the **Day 3 AI Show and Tell**. Prepare the team to share "raw" work early without fear of judgment.
- **Show, Don't Tell:** Visual demonstrations of AI solving real problems persuade skeptics far more effectively than strategic slide decks.

MIND MAP – CHANGE MANAGEMENT

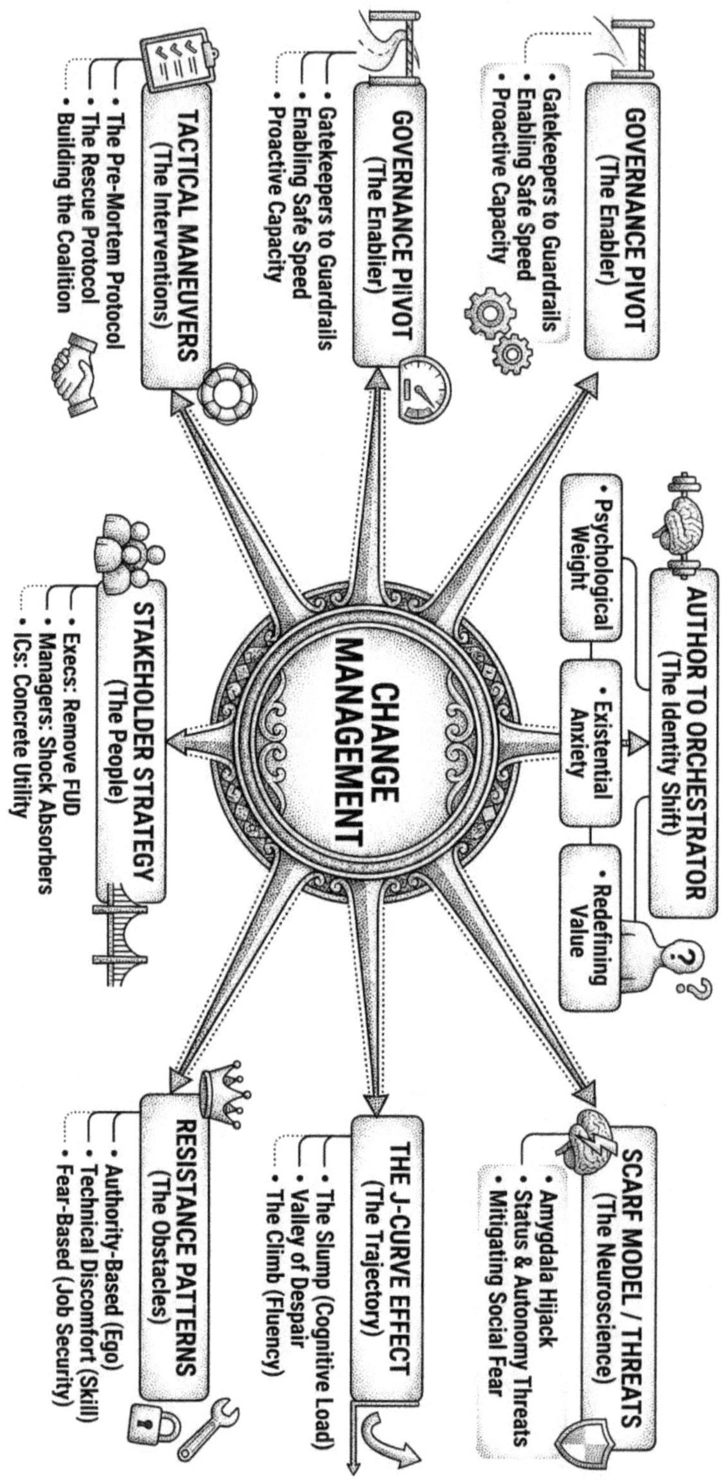

12

This comprehensive framework defines a quantitative model for evaluating organizational readiness across eight capabilities, ranging from governance to integration depth. It guides leaders to replace vanity metrics with rigorous data triangulation for honest benchmarking. Readers learn to sequence investments strategically, ensuring that speed and safety grow in balance to prevent structural failure during their transformation journey.

CHAPTER 12: MATURITY ASSESSMENT FRAMEWORK

Consider two organizations embarking on an AI-native agile transformation simultaneously.

The first organization, a logistics giant we will call "LogiCorp," charges forward. They purchase two thousand GitHub Copilot licenses, mandate a "prompt engineering" workshop, and declare their development teams "AI-Enabled." Their internal newsletters celebrate the modernization. Yet, six months later, their defect rate has climbed by 40 percent, their legal team is blocking releases due to murky IP provenance, and their most senior engineers are leaving because the noise-to-signal ratio in their codebases has become untenable.

The second organization, a mid-sized fintech firm we will call "FinStream," takes a slower, colder look at reality. They do not start with licenses; they start with a diagnostic. They systematically evaluate their current state across governance, testing automation, and cultural readiness. They identify a critical gap: their automated testing suite covers only 60 percent of their legacy code, a dangerous vulnerability if they introduce high-velocity AI code generation. They spend the first quarter closing that gap before rolling out generative agents.

Twelve months later, LogiCorp is conducting a "reset," rolling back permissions and explaining a data leak to their board. FinStream demonstrates a 30 percent reduction in cycle time and a fully compliant, AI-augmented workflow.

The difference is not ambition. It is structural maturity.

Capability development follows predictable physics. Whether implementing Six Sigma in manufacturing or DevSecOps in cloud infrastructure, organizations advance through recognizable stages. They move from heroic chaos to inconsistent standardization, then to optimized predictability, and finally to generative innovation.

Traditional agile maturity models fail in the current era because they measure the wrong constraints. They focus on sprint cadence, standup attendance, and velocity stability. But as we established in Chapter 1, AI shatters the constraints of coding speed. A team can be "mature" in traditional Scrum—holding perfect ceremonies and burning down story points reliably—while creating a disaster of unverified, hallucinated code.

This chapter presents the **AI/Agile Maturity Model**. It integrates the mechanical discipline of software delivery with the probabilistic complexity of AI. It provides the rubric for determining where you are, where you are going, and the specific investments required to bridge that gap.

WHAT THIS CHAPTER COVERS

- **The Five-Level Progression:** Moving from "Initial" chaos to "Transformative" industry leadership.
- **The Eight Capability Dimensions:** A detailed breakdown of the specific vectors of change, identifying the "Catalyst Actions" required to move up.
- **The Assessment Workshop Protocol:** A facilitator's guide to running the evaluation session without succumbing to groupthink.
- **Strategic Roadmapping:** How to sequence investments based on industry context (e.g., why banks prioritize governance while startups prioritize integration).

- **Success Metrics:** Leading and lagging indicators that prove maturity is actually improving.

KEY TERMS AND CONCEPTS

- **Maturity Level:** A qualitative shift in organizational capability, not merely an incremental improvement. Moving from Level 2 to Level 3 requires a fundamental change in *how* work is done, not just *how fast* it is done.
- **Capability Dimension:** A vertical slice of the organization (e.g., Observability) requiring independent assessment. High maturity in one dimension does not imply high maturity in others.
- **The Capability Gap:** The delta between current state and strategic necessity. This gap defines the investment roadmap.
- **Balanced Growth:** A transformation strategy that advances checking capabilities (Governance, Validation) in lockstep with generating capabilities (AI Integration, Process Speed).
- **Institutional Amnesia:** A Level 1 Knowledge Management state where insights reside exclusively in individual brains, rendering the organization incapable of learning from AI interactions.
- **Cowboy Agile:** A dysfunctional delivery pattern where teams use the flexibility of Agile as an excuse to bypass essential discipline. Characterized by the systematic omission of feedback loops (specifically Retrospectives) and the truncation of planning rigor under the guise of "efficiency." While often justified by the claim that "Scrum is just a framework," this approach prioritizes immediate output over sustainable velocity. In an AI context, Cowboy Agile is catastrophic, as it accelerates the creation of AI Debt without the safety mechanisms required to detect it.

MATURITY MODEL STRUCTURE AND PHILOSOPHY

The AI/Agile Maturity Model defines five progressive levels across eight capability dimensions. It acknowledges that transformation is inherently non-linear. Organizations operate at different levels simultaneously. A specific team might demonstrate Level 4 AI Integration depth, utilizing agentic workflows, while operating at Level 2 Governance, bypassing risk checks. This unevenness creates "structural tension," a precursor to failure.

Understanding these levels allows leaders to diagnose that tension before it snaps.

Figure 12.1: The Dunning-Kruger Gap. *Organizations often mistake the purchase of tools for the maturity of capability, creating a dangerous blind spot in their transformation strategy.*

Level 1: Initial (Ad Hoc and Heroic)

At the initial level, capabilities are ad hoc, inconsistent, and dependent on individual heroics. Success is a person, not a process.

In an AI context, Level 1 is characterized by "Shadow AI." Developers use personal ChatGPT accounts to generate code. There are no guidelines, no approved tools, and no shared prompts. One engineer might achieve a 50 percent productivity gain, while the person next to them struggles to debug hallucinated libraries.

Organizations at this level often suffer from the Dunning-Kruger effect. They believe they are "using AI" because they see the output. They do not see the massive, unmanaged risk accumulating in their codebase. When the "hero" developer leaves, the capability leaves with them.

Level 2: Developing (Establishing Consistency)

Level 2 represents the move from individual art to team habit. Basic processes establish consistency within teams. The organization purchases enterprise licenses. It publishes a policy on acceptable use.

However, implementation varies wildly. Some teams adopt the tools rigorously; others ignore them. The "Tower of Babel" effect is common here: different teams use different prompts, different validation standards, and different definitions of "done." The organization recognizes the need for a system but has not yet enforced it.

Consider a mid-sized software company that deploys GitHub Copilot. Some teams achieve 30–40% productivity gains, while others see minimal benefit. Leadership investigates and discovers that the high-performing teams have developed prompt engineering expertise, while the low performers treat AI as a glorified spell-checker. This variance is the hallmark of Level 2.

Level 3: Operational (Achieving Standardization)

This is the tipping point. At Level 3, the organization moves from "guidelines" to "guardrails." Governance is no longer a document; it is a checkpoint in the workflow. AI integration is systematic.

Success is organizational. New hires onboard into a system that makes them effective immediately. Outcome metrics replace velocity metrics. The organization has established a "baseline of competence" that exists independent of any single employee. This is the minimum viable state for safe enterprise AI adoption.

A Level 3 organization has formalized the *AI Collaboration Log*. They do not just talk about human-AI partnership; they record it, analyze it, and improve it.

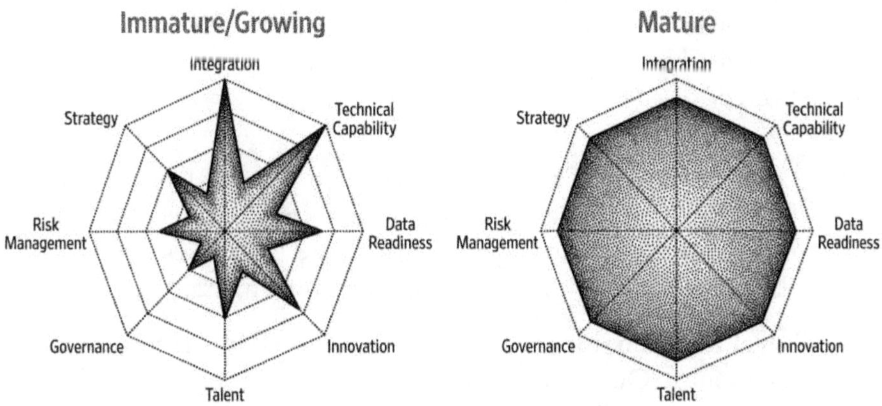

Figure 12.2: The Structural Tension of Immaturity. *An organization that spikes in AI Integration but lags in Governance (the jagged shape) is technically capable of generating code faster than it can safely deploy it.*

Level 4: Optimizing (Continuous Improvement)

Level 4 shifts the focus from *execution* to *optimization*. The organization uses data to refine its own processes.

At Level 3, a team uses an AI Collaboration Log. At Level 4, the organization aggregates data from *all* collaboration logs to identify which prompt patterns yield the lowest defect rates, then pushes those patterns back to the teams. Governance shifts from a gate to a flow, utilizing predictive risk analysis. The system learns from itself.

Level 5: Transformative (Industry Leadership)

Level 5 is the domain of market leaders. Here, AI-native practices become competitive differentiators. The organization does not just use tools; it builds proprietary capabilities. They shape the regulatory environment. They attract top talent specifically because their way of working allows engineers to focus entirely on high-value creative problem solving.

THE EIGHT CAPABILITY DIMENSIONS: A DEEP DIVE

An overall maturity score is meaningless. A doctor does not give a patient a single "Health Score"; they measure blood pressure, heart rate, and cholesterol separately. Similarly, we assess eight distinct dimensions. Organizations must evaluate each independently to identify the "structural tension" described earlier.

For each dimension below, we identify the **Catalyst Action**—the specific intervention required to bridge the difficult gap from Level 2 to Level 3.

Capability Dimension 1: Agile Process Maturity

This dimension assesses the shift from "feature factory" mechanics to outcome-oriented delivery. It asks: Are we optimizing for tickets closed, or value delivered?

- **Level 1 (Process Chaos):** Teams cling to Waterfall or chaotic "Cowboy Agile." Velocity is untracked or manipulated. Some teams abandon velocity tracking entirely because "AI makes metrics obsolete," but fail to replace it with outcome metrics.
- **Level 2 (Mechanical Execution):** Teams adopt AI Scrum ceremonies (Outcome Alignment, AI Show & Tell) but perform them mechanically. They utilize the *forms* of the new method but still think in terms of output volume. They are "doing" AI Scrum, not "being" AI Scrum.
- **Level 3 (Systematic Implementation):** The "Outcome Over Output" principle takes root. Outcome Alignment Meetings replace Sprint Planning. Teams refuse to start work without clear validation criteria. The AI Collaboration Log is a standard artifact.

- **Level 4 (Process Optimization):** Process bottlenecks are identified via data. The organization measures the "Concept-to-Cash" cycle time and uses AI to compress specific phases, such as requirements gathering.
- **Level 5 (Methodology Innovation):** The organization contributes to the industry standard. They invent new ceremonies or roles.

The False Positive: A team that closes 100% of their Jira tickets every sprint but has a "Return to Backlog" rate of 25%. They are efficient at closing tickets, not at solving problems. **The Catalyst Action (L2 -> L3): Implement "Outcome Gates."** Configure Jira (or your tracking tool) to prevent a ticket from moving to "In Progress" unless the "Definition of Intent" field is populated and verified by a second human. This forces the shift from activity to value.

Capability Dimension 2: AI Integration Depth

This measures how deeply AI is woven into the actual work, beyond simple code completion.

- **Level 1 (Experimental Adoption):** AI is a curiosity. Usage is sporadic and limited to simple Q&A. Engineers paste code from ChatGPT manually.
- **Level 2 (Tool Deployment):** Tools like GitHub Copilot or Cursor are deployed, but usage is effectively "fancy autocomplete." Engineers lack prompt engineering skills.
- **Level 3 (Systematic Integration):** AI is a teammate. Teams use "Context Specifications" to prime models. RAG is used to ground AI in the company's codebase.
- **Level 4 (Partnership Optimization): AI Alignment Engineers** optimize the interface. Custom agents handle specific parts of the workflow (e.g., a dedicated "Refactoring Agent").
- **Level 5 (Partnership Revolution):** The organization builds proprietary models or fine-tunes open weights models on their own data.

The False Positive: High "acceptance rate" of AI suggestions in the IDE. This often means the AI is suggesting trivial boilerplate (L2) rather than solving complex logic (L3). **The Catalyst Action (L2 -> L3): Mandatory Context Specs.** Require that every complex task begins with a written

"Context Specification"—a paragraph explicitly describing the architectural constraints and intent—which is then fed to the AI. This moves usage from "guessing" to "instructing."

Figure 12.3: The Integration Shift. *At Level 1 and 2, AI is merely a toy or a tool. At Level 3, it occupies the "second chair," requiring communication protocols just like a human pair programmer.*

Capability Dimension 3: Governance and Risk Management

This dimension evaluates the transition from "Department of No" to "Automated Guardrails."

- **Level 1 (Governance Vacuum):** "Move fast and break things." No formal policy. Risks are invisible until they manifest as incidents.
- **Level 2 (Policy Establishment):** Policies are written but live in PDFs. Compliance is reactive. Teams view governance as a tax on speed.
- **Level 3 (Systematic Governance):** Governance is integrated into the sprint. The **Governance Coordinator** role is active. An AI Governance Board reviews high-risk initiatives *before* code is written.
- **Level 4 (Predictive Governance):** Governance is codified. Policy-as-Code checks run in the CI/CD pipeline, automatically rejecting non-compliant artifacts.

- **Level 5 (Strategic Leadership):** Governance enables speed. Robust safety systems allow the organization to take calculated risks.

The False Positive: A "clean" audit report that was generated by manually creating documentation *after* the project was finished. This is "Governance Theatre." **The Catalyst Action (L2 -> L3): The Governance Coordinator Role.** Assign a specific human in the Scrum Team the explicit role of Governance Coordinator (rotating or permanent). Their job is to flag IP and data risks during the Daily Scrum, bringing compliance into the daily rhythm.

Capability Dimension 4: Validation and Quality Engineering

This is the most critical dimension for avoiding the "Velocity Trap."

- **Level 1 (Manual Chaos):** Manual testing is the norm. AI generates code faster than humans can test it.
- **Level 2 (Basic Automation):** Unit tests are automated, but integration testing remains slow. The bottleneck shifts from coding to testing.
- **Level 3 (Comprehensive Validation):** "Validation-First Development." Acceptance criteria are machine-readable. AI generates test cases before it generates code.
- **Level 4 (Predictive Quality):** "Red Teaming" is continuous. AI agents actively try to break the system in lower environments.
- **Level 5 (Quality Innovation):** Zero-defect production is the expectation. Self-healing systems identify and patch faults.

The False Positive: High unit test coverage (e.g., 90%) where the tests assert trivialities (e.g., assert true == true) or were generated by the AI to match its own hallucinated code. **The Catalyst Action (L2 -> L3): Inverted Workflow (Test-Then-Code).** Enforce a strict policy where the *test file* must be committed to the repo before the *implementation file*. The CI pipeline should fail any commit that introduces code without a preceding test.

Figure 12.4: The Detection Efficiency Gap. *Manual testing is digging through haystacks; Level 3 validation is turning on the magnet to catch defects instantly.*

Capability Dimension 5: Observability and Telemetry

You cannot manage what you cannot see.

- **Level 1 (Minimal Visibility):** Logs are scattered. Troubleshooting is an archeological dig.
- **Level 2 (Basic Monitoring):** Standard dashboards exist. Alert fatigue is high. Teams see *that* something broke, but not *why* the AI broke it.
- **Level 3 (Comprehensive Observability):** Distributed tracing allows teams to see the flow of data through AI components. Outcome metrics are tracked.
- **Level 4 (Predictive Operations):** AIOps tools predict failures. Anomaly detection identifies model drift.
- **Level 5 (Competitive Intelligence):** Telemetry is a strategic asset driving product strategy.

The False Positive: A "Green Dashboard" that tracks infrastructure uptime (Server is up) while the AI application is silently returning 0 results to users. **The Catalyst Action (L2 -> L3): Chain of Thought Logging.** Configure your logging framework to capture not just the input and output of the AI, but the *retrieval context* (what RAG chunks were used) and the *system prompt version.* This turns "magic" errors into debuggable logic.

Capability Dimension 6: Knowledge Management

This evaluates whether the organization learns or merely repeats.

- **Level 1 (Institutional Amnesia):** Knowledge lives in Slack and DMs. If an engineer leaves, the team lobotomizes itself.
- **Level 2 (Basic Documentation):** A stale wiki exists. "Write-only" memory.
- **Level 3 (Systematic Capture):** AI summarizes meetings and updates documentation. Knowledge graphs enable semantic search.
- **Level 4 (Personalized Learning):** A Vector Database acts as a "Company Brain." Learning is personalized.
- **Level 5 (Knowledge Excellence):** Insights are instantly indexable and discoverable globally.

The False Positive: A massive Confluence space with 10,000 pages, none of which have been updated in 2025. This is data, not knowledge. **The Catalyst Action (L2 -> L3): The Vector Pipeline.** Implement an automated pipeline that ingests "Decisions" (e.g., Architecture Decision Records) into a vector database, making them queryable by the AI. This allows developers to ask, "Why did we choose Postgres over Mongo?" and get a cited answer.

Figure 12.5: From Amnesia to Intelligence. *Level 3 Knowledge Management turns documentation from a chore into a queryable asset, moving from "piles of files" to a connected graph.*

Capability Dimension 7: Organizational Culture

Culture is the operating system of the enterprise.

- **Level 1 (Cultural Resistance):** Fear of replacement. "Blame culture."
- **Level 2 (Cultural Acceptance):** Skeptical curiosity. Management encourages AI, but psychological safety is low.
- **Level 3 (Cultural Embrace):** High psychological safety. Teams discuss hallucinations without fear.
- **Level 4 (Cultural Advantage):** Radical candor. Experimentation is a duty.
- **Level 5 (Cultural Leadership):** The culture is a recruiting magnet.

The False Positive: An "AI Hackathon" where everyone smiles, but on Monday morning, engineers are reprimanded for spending time on "non-roadmap" items. **The Catalyst Action (L2 -> L3): The "Best Failure" Award.** Publicly reward a team that tried an AI experiment that *failed*, provided they documented the learning. This proves that safety is real.

Capability Dimension 8: Strategic Alignment

- **Level 1 (Strategic Disconnect):** Leadership demands "AI" but provides no budget.
- **Level 2 (Strategic Acknowledgment):** Pilot projects are funded but isolated.
- **Level 3 (Strategic Integration):** AI transformation is a top-3 objective. Infrastructure is funded.
- **Level 4 (Strategic Advantage):** Business model evolves based on AI.
- **Level 5 (Strategic Identity):** Organization redefines itself around AI.

The False Positive: A CEO who talks about AI in earnings calls but cuts the training budget for engineering. **The Catalyst Action (L2 -> L3): The Unified Budget.** Move AI costs from "Innovation Discretionary" to "Core Operational Expense." This signals that AI is not a hobby; it is how we run the business.

THE ASSESSMENT WORKSHOP PROTOCOL

Organizations often treat assessment as a survey: "Rate yourself 1 to 5." This fails because Level 1 teams often rate themselves as Level 4 because they do not know what good looks like. The self-serving bias blinds them to their own incompetence.

You cannot just send out a survey. You must convene a **Maturity Assessment Workshop**.

Figure 12.6: The Calibration Debate. *The value of the assessment lies in the debate, not the number. Discrepancies in voting reveal the hidden friction in the organization.*

Preparation (Pre-Work)

Do not walk into the room empty-handed. The facilitator (usually the Transformation Lead or an external consultant) must prepare the "Evidence Pack."

- **Pull the Data:** * Cycle Time trends (last 3 months).
    - Defect Density (production bugs vs. lines of code).
    - Test Coverage reports.
    - Change Failure Rate.

- **Pull the Artifacts:**
    - Three random Jira tickets (to check for outcome definitions).
    - Three random Pull Requests (to check for AI-generated code reviews).
    - The last three Retrospective notes.

The Workshop Agenda (4 Hours)

The agenda for the workshop is broken into four distinct portions:

Part 1: The Evidence Walk (60 Minutes)

Display the Evidence Pack on the screen. Do not discuss maturity scores yet. Just agree on the reality.

- *Facilitator:* "Our data shows test coverage dropped from 80% to 65% this quarter. Do we agree this is accurate?"
- *Goal:* Bust the delusion that "everything is fine."

Figure 12.7: Triangulated Assessment. *A score based only on opinion (Consensus) will collapse under pressure. A stable assessment requires the support of Data and Artifacts.*

Part 2: Dimension Scoring (90 Minutes)

Go through the 8 Dimensions one by one. Use "Planning Poker" cards (1–5) to vote anonymously, then reveal.

- *Scenario:* The Lead Dev votes "4" on Quality. The QA Lead votes "2."
- *Facilitator:* "QA Lead, why a 2? Lead Dev, why a 4?"
- *Discussion:* The gap between the votes is where the truth lies. The Lead Dev sees the speed of code generation (4); the QA Lead sees the bugs in the backlog (2).
- *Rule:* You cannot score a Level 3 or higher without pointing to a specific URL or document that proves the system exists.

Part 3: The Gap Analysis (60 Minutes)

Compare the Current State scores to the Target State (defined by strategy).

- *Facilitator:* "We are a Level 2 in Governance, but our Strategy requires Level 4 because we are handling medical data. This is a Critical Gap."

Part 4: Action Planning (30 Minutes)

Define the "Catalyst Actions" for the next quarter. Do not try to fix everything. Pick the top 3 gaps.

The Triangulation Method

To prevent "Grade Inflation," use the **Triangulation** rule. A score is only valid if it is supported by three legs:

1. **Tool Data:** What does the system say? (Hard metrics)
2. **Artifact Evidence:** What can you show me? (Documents, links)
3. **Consensus:** Do the Product Owner, Scrum Master, and Lead Developer agree?

COMMON PROGRESSION PATTERNS

Organizations should not attempt to move all eight sliders to the right simultaneously. That is a recipe for paralysis. Strategic context dictates the path.

Pattern A: The Governance-First Path (Regulated Industries)

Banks, healthcare providers, and defense contractors utilize this pattern.

- **Focus:** Drive Governance and Validation to Level 3 or 4 *before* pushing AI Integration past Level 2.
- **Logic:** The cost of a compliance failure (fines, jail, license revocation) exceeds the value of development speed.
- **Risk:** "Analysis Paralysis." The organization creates a perfect safety cage but never puts the tiger inside.

Pattern B: The Innovation-First Path (Startups/Tech)

SaaS companies and digital natives often choose this route.

- **Focus:** Push AI Integration and Agile Process to Level 4 rapidly. Governance trails at Level 2 (Minimum Viable Compliance).
- **Logic:** Speed is survival. They must capture market share before competitors.
- **Risk:** "Technical Bankruptcy." They build so fast they create a fragile, unmaintainable mess that collapses under scale or regulatory scrutiny.

Pattern C: The Balanced-Growth Path (Large Enterprise)

This is the recommended path for most established firms (Global 2000).

- **Focus:** Move the "whole fleet" together. No dimension is allowed to lag more than one level behind the others.

- **Logic:** This prevents structural tension. It ensures that as speed increases, the brakes (governance/testing) are upgraded to match.
- **Risk:** Slow progress. It feels like moving through molasses because every advance requires coordination.

Figure 12.8: Strategic Vehicles. *Startups often choose the rocket (speed over safety), regulated industries require the armored transport (safety over speed), while mature enterprises build the train (speed on rails).*

STRATEGIC PLANNING: USING THE MODEL

The assessment is not the destination; it is the map. Once the organization knows its current state, it must plan the journey.

Step 1: Define the Target State

Do not aim for Level 5 in everything. Level 5 is expensive, resource-intensive, and often unnecessary for non-core functions. Perfection is the enemy of execution.

- **The Internal HR Portal:** Does this require Level 5 AI Innovation? Likely not. Level 3 (Systematic Integration) is sufficient to drive efficiency.
- **The Core Trading Algorithm:** Does this require Level 5 Governance? Absolutely. The cost of failure here is existential.

Strategic leadership defines a "Target Maturity Profile" that allocates resources to competitive differentiators while maintaining safety standards elsewhere.

Step 2: Identify the Constraints

Look for the "lagging" dimensions that create structural tension. The risk lies in the gap between capabilities.

- **The Blind Spot:** If AI Integration is at Level 3 but Observability is at Level 1, the organization is flying blind at high speed.
- **The Intervention:** In this scenario, the immediate strategic action is not to hire more AI engineers, but to **freeze AI rollout**. Investment shifts entirely to Observability until it reaches parity (Level 2 or 3).

Step 3: Sequence the Investments

Transformation costs money and political capital. You cannot boil the ocean.

- **Phase 1 (Foundation):** Focus on Governance Level 2 and Validation Level 2. You cannot build a skyscraper on quicksand.
- **Phase 2 (Adoption):** Focus on AI Integration Level 2 and Process Level 3. Get the teams working effectively.
- **Phase 3 (Optimization):** Focus on Observability Level 3 and Culture Level 3. Refine and scale the success.

Measuring Success: The ROI of Maturity

Proving the value of maturity requires shifting the conversation from "activity" (how much code did we write?) to "unit economics" (how much did it cost to deliver value safely?). When executive leadership demands the ROI of the transformation, the answer lies in the correlation between leading behavior and lagging results.

Effective leaders manage success by monitoring the **Leading Indicators** as the pulse of the transformation. For instance, a spike in **Knowledge Search Volume** (querying the vector database) signals that developers are checking their work against institutional history rather than guessing. Simultaneously, the **Test Automation Ratio** acts as the governor; if AI-generated code volume rises while test coverage flatlines, the leader knows AI Debt is accumulating. These metrics allow for course correction *during* the sprint, rather than after the failure.

This discipline produces the **Lagging Indicators** that validate the strategy. The ultimate proof of AI maturity is not speed, but stability at speed. A plummeting **Change Failure Rate** combined with a shrinking **Mean Time to Restore (MTTR)** demonstrates that the organization has decoupled velocity from risk.

Tying It Together: Consider a scenario where a competitor releases a feature that forces you to pivot. A low-maturity organization freezes, fearful that rapid changes will break their fragile code. A high-maturity organization, one that prioritized Validation and Governance in Phase 1, can unleash its AI agents to refactor code and generate new features aggressively. They move fast *because* they trust their brakes. The ROI of maturity is the ability to survive and exploit market volatility that breaks less mature competitors.

Figure 12.9: The Transformation Bridge. *Optimization requires the pillars of Foundation and Adoption. Attempting to skip phases usually results in a collapse of the initiative.*

FROM THE TRENCHES: MATURITY ASSESSMENT BEST PRACTICES

Organizations implementing maturity assessment frameworks often struggle not with the logic of the model, but with the psychology of the assessment. Based on the patterns observed in successful versus unsuccessful transformations, there are five critical practices that distinguish effective assessments.

1. Start with an Honest Baseline Assessment The most common point of failure is "vanity scoring." Organizations often inflate their initial maturity scores to satisfy executive expectations or to disguise prior stagnation. This sets the transformation up for failure. An accurate baseline, however uncomfortable, is the only foundation that enables realistic planning. Leaders must accept that a "Level 1" score is not an indictment of their past performance, but an accurate map for their future investment. Self-deception serves no strategic purpose.

2. Match Assessment Rigor to Transformation Stage One size does not fit all. Early-stage transformations often benefit from lightweight, quarterly assessments that track rapid capability development and adoption rates. Conversely, mature organizations (Level 3+) require more rigorous, annual assessments to support complex strategic investment planning. The administrative overhead of the assessment should scale in proportion to the transformation's maturity and strategic importance.

3. Use Maturity Gaps to Prioritize, Not Punish Teams naturally fear that a low maturity score will be weaponized against them. If an organization treats "Level 1" as a failure to be punished rather than a gap to be closed, teams will game the metrics and hide their true state. The model exists to guide investment, not to judge performance. Leadership must explicitly frame the assessment as a tool for securing the resources teams need to improve, not for assigning blame.

4. Recognize That Regression Is Normal Maturity is not a straight line. When organizations undergo significant restructuring, leadership transitions, or major technology platform shifts, maturity scores often regress temporarily. This "J-curve" effect represents normal adaptation, not transformation failure. The key is to recognize regression early as a signal to reinvest in foundational capabilities, rather than panicking.

5. Balance Quantitative Metrics with Qualitative Insights Neither data nor opinion is sufficient on its own. Tool-based assessments provide objective data (e.g., test coverage, cycle time) but miss critical cultural nuances like psychological safety. Self-assessments capture the cultural context but often lack objectivity. The most effective assessments combine these approaches, using quantitative metrics to anchor the discussion in reality while using qualitative insights to explain the *why* behind the numbers.

WHAT COMES NEXT

With a clear understanding of *where* you are and *where* you need to go, the question becomes *how* to execute the move.

The following chapters provide the playbook. **Chapter 13** outlines the high-level Transformation Roadmap to help you with implementing AI Scrum. It blends a specific playbook for each organization type; experienced or greenfield.

- Small organization running a single traditional Scrum team migrating to AI Scrum
- Medium to large organization with multiple traditional Scrum teams (or variants) migrating to AI Scrum

- Small organization without agile experience, possibly running waterfall projects, wanting to adopt AI Scrum (greenfield organizations)
- Medium to large organization with multiple project teams, possibly running waterfall projects, wanting to adopt AI Scrum (greenfield organizations)

KEY TAKEAWAYS

- **Activity ≠ Capability:** Owning a tool is not the same as having the capability to use it. Assess outcomes, not purchases.
- **The Chain is Weakest at the Link:** Your overall risk is determined by your lowest maturity dimension. High AI speed with low Governance is a liability, not an asset.
- **Context is King:** Regulated industries must prioritize Governance maturity; innovation industries must prioritize Integration maturity. Blindly copying another company's roadmap is dangerous.
- **Evidence Over Opinion:** Use the **Triangulation Method** (Tools + Artifacts + Consensus) to anchor your assessment in reality.
- **The Catalyst Actions:** Focus on the specific interventions (Outcome Gates, Context Specs, Governance Coordinators) that force the jump from Level 2 to Level 3.

MIND MAP – MATURITY ASSESSMENT FRAMEWORK

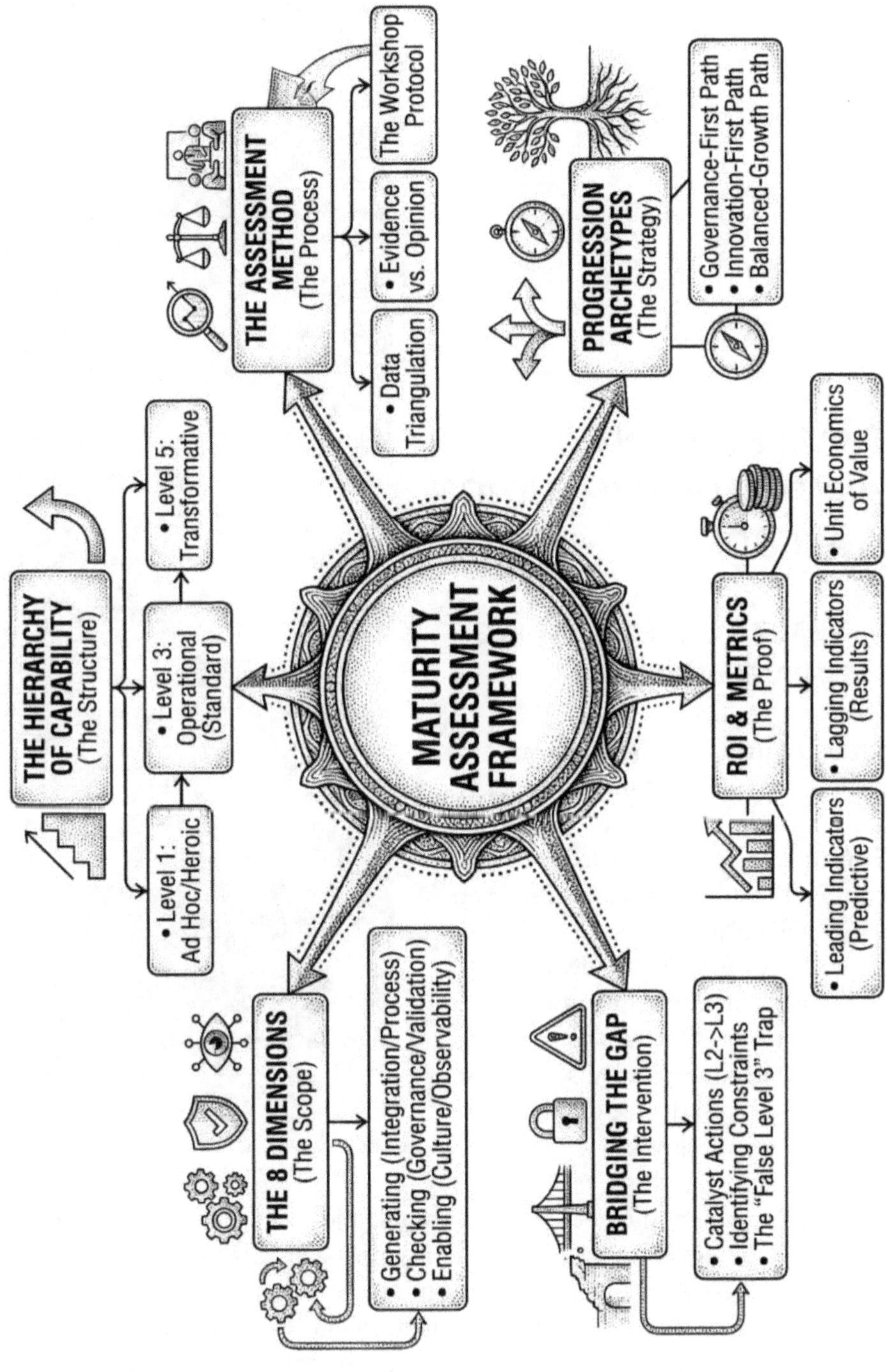

CHAPTER 13: THE AI SCRUM IMPLEMENTATION PLAYBOOK

This implementation playbook guides organizations through AI Scrum transformation across four scenarios: single or multiple teams, agile or non-agile starting points. It covers the nine-month preparation period, sprint-by-sprint execution through the J-curve productivity dip and recovery, budget models, risk management strategies, and timeline expectations ranging from nine to eighteen months post-preparation to achieve sustained excellence at 115%+ productivity.

——PART I: UNDERSTANDING THE FRAMEWORK——

The executive asks the same question in four different organizations across three months. "When can we start?" The answer depends entirely on context the executive cannot see yet. A single agile team migrating to AI Scrum faces fundamentally different constraints than multiple non-agile teams adopting the framework from scratch. Timelines, budgets, and risks all differ in both degree and kind across these scenarios.

Most transformation guides present a single pathway and expect readers to adapt the content to their situation. This approach fails because the differences between scenarios are not matters of degree but matters of kind. A single team makes autonomous decisions about timing, tooling, and process changes without cross-team coordination overhead. An agile team already possesses muscle memory for iterative development that non-agile teams must build from zero. Treating these as minor variations leads to failed transformations.

This chapter takes a different approach. We present one unified framework applicable to all implementation scenarios while explicitly calling out where your path diverges based on two critical dimensions. The first dimension is team count. Single teams move faster through transformation because they coordinate only with themselves. Multiple teams require wave-based rollouts and cross-team governance that single teams skip entirely. The second dimension is starting methodology. Agile teams already understand sprints, backlogs, and retrospectives. Non-agile teams must simultaneously learn agile principles and AI augmentation, creating a steeper learning curve.

From these dimensions, four distinct implementation vectors emerge from these dimensions. Scenario A covers single traditional agile Scrum teams migrating to AI Scrum. Scenario B addresses multiple traditional agile Scrum teams making the same migration. Scenario C involves single non-agile project teams adopting AI Scrum as their first agile framework. Scenario D handles multiple non-agile project teams moving to AI Scrum. Each scenario shares the same fundamental framework but diverges at specific decision points throughout the transformation journey.

The chapter begins with universal concepts including transformation philosophy, team architecture, infrastructure requirements, and the nine-month preparation period. These foundations apply across all scenarios. Later sections provide scenario specific guidance on budgets, timelines, stakeholder management, and backlog preparation. The chapter concludes with success measurement frameworks and common pitfalls organized by scenario type. When content applies universally, we present it once without qualification. When approaches diverge, we use clear labels like [Scenario A] or [B+D] to indicate which path you should follow.

The scenario-specific approach in this chapter addresses three critical transformation failures. First, teams waste months pursuing the wrong preparation activities because generic guides do not distinguish between scenarios. A single agile team does not need the wave planning that multi-team organizations require. Second, budget justifications fail because executives cannot connect generic cost estimates to their specific situation. Scenario C teams need larger training budgets than Scenario A teams. Third, stakeholder management strategies that work for one scenario actively harm another. The J-curve explanation that reassures stakeholders of a single team terrifies stakeholders of a multi-team organization who cannot afford organization-wide velocity drops.

This chapter eliminates that ambiguity. You will know exactly which sections apply to your scenario. You will build a timeline specific to your starting conditions. You will create a budget model that reflects your actual transformation costs. Most importantly, you will understand why certain decisions matter for your scenario even when they seem irrelevant to others.

Figure 13.1: The end-to-end process of implementation. *No matter which scenario your group falls under (A, B, C, or D), the process will be similar, although it may take longer due to unlearning old processes and mastering new ones. Each group is different so reaching sustained excellence could take a little shorter or longer than suggested.*

Before diving into the framework, we should clarify what this chapter deliberately excludes to avoid unrealistic scope expectations. A note on what this chapter deliberately excludes. Role evolution receives minimal coverage here because Chapter 4 already provides thorough treatment of how traditional roles (developers, testers, architects) transform into AI Scrum roles (AI Collaborator, Validation Lead, Prompt Engineer). We reference how Product Owners become Product Strategists and Scrum Masters evolve into Delivery Facilitators when relevant but do not repeat the content. Similarly, we focus on AI augmentation through large language models creating code, documentation, user stories, and presentations. More advanced implementations using AI agents and workflow automation tools like n8n are certainly possible, but we cannot provide expert guidance on every AI capability. Our focus remains on core

LLM-based augmentation that delivers consistent productivity improvements of 115% across all scenarios.

The transformation you are about to undertake is not a pilot program. It is not an experiment. It is a complete methodology migration that will fundamentally change how your teams work. Consider the timeline carefully: the preparation period alone spans nine months regardless of scenario. The path to sustained excellence ranges from nine to eighteen months depending on your starting point. This represents a significant organizational commitment.

Despite this investment, the payoff justifies the commitment. Based on observed implementations, teams completing this transformation typically achieve 115% of their previous baseline productivity—a 15% improvement—while maintaining or improving quality. Exceptional teams occasionally reach 120-125% productivity in optimal conditions. However, 115% represents a realistic sustainable outcome for most disciplined implementations. Teams deliver features faster and handle complexity with significantly less friction than before. They adapt to changing requirements with less friction. These outcomes are not theoretical. They are measurable and repeatable when organizations follow the framework presented in this chapter.

Your scenario determines your specific path, but the destination remains constant. You are building an elite AI-augmented team capable of significantly outperforming traditional teams. The question is not whether to make this transformation but how to execute it successfully for your specific starting conditions.

IMPLEMENTATION VECTOR FRAMEWORK

The implementation vector framework rests on two binary dimensions that create four distinct transformation pathways. Understanding where your organization sits on these dimensions determines every subsequent decision in your transformation journey. Getting this classification wrong means following guidance that does not apply to your situation, which inevitably leads to failed implementations.

The Two Critical Dimensions

Team Configuration determines coordination complexity. Single teams operate autonomously with complete control over transformation timing and decisions. Multiple teams require wave-based rollouts and cross-team governance to prevent organization-wide disruption.

Starting Methodology determines learning curve steepness. Agile teams augment existing practices with AI capabilities. Non-agile teams face dual transformation learning both agile principles and AI integration simultaneously.

The Four Implementation Scenarios

The four scenarios emerge from combining team configuration and starting methodology dimensions. Each scenario has distinct characteristics that determine transformation complexity and timeline.

Scenario	Team Count	Starting Method	Primary Challenge	Timeline to Excellence	Wave Approach
A	Single	Agile	AI integration only	9-12 months	Not needed
B	Multiple	Agile	Coordination overhead	12-18 months	Mandatory
C	Single	Non-Agile	Dual learning curve	9-12 months	Not needed
D	Multiple	Non-Agile	Dual learning + coordination	12-18 months	Mandatory

- **Scenario A** teams focus purely on AI integration without organizational complexity or methodology learning. This represents the fastest, lowest-risk path.
- **Scenario B** teams coordinate governance frameworks, infrastructure decisions, and rollout timing across multiple teams while building on existing agile foundation.
- **Scenario C** teams face dual learning of agile fundamentals and AI augmentation but operate autonomously without cross-team dependencies.
- **Scenario D** teams navigate the most complex transformation combining dual learning curves with organizational coordination and wave-based risk management.

Decision Framework

Answer three critical questions to determine your path:

- **Need wave planning?** Yes for B and D (multiple teams). No for A and C (single team).
- **Need agile training?** Yes for C and D (non-agile). No for A and B (already agile).

- **Timeline expectation?** 9-12 months for A and C. 12-18 months for B and D.

When organizations skip this classification, they and jump into transformation activities invariably fail. Scenario D organizations following Scenario A guidance launch all teams simultaneously, overwhelming infrastructure and forcing abandonment. Scenario A teams implementing unnecessary wave planning waste months on coordination overhead that provides no value for autonomous teams.

WHY THIS PLAYBOOK SKIPS EXPERIMENTATION

AI Scrum implementations do not use pilot programs. This generally applies across all four scenarios regardless of team count or starting methodology. The decision to skip pilots is not a shortcut or an oversight. It reflects fundamental differences between AI Scrum and traditional software transformations that make pilots actively counterproductive.

Historically, traditional transformations use pilots because the new methodology might not work. Organizations test whether Scrum improves velocity or whether continuous integration reduces defects. The pilot provides data that justifies broader adoption or reveals that the approach fails for the organization's specific context. This caution makes sense when the methodology's effectiveness remains uncertain.

However, AI Scrum does not face this uncertainty. The framework combines proven agile mechanics with AI augmentation that demonstrably accelerates work across software development contexts. The question is not whether AI Scrum works but whether your organization can execute the transformation successfully. That question cannot be answered by a pilot team.

The Pilot Paradox

In typical scenarios, pilot teams operate under artificial conditions that make their results irrelevant to full-scale implementation. The organization gives pilot teams extra support, removes bureaucratic obstacles, and protects them from standard organizational friction. Leadership pays unusual attention to the pilot's progress. Stakeholders grant flexibility they would never extend to regular teams. These conditions create temporary success that vanishes when the organization attempts broader rollout.

Additionally, the pilot creates a false choice between the old and new approaches. Teams watching the pilot can dismiss poor early results as evidence that AI Scrum does not work. They can also dismiss strong results as artifacts of the special treatment pilots receive. Either way, the pilot gives skeptics ammunition to resist transformation rather than building organizational commitment.

Perhaps more fundamentally, AI Scrum transformation is not reversible once started. Partial adoption of prompt engineering standards proves impossible, despite what teams initially believe or selectively use AI governance frameworks. The methodology requires complete integration into how teams think about work, manage context, and validate outputs. A pilot team either commits fully to the transformation or produces meaningless results. Once that commitment happens, reverting to traditional Scrum wastes the investment in training, infrastructure, and capability development.

The Wave Alternative

Organizations with multiple teams address risk through waves instead of pilots. Wave implementation differs from pilot programs in several ways in three critical ways. First, waves commit to permanent transformation rather than temporary experimentation. Wave 1 teams know they are pioneering the permanent approach for the entire organization. Second, waves receive the same support and constraints as subsequent waves. The organization does not create artificial success conditions. Third, waves inform subsequent implementation without creating an accept-reject decision point. The question is never whether to continue but how to improve execution for the next wave.

Scenario B and D organizations implement waves because multiple teams cannot transform simultaneously without overwhelming infrastructure and governance capacity. Wave 1 typically includes one team that runs for three to four sprints. This duration allows the team to work through the cognitive load spike and reach initial stabilization. Wave 2 launches at the beginning of sprint four or five with one to three additional teams. Wave 3 follows at sprint six or seven with another one to three teams. The staging prevents the organization from having all teams in the cognitive load spike simultaneously.

By contrast, single team scenarios skip waves entirely. Scenario A and C organizations have no coordination complexity to manage. The single team transforms completely during the nine-month preparation period and launches into AI Scrum on day one. There is no second team waiting to learn from experience or infrastructure to protect from overload. The transformation is binary: the team either executes AI Scrum or continues with traditional methods.

Making the Commitment

The no-pilot approach demands different organizational readiness. Executives must commit to transformation before seeing results from their specific teams in their specific context. This commitment requires understanding that AI Scrum works based on broader evidence rather than local proof. It requires trusting the preparation process to build necessary capabilities before sprint one begins. Most challenging, it requires accepting the J-curve where velocity drops before it rises.

Organizations unable to commit at this level should not attempt AI Scrum transformation. Importantly, the framework does not offer a low-risk experimentation path. It offers a high-return transformation that requires genuine organizational commitment and proper execution. Pilots cannot provide the confidence that commitment requires. Only leadership conviction combined with thorough preparation produces successful transformations.

The preceding sections provide the strategic context for this commitment decision. They explain why organizations across all four scenarios must commit fully to transformation rather than testing incrementally. This chapter assumes you have made that commitment and focuses on execution specifics for your scenario.

WHAT THIS CHAPTER COVERS

The following sections provide complete implementation guidance for transforming to AI Scrum across all four organizational scenarios. The content follows a universal-then-specific pattern that minimizes repetition while ensuring you receive actionable guidance for your exact situation.

Universal Framework and Infrastructure

The chapter begins with foundational elements that every organization needs regardless of scenario. These universal components include:

- AI Scrum constructs and terminology that apply across all scenarios
- The 11-person elite squad architecture that defines optimal team composition
- Risk management frameworks for operational continuity and psychological safety
- AI Sandbox technical infrastructure for experimentation and learning
- Governance Canvas and Checkpoint ceremony for maintaining quality and compliance

- Success measurement frameworks that track transformation progress
- The complete nine-month preparation period broken into three phases

Scenario-Specific Playbooks

After establishing universal foundations, the chapter provides detailed guidance tailored to each scenario. The scenario-specific content includes:

- Budget models with direct transformation costs, opportunity costs, and ROI timelines for each scenario
- Complete transformation timelines from preparation through sustained excellence
- Wave planning and rollout strategies for multi-team scenarios
- Stakeholder alignment approaches including J-curve education and scope amnesty protocols
- Backlog preparation requirements and work intake modifications
- Velocity recovery expectations and stabilization maneuvers
- Scenario-specific anti-patterns and failure modes

Success Measurement and Course Correction

The final section provides frameworks for evaluating transformation progress and correcting course when needed. The measurement and intervention tools include:

- Chapter 9 transformation metrics for tracking AI collaboration quality, validation effectiveness, and governance understanding
- The 4Cs Remediation Protocol for addressing capability gaps
- Continuous monitoring approaches and early intervention triggers
- Common pitfalls organized by scenario type
- Limitations and context dependencies that affect transformation outcomes

What This Chapter Does Not Cover

Role evolution receives minimal treatment because Chapter 4 provides full coverage of how Product Owners evolve into Product Strategists and Scrum Masters become Delivery Facilitators. We reference these transformations when relevant to specific scenarios but do not duplicate the detailed capability frameworks, responsibility shifts, and mindset changes already documented.

Beyond role evolution, this chapter also limits its technology scope to maintain focus on core transformation mechanics. Similarly, this chapter focuses on core LLM-based AI augmentation for creating code, documentation, user stories, and presentations. More advanced AI agent implementations and workflow automation tools like n8n represent possible extensions but fall outside our expertise scope. The guidance provided ensures you achieve the 115% productivity floor through proven LLM integration regardless of whether you later adopt additional AI capabilities.

Enterprise-wide scaling beyond ten teams requires additional considerations. Organizations scaling across dozens or hundreds of teams, build centers of excellence, manage knowledge transfer, and sustain transformation momentum. This chapter assumes you are transforming between one and ten teams maximum and focuses on getting those initial teams to sustained excellence.

How to Navigate This Chapter

The opening sections help you identify your scenario based on team configuration and starting methodology. The middle sections present universal content on transformation philosophy, team architecture, infrastructure, and the nine-month preparation period. Read these sections completely regardless of your scenario because they establish the foundation everything else builds upon.

Later sections provide scenario-specific guidance clearly marked with labels. When you see [All], the content applies to everyone. When you see [Scenario A] or [B+D], the content applies only to those specific scenarios. Read all universal content plus your scenario-specific sections. You may find value in scanning other scenarios to understand implementation differences, but focus your detailed reading on your own path.

The final sections cover success measurement frameworks and common pitfalls, then provide scenario-specific anti-patterns you should actively avoid. Read the universal sections completely and pay close attention to the pitfalls for your specific scenario.

The chapter concludes with key takeaways synthesizing the most critical success factors. It includes a mind map visualizing the complete transformation journey from preparation through sustained excellence.

This structure allows you to read the chapter linearly while clearly understanding which guidance applies to your situation. You will not waste time on irrelevant preparation activities or implement unnecessary governance overhead. You will build exactly the infrastructure, processes, and capabilities your scenario requires to achieve successful AI Scrum transformation.

KEY TERMS AND CONCEPTS

AI Scrum: An evolution of traditional Scrum that integrates large language model capabilities into every aspect of the development lifecycle. Teams use AI to accelerate requirements analysis, architectural design, code generation, testing, documentation, and stakeholder communication while maintaining human oversight and validation at all critical decision points. AI Scrum preserves Scrum's iterative structure, time-boxed sprints, and empirical process control while fundamentally changing how teams execute work within those constraints.

AI Sandbox: A controlled technical environment where team members experiment with AI capabilities, test prompts, validate outputs, and develop AI collaboration skills without risking production systems or violating governance policies. The sandbox provides access to approved LLM providers, maintains isolated data storage, logs all interactions for learning purposes, and allows team members to fail safely while building expertise. Organizations typically implement sandboxes during the preparation period and maintain them throughout transformation for ongoing experimentation.

Context-Ready Definition of Ready: An enhanced version of traditional Definition of Ready that ensures work items contain sufficient context for effective AI collaboration. Context-ready items include explicit acceptance criteria, relevant technical constraints, architectural decisions that limit solution space, examples of similar completed work, and any domain knowledge required for accurate AI output. This standard prevents teams from starting work that would require excessive prompt iteration or produce outputs misaligned with stakeholder expectations.

Digital Backlog Hoarding: The retention of massive quantities of aged backlog items to maintain "visibility" despite missing sponsors or obsolete value. This anti-pattern forces teams to expend transformation energy on forensic cleanup rather than future-focused outcome definition. It serves as the primary driver for extended velocity recovery timelines, often pushing baseline restoration beyond ten months.

Governance Canvas: A visual framework that maps AI usage decisions across work categories, risk levels, validation requirements, and approval authorities. The canvas helps teams quickly determine appropriate AI collaboration approaches for specific work types without requiring repeated governance discussions. Teams update the canvas quarterly based on capability growth and lessons learned, gradually relaxing restrictions as confidence and expertise increase.

Governance Checkpoint: A recurring ceremony held every two to four sprints where the team reviews AI usage patterns, validates governance compliance, shares learning across team members, and adjusts AI collaboration approaches based on outcomes. The checkpoint typically consumes thirty to sixty minutes and includes the entire team plus any organizational AI governance representatives. This ceremony prevents governance drift while ensuring rules evolve with team capabilities.

Institutional Memory System: A structured repository that captures team decisions, architectural patterns, coding standards, prompt templates, validation approaches, and lessons learned in machine-readable formats optimized for AI consumption. The system allows AI tools to maintain consistency with past decisions without requiring team members to manually provide context in every interaction. Effective institutional memory dramatically reduces prompt engineering effort and improves AI output quality over time.

J-Curve: The predictable productivity pattern where team velocity drops significantly during initial AI Scrum sprints before recovering and eventually exceeding previous baseline performance. The curve reflects cognitive load from learning new workflows, overhead from establishing governance practices, and time invested in capability building. Most teams experience the velocity trough during sprints two through four. They begin recovery by sprint five, reach baseline by sprints eleven through eighteen, and achieve sustained excellence beyond sprint nineteen.

Prompt Engineering: The practice of crafting effective instructions, context, and constraints that guide AI tools toward desired outputs. Effective prompt engineering includes specifying output format, providing relevant examples, establishing quality criteria, constraining solution approaches, and building on institutional memory. Teams develop prompt engineering as a core skill during preparation and continue refining approaches throughout transformation. Individual team members often specialize in prompt engineering for specific work types like testing, documentation, or user story creation.

Validation Lead: The evolved role formerly known as QA Engineer that focuses on validating both human-generated and AI-generated outputs. Validation Leads design test strategies verifying AI contributions meet quality standards. They develop automated validation catching common AI failure modes and establish confidence levels for different output categories. This role becomes critical during transformation because traditional testing approaches often miss AI-specific failure patterns like plausible but incorrect code or documentation that sounds authoritative but contains subtle inaccuracies.

Wave Implementation: A staged rollout approach for multi-team organizations where teams adopt AI Scrum in sequential cohorts rather than simultaneously. Wave 1 typically includes one team running three to four sprints before Wave 2 launches with one to three additional teams. Wave 3 follows two to four sprints later with the remaining teams. This staging prevents overwhelming organizational infrastructure, allows later waves to learn from earlier experiences, and ensures the organization never has all teams in the cognitive load trough simultaneously. Single team scenarios do not use waves.

Preparation Period (Months -9 to 0): The nine-month period before Sprint 1 of AI Scrum where organizations build the infrastructure, capabilities, and readiness required for successful transformation. This period divides into three phases covering strategic alignment and approval, technical infrastructure and governance framework development, and tactical readiness including deep immersion training. All four scenarios require the full nine-month preparation regardless of team count or starting methodology. Organizations that attempt to compress this timeline typically fail during the first month of AI Scrum implementation.

Cognitive Load Spike (Sprints 1-4): The initial transformation phase where teams experience significantly reduced velocity due to learning new workflows, establishing AI collaboration patterns, and building governance muscle memory. Teams typically operate at 40-60% of previous baseline velocity during this period. The spike is predictable, temporary, and necessary for building sustainable AI Scrum capabilities. Organizations that panic during the cognitive load spike and revert to traditional practices waste their entire transformation investment.

Recovery Climb (Sprints 5-10): The phase where teams begin applying learned patterns with increasing confidence and velocity starts recovering toward previous baseline. Teams reduce time spent on governance decisions, develop more effective prompts with less iteration, and build institutional memory that accelerates future work. Velocity typically reaches 70-85% of previous baseline by sprint eight. Teams maintaining discipline during this phase set themselves up for strong parity point performance.

Parity Point (Sprints 11-18): The phase where team velocity returns to previous baseline performance levels while completing work with AI augmentation that would have taken the same time using traditional approaches. Reaching parity represents a critical milestone because it proves the team can match previous productivity using the new methodology. Most teams reach parity between sprints eleven and fifteen.

Scenario C and D organizations sometimes extend into sprint eighteen due to dual learning curves.

Sustained Excellence (Sprint 19+): The phase where teams consistently exceed previous baseline velocity by 115% or more while maintaining quality standards. Teams at sustained excellence have fully internalized AI collaboration patterns, developed sophisticated institutional memory, and eliminated most governance overhead through automated validation approaches. The 115% threshold represents the expected sustainable outcome for successful transformations. Exceptional teams occasionally achieve 120-125% productivity in optimal conditions while simultaneously improving code quality, reducing defects, and enhancing documentation completeness.

Operational Continuity Risk: The probability that AI Scrum transformation disrupts critical business operations or prevents the organization from meeting existing commitments. Single team scenarios face lower operational continuity risk because disruption remains contained within one team's blast radius. Multi-team scenarios require wave implementation specifically to manage operational continuity risk by preventing simultaneous velocity drops across all teams. Organizations with zero tolerance for delivery disruption should not attempt AI Scrum transformation regardless of scenario.

Psychological Safety Risk: The probability that team members disengage, resist transformation, or leave the organization due to stress, uncertainty, or fear of obsolescence during AI Scrum adoption. Psychological safety risk increases during the cognitive load spike when team members question their own competence and worry about job security. Effective risk management includes transparent J-curve communication, explicit job security assurances, learning milestone celebration, and safe spaces for expressing concerns.

Scope Amnesty: A time-limited stakeholder agreement where the organization forgives velocity shortfalls during the cognitive load spike and recovery climb in exchange for the team's commitment to reach sustained excellence. Scope amnesty typically covers sprints one through ten and includes explicit metrics that define acceptable performance during each phase. The amnesty prevents stakeholders from panicking when velocity drops and demanding return to traditional practices before transformation can complete.

The New Green: Revised success metrics that account for AI Scrum's different productivity patterns and value delivery mechanisms. Traditional green status typically measures story points completed or features delivered per sprint. The new green includes AI collaboration quality, governance compliance rates, institutional memory growth, and Chapter 9 transformation metrics alongside traditional velocity metrics. Teams transition to new green reporting during the recovery climb and maintain these metrics through sustained excellence to demonstrate transformation progress beyond simple velocity numbers.

TRANSFORMATION PHILOSOPHY

AI Scrum transformation rests on principles that differ fundamentally from traditional agile adoption or tool implementation projects. Understanding these philosophical foundations prevents organizations from treating AI Scrum as just another process improvement initiative. It actually represents a complete reconception of how software teams operate.

Capability Over Process

Traditional transformations focus on process compliance. Organizations teach teams to follow sprint ceremonies, maintain backlogs properly, and conduct retrospectives on schedule. Success means teams execute the prescribed process consistently. AI Scrum inverts this priority by emphasizing capability development over process adherence.

Figure 13.2: Capability Formation Over Process Compliance. *Traditional transformations measure ceremonial adherence while AI Scrum prioritizes genuine skill development. Teams succeed when members master prompt engineering, output validation, and governance judgment rather than merely executing prescribed ceremonies on schedule.*

Teams build genuine AI collaboration skills rather than simply following AI usage procedures. A team that mechanically runs governance checkpoints without developing actual judgment about AI output quality fails regardless of ceremonial compliance. A team that develops strong prompt engineering capabilities and validates AI contributions effectively succeeds. This holds true even if their governance checkpoint timing varies slightly from the prescribed schedule.

This capability-first philosophy means transformation success cannot be measured by observing ceremonies or reviewing documentation. Success requires assessing whether team members can craft effective prompts, validate AI outputs accurately, maintain institutional memory effectively, and make sound governance decisions under uncertainty. These capabilities develop through practice and coaching rather than through process enforcement.

Whole-Team Transformation

AI Scrum rejects the specialist approach where a few team members become AI experts while others continue traditional practices. Every team member must develop AI collaboration capabilities appropriate to their role. Developers must master prompt engineering for code generation. Product Strategists must learn how to use AI for requirements analysis and stakeholder communication. Validation Leads must build skills in validating AI-generated outputs across multiple work categories.

Figure 13.3: Collective AI Capability Development. *Every team member develops AI collaboration capabilities appropriate to their role. Organizations achieve transformative productivity improvements when entire teams integrate AI into daily work rather than designating a few specialists as AI experts.*

The whole-team requirement stems from AI Scrum's fundamental premise that AI augmentation applies to nearly every development activity. A team where only developers use AI while Product Owners continue traditional user story creation will achieve marginal productivity gains. A team where everyone integrates AI into their daily work while maintaining appropriate human oversight achieves transformative productivity improvements. This philosophy demands different training approaches than traditional transformations. Organizations cannot send a few champions to external training and expect knowledge transfer to suffice. Every team member requires direct skill building through the deep immersion week, ongoing micro-training, and supported practice during early sprints. The investment is larger but the capability building is genuine.

Governance as Guardrails Not Gates

Chapter 10 establishes the AI Scrum governance philosophy where guardrails enable autonomous team decisions within defined boundaries rather than requiring approval gates for individual choices. The Governance Canvas created during Phase 2 of preparation maps which AI usage patterns teams can execute independently versus which require consultation or approval.

Low-risk internal tool development operates with minimal governance oversight making most AI usage decisions autonomously. Regulated domains require tighter guardrails but still decide independently within those boundaries. Preparation time invested in defining risk categories enables and validation requirements upfront so teams rarely wait for governance approvals during sprints. This front-loaded governance framework development enables sustained velocity throughout transformation.

Permanent Not Pilot

The no-pilot philosophy described earlier reflects a broader commitment to permanence over experimentation. Teams that believe they are testing AI Scrum to decide whether to continue will subconsciously limit their investment in capability building. They will maintain dual competencies in traditional and AI-augmented approaches. They will hedge on infrastructure decisions and governance frameworks.

Teams that commit to permanent transformation invest differently. They phase out traditional approaches as AI-augmented alternatives prove effective. They build institutional memory aggressively because they know it will compound benefits over time. They develop governance expertise because they understand it will govern their work indefinitely.

RISK MANAGEMENT DURING TRANSFORMATION

AI Scrum transformation creates two distinct risk categories requiring active management throughout the journey: operational continuity risks and psychological safety risks.

Operational Continuity Risks

Operational continuity risk reflects probability that transformation disrupts critical business operations or prevents meeting existing commitments.

[A+C] Single Team Scenarios: Single teams face contained risk because disruption remains limited to one team's deliverables. However, teams delivering mission-critical capabilities or supporting time-sensitive opportunities still face significant operational consequences from velocity degradation.

Single teams manage operational continuity through stakeholder alignment and scope amnesty. The Product Strategist negotiates explicit expectations before Sprint 1. Stakeholders understand velocity will drop during sprints 1-4, recover during sprints 5-10, and exceed baseline by sprint 19. Organizations unwilling to grant this flexibility should not attempt transformation. The cognitive load spike is inevitable.

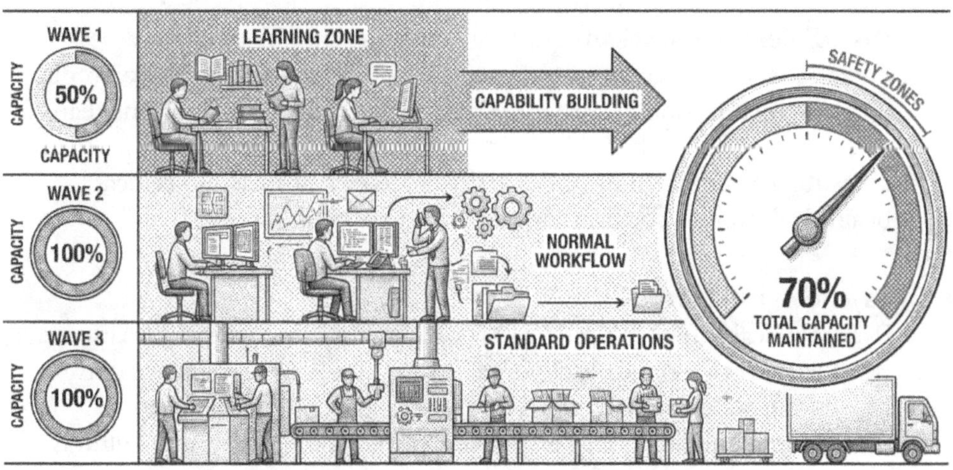

Figure 13.4: Wave-Based Risk Mitigation Strategy. *Multi-team organizations prevent delivery collapse by staging transformation across sequential waves. While Wave 1 operates at reduced capacity during capability building, remaining waves maintain normal productivity preventing organization-wide disruption.*

[B+D] Multi-Team Scenarios: Multi-team organizations face compounded risk because multiple teams experiencing simultaneous velocity drops creates organization-wide delivery disruption. Five teams hitting cognitive load spike together see delivery capacity drop 50-60% across entire product portfolio.

Wave implementation directly addresses this by staging transformation so only a subset of teams experiences velocity degradation simultaneously. Wave 1 includes one team running 3-4 sprints before Wave 2 launches. During sprints 1-4, four teams maintain normal velocity while one team works through cognitive load spike. Total organizational velocity drops approximately 10-12% rather than 50-60%.

Wave timing requires careful calibration. Launching waves too quickly means multiple cohorts hit cognitive load spike simultaneously. Launching too slowly extends total transformation duration. Most organizations find 2-4 sprint spacing between waves provides optimal balance.

Mitigation Strategies (All Scenarios)

- Maintain technical infrastructure redundancy during transformation for emergency coverage
- Defer major product launches or architectural migrations until teams reach at least sprint 8
- Establish clear escalation paths where transformation progress conflicts with critical business needs

Risk Tolerance Assessment: Organizations should explicitly assess operational continuity risk tolerance during preparation before committing. Can the organization absorb 40-60% velocity reduction from one team for four sprints? Will stakeholders maintain commitment when drops occur? Does the organization have flexibility to defer features? Do critical business deadlines conflict with transformation timeline?

Psychological Safety Risks

Psychological safety risk reflects probability that team members disengage, resist transformation, or leave organization due to stress, uncertainty, or fear during adoption. Chapter 11 addresses psychological safety management including fear responses, competence crises, and cultural prerequisites for successful change.

The Competence Crisis: Transformation triggers competence crisis during cognitive load spike when experienced professionals suddenly struggle with unfamiliar AI collaboration patterns.

Solution Protocol: Executive sponsors address this explicitly during Deep Immersion Week in Month -1 with specific messaging: "You are not being replaced. You are being amplified. The skills that made you excellent developers remain essential. AI collaboration is a new skill you will master like you mastered everything else in your career. We expect struggle during Sprints 1-8. Struggling means you are learning, not failing."

Delivery Facilitators reinforce this message weekly during retrospectives in Sprints 1-8. They explicitly ask "What did you struggle with this sprint?" before asking "What went well?" This normalizes struggle as expected rather than exceptional. "Struggle boards" emerge in team spaces to documenting common AI collaboration challenges and solutions discovered. This makes struggle visible and sharable rather than hidden and shameful.

Product Strategists protect teams from stakeholder pressure during competence crisis by enforcing scope amnesty strictly. When stakeholders complain about velocity drops, Product Strategists respond: "Our contract specifies 40-60% velocity during Sprints 1-4. The team is at 52% which meets expectations. Capability formation requires this investment period."

The Obsolescence Fear: Team members naturally question job security when AI generates code rapidly.

Solution Protocol: Executive sponsors communicate staffing commitment during Phase 1 Session 1 (Month -9) with explicit statement: "We are maintaining full team staffing through transformation and beyond. This investment aims to increase our team's capabilities, not reduce our team's size. If we intended workforce reduction, we would not invest in AI training and capability building."

Organizations planning workforce reductions should delay AI Scrum transformation until staffing decisions complete. Attempting transformation during organizational restructuring destroys psychological safety irreversibly.

Delivery Facilitators address obsolescence fear individually during one-on-ones in Month -1 and Sprint 1. They ask each team member directly, "What concerns do you have about AI and your role?" Team members expressing job security fears receive direct reassurance. The reassurance connects to executive messaging and points to organizational investment in their training as evidence of commitment.

Product Strategists demonstrate commitment through scope amnesty enforcement. Teams seeing stakeholders honor velocity contracts interpret organizational patience as genuine investment rather than temporary experiment before layoffs.

Failure Acceptance: AI Scrum requires extensive experimentation where approaches sometimes fail.

Solution Protocol: The AI Sandbox created during Phase 2 provides technical infrastructure for safe failure without production risk. Delivery Facilitators explicitly label sandbox work as "experiments where failure is expected and valuable" during Sprint Inception ceremonies.

Governance checkpoints create ceremony for sharing failures as learning opportunities. Delivery Facilitators facilitate checkpoint discussions starting with "What AI approaches failed this period and what did we learn?" before discussing successes. This signals that sharing failures is valued behavior rather than admission of incompetence.

Organizations with blame cultures require intensive intervention before transformation. Phase 1 should include psychological safety training for all team members and stakeholders covering blameless post-mortems, growth mindset, and learning from failure. Organizations unable to shift from blame to learning cultures should postpone AI Scrum transformation until cultural prerequisites exist.

Figure 13.5: Normalizing Struggle Through Psychological Safety.
Transformation triggers competence crises when experienced professionals struggle with unfamiliar patterns. Organizations succeed by explicitly framing struggle as expected learning rather than individual failure, creating safe environments for capability development.

The Silent Revert: Most common psychological safety failure occurs when team members claim AI usage while reverting to traditional practices because AI collaboration feels too uncertain.

Solution Protocol - Detection: Delivery Facilitators monitor for silent revert indicators during daily standups and sprint ceremonies:

- Developers completing work items faster than AI-assisted estimates suggest
- Team members providing vague responses about AI usage: "used it some" versus specific prompt patterns
- Commit histories showing traditional code patterns rather than AI-generated structures
- Sprint Telemetry Dashboard showing low AI tool engagement despite claimed usage

When Delivery Facilitators detect potential silent revert, they address it privately in one-on-ones rather than publicly in ceremonies. Conversation focuses on understanding obstacles: "I noticed you completed the API endpoint in two hours. Walk me through how you used AI for that work." Responses revealing minimal AI usage trigger support rather than criticism.

Solution Protocol - Intervention: Delivery Facilitators work with silent reverters to identify root causes. Common causes include:

- Inadequate prompt engineering skills requiring additional training
- Fear that AI-generated code is lower quality requiring validation skill building
- Belief that manual coding is faster than iterating with AI requiring pair programming with proficient AI collaborator
- Discomfort admitting AI struggles requiring psychological safety reinforcement

Intervention matches cause. Skills gaps get targeted training. Fear gets paired validation exercises showing AI quality. Speed beliefs get timed experiments comparing approaches. Psychological safety issues get explicit permission to struggle backed by concrete examples of other team members' challenges.

[C+D] Non-agile teams face elevated psychological safety risks because they simultaneously learn agile principles and AI augmentation. More intensive psychological safety management becomes necessary in these scenarios. Organizations should provide extended Deep Immersion Weeks

(5 days instead of 3) and weekly check-ins with all team members during Sprints 1-6 rather than as-needed interventions.

[A+B] Agile teams start from foundation of trusting iterative learning with existing psychological safety infrastructure. These teams still require explicit acknowledgment that AI collaboration introduces genuine uncertainty even for experienced agile practitioners.

INFRASTRUCTURE, TOOLING, AND GOVERNANCE SETUP

[All] AI Scrum requires specific technical infrastructure and governance frameworks established during Phase 2 of preparation (months -6 to -3). These foundational elements enable safe experimentation, maintain quality standards, and scale team capabilities throughout transformation.

The AI Sandbox: During Phase 2, implementation of the AI Sandbox following the infrastructure guidance in this chapter's preparation sections. The sandbox provides controlled environment for AI collaboration skill development separate from production systems. Organizations maintain sandbox permanently for ongoing learning and capability experimentation.

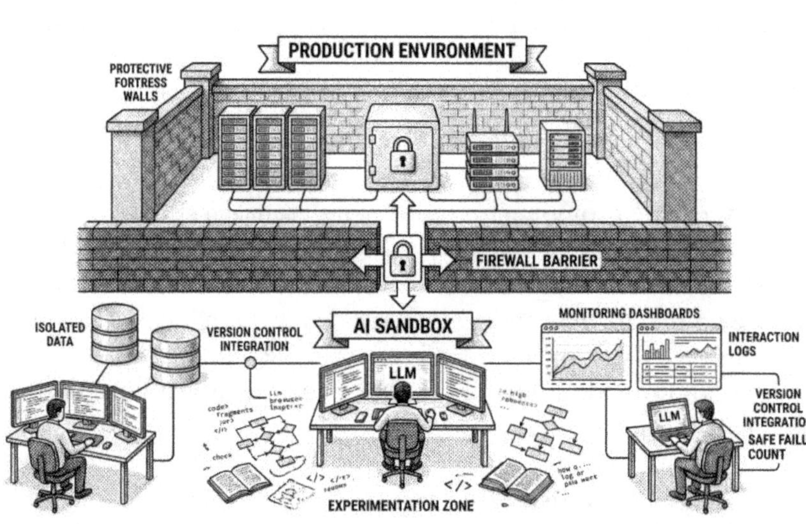

Figure 13.6: Isolated Learning Environment Architecture. *The AI Sandbox provides controlled experimentation space where team members develop collaboration skills without risking production systems. Organizations maintain sandboxes permanently for ongoing capability development and safe failure during learning.*

The Governance Canvas: During Phase 2, development of the initial Governance Canvas using the framework defined in Chapter 10. The canvas maps AI usage decisions across work categories and risk levels.

Teams refine governance quarterly through checkpoint ceremonies as capability scores improve.

The Governance Checkpoint Ceremony: Organizations establish the Governance Checkpoint as defined in Chapter 5. First governance checkpoints are scheduled for Sprint 2 or 3 with initial frequency of every two sprints during cognitive load spike and recovery climb. Chapter 5 provides complete ceremony structure, agenda, and facilitation guidance.

[B+D] Multi-team organizations align checkpoint timing across teams within each wave to facilitate knowledge sharing. Wave 1 teams should conduct checkpoints on consistent schedule before Wave 2 launches.

Delivery Platform and Institutional Memory: Delivery platform configuration occurs during Phase 2 to support AI collaboration patterns. The Infrastructure section of this chapter details platform requirements covering version control integration, CI/CD automation, testing frameworks, and monitoring. Institutional memory builds incrementally starting with basic architectural decisions and standards during preparation, expanding coverage throughout transformation sprints.

Figure 13.7: Risk-Based AI Usage Framework. *The Governance Canvas maps AI collaboration decisions across work categories and risk levels, enabling teams to operate autonomously within defined boundaries rather than requesting approval for each decision.*

[A+B] Agile teams augment existing delivery platforms with AI integration rather than building infrastructure from scratch. These teams typically have version control, CI/CD pipelines, and automated testing already operational from previous Scrum implementations. Phase 2 preparation focuses on integrating AI tools into existing workflows and configuring platforms to capture AI usage metrics. Institutional memory represents the primary new component these teams must build. Agile teams typically complete Phase 2 infrastructure setup within the full three-month window.

[C+D] Non-agile teams must build complete delivery platforms during Phase 2 while simultaneously establishing institutional memory systems. Traditional project management often relies on manual deployment processes and approval-based review gates. AI Scrum requires automated platforms that enable rapid experimentation without manual bottlenecks.

Teams implement version control systems that track every code change. They establish CI/CD pipelines that automatically test and deploy code. They configure testing frameworks that validate AI-generated outputs without human intervention for every change. The Infrastructure and Governance Setup section provides detailed guidance on platform selection, configuration, and integration patterns for teams building these capabilities from zero. Dual infrastructure requirements typically extend Phase 2 by four to six weeks beyond the standard three-month timeline. Additional consulting or training support should be budgeted during Phase 2 for non-agile teams building delivery automation for the first time.

[B+D] Multi-team organizations standardize delivery platforms across all teams to enable knowledge sharing and maintain governance consistency. Teams using different version control systems or CI/CD tools cannot easily share institutional memory, prompt libraries, or governance frameworks. Platform fragmentation creates coordination overhead that eliminates multi-team transformation benefits. Unified tooling stack selection becomes mandatory during Phase 1 strategic planning and implement standardized platforms across all teams during Phase 2. Standardization requirements apply regardless of current platform diversity. Teams with established platforms may need to migrate to the organization's selected standard during Phase 2.

SUCCESS MEASUREMENT FRAMEWORK

[All] AI Scrum transformation requires measurement approaches that differ from traditional agile metrics. Organizations track capability development alongside delivery metrics to understand transformation progress accurately. Chapter 9 establishes five core metrics for tracking AI Scrum transformation progress. These metrics collectively indicate whether teams are developing genuine AI collaboration capabilities or simply performing ceremonial activities without substance.

AI Collaboration Quality

This dimension measures how effectively team members integrate AI into their daily work. Strong AI collaboration means team members craft prompts that consistently generate high-quality outputs, provide appropriate context from institutional memory, and validate AI contributions efficiently. Weak AI collaboration means team members struggle with prompt engineering, waste time on excessive iteration, or avoid AI usage despite governance permissions.

Figure 13.8: Measuring AI Collaboration Effectiveness. *Collaboration quality reflects how efficiently team members generate desired outputs. Single-iteration success with well-crafted prompts demonstrates mastery while excessive iteration cycles indicate developing capabilities requiring improvement.*

Peer reviews provide the mechanism for assessing collaboration quality where Validation Leads examine AI-generated work products and the prompts that created them. A developer who generates a complex algorithm through a single well-crafted prompt demonstrates strong collaboration. A developer who requires fifteen prompt iterations to achieve the same result shows developing capability that needs improvement.

The dashboard scores collaboration quality on a scale from one to five. One represents minimal AI usage with poor results. Three represents competent usage with moderate iteration. Five represents expert usage with minimal iteration and consistently strong outputs. During the cognitive load spike, scores typically hover around two, three during recovery climb, four at parity point, and five at sustained excellence.

Validation Effectiveness

This dimension measures how accurately teams validate AI-generated outputs across different work categories. Effective validation means Validation Leads catch errors in AI-generated code, documentation, and tests while avoiding false positives that unnecessarily reject correct AI contributions. Ineffective validation means teams either let quality issues through or waste time rejecting valid AI outputs.

Figure 13.9: Systematic Output Validation Excellence. *Validation effectiveness measures how accurately teams catch AI-generated errors before production deployment. Strong validators identify subtle logical flaws and hallucinated dependencies while avoiding false positives that waste development time.*

Teams assess validation effectiveness by tracking defect escape rates for AI-generated work that passed validation versus defect discovery rates during validation itself. A Validation Lead who catches subtle logic errors during code review before tests reveal them demonstrates strong effectiveness. A Validation Lead who approves AI-generated code that later fails integration testing shows developing capability needing improvement.

The dashboard scores validation effectiveness similarly from one to five based on defect patterns and review thoroughness. Validation effectiveness improves through several approaches. They develop checklists for common AI failure modes and build automated validation tools that catch specific error patterns. They learn which work categories require more scrutiny versus which categories AI handles reliably.

Governance Understanding

This dimension measures how well team members understand AI usage decisions. Strong understanding includes knowing when to use AI versus when to avoid it and which validation approaches apply to different work types. It also includes making sound governance decisions under ambiguity. Strong governance understanding means teams navigate the governance canvas confidently and make appropriate risk-based decisions independently. Weak governance understanding means teams either violate governance policies accidentally or request excessive guidance for straightforward decisions.

Teams assess governance understanding through the governance checkpoint ceremony. They review recent decisions and evaluate whether those decisions aligned with both policy letter and policy spirit. A team that independently determines appropriate AI usage for a new work category demonstrates strong understanding. A team that routinely seeks approval for decisions the governance canvas clearly permits shows developing understanding.

The dashboard scores governance understanding based on decision quality and independence level. At level five on this dimension, governance representative input becomes rarely necessary. They have internalized the risk frameworks and decision criteria that guide appropriate AI usage.

Institutional Memory Development

This dimension measures how effectively teams build and maintain the knowledge repository that makes AI collaboration sustainable. Strong institutional memory development means teams consistently document validated patterns, maintain accurate architectural decisions, and update prompt libraries with proven approaches. Weak development means teams accumulate knowledge in individual heads rather than shared systems or maintain documentation that becomes stale and misleading.

Teams assess memory development by examining how often AI tools successfully reference institutional memory to maintain consistency. They also track how often team members must manually provide context instead. A codebase where AI-generated components automatically follow established patterns demonstrates strong memory. A codebase where AI frequently violates standards until team members provide explicit correction shows weak memory.

The dashboard scores memory development based on coverage completeness and AI tool success rates when referencing memory. Teams improve this dimension by designating specific team members to maintain memory sections and conducting quarterly memory audits to identify gaps. They also integrate memory updates into sprint workflows so knowledge capture happens continuously rather than as afterthought.

Capability Improvement Trajectories

The dashboard should show trajectory as well as absolute scores. A team at sprint six with scores averaging 2.5 but improving by 0.3 per sprint is on track for sustained excellence. A team at sprint twelve with scores averaging 3.0 but showing no improvement over the previous four sprints has plateaued and needs intervention.

Organizations should establish capability improvement targets based on scenario. Single agile teams should reach average scores of 4.0 by sprint twelve. Multiple agile teams should reach 4.0 by sprint fifteen due to coordination overhead. Non-agile teams should reach 4.0 by sprint fifteen for single teams and sprint eighteen for multiple teams due to the dual learning curve.

Teams failing to meet these trajectories require remediation through additional training or coaching from more advanced teams. They may also need governance policy adjustments that better match current capability levels. The dashboard makes capability problems visible early enough for intervention before teams develop bad habits or become discouraged by persistent struggle.

Integration with Traditional Metrics

Chapter 9 transformation metrics complement rather than replace traditional agile metrics like velocity, defect rates, and cycle time. Organizations should track both capability development and delivery outcomes to understand the complete transformation picture.

A team with strong transformation metric scores but declining velocity indicates process or tooling problems separate from AI collaboration capability. A team with strong velocity but weak capability scores is likely achieving gains through unsustainable shortcuts that will create future problems. A team with strong scores on both measures is executing AI Scrum successfully and building sustainable practices.

The measurement framework provides objective data for transformation decisions. When debating whether to relax governance restrictions, organizations, adjust wave timing, or increase training investment can reference capability scores rather than relying on subjective assessments.

Teams requesting additional tooling or infrastructure can demonstrate capability readiness through dashboard evidence. The framework transforms transformation management from opinion-based to data-driven.

——PART II: THE 9-MONTH PREPARATION——

THE PREP PERIOD: MONTHS -9 TO 0

[All] The nine-month preparation period represents the foundation upon which successful AI Scrum transformation rests. Organizations that skip this preparation do not merely risk failure—they guarantee it. Compressing this timeline or skipping preparation activities inevitably leads to failure. I recognize this sounds absolutist, but the pattern holds across every scenario we've examined. during the first month of actual AI Scrum implementation. The preparation is not optional regardless of scenario, team count, or starting methodology.

The prep period divides into three distinct phases covering strategic alignment, infrastructure development, and tactical readiness. Each phase builds on previous phases and cannot be reordered or executed in parallel without compromising transformation outcomes.

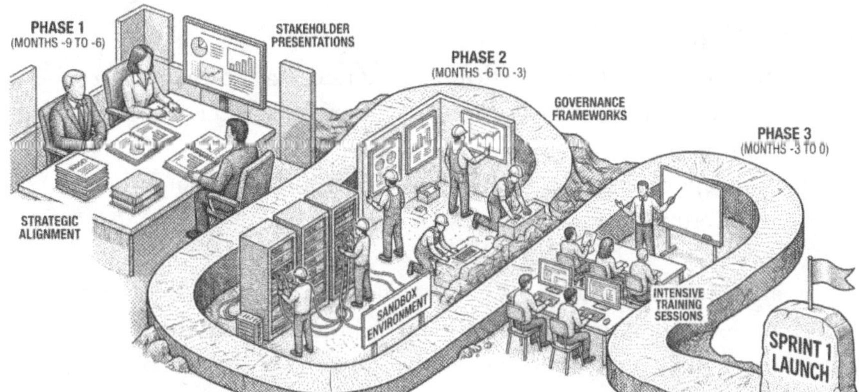

Figure 13.10: Three-Phase Preparation Journey. *The nine-month preparation period progresses from strategic alignment through infrastructure development to tactical readiness. Each phase builds essential foundations that enable successful Sprint 1 launch and sustained transformation capability.*

The Three Phases

Phase 1 spans months -9 to -6 and focuses on strategic alignment and financial approval. Executive commitment and budget model development represent key **Phase 1** outcomes specific to their scenario. Multi-team scenarios include team wave assignments during this phase. All scenarios include stakeholder education about the J-curve productivity pattern (the temporary velocity drop during capability formation followed by recovery to higher sustained performance). Phase 1 establishes the organizational commitment required to sustain transformation through the cognitive load spike.

Phase 2 covers months -6 to -3 and concentrates on technical infrastructure and governance framework development. Implementation of the AI Sandbox and creation of the Governance Canvas, build institutional memory systems, and configure delivery platforms. Multi-team scenarios also establish cross-team coordination mechanisms during this phase. The infrastructure created here enables safe AI experimentation and sustainable governance.

Phase 3 runs months -3 to 0 and addresses tactical readiness including role appointments, squad formation, and intensive training. The deep immersion week occurs during month -1 where they build hands-on AI collaboration skills. Readiness certification verifies teams possess minimum capabilities before Sprint 1. Phase 3 transforms abstract AI knowledge into practical skills team members can apply immediately.

Why Nine Months

Whether nine months of preparation is truly necessary remains a frequent question. Could organizations complete preparation faster? Perhaps—but the data suggests otherwise. The answer depends on understanding what the preparation period accomplishes beyond simple training and infrastructure setup.

The preparation period allows organizational skepticism to evolve into commitment—though exactly how this psychological transformation occurs varies considerably across organizations. Executives who initially view AI Scrum as risky experimentation develop confidence through education campaigns and engagement with the transformation framework. Stakeholders who fear velocity drops during transformation learn to expect and accept the J-curve pattern. Such psychological shifts cannot be rushed without creating commitment fragility that collapses under pressure.

The preparation period provides time for infrastructure procurement and deployment cycles that organizations cannot compress. Enterprise software purchases require vendor evaluation, contract negotiation, and procurement processes that span months. Cloud infrastructure setup involves security reviews, compliance validation, and capacity planning. Attempting to accelerate these organizational realities creates rushed decisions that undermine transformation quality.

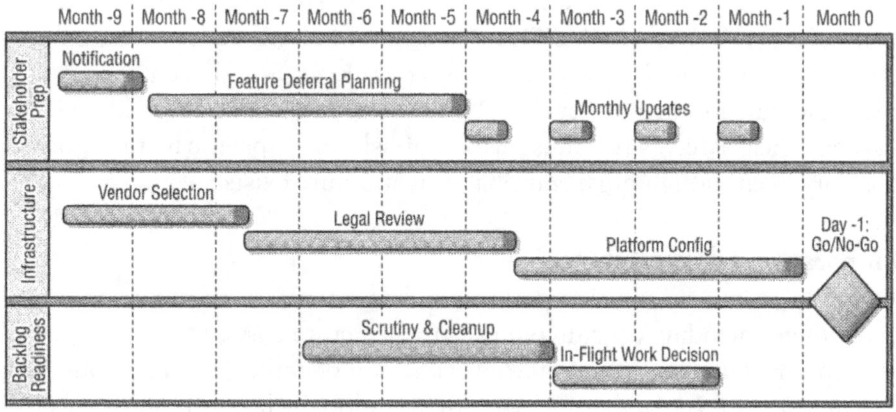

Figure 13.11: The Preparation Timeline. *Transition success is determined before Sprint 1 begins. This nine-month runway coordinates infrastructure, legal, and stakeholder readiness to ensure the team can focus entirely on execution once the operational clock starts.*

The preparation period spaces out decision-making and learning so teams avoid cognitive overload. Compressing nine months into three means teams must simultaneously learn AI capabilities, understand governance frameworks, build institutional memory, master new tools, and develop collaboration patterns. This cognitive burden exceeds what most team members can absorb effectively. The nine-month timeline distributes learning across manageable increments.

Scenario Variations

[A+C] Single team scenarios execute all three phases with focus on the specific team transforming. Phase 1 includes obtaining commitment from the team and their immediate stakeholders rather than organization-wide alignment. Phase 2 builds infrastructure for one team rather than multi-team coordination systems. Phase 3 trains eleven specific people rather than multiple cohorts. The timeline remains nine months but the scope stays contained.

[B+D] Multi-team scenarios execute the same three phases but with expanded scope. Phase 1 includes wave assignment decisions and cross-team stakeholder education. Phase 2 builds infrastructure that supports multiple teams and establishes governance consistency mechanisms. Phase 3 trains Wave 1 teams during months -3 to 0, then trains Wave 2 teams during their preparation window before their launch. Each wave gets its own preparation period aligned to its launch date.

Organizations sometimes propose that later waves can compress preparation because earlier waves have already established infrastructure and governance frameworks. This assumption fails because preparation is about capability building, not just infrastructure access. Wave 2 teams need the same nine months to develop AI collaboration skills and build psychological safety. They must also establish team-specific institutional memory even though organizational infrastructure exists.

Preparation as Investment

The nine-month preparation period requires significant organizational investment including executive time, infrastructure costs, training expenses, and opportunity cost from deferred AI Scrum benefits. Organizations view this investment as mandatory rather than optional when committing to transformation.

The investment pays returns that compound throughout transformation and beyond. Teams that complete thorough preparation reach sustained excellence faster, experience fewer setbacks during the cognitive load spike, and build more robust AI collaboration capabilities. Organizations that skimp on preparation extend their path to sustained excellence, experience more severe velocity drops, and often abandon transformation before reaching payback.

Phase 1: Strategic and Financial Readiness (Months -9 to -6)

Phase 1 establishes the organizational foundation required to sustain AI Scrum transformation through inevitable challenges. This phase focuses on securing genuine commitment rather than superficial approval and developing financial models that withstand executive scrutiny.

Executive Sponsorship and Commitment

[All] AI Scrum transformation requires executive sponsorship that extends beyond verbal support to active participation in key decisions. The executive sponsor must understand the J-curve productivity pattern and accept velocity drops during the cognitive load spike. They must commit to maintaining transformation investment when stakeholders demand explanations for decreased output.

Organizations should identify sponsors who possess budget authority for the transformation costs and influence over stakeholders who will experience velocity impacts. The sponsor must also have credibility with teams who will execute the transformation. The sponsor cannot be someone who merely endorses the initiative but rather someone willing to defend it during difficult moments.

The sponsor participates in developing the transformation budget and reviews wave assignments for multi-team scenarios. They approve the governance framework and intervene when organizational pressures threaten to derail capability building. This involvement requires approximately four to six hours per month during Phase 1, increasing to eight to ten hours monthly during early transformation sprints.

Figure 13.12: Executive Commitment Beyond Verbal Support. *Genuine sponsorship requires active participation in transformation decisions and acceptance of velocity drops during capability building. Sponsors must defend investment when stakeholders demand explanations for decreased output during the J-curve trough.*

Financial Model Development

[All] Organizations develop scenario-specific budget models during Phase 1 that account for direct transformation costs, opportunity costs from velocity impacts, and expected ROI timelines. Generic budget estimates fail during executive reviews because they cannot answer specific questions about the organization's actual financial commitment.

Direct transformation costs include AI tool subscriptions, infrastructure investments, training expenses, and any consulting support required. These costs vary substantially across scenarios based on team count, existing infrastructure maturity, and whether teams already practice agile methodologies.

Opportunity costs reflect the revenue or strategic value foregone during velocity degradation periods. A team operating at 50% capacity for four sprints represents real business impact that must be quantified and justified. Organizations that ignore opportunity costs face executive panic when stakeholders demand explanations for delayed features.

ROI timelines specify when productivity gains offset transformation investments. Most scenarios reach breakeven between months twelve and twenty-four post-preparation depending on baseline velocity, recovery trajectory, and sustained excellence performance levels. Organizations should model conservative, expected, and optimistic scenarios to establish realistic expectations.

[All] Phase 1 includes systematic stakeholder education about AI Scrum mechanics, the J-curve pattern, and transformation timelines. Stakeholders who understand why velocity drops temporarily prove far more supportive than stakeholders experiencing unexpected delivery delays without context.

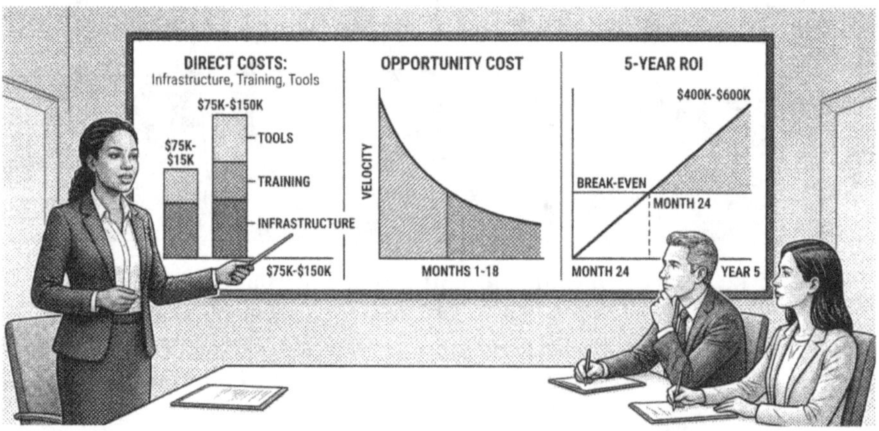

Figure 13.13: Comprehensive Transformation Investment Analysis. *Budget models account for direct transformation costs, opportunity costs from velocity impacts, and ROI timelines. Most scenarios reach breakeven within twenty-four to thirty-four months with cumulative five-year value exceeding four hundred thousand dollars per*

The education campaign should target three stakeholder groups with different messages. Product stakeholders need to understand how AI Scrum affects feature delivery timelines and why scope amnesty during early sprints serves long-term productivity goals. Executive stakeholders need to understand transformation costs, ROI timelines, and governance oversight mechanisms. Adjacent team stakeholders need to understand why the transforming team may need different processes or infrastructure than other teams.

Organizations typically conduct three to four education sessions during Phase 1 using presentations, case studies from similar organizations, and detailed Q&A discussions. The sessions should acknowledge real concerns about velocity drops and delivery delays rather than dismissing stakeholder worries as resistance to change.

Scenario-Specific Activities

[A+C] Single team scenarios focus Phase 1 activities on the specific team transforming and their immediate stakeholders. The executive sponsor works directly with the team's product stakeholders to establish scope amnesty agreements and realistic delivery expectations. Financial models focus on one team's costs and one team's ROI timeline.

[B+D] Multi-team scenarios expand Phase 1 to include wave assignment decisions and cross-team stakeholder coordination. Organizations must determine which teams join Wave 1, Wave 2, and Wave 3 based on strategic priorities, technical readiness, and stakeholder tolerance for disruption.

Wave assignments should consider team dependencies where one team's output feeds another team's work. Dependent teams should not both be in Wave 1 because simultaneous velocity drops could create cascading delays. Teams with more flexible stakeholder relationships make better Wave 1 candidates than teams operating under tight contractual deadlines.

The stakeholder education campaign for multi-team scenarios must address organization-wide concerns about coordinated velocity drops and explain how wave staging mitigates operational continuity risks. Executive stakeholders need to understand the total transformation timeline spanning all waves, not just Wave 1 duration.

Commitment Validation

Phase 1 concludes with formal commitment validation where the executive sponsor confirms readiness to proceed with preparation Phases 2 and 3. This validation is not ceremonial but rather a genuine assessment of whether organizational commitment exists to sustain transformation through challenges.

The validation should confirm that budget models have been approved with full understanding of costs and opportunity impacts. It should verify that stakeholders understand and accept the J-curve pattern. It should establish that wave assignments for multi-team scenarios reflect actual organizational priorities rather than political compromises.

Organizations failing validation should not proceed to Phase 2. Attempting transformation without genuine commitment wastes the Phase 2 infrastructure investment and Phase 3 training effort. Better to acknowledge insufficient readiness during Phase 1 than to discover commitment gaps during Sprint 3 when teams hit the velocity trough. Phase 1 also includes critical decisions about which teams transform and when they begin their AI Scrum journey. These decisions differ fundamentally between single team and multi-team scenarios but share the common requirement of genuine team member commitment.

[B+D] Wave Assignment Criteria

Multi-team organizations must determine wave composition during Phase 1 based on strategic priorities, technical capabilities, and organizational constraints. Wave 1 typically includes one team that serves as the organizational pioneer. Wave 2 and Wave 3 follow at staged intervals with one to three teams each depending on organizational capacity.

Wave 1 team selection should prioritize teams with strong technical capabilities who can navigate ambiguity without extensive support. These teams establish organizational patterns, validate governance frameworks, and demonstrate that AI Scrum delivers promised productivity gains. Selecting weak teams for Wave 1 increases failure probability and undermines organizational confidence in transformation viability.

Wave 1 teams should also have stakeholders willing to accept velocity drops during the cognitive load spike. Teams operating under inflexible contracts or tight regulatory deadlines make poor Wave 1 candidates because stakeholder pressure may force premature abandonment. Teams with internal stakeholders or flexible roadmaps provide better environments for working through early challenges.

Technical readiness matters for wave assignment. Teams already using modern development practices like continuous integration, automated testing, and version control adapt faster to AI Scrum than teams relying on manual processes. Teams with strong documentation cultures build institutional memory more effectively than teams operating primarily through tribal knowledge. These readiness factors should influence wave placement.

Wave 2 and Wave 3 assignments can include teams with less technical maturity because they benefit from Wave 1 learnings and established organizational infrastructure. Later waves receive refined governance frameworks, validated prompt libraries, and documented lessons that accelerate their capability building. Organizations should sequence waves so later teams learn from earlier experiences rather than repeating mistakes.

Team dependencies require careful consideration in wave assignments. Two teams with tight integration where one team's output feeds another team's work should not both be in Wave 1. Simultaneous velocity drops would compound delays and create stakeholder frustration that threatens transformation continuation. Stagger dependent teams across waves to maintain organizational delivery capacity.

[A+C] Team Commitment Building

Single team scenarios skip wave assignment but require deliberate team member commitment building during Phase 1. Organizations cannot mandate enthusiasm for transformation but they can create conditions where team members understand benefits and choose to engage fully.

Commitment building starts with transparent communication about what AI Scrum transformation entails. Team members should understand they will learn entirely new working methods, experience temporary productivity decreases, and invest significant effort in capability building before seeing benefits. Pretending transformation is easy or painless creates disillusionment when reality contradicts messaging.

Team members need opportunities to voice concerns without those concerns being interpreted as resistance. Someone worried about job security due to AI capabilities deserves honest conversation about organizational intentions. Someone skeptical about productivity claims deserves evidence from comparable transformations. Creating space for legitimate concerns builds trust that superficial reassurances cannot.

Organizations should allow team members who fundamentally oppose AI Scrum to opt out during Phase 1 rather than forcing participation. A single resistant team member can undermine transformation through silent reversion, negative influence on peers, or outright sabotage. Better to reassign reluctant members and find committed replacements than to carry resistance through nine months of preparation.

The commitment building process includes team members in key Phase 1 decisions about governance frameworks, tooling selections, and training approaches. Teams that shape their own transformation demonstrate higher engagement than teams having transformation imposed upon them. This participation requires more Phase 1 coordination effort but produces better long-term outcomes.

Commitment Validation Mechanisms

Both single and multi-team scenarios should validate team commitment before proceeding to Phase 2. This validation differs from executive sponsor commitment validation because it focuses on actual team member readiness rather than organizational approval.

Team commitment validation includes direct conversations between the Delivery Facilitator and each team member. These conversations cover their understanding of transformation requirements and their willingness to engage fully. The conversations happen individually to create psychological safety for honest responses. Team members who express reservations receive additional support or alternative placement rather than being pressured into insincere agreement.

Organizations should expect some team member turnover during Phase 1 as individuals decide AI Scrum transformation does not align with their career interests. This turnover is healthy when it happens early. Losing team members during Sprint 3 or 4 creates far more disruption than replacing them during preparation.

Beyond team member commitment, Phase 1 stakeholder education determines whether organizations can sustain transformation through the cognitive load spike when velocity drops create pressure to abandon AI Scrum. Stakeholders who understand the J-curve pattern before Sprint 1 tolerate velocity decreases far better than stakeholders experiencing unexpected delivery delays.

The J-Curve Education Campaign

The J-curve represents the predictable productivity pattern where velocity drops significantly during initial sprints, recovers gradually through middle sprints, and eventually exceeds previous baseline performance. Stakeholders must understand this pattern is not a sign of implementation failure but rather evidence of proper capability building.

Education sessions should present the J-curve with specific velocity projections based on the organization's scenario. Single agile teams typically bottom out at 50-60% of baseline during sprints two through four. Multiple agile teams experience similar individual drops but organizational capacity stays higher due to wave staging. Non-agile teams may reach slightly lower troughs due to dual learning curves.

The campaign must address stakeholder concerns directly rather than glossing over difficulties. Product stakeholders worried about feature delivery delays need realistic timelines showing when velocity recovers. Executive stakeholders concerned about competitive impacts need data about long-term productivity gains justifying short-term disruption. Customer-facing stakeholders need communication strategies for managing external expectations during transformation.

Figure 13.14: Transparent Velocity Pattern Education. *Stakeholder education about the J-curve pattern prevents panic when velocity drops occur. Understanding that forty to sixty percent capacity during early sprints represents normal capability formation rather than implementation failure maintains transformation commitment.*

Visual presentations prove more effective than verbal explanations for J-curve education. Graphs showing historical transformations with actual velocity data, sprint-by-sprint capability building milestones, and ROI crossover points help stakeholders internalize transformation realities. Organizations should develop scenario-specific visualizations rather than using generic templates.

Scenario-Specific Messaging

[A+B] Agile team stakeholders need education focused on how AI augmentation changes familiar Scrum practices. These stakeholders understand sprint mechanics and velocity tracking but may not grasp how AI collaboration affects those established patterns. Education should emphasize that AI Scrum preserves agile principles while fundamentally changing execution approaches.

[C+D] Non-agile team stakeholders require education about both agile methodologies and AI augmentation simultaneously. These stakeholders may not understand concepts like sprints, backlogs, or velocity measurement that agile stakeholders take for granted. Education sessions must cover foundational agile principles before layering on AI-specific transformations.

[B+D] Multi-team stakeholders need additional education about wave implementation mechanics and how staging mitigates organizational risk. These stakeholders must understand why all teams cannot transform simultaneously and how wave spacing maintains delivery capacity during transformation.

Scope Amnesty Negotiation

Phase 1 education campaigns should culminate in explicit scope amnesty agreements where stakeholders commit to forgiving velocity shortfalls during defined transformation periods. These agreements typically cover sprints one through ten and specify acceptable performance levels for each phase.

Scope amnesty is not unlimited forgiveness for poor performance but rather negotiated flexibility based on realistic transformation expectations. A team expected to operate at 50% capacity during sprints one through four should still deliver approximately half their normal output. Stakeholders grant amnesty for the missing half while teams maintain accountability for achievable work.

The amnesty agreement should specify conditions that would invalidate the arrangement. Teams that violate governance policies, skip capability building activities, or fail to show progressive improvement lose amnesty protection. This conditionality ensures teams take transformation seriously rather than treating amnesty as permission to underperform indefinitely.

Documentation and Alignment

Organizations should document stakeholder education outcomes including who attended sessions, what concerns were raised, how those concerns were addressed, and what commitments were made. This documentation prevents later disputes about whether stakeholders understood transformation implications or agreed to velocity impacts.

Multi-team organizations must ensure stakeholder education reaches all relevant parties across organizational boundaries. A Wave 1 team's transformation can affect adjacent teams, shared infrastructure, or downstream processes. Education should extend beyond direct stakeholders to anyone whose work intersects with transforming teams.

Phase 2: Technical Stack and Governance (Months -6 to -3)

Phase 2 shifts focus from strategic alignment to concrete infrastructure and governance implementation. Organizations build the technical foundations and policy frameworks that enable safe AI experimentation while maintaining quality standards.

AI Sandbox Implementation

[All] The AI Sandbox provides controlled environment where team members develop AI collaboration skills without risking production systems. Organizations complete sandbox implementation by month -5 allowing at least two months of experimentation before deep immersion training.

Technical Components: Sandbox includes approved LLM providers (typically Anthropic Claude, OpenAI GPT-4, or Google Gemini rather than on-premises models), isolated data storage for code samples and documentation, interaction logging for learning purposes, and version control integration for safe experimentation. Organizations with regulatory requirements implement data anonymization, access logging, and retention policies aligned with compliance mandates.

Access and Usage: Sandbox access should be unrestricted for approved team members to encourage experimentation. Organizations establish usage policies including acceptable use guidelines, prohibited activities, and cost controls (rate limits and spending caps). Team members can generate and test code freely but cannot deploy to production without proper validation.

Integration Requirements: Integration with development environments (IDEs, command-line interfaces) proves critical for effectiveness. This integration reduces friction and encourages frequent experimentation that builds muscle memory. [B+D] Multi-team organizations require shared sandbox configurations maintaining consistency across teams.

Governance Canvas Development

[All] The Governance Canvas maps AI usage decisions across work categories, risk levels, validation requirements, and approval authorities. Teams develop initial canvases during Phase 2 through collaborative workshops involving technical leads, compliance representatives, and executive sponsors.

Work Categorization: Canvas development starts with categorizing work types by risk and complexity. Low-risk categories include internal documentation, test case generation, code comments, and developer tooling. Medium-risk categories include business logic implementation, API development, and database queries. High-risk categories include security controls, payment processing, authentication mechanisms, and regulated data handling.

Figure 13.15: Controlled AI Experimentation Infrastructure. *The AI Sandbox provides isolated environment with approved LLM providers, secure data storage, comprehensive interaction logging, and version control integration. Teams experiment safely without risking production systems while building essential collaboration capabilities.*

Permission Levels: For each category, canvas specifies whether AI usage is permitted, prohibited, or conditional during different transformation phases. Most organizations prohibit AI for high-risk categories during sprints 1-4. They allow conditional usage during sprints 5-10 with enhanced validation. Standard usage is permitted after teams demonstrate capability through dashboard scores.

Validation Requirements: Permitted usage requires standard peer review. Conditional usage requires Validation Lead approval plus architectural review for significant decisions. Prohibited categories use traditional approaches exclusively.

Canvas Evolution: The canvas is explicitly temporary in its initial restrictive form. As transformation metrics improve through governance checkpoint reviews, canvas relaxes restrictions and expands permitted categories. Teams base relaxation on demonstrated capability rather than time elapsed. Teams reaching sustained excellence operate with minimal governance overhead.

Team Engagement: Canvas development should engage team members who will operate under policies rather than having compliance teams create frameworks in isolation. Team input ensures policies reflect actual work patterns and remain practical to implement.

Figure 13.16: Collaborative Governance Framework Design. *Governance canvas development engages technical teams and compliance representatives in mapping AI usage boundaries. Initial restrictive frameworks protect quality during capability formation then relax systematically as teams demonstrate proficiency through transformation metrics.*

[A+B] Agile Team Governance Adaptations

Agile teams typically have existing code review processes, quality standards, and architectural guidelines adaptable for AI governance. Phase 2 work extends existing frameworks to cover AI-specific considerations rather than building governance from scratch. Key adaptation involves defining how existing quality gates apply to AI-generated outputs, including checking for hallucinated dependencies, verifying logical correctness, and confirming alignment with architectural patterns. [B] Multi-team organizations must maintain consistent governance canvases across all teams within a wave.

[C+D] Non-Agile Team Governance Creation

Non-agile teams must establish complete governance frameworks during Phase 2 because they lack existing quality processes to adapt. Organizations should implement basic agile governance practices (peer review, automated testing gates, architectural oversight) before layering AI-specific policies. Governance frameworks should emphasize simplicity and clarity because team members learn governance concepts simultaneously with AI collaboration skills. Start with straightforward permitted-prohibited categories and refine based on experience. [D] Multi-team organizations coordinate governance across waves.

Infrastructure Configuration

Phase 2 includes configuring delivery platforms to support AI collaboration patterns. Organizations should ensure version control systems, CI/CD pipelines, and deployment automation accommodate increased code volume from AI generation without creating bottlenecks.

Automated testing infrastructure becomes particularly critical because AI generates more code than traditional approaches and manual review alone cannot catch all issues. Teams should implement thorough unit tests, integration tests, and quality checks that run on every commit. AI-generated code failing these automated gates receives immediate feedback without consuming Validation Lead capacity.

Monitoring and observability systems should track AI usage patterns across multiple dimensions. These include which tools teams use most frequently, where AI contributions concentrate in the codebase, and what error patterns emerge from AI-generated code. These metrics inform governance refinements and capability assessments throughout transformation.

Additionally, Phase 2 includes establishing the knowledge repositories and coordination mechanisms that enable sustainable AI collaboration. The scope and complexity of this work varies significantly between single team and multi-team scenarios.

[All] Institutional Memory Systems

Institutional memory captures team decisions, architectural patterns, coding standards, and validated prompt templates in structured formats optimized for AI consumption. Organizations should implement basic memory systems during Phase 2 with the understanding that content accumulates over time rather than being created entirely upfront.

Memory systems should use machine-readable formats like JSON, YAML, or structured markdown rather than free-text documents. An architectural decision record specifying microservices patterns, naming conventions, and communication protocols allows AI tools to maintain consistency automatically. Free-text documentation describing the same decisions requires manual extraction and interpretation that slows AI collaboration.

The memory system architecture should support versioning alongside code so decisions and implementations stay synchronized. When teams change architectural standards, both code and memory update together. This prevents AI tools from following outdated patterns that no longer align with current practices.

Organizations should resist the temptation to build custom memory platforms during Phase 2. File-based systems using structured markdown in version control work effectively for most teams. More sophisticated organizations might use database-backed documentation platforms, but custom development creates maintenance burdens that distract from transformation priorities.

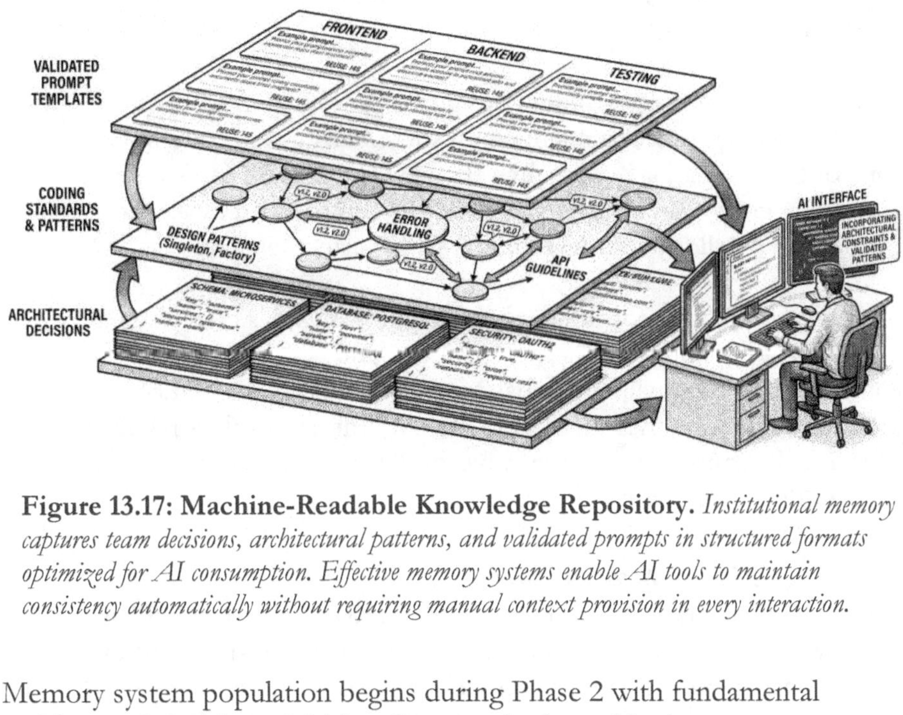

Figure 13.17: Machine-Readable Knowledge Repository. *Institutional memory captures team decisions, architectural patterns, and validated prompts in structured formats optimized for AI consumption. Effective memory systems enable AI tools to maintain consistency automatically without requiring manual context provision in every interaction.*

Memory system population begins during Phase 2 with fundamental architectural decisions, initial coding standards, and basic prompt patterns discovered during sandbox experimentation. Teams expand memory incrementally throughout transformation as they validate new patterns and encounter new work categories. Attempting detailed memory documentation before Sprint 1 wastes effort on hypothetical scenarios rather than capturing actual validated knowledge.

[B+D] Cross-Team Coordination Infrastructure

Multi-team scenarios require coordination mechanisms during Phase 2 that single team scenarios skip entirely. These mechanisms ensure governance consistency, facilitate knowledge sharing, and prevent teams from developing incompatible approaches.

Organizations should establish regular cross-team synchronization meetings during Phase 2 where Wave 1 teams align on governance frameworks, share sandbox learnings, and coordinate infrastructure decisions. These meetings typically occur biweekly during Phase 2 and continue through early transformation sprints.

The synchronization meetings should have clear agendas focused on decision-making rather than general updates. Teams might decide on shared prompt library organization, agree on institutional memory formats, or establish consistent validation criteria for specific work categories. Meetings that become status reports waste time without producing alignment.

Multi-team organizations should designate coordination roles during Phase 2. Roles include someone maintaining shared institutional memory, someone coordinating governance framework updates, and someone facilitating cross-team learning. These roles can be filled by existing team members with allocated time rather than requiring new hires.

Shared infrastructure for multi-team scenarios includes centralized institutional memory accessible to all teams and common AI Sandbox configurations that maintain consistency. It also includes unified monitoring systems that track AI usage patterns across the organization. This infrastructure enables later waves to leverage earlier wave learnings without duplicating discovery effort.

[A+C] Single Team Infrastructure

Single team scenarios implement simpler infrastructure during Phase 2 because no cross-team coordination is required. The team builds one institutional memory system serving eleven people rather than coordinating multiple team repositories.

Single teams have more flexibility in tooling and process choices because they only need internal consistency rather than organizational alignment. A single team might select specific AI providers, documentation formats, or validation approaches based purely on team preferences. Multi-team organizations must standardize these choices to enable knowledge sharing.

The simplicity advantage for single teams means Phase 2 infrastructure work typically completes faster than multi-team scenarios. Single teams can finish core infrastructure in six to eight weeks. Multi-team organizations often need the full twelve weeks to establish coordination mechanisms and shared systems.

Infrastructure Validation

Phase 2 should conclude with infrastructure validation confirming that all systems work correctly before Phase 3 training begins. Validation includes verifying the AI Sandbox allows team members to experiment freely. It confirms institutional memory systems integrate with development workflows and tests that governance frameworks cover relevant work categories.

Multi-team validations should include cross-team exercises where teams access shared institutional memory, use common governance frameworks, and coordinate through established synchronization mechanisms. These exercises reveal integration issues that individual team testing would miss.

Organizations discovering infrastructure gaps during validation should delay Phase 3 until issues resolve. Starting deep immersion training with broken infrastructure frustrates team members and undermines confidence in transformation planning. Better to extend Phase 2 by a few weeks than to proceed with inadequate technical foundations.

Phase 3: Roles and Squad Formation (Months -3 to 0)

Phase 3 shifts focus to tactical readiness including appointing individuals to evolved roles, forming the eleven-person squad, and building hands-on AI collaboration capabilities. This phase transforms abstract preparation into concrete team readiness.

Role Appointments

[All] Organizations should appoint specific individuals to the four AI Scrum roles during early Phase 3. The appointments should complete by month -2 to allow time for role-specific training before the deep immersion week in month -1.

The Product Strategist appointment requires identifying someone who understands product vision, maintains stakeholder relationships, and can learn AI-assisted requirements analysis. This person typically comes from existing Product Owner roles in agile scenarios or from business analyst or project manager roles in non-agile scenarios. Chapter 4 provides detailed guidance on Product Strategist capabilities and responsibilities.

The Delivery Facilitator appointment requires someone skilled in team coordination, impediment removal, and coaching. Existing Scrum Masters transition naturally to this role in agile scenarios. Non-agile scenarios must identify someone with facilitation skills even if they lack formal Scrum experience because Phase 3 training covers agile fundamentals alongside AI collaboration.

Developer appointments focus on technical capability and willingness to learn AI collaboration patterns. Organizations should prioritize developers with strong fundamentals over those with specific framework expertise because AI augmentation changes how teams apply technical knowledge. Seven developers form the optimal squad size regardless of scenario.

Validation Lead appointments require individuals with quality mindset and attention to detail. Existing QA Engineers fit naturally but organizations can also appoint developers with testing expertise to these roles. Two Validation Leads provide adequate coverage for reviewing AI-generated outputs without creating bottlenecks.

[A+B] Agile Team Appointments

Agile teams typically have existing role holders who transition to evolved AI Scrum roles. The Product Owner becomes the Product Strategist with expanded responsibilities for AI-assisted analysis. The Scrum Master evolves into the Delivery Facilitator with added focus on AI collaboration coaching.

These transitions require explicit role redefinition rather than assuming existing role holders automatically gain new capabilities. Organizations should document how responsibilities change, what new skills role holders must develop, and how success metrics evolve for AI Scrum contexts.

[C+D] Non-Agile Team Appointments

Non-agile teams must appoint people to roles they may not fully understand yet because they lack agile experience. Organizations should emphasize that Phase 3 training will develop necessary capabilities rather than expecting appointees to arrive fully formed.

Role appointments in non-agile scenarios often involve more significant organizational change. A project manager might become the Product Strategist but must shift from detailed planning mindsets to adaptive prioritization approaches. A team lead might become the Delivery Facilitator but must learn facilitation techniques beyond directive management.

Squad Formation and Team Dynamics

The eleven appointed individuals must form into a cohesive squad during Phase 3. Organizations should not assume that putting eleven capable people together automatically creates an effective team. Deliberate team formation activities build relationships and establish working norms.

Squad formation includes initial team meetings where members introduce backgrounds, discuss expectations for collaboration, and identify potential working style conflicts. These meetings surface interpersonal dynamics that might create problems during transformation stress.

Organizations should facilitate explicit discussions about how the team will handle disagreements, make decisions under uncertainty, and support each other through the cognitive load spike. Teams that establish psychological safety during Phase 3 navigate transformation challenges more effectively than teams that defer these conversations.

Backup Planning

Phase 3 should include backup planning for key roles in case appointed individuals leave during transformation. Organizations should identify potential replacements for the Product Strategist and Delivery Facilitator because these roles prove most difficult to fill mid-transformation.

Figure 13.18: Intensive Capability Formation Week. *Deep Immersion Week builds hands-on AI collaboration skills through structured progression from capability exploration to practical application. Teams develop functional competencies in prompt engineering, output validation, and governance navigation preparing them for Sprint 1 execution.*

The backup planning should not involve training multiple people for the same role because this dilutes focus and wastes training investment. Rather, organizations should identify individuals who could step into roles with accelerated training if primary appointees become unavailable.

The deep immersion week represents the most intensive preparation activity where team members build hands-on AI collaboration skills through structured exercises and real work simulations. This week occurs during month -1 and requires full-time participation from all eleven squad members.

Immersion Week Duration by Scenario

Immersion week duration varies by scenario based on whether teams need agile fundamentals training before AI collaboration content. Non-agile teams require an additional Day 0 to establish basic agile concepts before diving into AI-augmented practices.

Scenario	Duration	Additional Content
A (Single Agile)	5 days	AI collaboration only
B (Multiple Agile)	5 days	AI collaboration only
C (Single Non-Agile)	6 days	Day 0: Agile fundamentals
D (Multiple Non-Agile)	6 days	Day 0: Agile fundamentals

Core Curriculum (All Scenarios, Days 1-5)

- **Day 1** introduces AI capabilities and limitations through sandbox demonstrations and experimentation. Team members develop realistic expectations about what AI can accomplish reliably versus where it fails.
- **Day 2** focuses on prompt engineering techniques specific to different work types and roles. Each team member builds initial prompt libraries for their role-specific work.
- **Day 3** covers validation approaches for AI-generated outputs including spotting hallucinated dependencies, plausible but incorrect logic, and outputs violating implicit constraints. Validation Leads receive additional training on systematic review techniques.
- **Day 4** addresses governance framework application through scenario-based exercises spanning different risk categories. Teams resolve governance ambiguities before encountering them under sprint pressure.
- **Day 5** integrates all skills through simulated sprint work completing realistic work items using AI collaboration under time constraints. Teams learn from failures and build confidence in emerging capabilities.

Role-Specific Training: Afternoon sessions provide targeted training for Product Strategists (stakeholder analysis, requirements documentation), Delivery Facilitators (coaching techniques), Developers (language-specific prompt engineering), and Validation Leads (failure mode detection, automated validation configuration).

[C+D] Agile Fundamentals (Day 0): Non-agile teams require additional day covering sprint structure, backlog management, ceremony purposes, and empirical process control. Training uses AI-assisted examples so teams simultaneously learn both agile principles and AI collaboration possibilities.

Psychological Preparation: Week explicitly addresses transformation psychology including competence concerns, obsolescence fears, and cognitive load spike reality. Trainers provide transparent information about velocity patterns, capability building timelines, and common emotional challenges. Teams establish peer support commitments.

Expected Outcomes: Teams exit with functional but developing AI collaboration skills, realistic expectations about transformation challenges, initial prompt libraries, validated governance interpretations, and documented lessons feeding institutional memory systems.

Following immersion week, Phase 3 concludes with formal readiness certification that validates teams possess minimum capabilities required for Sprint 1 launch. Organizations should not proceed to AI Scrum implementation without confirmed readiness regardless of schedule pressures.

Readiness Certification Gates

[All] Certification assesses team capabilities across multiple dimensions including technical infrastructure, governance understanding, AI collaboration basics, and psychological readiness. Teams must meet minimum thresholds on all dimensions because weakness in any area creates transformation vulnerability.

Technical infrastructure certification confirms the AI Sandbox functions correctly and institutional memory systems integrate with workflows. It verifies delivery platforms support AI collaboration patterns and monitoring systems track usage appropriately. Organizations verify these capabilities through hands-on exercises where team members complete realistic tasks using production infrastructure.

Governance understanding certification tests whether team members can interpret the Governance Canvas correctly and make sound decisions about AI usage for different work categories. It also assesses their understanding of validation requirements for various permission levels. The assessment uses scenario-based questions where team members explain their reasoning rather than simple multiple choice tests.

Figure 13.19: Multi-Dimensional Readiness Validation. *Certification assesses teams across technical infrastructure, governance understanding, AI collaboration capabilities, and psychological readiness. Teams must demonstrate minimum competency in all dimensions before Sprint 1 launch because weakness in any area creates transformation vulnerability.*

AI collaboration certification evaluates basic prompt engineering skills, output validation capabilities, and institutional memory usage. Team members demonstrate they can generate acceptable code or documentation through AI assistance, identify common failure modes in AI outputs, and reference institutional memory appropriately. The bar is functional competence rather than expertise.

Psychological readiness certification confirms team members understand the J-curve pattern, accept that velocity will drop during early sprints, and commit to capability building despite frustration. The Delivery Facilitator conducts individual conversations assessing whether team members have realistic expectations and genuine willingness to persist through challenges.

Certification Failure Handling

Teams failing any certification dimension should not proceed to Sprint 1 on schedule. Organizations face a critical decision between delaying transformation to address gaps or proceeding with known weaknesses that will likely cause failure.

The responsible decision is delay. Launching with inadequate infrastructure creates technical debt that compounds through transformation. Launching with poor governance understanding leads to quality problems or excessive risk-taking. Launching with weak AI collaboration skills produces frustration that undermines psychological safety. Launching with fragile commitment almost guarantees abandonment during the cognitive load spike.

Organizations should extend Phase 3 by two to four weeks when teams fail certification and provide targeted remediation addressing specific gaps. A team failing governance certification needs additional scenario practice and policy clarification. A team failing technical certification needs infrastructure fixes and additional testing.

Ongoing Micro-Training

Readiness certification confirms minimum capabilities but teams continue learning throughout transformation. Organizations should implement ongoing micro-training that delivers targeted skill building in fifteen to thirty minute sessions during sprints.

Micro-training topics emerge from governance checkpoint discussions, Chapter 9 metric assessments, and Delivery Facilitator observations about team struggles. A team showing weak prompt engineering for test generation receives focused training on that specific skill. A team making governance errors around conditional AI usage gets clarification on decision criteria.

The micro-training approach differs from traditional training programs by delivering just-in-time learning exactly when teams need specific capabilities. Rather than front-loading thorough training that teams partially forget, micro-training provides frequent reinforcement at points of application.

Micro-training should be led by team members who demonstrate strength in the relevant area rather than always relying on external trainers. A developer who masters effective prompting for refactoring shares techniques with teammates. A Validation Lead who develops strong code review checklists trains others on their approach. This peer training builds team capability while reducing dependence on external support.

[B+D] Wave Staggering and Knowledge Transfer

Multi-team organizations must manage readiness certification across waves with later waves learning from earlier experiences. Wave 1 teams complete full certification during their Phase 3. Wave 2 teams receive Wave 1 lessons integrated into their deep immersion week and certification process.

Organizations should document Wave 1 certification outcomes including which areas proved most challenging, what remediation approaches worked effectively, and how certification criteria might improve. This documentation informs Wave 2 and Wave 3 preparation without requiring later waves to repeat earlier mistakes.

Wave 2 and Wave 3 teams benefit from refined training materials, validated governance frameworks, and proven prompt libraries that Wave 1 teams developed. However, these teams must still demonstrate actual capability through certification rather than assuming inherited artifacts substitute for skill building.

Certification as Commitment Signal

Readiness certification serves psychological purposes beyond capability validation. Successfully completing certification builds team confidence that they can navigate transformation challenges. The certification process demonstrates organizational seriousness about proper preparation rather than rushing into implementation.

Teams that invest effort in meeting certification standards develop ownership of transformation success. They understand readiness requirements exist to support them rather than create bureaucratic obstacles. This ownership proves valuable during early sprints when maintaining capability building discipline requires intrinsic motivation. Phase 3 concludes with executive validation that all preparation activities completed successfully and backlog preparation ensuring Sprint 1 has appropriate work ready.

Executive Readiness Validation

[All] The executive sponsor should conduct formal readiness validation covering strategic, technical, and operational dimensions before authorizing Sprint 1 launch. This validation differs from team readiness certification by focusing on organizational preparedness rather than individual capabilities.

Strategic readiness confirms budget approval remains current and stakeholder education produced genuine understanding. It verifies scope amnesty agreements are documented and accepted. For multi-team scenarios, it confirms wave assignments reflect actual priorities. The sponsor verifies these elements through stakeholder conversations rather than document review alone.

Technical readiness validates infrastructure deployment completion, governance framework documentation and team understanding, institutional memory system functionality, and monitoring capabilities for tracking transformation progress. The sponsor should see demonstrations of working systems rather than accepting status reports without verification.

Operational readiness assesses whether the organization can sustain transformation through the cognitive load spike. Adequate delivery capacity must be maintained and support team members through capability building. It also covers managing stakeholder expectations during velocity drops. The sponsor evaluates organizational resolve through frank discussions about what happens when transformation gets difficult.

Checklist Components

The executive checklist should include specific yes-no questions that force honest assessment rather than optimistic assumptions. Can the organization afford nine to eighteen months of reduced velocity from transforming teams? Will stakeholders maintain commitment when features get delayed? Does the team demonstrate genuine readiness through certification rather than claiming preparedness?

Organizations should require evidence for each checklist item rather than accepting verbal confirmation. Budget approval means signed authorization documents not verbal agreement. Stakeholder education means documented session attendance and recorded commitments not claims that conversations occurred.

The checklist should identify decision authority for proceeding versus delaying. The executive sponsor makes the final call but should consult with finance, impacted stakeholders, and team leadership before authorizing launch. Distributed decision-making creates organizational buy-in that concentrated authority cannot achieve.

Backlog Preparation

[All] The Product Strategist must prepare the Sprint 1 backlog during final Phase 3 weeks ensuring work items meet context-ready definition of ready standards. Sprint 1 backlog preparation proves more critical than typical sprint planning because teams lack AI collaboration experience to overcome poorly defined work.

Context-ready work items include explicit acceptance criteria specifying what constitutes completion and relevant technical constraints limiting solution approaches. They contain architectural decisions that guide implementation and examples of similar completed work for AI reference. They also include domain knowledge required for accurate outputs. Items missing these elements create excessive prompt iteration and frustration during Sprint 1.

The backlog should emphasize lower-risk work categories during Sprint 1 to build team confidence. Internal documentation, developer tooling, and test automation provide good initial targets because failures have contained impact. Mission-critical features or customer-facing functionality should wait until Sprint 2 or later when teams demonstrate basic capability.

Work items should offer learning opportunities across different AI collaboration patterns. Sprint 1 might include code generation tasks, documentation creation, test case development, and requirements analysis to give different team members relevant practice. Concentrating all work in single categories leaves some roles underutilized during critical capability building periods.

[C+D] Backlog Considerations for Non-Agile Teams

Non-agile teams require additional backlog preparation because they are simultaneously learning sprint mechanics and AI augmentation. The Product Strategist should create smaller work items than agile teams would use to enable more frequent completion and feedback cycles.

The backlog should include explicit learning objectives alongside delivery objectives. Sprint 1 goals might state teams will deliver three user stories while demonstrating competent prompt engineering and appropriate governance decisions. Making learning explicit prevents teams from feeling they failed if velocity runs low during capability building.

Launch Authorization

The executive sponsor should provide explicit written authorization to proceed to Sprint 1 after reviewing the complete readiness checklist and backlog preparation. This formality ensures the sponsor consciously accepts transformation risks and commits organizational resources rather than drifting into implementation.

The authorization should specify conditions under which the organization would pause or terminate transformation. These include persistent governance violations, failure to show capability improvement, or stakeholder revolt despite education efforts. Defining failure conditions in advance prevents reactive decisions during difficult moments.

The following sections provide scenario-specific implementation guidance. Content marked [All] applies universally. Content marked [A], [B], [C], [D], or combinations like [B+D] applies only to those specific scenarios. Read universal content plus your scenario-specific sections for complete guidance.

Organizational Starting Conditions

Understanding your starting conditions determines which scenario applies and what specific challenges you will face during transformation. The four scenarios reflect fundamentally different organizational realities that require adapted approaches.

[A] Single Traditional Agile Scrum Team

Scenario A organizations operate one Scrum team using traditional agile practices. The team runs two-week sprints, maintains a product backlog, conducts sprint planning and retrospectives, and delivers working software iteratively. The Product Owner prioritizes work and accepts completed features. The Scrum Master facilitates ceremonies and removes impediments. Developers write code, conduct peer reviews, and participate in sprint activities.

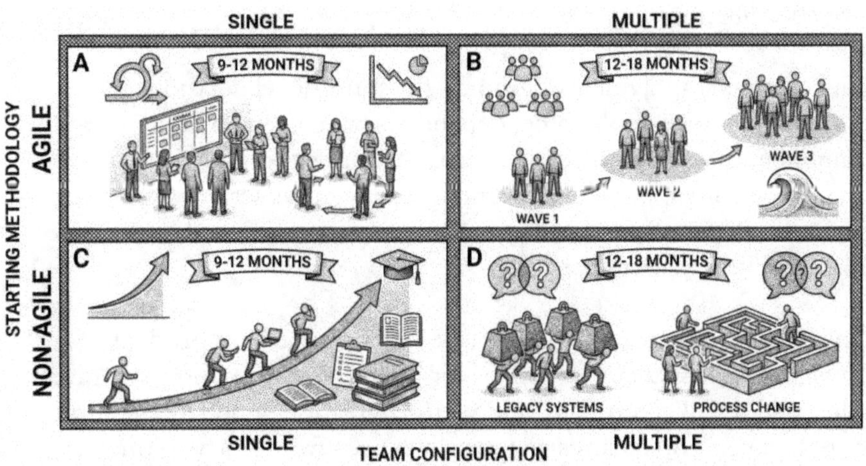

Figure 13.20: Four Transformation Pathway Matrix. *Team configuration and starting methodology create four distinct implementation scenarios. Each scenario has unique characteristics determining transformation complexity, timeline, and required organizational support despite sharing the same fundamental AI Scrum framework.*

The team exists within a larger organization but operates with significant autonomy. The team controls its own velocity, makes technical decisions without extensive approval processes, and adapts practices based on retrospective insights. Adjacent teams may use different methodologies without creating coordination problems.

Common organizational characteristics include product development companies with small to medium codebases and technology teams within larger enterprises that have adopted agile practices. Independent software vendors with focused product offerings also fit this scenario. Team sizes typically range from five to nine people before AI Scrum adoption requires expansion to eleven.

The team demonstrates agile maturity through consistent sprint execution, meaningful retrospectives that drive improvement, and stakeholder trust built through reliable delivery. The organization understands agile principles even if not all teams practice them.

[B] Multiple Traditional Agile Scrum Teams

Scenario B organizations operate three to ten Scrum teams all using traditional agile practices. Teams share common infrastructure, coordinate on architectural decisions, and contribute to the same product or related product suite. The organization has invested in agile coaching, established communities of practice, and developed organizational patterns for multi-team coordination.

Teams maintain individual autonomy for sprint-level decisions but coordinate on release planning, dependency management, and shared component development. The organization uses frameworks like SAFe, LeSS, or custom approaches for multi-team agile at scale. Cross-team synchronization happens through regular meetings, shared backlogs, or program-level planning events.

Common organizational characteristics include software product companies with substantial codebases, digital transformation initiatives within traditional enterprises, and platform teams supporting multiple business units. Total team size typically ranges from thirty to one hundred people across all teams.

The organization demonstrates agile maturity across teams through consistent ceremonies, shared understanding of agile principles, and established patterns for resolving cross-team conflicts. Leadership understands agile economics and supports iterative development approaches.

[C] Single Non-Agile Project Team

Scenario C organizations operate one team using waterfall, proprietary project methodologies, or ad hoc approaches that do not follow agile principles. The team plans work in extended phases, maintains detailed upfront specifications, and uses stage-gate approval processes. Project managers create complete plans, track tasks against schedules, and report status through traditional project management tools.

The team may use some modern development practices like version control or continuous integration but does not organize work into sprints or maintain product backlogs. Planning happens at project inception rather than iteratively. Changes require formal change control processes rather than adaptive prioritization.

Common organizational characteristics include IT departments within traditional industries, custom software development for internal business processes, and professional services firms delivering client projects. Team sizes vary widely but typically range from five to twelve people depending on project scope.

The team demonstrates technical capability through successful project delivery but lacks experience with iterative development, adaptive planning, or empirical process control. The organization may view agile skeptically or simply never considered it for their context.

[D] Multiple Non-Agile Project Teams

Scenario D organizations operate three to ten teams using non-agile methodologies. Teams work on related projects or programs coordinated through traditional program management approaches. The organization uses waterfall, V-model, or proprietary methodologies consistently across teams with established governance for multi-project coordination.

Teams share resources, coordinate on dependencies through detailed planning, and integrate deliverables at defined milestones. The organization has project management offices, standardized templates, and formal processes for scope control, risk management, and stakeholder communication.

Common organizational characteristics include large IT organizations within regulated industries, government agencies delivering technology projects, and systems integrators managing complex implementations. Total team size typically ranges from thirty to one hundred twenty people across all teams.

The organization demonstrates project management maturity through established processes, successful multi-project delivery, and governance frameworks that maintain coordination. Leadership understands traditional project economics but may lack familiarity with agile principles or iterative delivery approaches.

Common Elements Across Scenarios

All scenarios share certain characteristics that make AI Scrum transformation viable. Teams possess technical capability to deliver software successfully using their current methodologies. Organizations have executive support sufficient to fund transformation investments. Stakeholders value productivity improvements enough to tolerate transformation disruption.

Teams demonstrate functional collaboration even if methodologies differ. Developers work together effectively, communicate clearly, and solve problems collaboratively. Organizations maintain basic technical infrastructure including version control, testing capabilities, and deployment processes even if not fully automated.

Leadership exhibits willingness to change processes and invest in capability building. Organizations are not merely seeking incremental improvements but rather transformational productivity gains that justify significant investment and risk-taking.

——PART III: OPERATIONAL PLANNING——

BUDGET FRAMEWORK

AI Scrum transformation requires significant financial investment across infrastructure, training, and opportunity costs during velocity recovery. Executive stakeholders need clear justification showing why the investment produces returns that exceed costs. The budget framework presents ROI before discussing expenses, helping stakeholders understand value creation before considering resource allocation.

The Value Proposition: ROI Before Costs

AI Scrum transformation delivers 115-125% of baseline productivity at sustained excellence (15-25% improvement over original velocity). Organizations should evaluate transformation as strategic investment with measurable return rather than expense requiring justification.

Figure 13.21: Investment Recovery Timeline Visualization. *Most scenarios reach ROI breakeven within 24 to 34 months as sustained productivity gains offset transformation costs. Five-year cumulative value exceeds $400,000 per team under conservative performance assumptions.*

Conservative ROI Scenario (115% productivity)

- Single team investment: $75,000-$150,000 total
- Recovery timeline: 24-34 months
- Annual value after recovery: $100,000-$150,000 ongoing
- Five-year cumulative value: $400,000-$600,000

Expected ROI Scenario (120% productivity)

- Single team investment: Same $75,000-$150,000
- Recovery timeline: 20-28 months
- Annual value after recovery: $140,000-$180,000 ongoing
- Five-year cumulative value: $500,000-$700,000

Organizations already using AI tools and possessing agile infrastructure can achieve these returns with investments toward the lower end of ranges. Organizations building infrastructure from zero should budget toward higher end while still achieving substantial returns within 24-36 months.

Investment Components and Timing

Most transformation costs concentrate in the preparation period (Months – 6 to 0) rather than spreading across full 18-27 month timeline. Organizations make majority of investment decisions upfront enabling clear budgeting and approval process.

Core Investment Categories:

AI Tool Subscriptions ($18,000-$36,000 for single team): Organizations need licenses for AI coding assistants, documentation tools, and collaboration platforms. Timeline duration is 21-27 months including preparation and transformation phases.

[A+B] Many agile teams already use GitHub Copilot, Cursor, or similar tools. Organizations continuing with existing tools subtract this category from incremental transformation costs.

[C+D] Non-agile teams typically need new tool subscriptions representing genuine incremental cost.

Infrastructure Investment ($8,000-$35,000 for single team): Organizations need AI Sandbox environment, institutional memory systems, and delivery platform enhancements. Costs vary dramatically based on starting point.

[A+B] Agile teams with existing version control, CI/CD pipelines, and testing frameworks need only AI Sandbox and institutional memory systems ($8,000-$15,000). Infrastructure investment occurs during Phase 2 (Months -6 to -3).

[C+D] Non-agile teams building complete delivery platforms from zero require full infrastructure investment ($15,000-$35,000). This investment delivers value even without AI Scrum adoption by establishing modern development practices.

Training Investment ($12,000-$35,000 for single team): Teams require Deep Immersion Week training in Month -1 covering AI collaboration fundamentals, prompt engineering, and governance frameworks.

[A+B] Agile teams need 5-day training focused on AI augmentation ($12,000-$20,000).

[C+D] Non-agile teams need 6-day training covering agile foundations plus AI augmentation ($20,000-$35,000).

Organizations with internal trainers possessing genuine AI collaboration expertise can reduce training costs 40-50% by conducting Deep Immersion Week internally. Organizations should validate trainer capability by demonstrating actual prompt engineering skills rather than theoretical knowledge.

Consulting Support (Optional: $0-$60,000 for single team): Organizations may choose to engage external transformation coaches for additional support during preparation (Months -6 to 0) and early transformation (Sprints 1-8). This consulting support is optional rather than required. Teams with strong internal agile coaching capabilities or prior AI Scrum experience often proceed successfully without external consultants.

Organizations choosing external support typically invest based on internal capability and transformation complexity:

- Strong internal capability with AI experience: $0 (no external support needed)
- Moderate internal capability: $10,000-$20,000 for governance framework design and initial training review
- Limited AI Scrum experience: $40,000-$60,000 for intensive coaching through preparation and early sprints

External coaches primarily support governance framework design, Deep Immersion Week training delivery, and early sprint troubleshooting. Organizations should evaluate whether internal staff can fulfill these functions before budgeting external support.

Change Management ($0-$20,000 for single team): Single teams typically absorb change management within existing organizational functions. Multi-team implementations and non-agile transformations require dedicated change management support.

- [A] Single agile teams: No incremental change management investment
- [B] Multiple agile teams: $15,000-$35,000 for cross-team coordination
- [C] Single non-agile teams: $10,000-$20,000 for cultural transition support
- [D] Multiple non-agile teams: $35,000-$70,000 for full change management

Calculating Your Organization's Investment:
Product Strategists should estimate transformation investment during Month -9 using this framework:

1. **Start with scenario baseline:** Identify whether you are [A], [B], [C], or [D]
2. **Subtract existing investments:** Remove categories where infrastructure or tools already exist
3. **Adjust for internal capability:** Reduce consulting and training if strong internal expertise exists
4. **Add opportunity cost:** Calculate velocity shortfall during transformation (see methodology below)
5. **Present total investment:** Direct costs + Opportunity costs = Total transformation investment

Example for [A] scenario (single agile team) with existing tools

- AI Tool Subscriptions: $0 (already using GitHub Copilot)
- Infrastructure: $10,000 (AI Sandbox + institutional memory)
- Training: $15,000 (external 5-day Deep Immersion Week)
- Consulting: $0 (strong internal agile coaching, no external support needed)
- Change Management: $0 (absorbed in existing functions)
- Miscellaneous: $5,000 (documentation, licenses, etc.)
- **Direct Costs: $30,000**
- Opportunity Cost: $210,000 (calculation below)
- **Total Investment: $240,000**
- **Recovery Timeline: 20-24 months at 120% productivity**

If this same organization chose to engage external coaching support for governance design and early sprint troubleshooting, consulting would add $15,000-$20,000. This brings direct costs to $45,000-$50,000 and total investment to $255,000-$260,000.

Opportunity Cost: Calculating Velocity Shortfall

Opportunity cost represents business value foregone during velocity reduction in Sprints 1-18. Organizations should include opportunity cost in total investment calculations for realistic ROI projections.

Calculation Methodology:

1. **Establish baseline velocity:** Use previous 6-12 sprints average under current framework (example: 40 story points per sprint)
2. **Project transformation velocity:** Apply J-curve pattern to estimate actual delivery:
    - Sprints 1-4: 40-60% of baseline (example: 24 points per sprint)
    - Sprints 5-10: 60-90% of baseline (example: 30 points per sprint)

    - Sprints 11-18: 95-110% of baseline (example: 38 points per sprint)
3. **Calculate story point shortfall:** Sum the difference between baseline and projected for each sprint phase
4. **Apply business value per story point:** Typical range $1,000-$5,000 depending on industry and product value

Example Calculation (40 point baseline, $1,500 per point)

- Sprints 1-4: (40 - 24) × 4 sprints = 64 points foregone
- Sprints 5-10: (40 - 30) × 6 sprints = 60 points foregone
- Sprints 11-18: (40 - 38) × 8 sprints = 16 points foregone
- **Total shortfall: 140 story points**
- **Opportunity cost: 140 × $1,500 = $210,000**

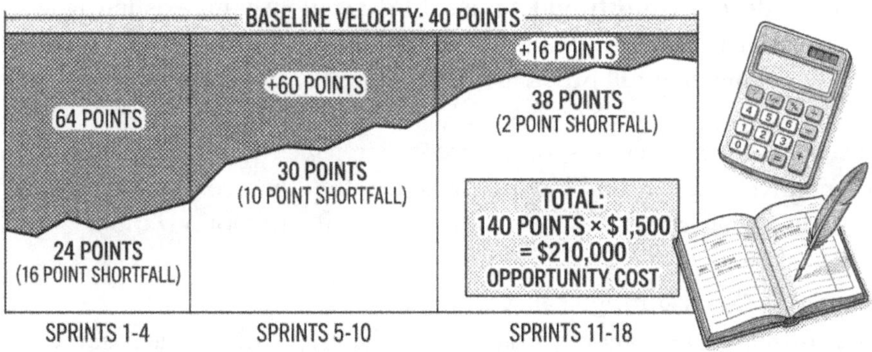

Figure 13.22: Quantifying Velocity Shortfall Impact. *Opportunity cost represents business value foregone during velocity reduction across eighteen sprints. Organizations calculate total story point shortfall and multiply by business value per point to determine complete transformation investment including foregone delivery capacity.*

Organizations with higher business value per story point face larger opportunity costs but also capture larger productivity gains after reaching sustained excellence. Software-as-a-service products often show $3,000-$5,000 per story point while internal tools may show $1,000-$2,000 per point.

Presenting Investment to Executive Stakeholders

Product Strategists should structure financial presentations during Month -9 using value-first messaging:

Recommended Presentation Flow:

1. **Frame the competitive necessity:** "Competitors adopting AI Scrum gain compounding productivity advantages. Delaying transformation costs more than executing it."

2. **Present productivity improvement potential:** "AI Scrum delivers 115-125% of current productivity at sustained excellence. Conservative scenario shows 115% of baseline (15% improvement)."
3. **Show investment required:** "Total investment including opportunity cost ranges from $255,000 to $425,000 for single team depending on existing infrastructure."
4. **Demonstrate recovery timeline:** "Conservative scenario recovers investment in 24-34 months. Expected scenario recovers in 20-28 months. Organization gains positive cash flow for decades after recovery."
5. **Emphasize five-year value:** "Five-year cumulative value exceeds $400,000 under conservative assumptions. This represents 150%+ ROI on transformation investment."
6. **Address conditional costs:** "Organizations with existing agile infrastructure and AI tools can reduce direct costs 30-40% while maintaining full productivity benefits."

Executive stakeholders respond better to strategic framing emphasizing competitive necessity and long-term value rather than cost minimization messaging suggesting transformation is discretionary optimization.

Managing Costs During Transformation

Transformation costs concentrate in specific preparation phases rather than spreading evenly across the timeline. Understanding when expenses peak helps organizations plan cash flow and budget approvals effectively.

Cost Concentration Timeline:

Most transformation costs occur during specific phases rather than spreading evenly:

- **Months -6 to -3 (Phase 2):** Infrastructure investment, platform setup, governance framework design (30-40% of direct costs)
- **Month -1:** Deep Immersion Week training (15-25% of direct costs)
- **Months 0-12:** Consulting support, tool subscriptions, ongoing training (40-50% of direct costs)
- **Months 13-21:** Primarily tool subscriptions and minor consulting (10-15% of direct costs)

Organizations can budget across fiscal years by timing transformation start to align major expenditures with budget availability.

Monthly Cost Monitoring:
Product Strategists should track actual costs against budget monthly to detect overruns before they become significant:

- AI tool subscriptions: Predictable fixed costs requiring minimal monitoring
- Infrastructure: Front-loaded in Phase 2, monitor for scope creep during buildout
- Training: Concentrated in Month -1, manage consultant scope to prevent session expansion
- Consulting: Highest risk category, track hours closely against original engagement scope

When actual costs exceed budget by 10% in any category, Product Strategist escalates to executive sponsor to determine whether budget increase is warranted or spending reduction is necessary.

Preventing Premature Abandonment:
Organizations experiencing velocity trough in Sprints 4-8 sometimes consider abandoning transformation to restore immediate productivity. This represents worst possible financial outcome because organization has invested 60-70% of total costs while capturing zero productivity benefits.

When executives raise abandonment concerns, Product Strategists should present:

1. **Sunk cost calculation:** Show investment already committed through preparation and early sprints
2. **Recovery timeline:** Project specific sprint numbers when velocity returns to baseline and exceeds it
3. **Abandonment cost:** Demonstrate that stopping wastes all investment with no return
4. **Continuation cost:** Show modest additional investment required to capture full ROI
5. **Competitive impact:** Quantify advantage competitors gain if organization abandons while they persist

Organizations that persist through velocity trough reach sustained excellence capturing full ROI. Organizations that abandon waste preparation investment and must restart from zero if attempting transformation again later.

Scenario-Specific Cost Considerations:
[A] Single agile teams can reduce costs through internal training delivery if trainers possess genuine AI collaboration expertise. Validate trainer capability through demonstrations rather than credentials before committing to internal delivery. Many single agile teams with strong internal coaching capabilities proceed without external consulting support, reducing total direct costs to $30,000-$50,000 range depending on existing infrastructure.

[B+D] Multi-team organizations should spread costs across fiscal years aligned with wave timing. Wave 1 costs concentrate in year one. Later wave costs span years one and two. Organizations can pause between waves for budget management though pauses exceeding four months risk institutional memory degradation. Multi-team organizations more frequently benefit from external consulting to establish consistent governance frameworks across waves, though this remains optional based on internal capability.

[C+D] Non-agile teams should present infrastructure investment as delivering dual benefits: enabling AI Scrum transformation and establishing modern development practices valuable regardless of AI adoption. This framing reduces perceived AI-specific risk when seeking executive approval. Non-agile teams learning both agile principles and AI augmentation simultaneously may find external coaching valuable. Evaluation against internal capability should occur rather than assuming external support is required.

Organizations should resist pressure to reduce costs below scenario-appropriate ranges. Underfunded transformations fail more frequently than adequately funded transformations. Organizations needing lower investment should delay transformation until budget becomes available rather than proceeding with insufficient resources.

WHAT NOT TO AI-AUGMENT DURING INITIAL SPRINTS

[All] AI Scrum requires deliberate restraint about where teams use AI during early transformation sprints. Organizations that allow unrestricted AI usage across all work categories risk quality failures, compliance violations, and catastrophic production incidents that undermine stakeholder confidence.

Prohibition Philosophy

The prohibition approach recognizes that teams must develop AI collaboration capabilities before using AI for high-stakes work. Early sprint teams lack the validation skills, governance judgment, and pattern recognition required to catch subtle AI failures in critical systems.

Prohibitions serve learning purposes beyond risk mitigation. Teams building AI collaboration skills on low-risk work develop competence without facing consequences from failures. A team learning prompt engineering through internal documentation can iterate freely. A team learning on payment processing faces unacceptable risk from each iteration.

Prohibitions are explicitly temporary and capability-dependent rather than permanent or time-based. Teams demonstrate capability through governance checkpoint assessments and Chapter 9 transformation metrics. As scores improve, prohibitions relax systematically. High-performing teams eventually use AI across nearly all work categories with appropriate validation.

The prohibition framework balances two competing needs. Organizations need confidence that AI usage will not cause production failures or compliance violations. Teams need sufficient AI usage opportunities to build genuine capabilities rather than theoretical knowledge. Prohibitions too restrictive prevent capability building. Prohibitions too permissive create unacceptable risks.

High-Risk Work Categories

Chapter 10 defines governance risk categories and validation requirements for AI usage across different work types. Organizations should prohibit AI usage for high-risk categories during sprints 1-4 regardless of scenario. These universal prohibitions include:

Prohibited During Sprints 1-4

- Security controls and authentication logic
- Payment processing and financial transactions
- Regulated data handling (PII, PHI, compliance-sensitive)
- Production database schema modifications
- Production infrastructure and deployment configurations
- Customer-facing critical paths (checkout, registration)

Enforcement During Sprint Planning: Product Strategists identify prohibited work during backlog refinement before sprint planning begins. Work items requiring prohibited AI usage get tagged with governance markers in the delivery platform. Teams either defer prohibited items to later sprints when capabilities mature or implement them manually during early sprints. The Governance Coordinator reviews sprint commitments during Sprint Inception ceremony to verify no prohibited work entered the sprint inappropriately.

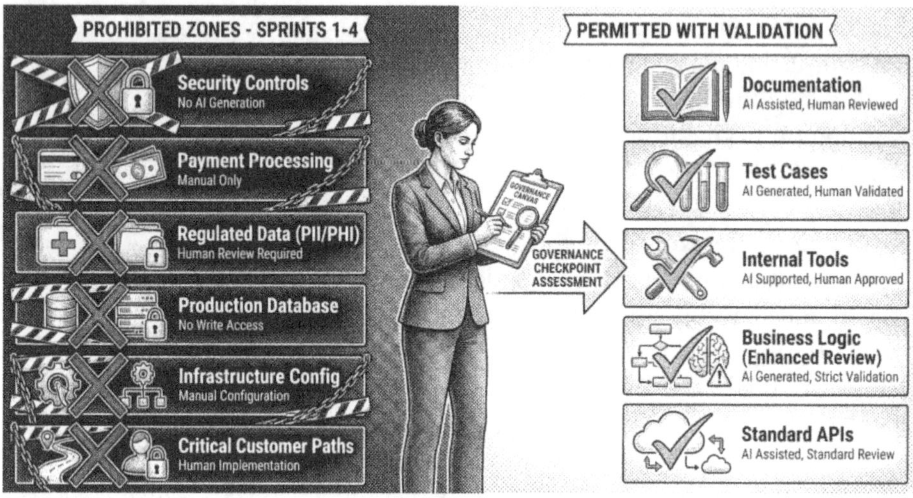

Figure 13.23: Risk-Based AI Usage Boundaries. *Organizations prohibit AI usage for high-risk work categories during early sprints when teams lack validation capabilities to catch subtle failures. Restrictions relax systematically as teams demonstrate competency through governance checkpoint assessments and transformation metrics.*

Handling Violations: Teams discovering accidental AI usage on prohibited work during sprint execution must immediately halt the work item. The Validation Lead conducts detailed security review of any AI-generated code even if not deployed. The Governance Coordinator documents the violation during the next governance checkpoint to identify process gaps preventing recurrence. Repeated violations indicate inadequate sprint planning governance or stakeholder pressure requiring executive intervention.

Chapter 10 provides detailed risk analysis and validation requirements for each category. Teams implement prohibited components manually during early sprints while documenting approaches in institutional memory. This documentation becomes valuable context for later AI-assisted work when teams possess adequate validation capabilities.

Scenario-Specific Considerations

[C+D] Non-agile teams should apply stricter prohibitions than agile teams during early sprints because they are simultaneously learning agile practices and AI collaboration. These teams should extend prohibitions through sprint six rather than sprint four for high-risk categories.

[B+D] Multi-team organizations must maintain consistent prohibitions across all teams within a wave. Individual teams cannot decide to relax prohibitions independently because this creates governance inconsistency and knowledge sharing challenges. Wave-level prohibition policies ensure all teams operate under aligned constraints.

Organizations in highly regulated industries should extend prohibitions beyond sprint four regardless of scenario. Healthcare, financial services, and government organizations face compliance requirements that demand additional caution. These organizations might maintain prohibitions through sprint eight or ten while building complete validation capabilities.

Conditional AI Usage Rules

Chapter 10 establishes conditional usage policies for moderate-risk work categories permitting AI assistance with enhanced validation during early sprints:

Permitted with Enhanced Validation (Sprints 1-4)

- Business logic implementation: Requires two-developer plus Validation Lead review
- Standard CRUD API endpoints: Validation matching business logic requirements
- Read-only database queries: Validation Lead review mandatory
- Test case generation: Extensive usage permitted with coverage verification
- Documentation and comments: Light validation standards acceptable

Enhanced Validation in Practice: Two-developer review means the AI-generating developer plus one peer developer both examine AI outputs independently before Validation Lead final review. The first peer developer verifies logical correctness and adherence to business requirements. The second review by Validation Lead checks security implications, performance characteristics, and integration patterns. Each reviewer signs off in pull request comments documenting specific aspects validated.

Effective enhanced validation catches subtle errors AI introduces. A reviewer might notice that AI-generated business logic handles typical customer scenarios correctly but fails edge cases like zero-quantity orders or negative inventory adjustments. Poor validation simply confirms code compiles and basic tests pass without examining business logic accuracy or edge case handling.

Teams document validation checklists in institutional memory covering common AI failure modes for each work category. Business logic checklists verify edge case handling, error conditions, and boundary value testing. API endpoint checklists confirm authentication, input validation, error response formats, and rate limiting. Query checklists validate join correctness, filtering logic, and performance against production data volumes.

Organizations reference Chapter 10 for complete validation requirements and governance policy frameworks for each category.

Restriction Relaxation Timeline

Teams relax prohibitions based on demonstrated capability through Chapter 9 metrics rather than rigid time schedules. Organizations establish specific metric thresholds triggering restriction relaxation:

Relaxation Trigger Criteria

- AI First-Pass Acceptance Rate sustained above 60% for three consecutive sprints
- Validation Lead Time consistently under 3 days for four weeks
- Cognitive Load Score below 50 for three consecutive assessments
- Zero governance violations during previous two governance checkpoints
- Sprint Outcome Achievement meeting or exceeding baseline from previous framework

Teams meeting these thresholds submit relaxation requests to the Governance Coordinator during governance checkpoint ceremonies. The Governance Coordinator verifies metric evidence and consults with executive sponsors before approving policy changes. Relaxation proceeds incrementally starting with lowest-risk prohibited categories before advancing to higher-risk work.

Typical Relaxation Progression

- Sprints 5-10: Security and financial work with mandatory security-focused code review and penetration testing validation
- Sprints 8-12: Database schema modifications with database administrator oversight and staged deployment
- Sprints 10-15: Production infrastructure changes with extensive testing and gradual rollout
- Sprint 15+: Critical customer paths with thorough testing and phased user exposure

Teams failing to meet metric thresholds maintain prohibitions beyond standard timelines. A team at Sprint 8 with acceptance rates below 55% continues prohibited work restrictions until metrics improve. Organizations prioritize capability development over artificial timeline adherence.

- [C+D] Non-agile teams should extend prohibitions through Sprint 6 rather than Sprint 4 for high-risk categories due to dual learning curves.

- [B+D] Multi-team organizations must maintain consistent prohibition policies across all teams within each wave to ensure governance alignment and enable knowledge sharing.

Organizations in highly regulated industries may extend prohibitions through Sprint 8-10 while building full validation capabilities regardless of scenario.

Managing Pressure for Early Relaxation

Teams facing delivery pressure often request early prohibition relaxation to use AI for high-risk work before demonstrating adequate capabilities. Organizations must resist this pressure because premature relaxation creates exactly the failures that undermine transformation credibility.

Delivery Facilitators should help teams understand that prohibitions protect transformation success rather than impede productivity. Teams that violate prohibitions and cause production incidents face stakeholder backlash that threatens transformation continuation.

Organizations should establish clear escalation paths for cases where prohibitions genuinely prevent critical work completion. Executive sponsors can grant one-time exceptions with enhanced validation requirements and explicit risk acceptance rather than broadly relaxing prohibitions.

The escalation process should require written justification explaining why the work cannot wait until teams develop adequate capabilities and what additional validation will mitigate risks. Most requests for early relaxation fail to meet these criteria when examined rigorously.

TRANSFORMATION TIMELINE FRAMEWORK

[All] AI Scrum transformation follows predictable phases from preparation through sustained excellence. Understanding timeline expectations prevents premature abandonment when teams experience temporary velocity drops and enables realistic stakeholder communication.

Universal Timeline Components

All scenarios include the nine-month preparation period from months minus nine through zero. This duration cannot be compressed without compromising transformation foundations. Organizations attempting shorter preparation face higher failure rates during implementation.

All scenarios experience the cognitive load spike during initial sprints where velocity drops significantly as teams learn new workflows. The depth and duration of this spike vary by scenario but the pattern remains consistent.

All scenarios require recovery climb periods where velocity gradually returns toward baseline. Teams apply learned patterns with increasing confidence while continuing to build capabilities.

All scenarios must reach parity point where velocity returns to previous baseline before claiming transformation success. Sustained excellence follows parity point as teams exceed baseline performance consistently.

Timeline Phases Defined

The cognitive load spike represents sprints one through four for most scenarios. Teams operate at 40-60% of baseline velocity depending on starting conditions. This phase proves psychologically challenging because teams feel incompetent despite investing significant preparation effort.

The recovery climb spans sprints five through ten where teams apply learned patterns with decreasing struggle. Velocity gradually improves from the trough toward baseline. Teams begin experiencing benefits from AI collaboration while still below previous productivity levels.

The parity point occurs between sprints eleven and eighteen when velocity returns to baseline levels. Exact timing depends on team capability development, organizational support quality, and how aggressively teams push AI adoption.

Sustained excellence begins after teams consistently exceed baseline productivity by 115% or more for at least four consecutive sprints. This threshold typically occurs between sprints nineteen and twenty-four depending on scenario.

Figure 13.24: Complete Transformation Journey Timeline. *AI Scrum transformation follows predictable phases from nine-month preparation through cognitive load spike, recovery climb, parity achievement, and sustained excellence. Understanding phase durations enables realistic stakeholder communication and prevents premature transformation abandonment.*

Timeline Dependencies

Timeline achievement depends on several critical factors beyond simple time passage. Team capability development through deliberate practice determines how quickly teams progress through phases. Teams that practice AI collaboration extensively reach parity faster than teams that avoid AI usage.

Organizational support quality affects timeline through stakeholder patience, resource availability, and psychological safety maintenance. Teams facing stakeholder pressure during the cognitive load spike often revert to traditional practices delaying capability building.

Infrastructure stability influences timeline through friction reduction or friction creation. Teams with reliable sandboxes, functional institutional memory, and smooth delivery platforms progress faster than teams fighting technical issues.

Governance framework appropriateness matters because overly restrictive policies prevent adequate AI usage while overly permissive policies create quality issues requiring remediation. Governance frameworks that match team capabilities enable optimal learning.

Timeline Comparison by Scenario

Each scenario follows the same preparation and transformation phases but with different durations and activities. This comparison shows how timelines vary based on team count and starting methodology.

Phase	[A] Single Agile	[B] Multiple Agile	[C] Single Non-Agile	[D] Multiple Non-Agile
Preparation	9 months	9 months	9 months	9 months
Cognitive Load Spike	Sprints 1-4 (50-60% velocity)	Wave-dependent	Sprints 1-6 (40-50% velocity)	Wave-dependent
Recovery Climb	Sprints 5-10 (60-95% velocity)	Wave-dependent	Sprints 7-12 (60-95% velocity)	Wave-dependent
Parity Point	Sprints 11-18 (95-110% velocity)	Wave-dependent	Sprints 13-18 (95-110% velocity)	Wave-dependent
Sustained Excellence	Sprints 19-22 (115%+ velocity)	Sprints 19-24 (115%+ velocity)	Sprints 19-24 (115%+ velocity)	Sprints 19-30 (115%+ velocity)
Post-Prep Duration	9-12 months	12-18 months	9-12 months	12-18 months
Total Duration	18-21 months	21-27 months	18-21 months	21-27 months

Preparation Phase (All Scenarios: Months -9 to 0)

The nine-month preparation period divides into three phases with distinct activities and deliverables. All four scenarios follow the same preparation timeline, though content varies by team configuration and starting methodology.

Phase 1 (Months -9 to -6): Strategic alignment and budget approval. Executive sponsor secures commitment, develops financial models, conducts stakeholder education.

Phase 2 (Months -6 to -3): Infrastructure and governance implementation. Organizations deploy AI Sandbox, create Governance Canvas, build institutional memory systems, enhance delivery platforms. [B+D] Multi-team organizations establish cross-team coordination mechanisms.

Phase 3 (Months -3 to 0): Tactical readiness. Teams participate in deep immersion week during month -1 ([A+B] five days, [C+D] six days), complete readiness certification, prepare Sprint 1 backlog. [B+D] Only

Wave 1 teams train during initial Phase 3; later waves train before their respective launches.

Transformation Phases (Sprint-by-Sprint)

AI Scrum transformation follows predictable sprint-by-sprint patterns across four distinct phases. Teams progress from cognitive load spike through recovery climb to parity point and sustained excellence. Understanding these phases helps organizations set realistic expectations and identify when teams deviate from expected patterns.

Cognitive Load Spike (Sprints 1-4)

- Sprint 1: Velocity drops to 50-60% ([A+B]) or 40-50% ([C+D]) of baseline
- Sprint 2: Velocity trough at 45-55% ([A+B]) or 35-45% ([C+D])
- Sprint 3: Initial improvement to 50-60% ([A+B]) or 40-50% ([C+D])
- Sprint 4: Cognitive load spike ends at 55-65% ([A+B]) or 50-60% ([C+D])

Recovery Climb

- [A+B] Sprints 5-10: Progressive recovery from 60-70% to 85-95%
- [C+D] Sprints 5-12: Extended recovery from 60-70% to 85-95% due to dual learning curve
- First governance checkpoint typically occurs Sprint 5 or 6
- Second governance checkpoint Sprint 9 or 10

Parity Point

- [A+B] Sprints 11-18: Teams achieve 95-105% baseline, then consistently exceed at 110-120%
- [C+D] Sprints 13-18: Similar pattern with delayed entry due to extended recovery
- Psychological significance: Teams match previous productivity using AI-augmented approaches

Sustained Excellence

- [A+C] Sprints 19-22: Single teams reach 115%+ productivity
- [B] Sprints 21-24: Multiple agile teams complete (wave-dependent)
- [D] Sprints 24-30: Multiple non-agile teams complete (wave-dependent, highest variability)
- Transformation ends when teams operate at sustained excellence without special support

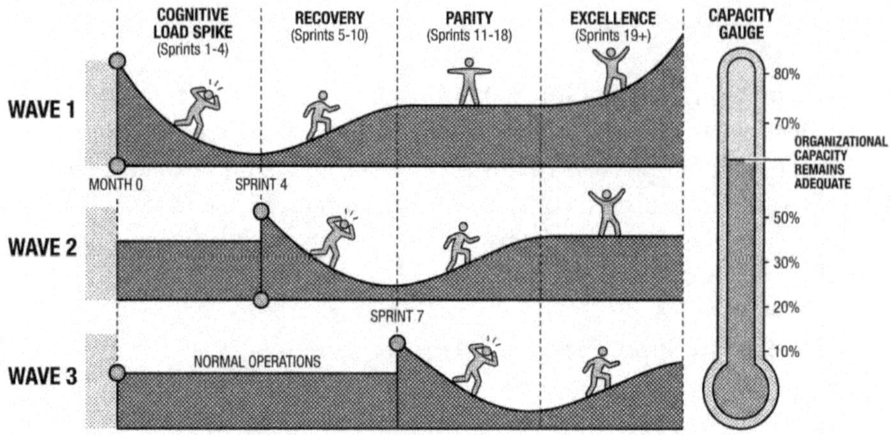

Figure 13.25: Wave-Staggered Transformation Coordination. *Multi-team organizations launch waves sequentially preventing simultaneous velocity drops across all teams. Staggered timing ensures organizational capacity remains adequate while each wave progresses through its capability formation journey toward sustained excellence.*

Wave-Based Timeline Adjustments ([B+D] Only)

Multi-team organizations stage transformation across waves to prevent simultaneous velocity drops that would cripple organizational capacity. Each wave follows the same phase progression but launches at different times to maintain overall delivery capability.

Wave 1: Follows timeline patterns above based on agile vs. non-agile starting point

Wave 2: Launches at Sprint 4 or 5 with one to three additional teams. Wave 2 enters extended cognitive load spike while Wave 1 begins recovery. Total organizational capacity drops by approximately 30% as multiple teams operate significantly below baseline simultaneously. This represents the maximum organizational capacity impact during transformation.

Wave 3: Launches at Sprint 6 or 7 with one to three additional teams. Wave 3 enters extended cognitive load spike while Wave 1 approaches parity and Wave 2 begins recovery. Total capacity reduction moderates to approximately 25% because earlier waves offset later wave drops. The staging prevents all teams from experiencing the cognitive load spike simultaneously.

Overall Wave Timeline: Organizational capacity gradually improves as all waves sequentially reach parity then sustained excellence. Wave 1 reaches sustained excellence around Sprint 19, Wave 2 around Sprint 23-24, and

Wave 3 around Sprint 25-26. Organizations experience meaningful capacity constraints through approximately Sprint 15, requiring careful stakeholder management and realistic delivery commitments.

Agile Adoption Complexity

Multiple non-agile teams adopting agile simultaneously create cultural change challenges beyond individual team transformations. Organizations must establish agile governance suitable for novices, develop cross-team coordination patterns without shared agile vocabulary, and maintain stakeholder confidence in iterative delivery.

Agile coaches prove essential for Scenario D success. Organizations should engage experienced coaches familiar with large-scale agile adoption to supplement AI collaboration consultants. The coaching requirement extends through at least Wave 2 completion.

Cross-team agile practices including program increment planning, dependency management, and release coordination must accommodate teams learning agile fundamentals. Standard scaled agile frameworks assume team-level agile maturity that Scenario D teams lack initially.

Change Management Requirements

Scenario D demands complete change management addressing resistance to both agile adoption and AI integration. Non-agile organizations often have cultures emphasizing detailed planning, thorough documentation, and phase-gate approvals that conflict with agile principles.

Change management should address concerns proactively through stakeholder education, pilot successes from Wave 1, and transparent communication about transformation progress. Organizations should allocate dedicated change management resources through at least month eighteen.

Leadership alignment proves critical because Scenario D transformation spans multiple organizational levels. Executive sponsors must maintain commitment through extended periods of reduced capacity. Middle management must support teams through methodology changes. Team members must embrace entirely new working approaches.

Timeline Variability and Risk

Scenario D faces highest timeline variability among all scenarios. Extended cognitive load spikes can persist through sprint eight rather than sprint six if teams struggle with dual learning curves. Recovery climbs can extend to sprint sixteen rather than sprint twelve if organizational support proves inadequate.

Organizations should build twelve-month contingency into Scenario D timelines. Planning for thirty-three months total with possibility of extending to forty-five months creates realistic expectations and prevents premature transformation abandonment.

The complexity of Scenario D makes it unsuitable for organizations lacking strong executive commitment, substantial transformation budgets, and tolerance for extended capacity constraints. Organizations questioning their ability to sustain three-year transformation efforts should reconsider whether Scenario D transformation aligns with their strategic priorities and organizational capabilities.

STAKEHOLDER ALIGNMENT STRATEGY

[All] Stakeholder alignment determines whether organizations can sustain transformation through the cognitive load spike when velocity drops create pressure to abandon AI Scrum. Effective alignment requires transparent communication about transformation realities, explicit agreements about acceptable performance, and consistent reporting approaches.

The J-Curve Education Campaign

Chapter 11 addresses stakeholder education strategy for transformation initiatives. The J-curve education campaign represents implementation-specific application of those principles to AI Scrum velocity patterns.

Campaign Timing and Structure: Organizations conduct J-curve education through four sessions during Phase 1 preparation (Months -9 to -6). Multiple sessions spaced across three months provide adequate reinforcement for stakeholders to internalize transformation implications. Single presentations prove insufficient because stakeholders need repeated exposure to accept velocity degradation as necessary transformation cost.

Session 1 (Month -9): Introduces AI Scrum concepts and productivity improvement potential. Stakeholders learn what AI augmentation means and why organizations pursue transformation. This session establishes context for subsequent detailed discussions about velocity impacts.

Session 2 (Month -8): Presents the J-curve pattern with scenario-specific velocity projections using actual story point estimates. Stakeholders see concrete numbers showing velocity at baseline (current), during trough (40-60% of baseline in Sprints 1-4), through recovery (60-90% in Sprints 5-10), and at sustained excellence (115%+ at Sprint 19+). Visual graphs prove more effective than verbal descriptions.

Product Strategists present graphs showing two lines: baseline velocity (flat horizontal line at current performance) and projected AI Scrum velocity (J-curve showing dip, recovery, then exceeding baseline). Graphs include specific sprint numbers on x-axis and story points on y-axis. Stakeholders need concrete visualization: "You currently deliver 60 story points per sprint. During Sprints 1-4, expect 24-36 points. By Sprint 12, we return to 60 points. By Sprint 20, we deliver 70-75 points sustainably."

Session 3 (Month -7): Covers organizational impacts including resource allocation, delivery timeline adjustments, and stakeholder role changes during transformation. Stakeholders understand how transformation affects their responsibilities and what support teams need during capability building.

Session 4 (Month -6): Finalizes scope amnesty agreements documented in stakeholder contracts. Stakeholders commit to specific performance expectations through each transformation phase and agree on reporting approaches teams will use.

Handling Stakeholder Resistance During Campaign:
Some stakeholders reject J-curve education even after four sessions. Common objections include "We cannot afford velocity drops," "Competitors will gain advantage," or "Customers have commitments we must meet."

Solution Protocol - Affordability Objections: Product Strategists present total cost analysis showing that delaying transformation costs more than accepting short-term velocity reduction. Calculate competitor advantage if they adopt AI Scrum while organization delays. Show compounding productivity gap: "If competitors transform now, they will deliver 115% productivity while we remain at 100%. Over 12 months, that 15% gap means they deliver 9 additional sprints of work. We cannot catch up without transforming."

Solution Protocol - Competitive Concerns: Executive sponsors address competitive timing by reframing transformation as competitive necessity rather than optional improvement. Present market research showing AI adoption rates in industry. Position transformation as defensive move preventing competitive disadvantage rather than optional productivity optimization.

Solution Protocol - Customer Commitment Conflicts: Product Strategists work with stakeholders to identify which customer commitments are genuinely immovable versus which can be renegotiated. For immovable commitments, consider:

- Delaying transformation start until after critical delivery milestone
- Staffing temporary traditional team for critical work while main team transforms
- Reducing transformation team to partial squad working critical path while remaining organization transforms

If stakeholders remain opposed after Session 4, executive sponsors must make explicit go/no-go decision. Proceeding with transformation despite stakeholder opposition guarantees failure when stakeholders undermine scope amnesty during velocity trough.

Managing Mid-Transformation Stakeholder Panic:
Even stakeholders who agreed to J-curve contracts may panic when experiencing actual velocity drops at Sprint 4-6.

Solution Protocol - Week 6 Panic Response: Product Strategists anticipate Week 6 panic by proactively scheduling stakeholder check-in during Sprint 5 or 6. Check-in agenda includes:

1. Review stakeholder contract velocity projections showing current performance matches expectations
2. Present Chapter 9 metrics showing upward trajectory in AI First-Pass Acceptance Rate and other capability indicators
3. Show institutional memory growth and governance checkpoint improvements demonstrating capability formation progress
4. Reinforce that Sprint 8-12 recovery phase begins soon with specific projections

When stakeholders demand immediate productivity recovery despite contract terms, executive sponsors intervene using specific language: "Our contract specified 40-60% velocity during this phase. The team is performing at 48% which meets our agreement. The metrics show capability formation is progressing. Demanding immediate recovery would force the team to abandon AI collaboration and revert to traditional practices, destroying our transformation investment. We honor our contract and maintain course."

Solution Protocol - Feature Demand Management: Stakeholders may submit critical feature requests during scope amnesty period despite contract prohibitions. Product Strategists evaluate each request using triage criteria:

- Business-critical (revenue loss, compliance violation, security breach): Escalate to executive sponsor for potential exception
- Important but not critical (customer satisfaction, competitive feature): Defer to post-amnesty backlog
- Nice-to-have: Decline explicitly with reference to scope amnesty contract

For business-critical exceptions, executive sponsor approves only after verifying genuine criticality. Approved exceptions get assigned to team members outside sprint commitment as interrupt work, not added to sprint backlog. Team maintains focus on capability building while critical work receives coverage.

[A+C] Single team stakeholders receive education focused on one team's transformation journey emphasizing contained impact where only this team experiences velocity changes.

[B+D] Multi-team stakeholders require additional education about wave staging and how wave implementation prevents simultaneous velocity drops across all teams maintaining adequate organizational capacity. Product Strategists show organizational capacity graphs demonstrating that Wave 1 transformation affects only 20-25% of total capacity while Waves 2-4 remain at full productivity.

[C+D] Non-agile team stakeholders need additional Session 0 (Month -10) covering agile fundamentals before addressing AI augmentation. This session introduces iterative development, adaptive planning, and empirical process control providing foundation for understanding why velocity fluctuates during transformation.

The Stakeholder Contract

Chapter 11 establishes stakeholder alignment principles for transformation initiatives. The stakeholder contract represents formal documentation of those commitments specific to AI Scrum transformation.

Contract Components:

Organizations create written stakeholder contracts during Month -6 of Phase 1 preparation following J-curve education campaign completion.

Written documentation provides accountability that verbal agreements cannot sustain through transformation difficulties.

Velocity Expectations: Contract specifies velocity targets through each transformation phase. Sprints 1-4 operate at 40-60% of baseline. Sprints 5-10 recover to 60-90% of baseline. Sprints 11-18 reach parity at 95-110% of baseline. Sprint 19+ achieves sustained excellence at 115%+ of baseline. These percentages should reflect scenario-specific expectations where [C+D] teams may require extended timelines.

Scope Amnesty Periods: Contract defines sprints where stakeholders forgive velocity shortfalls without demanding explanations or corrective action. Amnesty typically covers Sprints 1-10 when teams build capabilities rather than deliver maximum productivity. Contract explicitly prohibits new feature requests during amnesty periods except critical business needs requiring executive sponsor approval.

Reporting Requirements: Contract establishes weekly or biweekly status updates during transformation. Reports include velocity actuals versus projections from stakeholder contract, Chapter 9 metric trajectories (AI First-Pass Acceptance Rate, Validation Lead Time, Cognitive Load Score), and governance checkpoint outcomes. Reporting format should match stakeholder preferences balancing transparency with overhead minimization.

Escalation Paths: Contract specifies how teams raise transformation challenges and how stakeholders provide support. Teams commit to proactive issue reporting during sprint reviews when metrics trend negatively. Stakeholders commit to support provision rather than demanding immediate fixes undermining capability building.

Success Criteria: Contract defines minimum acceptable performance triggering escalation discussions. Thresholds include reaching 70% velocity by Sprint 8, 85% velocity by Sprint 12, and parity by Sprint 18. Teams missing thresholds trigger 4Cs Remediation Protocol investigations rather than automatic transformation abandonment.

Contract Negotiation and Enforcement: Contracts require genuine negotiation rather than unilateral imposition. Product Strategists lead contract negotiations during Month -6 addressing stakeholder-specific concerns explicitly. Stakeholders worried about customer commitments need concrete assurance about delivery timeline impacts. Stakeholders concerned about budgets need transformation cost transparency.

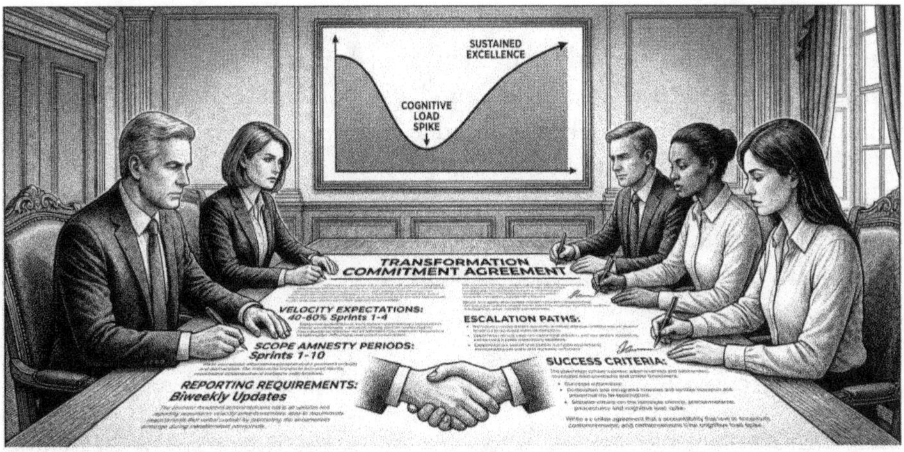

Figure 13.26: Formalized Transformation Commitment Agreement. *Stakeholder contracts document velocity expectations, scope amnesty periods, reporting requirements, and escalation protocols. Written agreements create accountability that verbal commitments cannot sustain when transformation difficulties inevitably emerge during the cognitive load spike.*

Negotiation Protocol: Product Strategists prepare for negotiations by identifying each stakeholder's primary concerns through pre-negotiation one-on-ones. Common concern categories include customer commitments, competitive timing, budget constraints, and career risk if transformation fails.

For customer commitment concerns, Product Strategists create delivery timeline adjustments showing how scope amnesty affects specific customer deliverables. Present options: "We can defer Feature X to Q3 allowing transformation to proceed, or we can staff traditional team for Feature X while main team transforms."

For competitive timing concerns, present market analysis showing competitor AI adoption rates and productivity gap projections if organization delays transformation. Quantify cost of waiting: "Delaying six months means competitors gain 9 sprints of productivity advantage we cannot recover."

For budget concerns, present total cost of ownership comparison showing transformation investment versus ongoing productivity deficit from not transforming. Include opportunity cost calculations for features not delivered due to lower productivity.

For career risk concerns, executive sponsors must explicitly assume transformation accountability rather than placing risk on Product Strategist. Executive messaging should state: "This transformation has my full support and accountability. Success or challenges reflect on organizational commitment, not individual performance."

Contract Enforcement - Team Violations: Teams violate contracts by reverting to traditional practices to boost short-term velocity or by failing to demonstrate genuine capability building progress despite adequate support.

Detection: Executive sponsors monitor for team violation indicators:

- Velocity recovering faster than contract projections without corresponding Chapter 9 metric improvements
- Sprint Telemetry Dashboard showing declining AI tool usage
- Governance checkpoint reports showing decreased AI collaboration attempts
- Team members admitting during retrospectives that they "went back to manual coding to hit deadlines"

Response Protocol: When executive sponsors detect team violations, they immediately schedule stakeholder contract review session including Product Strategist, Delivery Facilitator, and team representatives. Session agenda includes:

1. Present evidence of contract violation
2. Understand root cause using 4Cs Remediation Protocol framework
3. Determine whether violation stems from inadequate support, external pressure, or team resistance
4. Implement targeted correction matching cause
5. Reset expectations and timeline if necessary

Common causes and corrections:

- Inadequate training: Provide additional prompt engineering workshops and pair programming with AI-proficient coaches
- External pressure: Executive sponsor reinforces scope amnesty with stakeholders violating contract
- Team resistance: Address through psychological safety interventions and potentially team changes if resistance persists
- Wrong-fit scenario: Acknowledge transformation may not suit organization and plan explicit opt-out using Chapter 12 guidance

Contract Enforcement - Stakeholder Violations: Stakeholders violate contracts by demanding feature delivery during scope amnesty periods, pressuring teams for immediate velocity recovery, or bypassing Product Strategist to assign work directly to developers.

Detection: Product Strategists monitor for stakeholder violation indicators:

- New feature requests submitted during Sprints 1-10 without executive approval
- Stakeholders questioning velocity in sprint reviews despite meeting contract terms
- Stakeholders contacting team members directly to request work
- Stakeholders threatening transformation cancellation when metrics match contract expectations

Response Protocol: When Product Strategists detect stakeholder violations, they immediately escalate to executive sponsor rather than attempting direct enforcement. Executive sponsor schedules contract violation review with offending stakeholder using specific protocol:

1. **Private meeting:** Executive sponsor meets stakeholder privately rather than publicly to avoid embarrassment that hardens resistance
2. **Contract review:** Present original contract showing specific clause violated and current team performance meeting contract terms
3. **Impact explanation:** Explain how violation undermines capability building and threatens transformation success
4. **Recommitment:** Request stakeholder recommit to contract terms explicitly or escalate to transformation continuation decision
5. **Support offering:** Identify what additional support stakeholder needs to maintain contract (more frequent updates, different metrics, clearer timeline visibility)

If stakeholder refuses recommitment, executive sponsor must make explicit decision. Options are continuing transformation with reduced stakeholder support (risky) or pausing transformation until stakeholder alignment can be rebuilt (delay but preserve viability).

Example Enforcement Dialogue:

Stakeholder Violation - Feature Demand During Amnesty:

Stakeholder: "We need the customer portal enhancement in Sprint 6 for the board presentation."

Product Strategist: "Our contract specifies scope amnesty through Sprint 10. Is this feature business-critical requiring executive sponsor exception approval?"

Stakeholder: "It's important for the presentation but not business-critical."

Product Strategist: "I'll add it to the post-amnesty backlog for Sprint 11 planning. For the Sprint 6 presentation, we can show the Chapter 9 metrics demonstrating capability formation progress and projected productivity improvements supporting the strategic value of our transformation investment."

Team Violation - Velocity Manipulation:

Executive Sponsor (in contract review session): "Velocity recovered to 75% in Sprint 5. However, AI First-Pass Acceptance Rate remains at 38% and Sprint Telemetry shows 40% decrease in AI tool usage. This suggests the team is reverting to traditional coding rather than building AI collaboration capabilities."

Delivery Facilitator: "Team members reported feeling pressure to 'show progress' after stakeholder comments in Sprint 4 review. They reduced AI usage to complete more story points."

Executive Sponsor: "I will address the stakeholder pressure directly. The team's job is capability building, not velocity optimization during amnesty. We honor our contract commitments. I need the team to recommit to AI collaboration even if velocity drops further short-term. Capability formation is the priority."

Contracts include quarterly review points at Sprint 6, Sprint 12, and Sprint 18 aligned with transformation phases. Reviews examine whether teams progress as expected, whether stakeholders maintain commitments, and whether contract terms remain appropriate. Contract adjustments require mutual agreement maintaining collaborative stakeholder relationships.

[B+D] Multi-team scenarios require contracts covering all waves with wave-specific terms. Wave 1 contracts establish patterns that later waves adapt based on actual Wave 1 experience. Later wave contracts may offer tighter velocity targets reflecting organizational learning and improved support systems. For example, Wave 1 contracts might specify 70% velocity by Sprint 8. Wave 2 contracts could specify 75% velocity by Sprint

8 based on refined governance and extensive institutional memory availability.

Scope Amnesty Protocol

[All] Scope amnesty represents the formal agreement where stakeholders accept reduced velocity during transformation in exchange for long-term productivity gains. The protocol establishes clear boundaries around what constitutes acceptable performance, what triggers concern, and how teams maintain stakeholder confidence despite temporary productivity reduction.

Figure 13.27: Protected Capability Formation Period. *Scope amnesty creates time-bounded protection where stakeholders accept reduced velocity without demanding immediate corrections. Teams focus on capability building rather than feature delivery during sprints one through ten, investing in skills that enable sustained excellence.*

Amnesty Boundaries

Scope amnesty covers sprints one through ten for most scenarios representing the cognitive load spike and recovery climb periods. Stakeholders agree to forgive velocity shortfalls within expected ranges without demanding explanations or immediate corrective action.

The amnesty does not excuse complete productivity collapse or abandonment of delivery commitments. Teams operating at 20% of baseline when 50% is expected have violated amnesty terms. Teams delivering zero value because they are only experimenting with AI violate stakeholder trust.

Amnesty boundaries should specify minimum acceptable velocity for each transformation phase. Sprint 1 through 4 minimum might be 40% of baseline. Sprint 5 through 7 minimum might be 55% of baseline. Sprint 8 through 10 minimum might be 70% of baseline. Teams falling below minimums trigger escalation discussions.

The boundaries should also define what work receives priority during amnesty periods. Critical production fixes, urgent customer commitments, and regulatory compliance work typically fall outside amnesty scope. Teams must maintain ability to address true emergencies regardless of transformation status.

Amnesty Communication

Teams should communicate proactively about velocity throughout amnesty periods rather than surprising stakeholders with poor sprint outcomes. Weekly updates showing velocity actuals against projections help stakeholders track progress and maintain confidence in eventual recovery.

Communication should celebrate capability building milestones alongside delivery metrics. A team completing sprint two at 45% velocity while demonstrating strong prompt engineering skills shows genuine progress. Stakeholders need to understand that capability development represents investment in future productivity.

Teams should acknowledge when velocity falls below expectations and explain contributing factors. Honest communication about struggles builds trust more effectively than minimizing difficulties or making excuses. Stakeholders appreciate transparency about challenges and plans for addressing them.

Amnesty Termination

Scope amnesty ends when teams reach parity point and demonstrate sustained baseline productivity. Most teams exit amnesty between sprints eleven and fifteen depending on scenario and performance.

Amnesty termination should occur through explicit agreement rather than automatic expiration. Teams achieving parity at sprint twelve might request amnesty termination demonstrating confidence in sustained performance. Stakeholders grant termination acknowledging transformation success.

Early amnesty termination benefits teams by removing the psychological burden of operating under special conditions. Teams performing at baseline want recognition of normal operational status rather than continued transformation labeling.

Organizations should celebrate amnesty termination as major transformation milestone. The achievement validates that teams successfully navigated the cognitive load spike and built sustainable AI collaboration capabilities.

Scenario-Specific Challenges

[A] Single agile team stakeholders face contained impact where only one team operates under scope amnesty. The primary challenge involves maintaining stakeholder patience when they see adjacent non-transforming teams delivering normally. Stakeholders may question why transformation is necessary if other teams function adequately without AI.

Organizations should address this challenge by emphasizing competitive advantages and future productivity gains. The transforming team is building capabilities that will eventually make them vastly more productive than adjacent teams. Stakeholder patience during six months of reduced velocity earns years of superior performance.

[B] Multiple agile team stakeholders manage organizational capacity concerns as multiple teams operate under scope amnesty across waves. The primary challenge involves coordinating delivery commitments across transforming and non-transforming teams to maintain adequate organizational capacity.

Organizations should address this through careful wave planning that ensures no more than 30% of teams experience cognitive load spike simultaneously. Stakeholders see that organizational delivery capacity remains adequate even while some teams build capabilities.

[C] Single non-agile team stakeholders face extended amnesty periods due to dual learning curves. The primary challenge involves maintaining confidence through twelve to fifteen sprints of below-baseline performance. Stakeholders accustomed to waterfall predictability may struggle with extended uncertainty about when productivity recovers.

Organizations should address this by emphasizing dual transformation benefits. Teams are not just adopting AI but also gaining agile capabilities valuable regardless of AI outcomes. Stakeholders receive two methodology improvements for the cost of extended velocity reduction.

[D] Multiple non-agile team stakeholders manage the most complex challenge combining extended amnesty periods with organizational capacity constraints. The primary challenge involves sustaining executive commitment through two to three years of transformation activity and periodic capacity constraints.

Organizations should address this through staged commitments where executives approve each wave separately based on previous wave success. This approach reduces total at-risk commitment while maintaining transformation viability.

Managing Stakeholder Pressure

Stakeholders inevitably apply pressure for faster velocity recovery despite amnesty agreements. This pressure intensifies during sprint three or four when teams hit the velocity trough and stakeholder anxiety peaks.

Pressure Patterns

Common pressure patterns include stakeholders requesting specific features "just this once" outside normal prioritization. These requests attempt to bypass amnesty protections by framing individual items as exceptions. Teams must resist because exceptions accumulate undermining transformation entirely.

Another pattern involves stakeholders questioning transformation value based on early struggles. Stakeholders see reduced velocity and question whether eventual productivity gains justify current disruption. This questioning requires reinforcement of transformation rationale and long-term value.

Stakeholders sometimes suggest reverting to traditional practices temporarily "until the team gets their feet under them." This suggestion fundamentally misunderstands transformation mechanics. Reverting prevents capability building that enables eventual productivity gains.

Pressure Response Strategies

Teams should acknowledge stakeholder concerns as legitimate rather than dismissing them as resistance. Reduced velocity creates real business impact that stakeholders rightfully worry about. Validation of concerns builds trust while maintaining transformation commitment.

Teams should reference the stakeholder contract reminding stakeholders of agreed expectations. Sprint three velocity at 45% of baseline aligns with contracted 40-60% range. The team is performing as expected rather than underperforming.

Teams should share capability building evidence demonstrating genuine progress. Improving transformation metrics, expanding prompt libraries, and growing institutional memory show teams are investing transformation time productively.

The executive sponsor should intervene when stakeholder pressure threatens transformation continuation. Sponsor reinforcement of amnesty commitments and transformation value helps stakeholders maintain perspective during difficult periods.

Rebuilding Stakeholder Confidence

Organizations sometimes need to rebuild stakeholder confidence when early transformation results disappoint or external events create additional pressure. Confidence rebuilding requires acknowledging problems, demonstrating corrective action, and providing evidence of improvement.

Teams should conduct stakeholder reviews at governance checkpoint timing showing transformation progress across multiple dimensions. Velocity recovery matters but capability development, governance maturity, and institutional memory growth also demonstrate transformation success.

Organizations might invite stakeholders to observe governance checkpoint ceremonies where teams demonstrate AI collaboration capabilities. Seeing teams effectively use AI builds confidence more than abstract velocity reports.

Transparency about struggles and solutions builds credibility. Teams admitting that prompt engineering proved harder than expected but showing improved approaches developed through practice demonstrate honest assessment and genuine learning.

BACKLOG AND WORK INTAKE READINESS

[All] Sprint 1 success depends heavily on backlog quality and work intake processes that support AI collaboration. Teams cannot overcome poorly defined work items through skilled prompt engineering. Organizations must prepare backlogs specifically for AI-augmented development.

The Context-Ready Definition of Ready

Traditional Definition of Ready criteria focus on ensuring work items have clear acceptance criteria, are appropriately sized, and have identified dependencies. AI Scrum adds context-ready requirements ensuring work items contain information AI tools need to generate accurate outputs.

Context-Ready Requirements

Work items must include explicit acceptance criteria specifying what constitutes completion. Vague criteria like "improve performance" provide insufficient direction for AI code generation. Specific criteria like "reduce API response time from 300ms to under 100ms for 95th percentile requests" enable AI to generate targeted solutions.

Figure 13.28: Context Quality Determines AI Success. *Context-ready work items contain explicit acceptance criteria, technical constraints, architectural decisions, similar work examples, and required domain knowledge. Adequate context enables AI to generate accurate outputs while inadequate context creates excessive iteration frustration.*

Items must document relevant technical constraints limiting solution approaches. A requirement to "add user search functionality" leaves implementation completely open. Adding constraints like "use existing Elasticsearch infrastructure, maintain sub-second response times, support fuzzy matching" narrows solution space appropriately.

Items should reference architectural decisions affecting implementation. A task to "create new microservice" needs context about naming conventions, communication patterns, deployment approaches, and monitoring standards. This context might live in institutional memory but items should explicitly reference relevant decision records.

Items benefit from examples of similar completed work. A requirement to "implement OAuth integration" becomes more context-ready when referencing previous OAuth implementations the team completed. AI can analyze those examples to maintain consistency with established patterns.

Items should include domain knowledge required for accurate outputs. A healthcare feature requiring HIPAA compliance needs explicit compliance requirements in the work item. AI cannot infer regulatory requirements from functional descriptions alone.

Definition of Ready Enforcement

Teams should enforce context-ready standards during backlog refinement preventing inadequate items from reaching sprint planning. The Product Strategist reviews items against context-ready criteria before marking them ready for sprint commitment.

Items failing context-ready standards return to the Product Strategist for enhancement rather than entering sprint backlog with quality issues. Teams should not attempt to fix inadequate items during sprint execution because prompt iteration consumes time better spent on properly prepared work.

Enforcement requires Product Strategist capability development in providing AI-appropriate context. The role evolution from Product Owner to Product Strategist includes learning what information AI tools need to generate quality outputs. Chapter 4 covers this capability development in detail.

The Backlog Purge

[All] Digital Backlog Hoarding—the retention of massive quantities of aged backlog items to maintain "visibility" despite missing sponsors or obsolete value—represents the most common pre-transformation mistake organizations make. This anti-pattern forces teams to expend transformation energy on forensic cleanup rather than future-focused outcome definition, serving as the primary driver for extended velocity recovery timelines that often push baseline restoration beyond ten months instead of the standard Sprint 11 parity point.

Teams should purge old backlog items during preparation Phase 3 eliminating work lacking adequate context or questionable current relevance. Many organizations accumulate extensive backlogs of items requested months or years ago that no longer align with product strategy.

Purge Criteria

Items older than three months should face scrutiny about continued relevance. Business priorities change rapidly making old requests potentially obsolete. The Product Strategist should verify each old item still provides value before preserving it.

Items lacking sufficient context for AI collaboration should be enhanced or removed. Attempting to improve dozens of inadequate backlog items during transformation creates work preventing proper Sprint 1 preparation. Better to purge items and recreate them properly if they prove genuinely valuable.

Items with unclear business value should be eliminated. Many backlog items exist because someone requested them once rather than because they deliver meaningful value. The purge provides opportunity to remove low-value work.

Figure 13.29: Eliminating Digital Backlog Hoarding. *Organizations maintaining hundreds of aged backlog items force teams to waste transformation energy on forensic cleanup. Systematic purging during preparation removes obsolete work focusing capability formation on valuable context-ready items, preventing extended velocity recovery timelines.*

The purge typically reduces backlog size by 40-60% as teams eliminate obsolete, low-value, and inadequately defined work. This reduction benefits teams by focusing effort on genuinely valuable properly prepared items.

Purge Communication

Teams should communicate backlog purges to stakeholders explaining rationale and inviting objections. A stakeholder might care deeply about an item the team considers low-value. The communication prevents surprise when stakeholders discover their requested features were removed.

The purge message should emphasize quality over quantity. Organizations benefit more from twenty well-prepared high-value items than one hundred poorly defined uncertain-value items. Stakeholders should understand that smaller focused backlogs enable better delivery.

Work Intake Triage

[All] Organizations should modify work intake processes during transformation to ensure new requests meet context-ready standards before entering backlog. The triage process prevents accumulation of inadequate items requiring later enhancement.

Triage Process

New requests enter triage where the Product Strategist evaluates context completeness, value clarity, and priority relative to existing work. Requests lacking adequate context return to requesters for enhancement before backlog entry.

The triage should happen weekly or biweekly preventing large batches of requests accumulating without review. Regular triage maintains backlog quality throughout transformation.

Triage criteria should be documented and shared with frequent requesters. Stakeholders understanding context-ready standards can provide better initial requests reducing iteration required during triage.

Scenario-Specific Approaches

[A+C] Single team triage remains relatively simple with the Product Strategist managing intake directly. Weekly triage sessions consuming thirty to sixty minutes typically suffice for maintaining backlog quality.

[B+D] Multi-team triage requires coordination to prevent duplicate work and maintain consistent context standards. Organizations should establish cross-team triage forums where Product Strategists align on shared work and context expectations.

[C+D] Non-agile teams transitioning from waterfall often face challenges with triage cadence. Traditional project planning happens infrequently while agile work intake occurs continuously. Teams must develop discipline around regular triage sessions.

Sprint 1 Backlog Composition

Sprint 1 backlog should emphasize lower-risk work categories where AI usage is permitted without extensive restrictions. Internal documentation, developer tooling, and test automation provide good initial targets.

The backlog should include variety across different AI collaboration patterns. Sprint 1 might include code generation tasks, documentation creation, test case development, and requirements analysis to give different roles relevant practice.

Items should be sized slightly smaller than typical sprint work to account for cognitive overhead from learning new workflows. A team normally handling five story points per item might use three story points for Sprint 1 items ensuring completion despite reduced velocity.

The backlog should contain 50-75% more items than normal sprint capacity to provide flexibility when teams struggle with specific work types. Excess backlog prevents sprint failure if teams cannot complete specific items but can substitute alternatives.

Sprint 1 Success Criteria

Sprint 1 success should be measured by capability building and governance compliance rather than velocity achievement. A team completing three items at 50% velocity while demonstrating competent AI collaboration achieves more. This succeeds more than a team completing eight items by avoiding AI usage entirely.

Success criteria should include teams using AI for permitted work categories, validating outputs appropriately, making sound governance decisions, and contributing to institutional memory. These capability indicators matter more than story points delivered.

Teams should celebrate Sprint 1 completion regardless of velocity because finishing a sprint using AI Scrum represents genuine achievement. The ceremony and reflection build team confidence supporting continued capability development through subsequent sprints.

——PART IV: NAVIGATING THE TRANSFORMATION——

VELOCITY RECOVERY ACROSS TRANSFORMATION

Teams typically experience predictable velocity patterns during transformation regardless of their starting methodology or team count. Understanding these patterns helps stakeholders maintain confidence when velocity drops and recognize normal progress markers.

[All Scenarios]

Velocity recovery follows a predictable pattern across the first twelve to twenty sprints. Teams do not maintain baseline productivity during the formation window. Organizations that understand this trajectory can distinguish normal struggle from systemic failure. Those that do not often trigger premature rollback decisions during the most vulnerable phase.

The Wobble: Sprints 1-3

[All] Velocity becomes unstable during the first three sprints as teams master AI tools and governance protocols. Sprint-to-sprint variance often exceeds 30 percentage points. This volatility creates stakeholder alarm and executive pressure to abandon the transformation before capability investment matures.

Figure 13.30: Expected Velocity Instability Pattern. *Velocity becomes unstable during first three sprints as teams experiment with AI integration patterns. Sprint-to-sprint variance often exceeds thirty percentage points representing normal learning curves rather than implementation failure requiring immediate correction.*

The wobble represents the natural learning curve as teams experiment with AI integration patterns. Sprint 1 typically shows over-reliance on generative models. Sprint 2 swings toward excessive skepticism. Sprint 3 reveals pattern recognition as teams identify where AI provides genuine leverage.

- **[A]** Single agile teams experience shallower wobbles, bottoming at 55% to 65% of baseline during Sprint 2. Existing sprint discipline accelerates learning.
- **[B]** Multi-team agile organizations face coordination overhead. Wave 1 teams follow Scenario A patterns, while Wave 2 benefits from institutional memory transfer.
- **[C]** Single greenfield teams encounter deeper wobbles, bottoming at 40% to 50% during Sprint 3 as they learn Scrum and AI simultaneously.

- **[D]** Multi-team greenfield organizations experience the most severe wobbles, potentially dropping to 35% of baseline during Sprint 4.

The wobble phase is where stakeholders panic and demand course correction. This is why the J-Curve Education Campaign and Scope Amnesty Protocol are mandatory pre-transformation investments. The Product Strategist must defend the timeline and resist pressure to accelerate delivery through governance shortcuts.

The Stabilization: Sprints 4-8

[All] Stabilization begins when the Prompt Rejection Rate drops below 15% for two consecutive sprints. Teams develop consistent workflows and muscle memory around the Governance Checkpoint. Velocity remains below baseline but climbs 5 to 10 percentage points per sprint as institutional memory accumulates.

Chapter 9 transformation metrics reflect measurable progress. AI First-Pass Acceptance Rate improves from 35% to 65%. Validation Lead Time decreases from 4 hours to 90 minutes. Governance Cycle Time drops from 2 days to 4 hours. Sprint Outcome Achievement climbs from 60% to 85%.

- **[A]** Single agile teams reach 85% of baseline by Sprint 6 or Sprint 7.
- **[B]** Multi-team agile organizations see Wave 1 stabilize by Sprint 7, while Wave 2 achieves stabilization by Sprint 5.
- **[C]** Single greenfield teams reach 80% of baseline by Sprint 9 or Sprint 10.
- **[D]** Multi-team greenfield organizations reach 75% of baseline by Sprint 10 to Sprint 12.

Tactical Stabilization Maneuvers

Mandatory Actions: Celebrate micro-wins in every retrospective. Document reusable prompt patterns. Recognize contributors publicly. Maintain absolute governance discipline even when it constrains velocity. The Governance Checkpoint remains mandatory for all high-risk work categories. Invest aggressively in institutional memory. Capture every successful AI interaction in the SharePoint collaboration log with full prompt lineage.

Prohibited Actions: Forcing velocity through governance shortcuts creates rework loops that consume more time than manual implementation. Reverting to traditional workflows when AI feels difficult abandons the capability investment. Manual coding feels familiar, but reversion resets the learning curve. Skipping validation ceremonies erodes stakeholder trust. Validation is the primary mechanism for catching non-deterministic errors before production deployment.

Scenario-Specific Recovery Patterns

- **[A]** Single agile teams achieve the fastest recovery. J-Curve bottoms at Sprint 2 or Sprint 3 (55% to 65% velocity). Parity arrives by Sprint 11 to Sprint 14. Sustained excellence (115%+ baseline) by Sprint 19 to Sprint 22.

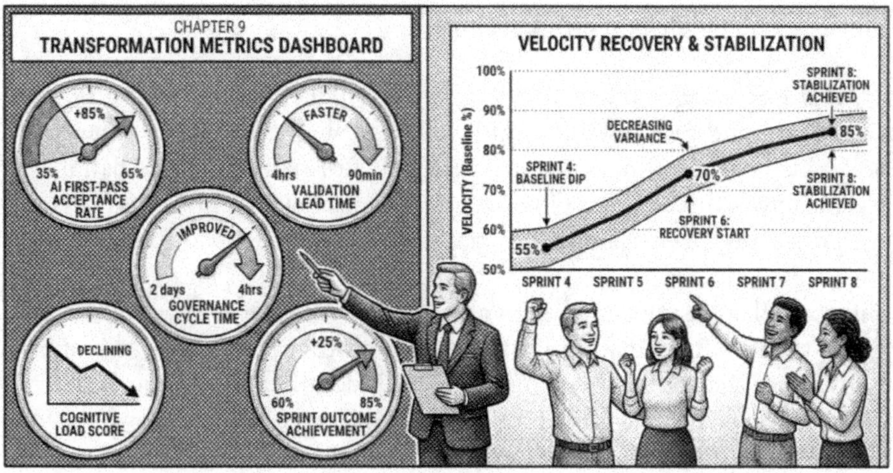

Figure 13.31: Capability Formation Through Measured Progress.
Stabilization emerges when transformation metrics show consistent improvement across four to eight sprints. Teams develop workflow muscle memory, build institutional memory, and demonstrate measurable skill advancement despite operating below previous baseline velocity.

- **[B]** Multi-team agile organizations experience differentiated patterns. Wave 1 follows Scenario A but with coordination overhead extending stabilization by one to two sprints. Wave 2 achieves stabilization faster due to institutional memory transfer. Wave 3 can reach parity three to five sprints ahead of Wave 1.

- **[C]** Single non-agile teams experience deeper recovery windows. J-Curve bottoms at Sprint 3 or Sprint 4 (40% to 50% velocity). Parity by Sprint 13 to Sprint 18. Sustained excellence by Sprint 21 to Sprint 24. Non-agile teams often demonstrate strong long-term performance once they internalize both agile principles and AI collaboration patterns.

- **[D]** Multi-team non-agile organizations face the most demanding journey. Wave 1 bottoms at Sprint 4 or Sprint 5 (35% to 45% velocity). Parity by Sprint 16 to Sprint 22. Sustained excellence materializes by Sprint 24 to Sprint 30 depending on wave progression and organizational support.

When to Worry vs Normal Struggle

Normal Struggle: Team members express frustration during retrospectives. Frequent questions emerge during standups. They struggle with stories that would have been trivial manually. Velocity shows high variance but demonstrates positive trend lines across four-sprint rolling windows. Governance violations decrease across sprints. Chapter 9 metrics improve incrementally but not uniformly.

Critical Red Flags: Team morale collapses. Developers stop contributing to standups and retrospectives. Individuals begin job hunting. Governance violations increase rather than decrease. transformation metrics flatline or decline across three consecutive sprints. Velocity shows no recovery trend by Sprint 6 or Sprint 7, remaining trapped at 50% to 60% of baseline.

When red flags manifest, the Governance Coordinator must immediately trigger the 4Cs Remediation Protocol. The protocol requires systematic diagnosis of the root cause (Condition) and identification of the underlying failure mode (Cause). Teams then implement targeted corrections (Correction) and verify resolution (Confirm). Early intervention prevents transformation collapse.

THE 4CS REMEDIATION PROTOCOL

Organizations that complete the Chapter 12 assessment do not abandon transformations when challenges arise. They diagnose problems systematically and implement targeted corrections. The 4Cs Remediation Protocol provides structured diagnostic approach for addressing transformation challenges.

The 4Cs framework provides systematic problem-solving methodology:

- **Condition**: The specific metric or symptom indicating a problem exists
- **Cause**: Root cause analysis to identify why the condition occurred
- **Correction**: Targeted intervention to address the root cause
- **Confirm**: Verification method to ensure the correction resolved the condition

This framework prevents emotional reactions during difficult transformation periods by requiring teams to diagnose problems systematically before implementing solutions. The protocol recognizes that transformation challenges stem from correctable causes rather than fundamental approach failures.

Example: High Prompt Rejection Rate

If Prompt Rejection Rate exceeds 25% for three consecutive sprints:

- **Condition**: High rejection rate indicates ineffective context blocks or inappropriate AI task application
- **Cause**: Investigate whether prompts lack architectural detail, governance boundaries are violated, or training was insufficient
- **Correction**: Implement intensive prompt engineering workshops, expand prohibited work categories, or provide additional coaching
- **Confirm**: Monitor rejection rate for two sprints post-correction to verify improvement below 15%

Figure 13.32: Systematic Problem Diagnosis Framework. *The 4Cs Remediation Protocol prevents emotional reactions during transformation challenges by requiring systematic diagnosis. Teams identify specific conditions, analyze root causes, implement targeted corrections, and verify resolution through metric validation.*

Example: Elevated Defect Escape Rate

If defect escape rate for AI-generated work exceeds 2x the manual baseline:

- **Condition**: Quality gates failing to catch AI-generated errors before production
- **Cause**: Determine whether Governance Canvas is incomplete, validation is being skipped, or Governance Checkpoint is rushed

- **Correction**: Expand Governance Canvas templates, enforce stricter validation protocols, or extend checkpoint ceremony duration
- **Confirm**: Track defect escape rate for four sprints to verify return to baseline or better

Example: Team Burnout Indicators

If team health indicators show sustained burnout or attrition threats:

- **Condition**: Team members show exhaustion, morale degradation, or key personnel threaten resignation
- **Cause**: Assess whether scope amnesty is violated, roles are understaffed, or expectations are unrealistic
- **Correction**: Enforce stricter scope amnesty boundaries, hire additional specialized roles, or extend cognitive load spike timeline
- **Confirm**: Conduct anonymous team surveys monthly until health indicators stabilize

Example: Stakeholder Trust Erosion

If stakeholder trust erodes to the point where J-Curve Contract is violated:

- **Condition**: Executives bypass scope amnesty protocols or demand premature acceleration through cognitive load spike
- **Cause**: Determine whether stakeholder communication has been inadequate, transformation metrics are being ignored, or external market pressure is overriding transformation commitment
- **Correction**: Schedule emergency stakeholder alignment sessions, present leading indicators proving capability formation progress, or renegotiate realistic timelines based on actual team performance
- **Confirm**: Stakeholders recommit to J-Curve Contract in writing and demonstrate respect for scope amnesty for four sprints

The transformation coach and Product Strategist monitor remediation triggers through Chapter 9 metrics. Early warning signals (such as metrics trending toward trigger thresholds) prompt proactive intervention before formal remediation is required. The goal is continuous course correction rather than binary success-failure outcomes.

TRANSFORMATION METRICS IN PRACTICE

The Chapter 9 metrics provide numerical scores but require interpretation within transformation context. Teams and stakeholders need guidance on what scores mean during different phases and when scores indicate problems requiring intervention.

[All Scenarios]
The 4Cs Remediation Protocol provides structured intervention when transformation metrics indicate teams are struggling. The protocol addresses four common failure modes: Capability regression, Context neglect, Compliance violations, and Communication breakdowns. Each failure mode has specific diagnostic triggers from Chapter 9 metrics and corresponding remediation actions.

The Delivery Facilitator updates scores weekly based on observable behaviors and measurable outcomes. The Governance Coordinator uses these scores to identify teams requiring intensive support versus those ready for advanced techniques. Organizations that master dashboard interpretation can diagnose problems early and intervene before minor friction escalates into transformation failure.

Interpreting Scores: Months 1-3

[All] During the formation window, scores reflect capability installation rather than execution proficiency. Month 1 scores typically range from 1.0 to 2.0 across all dimensions as teams struggle with basic protocols. This is expected baseline as developers transition from creation-centric to validation-centric workflows.

Collaboration scores between 1.5 and 2.0 indicate teams are asking questions but not yet sharing solutions. Context scores at this level reveal sporadic context block creation during sprint execution rather than proactive refinement. Compliance scores below 2.0 show frequent governance violations as teams learn boundaries through operational friction.

By Month 2, teams should demonstrate incremental progress. Collaboration climbing to 2.5 indicates developers are sharing prompt patterns. Context improving to 2.5 signals consistent context block creation. Compliance reaching 2.5 shows governance violations decreasing from six per sprint to three or four.

[A+B] Agile teams show faster Collaboration improvement because existing retrospective disciplines accelerate knowledge sharing.

[C+D] Greenfield teams often lag in Collaboration through Month 2 as they build team cohesion and AI proficiency simultaneously.

Month 3 represents the first validation checkpoint. Teams should reach 2.5 to 3.0 across most dimensions. Scores below 2.0 in any dimension by

Month 3 trigger immediate remediation. Scores flatlined at 1.5 for three consecutive months indicate fundamental capability gaps.

Interpreting Scores: Months 4-8

[All] During stabilization, scores transition from measuring protocol adherence to measuring execution efficiency. Teams should demonstrate consistent upward trajectories toward 3.5 to 4.0. Scores between 3.0 and 3.5 during Months 4-5 indicate teams have internalized governance protocols.

Key metric thresholds during stabilization:

- Collaboration scores above 3.0 reveal active institutional memory contribution
- Context scores at 3.5 show high-fidelity context blocks created during refinement
- Compliance above 3.5 indicates governance violations are rare exceptions
- Capability at 4.0 reflects confident operation across most AI augmentation categories

Scenario-specific timelines for reaching 3.5-4.0 thresholds:

- **[A]** Single agile teams: Month 6
- **[B]** Multi-team agile organizations: Wave 1 by Month 7, Wave 2 by Month 5
- **[C]** Single greenfield teams: Month 8
- **[D]** Multi-team greenfield organizations: Month 10

Intervention Triggers and Actions

[All] Scores declining across two consecutive assessments trigger immediate diagnostic review using the 4Cs Remediation Protocol. Scores flatlined for three assessments indicate the team needs intensive coaching or methodology adjustment.

Collaboration Interventions: Scores stuck below 2.5 by Month 3 require mandatory knowledge-sharing rituals. The Delivery Facilitator institutes daily prompt pattern reviews. Scores declining after Month 3 suggest team conflict requiring psychological safety work.

Context Interventions: Scores below 2.0 by Month 2 indicate AI Alignment Engineers need additional training or support. The team may lack architectural documentation required for high-fidelity context blocks. Scores improving slowly suggest work item quality issues requiring refinement process correction.

Compliance Interventions: Scores below 2.5 by Month 3 reveal teams are bypassing governance under delivery pressure. The Product Strategist must reinforce Scope Amnesty and defend against stakeholder demands. Declining scores during stabilization indicate governance fatigue requiring process streamlining.

Capability Interventions: Scores stalled below 2.0 suggest insufficient Deep Immersion Week training or wrong tool selection. Teams may need additional workshops or supplementary certifications. Scores improving unevenly across members reveal skill distribution problems requiring targeted coaching.

Scenario-Specific Interpretation

- **[A+B]** Agile teams should treat scores below 3.0 by Month 4 as warning signals. Existing process maturity should accelerate capability formation. Persistent underperformance suggests teams are reverting to Traditional Agile habits rather than adopting AI Scrum protocols.
- **[C+D]** Greenfield teams should expect scores to lag one to two months compared to agile counterparts. Scores below 2.5 by Month 4 are concerning but not critical. Scores below 2.0 by Month 5 require intensive remediation. The dual learning curve justifies extended timelines but not permanent underperformance.

- **[B+D]** Multi-team implementations require cross-wave score comparison. Wave 2 teams scoring below Wave 1 at equivalent timeframes indicate institutional memory transfer failures. The Governance Coordinator must investigate whether knowledge capture processes are functioning properly.

Continuous Monitoring and Early Intervention

Successful transformations require active monitoring for warning signs and rapid intervention when problems emerge. Waiting until formal metrics show issues often means problems have compounded beyond easy correction.

[All Scenarios]

Continuous monitoring prevents small problems from becoming transformation failures. The Governance Coordinator and Delivery Facilitator establish surveillance systems that detect warning signs before they manifest as dashboard score declines. Early intervention requires less effort and produces better outcomes than crisis remediation.

Monitoring operates across three time horizons. Daily signals come from standup observations and Slack activity patterns. Weekly signals emerge from sprint ceremonies and governance checkpoint metrics. Monthly signals appear in dashboard score trends and velocity trajectories.

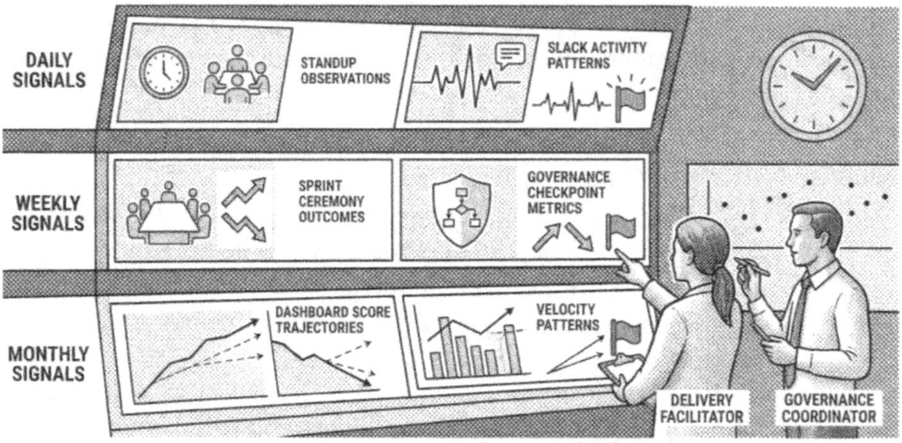

Figure 13.33: Multi-Horizon Transformation Surveillance. *Continuous monitoring operates across daily, weekly, and monthly time horizons detecting warning signs before they manifest as transformation failures. Early intervention based on leading indicators requires less effort and produces better outcomes than crisis remediation.*

Warning Signs Per Phase

[All] During the wobble phase (Sprints 1-3), normal struggle looks like confusion and frequent questions. Warning signs include silence rather than engagement. Teams that stop asking questions are not thriving; they're withdrawing. Developers who skip retrospectives or provide minimal standup updates signal psychological safety failures.

The Prompt Rejection Rate should be high during the wobble but trending downward. Rejection rates that increase across consecutive sprints indicate teams are generating worse prompts over time. This suggests training deficiencies or documentation gaps requiring immediate attention.

During stabilization (Sprints 4-8), teams should demonstrate increasing autonomy. Warning signs include persistent dependency on the

Governance Coordinator for basic decisions. Teams that cannot self-diagnose simple problems after four sprints lack fundamental capability formation.

Governance violations should decline steadily during stabilization. Violations that plateau or increase signal teams are under excessive delivery pressure. The Product Strategist must verify that Scope Amnesty remains in force.

- **[A+B]** Agile teams showing these warning signs by Sprint 4 require immediate intervention.
- **[C+D]** Greenfield teams get additional tolerance through Sprint 6, but persistent warning signs beyond that point trigger remediation.

Early Intervention Triggers

[All] Specific metrics trigger automatic intervention regardless of subjective assessment. Dashboard scores declining across two consecutive weeks require diagnostic review within 48 hours. The Governance Coordinator schedules emergency sessions to identify root causes before trends become entrenched.

Velocity showing no improvement across four consecutive sprints triggers executive escalation. The transformation may face structural impediments beyond team control. Budget constraints, stakeholder interference, or technical infrastructure failures require leadership intervention.

Individual developers consistently underperforming on Capability assessments trigger one-on-one coaching. The Delivery Facilitator determines whether additional training, role adjustment, or mentorship can close gaps. Waiting until Month 3 to address individual struggles wastes critical formation time.

Governance Checkpoint duration increasing rather than decreasing signals process inefficiency. The ceremony should become faster as teams internalize protocols. Increasing duration suggests teams are arguing about fundamentals rather than executing established standards.

Scenario-Specific Red Flags

- **[A]** Single agile teams experiencing cross-functional dependency breakdowns signal architectural problems. Their small size should prevent coordination failures. Persistent blockers indicate technical debt requiring architectural refactoring.

- **[B]** Multi-team agile organizations showing Wave 2 performance below Wave 1 at equivalent timeframes indicate institutional memory transfer failures. The Governance Coordinator must audit SharePoint documentation quality and accessibility.
- **[C]** Single greenfield teams reverting to ad-hoc workflows during pressure situations signal insufficient Scrum foundation. The Delivery Facilitator may need to slow AI integration to solidify fundamental agile practices.
- **[D]** Multi-team greenfield organizations experiencing coordination chaos beyond Sprint 6 require governance simplification. The protocols may be too complex for teams simultaneously learning Scrum and AI. The Governance Coordinator must identify which ceremonies can be streamlined without compromising validation rigor.

Escalation Paths

[All] Minor issues are resolved at the team level through retrospectives and daily coaching. Persistent issues escalate to the Governance Coordinator for 4Cs Protocol application. Structural issues escalate to executive sponsors for resource allocation or stakeholder management.

Early intervention converts potential failures into learning opportunities. Teams that receive timely support build confidence and resilience. Teams that struggle without intervention develop learned helplessness and abandon the transformation mentally before officially reverting.

UNIVERSAL ANTI-PATTERNS

Certain failure patterns emerge consistently across all transformation scenarios regardless of team count or starting methodology. Recognizing these anti-patterns early allows teams to correct course before problems become systemic.

[All Scenarios]

Anti-patterns emerge with predictable consistency across AI Scrum transformations. These behaviors sabotage capability formation regardless of scenario archetype or organizational maturity. Teams that recognize these patterns early can intervene before damage becomes permanent.

The Silent Revert

What It Looks Like: Developers quietly abandon AI tools for manual workflows during sprint execution. They attend governance checkpoints with AI-generated code but actually wrote everything manually. The institutional memory library receives no contributions despite weeks of claimed AI usage.

Why It Happens: The cognitive load of learning AI tools while maintaining velocity feels overwhelming. Manual coding provides psychological comfort. Developers fear judgment for struggling with tools that marketing claims are "intuitive."

How to Fix: The Delivery Facilitator institutes anonymous retrospective feedback mechanisms. The Governance Coordinator reviews actual LLM usage logs rather than trusting self-reported metrics. The organization celebrates struggle as evidence of learning rather than incompetence.

Figure 13.34: Silent Reversion to Manual Workflows. *Developers quietly abandon AI tools for manual coding while claiming AI usage at governance checkpoints. Detection requires monitoring actual tool engagement logs rather than trusting self-reported metrics, with intervention addressing root causes through targeted training or psychological safety reinforcement.*

The Prompt Hoarder

What It Looks Like: Individual developers discover breakthrough techniques but refuse to share them. High performers maintain personal prompt libraries on local machines rather than contributing to SharePoint. Transformation metrics show extreme variance across team members.

Why It Happens: Traditional performance management rewards individual productivity. Developers fear that sharing techniques will eliminate their competitive advantage during review cycles. Star performers protect their status by hoarding institutional knowledge.

How to Fix: The Product Strategist revises performance evaluation criteria to emphasize knowledge contribution over individual velocity. The Governance Coordinator institutes "prompt bounties" where developers earn recognition for high-value library contributions. Knowledge sharing becomes a promotion prerequisite.

Figure 13.35: Knowledge Hoarding Sabotages Collective Progress. *High performers who discover effective techniques but refuse to share them create extreme variance across team members. Organizations must revise performance evaluation criteria to reward knowledge contribution over individual productivity, making institutional memory sharing a promotion prerequisite.*

Executive Panic at Week 6

What It Looks Like: Leadership demands immediate course correction when velocity drops during the wobble phase. Executives question the transformation's validity despite pre-approved J-Curve expectations. Stakeholders pressure teams to abandon governance protocols. The Product Strategist receives mandates to accelerate roadmap delivery regardless of capability formation status.

Why It Happens: Executives intellectually accept the J-Curve during planning but emotionally panic when experiencing it operationally. Board pressure or competitive threats create urgency that overrides transformation discipline. Leaders who didn't genuinely commit during Chapter 12 use the wobble as abandonment justification.

How to Fix: The Governance Coordinator presents dashboard metrics showing capability formation progress despite velocity decline. Executive sponsors reinforce the original transformation contract and defend the timeline. If panic persists, the organization conducts honest cost-benefit reassessment rather than forcing dysfunctional acceleration.

The "Just This Once" Exception

What It Looks Like: Teams bypass governance protocols for "urgent" features that cannot wait for validation. Sanctuary systems receive AI augmentation because stakeholders demand immediate delivery. The double-blind validation rule is suspended during critical sprints. Prompt lineage documentation is skipped to save time.

Why It Happens: External pressure creates rationalization that governance is optional during emergencies. Teams have not internalized that protocols prevent disasters rather than slow delivery. One successful shortcut creates precedent that erodes discipline.

Figure 13.36: Stakeholder Panic During Cognitive Load Trough. *Executives who intellectually accepted the J-curve during planning often panic when experiencing operational velocity drops. Product strategists defend transformation by presenting contract terms showing current performance meets expectations and capability metrics demonstrate genuine progress.*

How to Fix: The Product Strategist establishes absolute governance non-negotiables during the prep period. The Delivery Facilitator helps teams understand that "just this once" is how production failures occur. The Governance Coordinator receives executive backing to veto exception requests without career risk.

The Capability Illusion

What It Looks Like: Dashboard scores show improvement, but underlying capability formation is hollow. Teams game metrics by focusing on easily measurable behaviors while ignoring substantive skill development. Collaboration scores climb because developers attend sessions, but no actual learning occurs.

Why It Happens: Measurement systems create incentives to optimize for scores rather than genuine capability. Teams under pressure choose performative compliance over substantive development. Organizations mistake activity for progress.

How to Fix: The Delivery Facilitator conducts spot-check audits where teams demonstrate actual capabilities rather than showing documentation. Developers must live-code using AI tools during randomized assessments. The dashboard incorporates outcome metrics like defect rates alongside behavioral scores.

Figure 13.37: Superficial Versus Substantive Code Validation. *Validation theater occurs when governance checkpoints become pro forma rubber-stamping rather than genuine risk management. Effective validation requires adequate ceremony time, validation lead training on hallucination detection, and mandatory defect discovery targets ensuring scrutiny remains substantive.*

The Validation Theater

What It Looks Like: Governance checkpoints occur on schedule, but validation is superficial. Validation Leads rubber-stamp AI-generated code without genuine review. The ceremony becomes a 15-minute formality. Hallucinations slip into production because nobody actually verified logic.

Why It Happens: Teams view validation as bureaucratic overhead rather than critical safety mechanism. Validation Leads lack training on what constitutes adequate verification. Organizations treat the ceremony as compliance checkbox rather than genuine risk management.

How to Fix: The Governance Coordinator reschedules checkpoints to optimal times when teams are mentally fresh. Validation Leads receive intensive training on hallucination detection techniques. The ceremony includes mandatory defect discovery targets; sessions without findings trigger deeper investigation.

The Documentation Void

What It Looks Like: The institutional memory library remains empty or filled with useless documentation. Prompt lineage tracking is incomplete or fabricated. Context blocks lack architectural detail required for effective AI augmentation.

Why It Happens: Documentation feels like non-value-added overhead when teams are behind schedule. The SharePoint interface is clunky and discourages contribution. Nobody is explicitly responsible for documentation quality.

How to Fix: The AI Alignment Engineers make documentation contribution a required Definition of Done criterion. The Governance Coordinator showcases velocity improvements directly attributable to reusable context blocks. Platform infrastructure is streamlined to minimize contribution friction.

The Premature Optimization

What It Looks Like: Teams attempt advanced AI techniques before mastering fundamentals. They deploy agent workflows during Sprint 2 despite struggling with basic prompting. Developers experiment with novel LLM features instead of solidifying core validation protocols.

Why It Happens: Developers are excited by advanced possibilities and bored by fundamental discipline. Marketing materials showcase impressive demos that teams want to replicate immediately. The organization rewards innovation theater over operational stability.

How to Fix: The Governance Coordinator enforces a staged capability unlock system. Advanced techniques become available only after teams demonstrate mastery at current levels. The Delivery Facilitator helps teams appreciate that excellence at fundamentals produces better outcomes than mediocrity at advanced techniques.

Digital Backlog Hoarding

What It Looks Like: Organizations maintain backlogs containing 200-500+ items including work requested years ago by departed stakeholders. Teams spend hours in backlog refinement sessions trying to determine relevance of obsolete requests. Product Strategists cannot articulate business value for 40-60% of backlog items yet resist deletion to maintain "visibility."

Why It Happens: Organizations confuse large backlogs with product vision. They fear deleting items that might prove valuable later. Waterfall heritage creates attachment to "complete requirements documentation." Teams lack courage to tell stakeholders their old requests were eliminated.

Figure 13.38: Backlog Hoarding Extends Recovery Timelines.
Organizations maintaining hundreds of aged backlog items force teams to expend transformation energy on forensic archaeology rather than capability building. Systematic purging during preparation removes obsolete work, enabling teams to focus on valuable outcomes and achieve standard recovery timelines.

How to Fix: Execute systematic backlog purge during Phase 3 (Months -3 to 0) eliminating all items older than three months unless sponsor actively confirms current value. For items stakeholders wish to retain but cannot prioritize for immediate work, establish a stakeholder-managed "parking lot" separate from the team's active backlog. Stakeholders maintain ownership of aged items and reprioritize them into the team backlog only when genuine business value emerges and resources become available. This transfers maintenance burden from the transformation team to the requesting stakeholders, preventing cognitive overhead from forensic requirement archaeology. Reduce active team backlog to 60-80 context-ready items maximum. Communicate purge rationale emphasizing quality over quantity and stakeholder ownership of unprioritized requests.

SCENARIO A+B SPECIFIC ANTI-PATTERNS

Organizations with existing agile maturity face unique anti-patterns that do not emerge in greenfield scenarios. Teams with deep Traditional Agile fluency often struggle more than expected because their ingrained habits conflict with AI Scrum principles.

[Agile Organizations Only]

Organizations with existing agile maturity face unique anti-patterns that do not emerge in greenfield scenarios. Teams with deep Traditional Agile fluency often struggle more than expected because their ingrained habits conflict with AI Scrum principles.

The "We Already Know Agile" Trap

What It Looks Like: Teams dismiss AI Scrum training as unnecessary because they've run Scrum for years. Developers skip Deep Immersion Week sessions claiming they "just need the AI tools training." The Product Owner refuses to transition to Product Strategist role. They insist the responsibilities are identical.

Why It Happens: Experienced agile practitioners view AI as merely a new tool within their existing framework. They don't recognize that AI fundamentally inverts the agile model from creation-centric to validation-centric. Confidence in existing process mastery creates blindness to necessary adaptations.

How to Fix: The Governance Coordinator presents specific examples of Traditional Agile practices that actively harm AI Scrum adoption. Executive sponsors mandate full participation in Deep Immersion Week regardless of agile tenure. The organization frames transformation as "Agile 2.0" rather than incremental tool addition.

The Velocity Obsession

What It Looks Like: Teams continue measuring success by story points completed per sprint despite velocity being obsolete in AI contexts. The Product Strategist pressures developers to maintain pre-transformation throughput during the wobble phase. Dashboard metrics are ignored in favor of traditional burndown charts.

Why It Happens: Traditional Agile deeply conditions teams to equate velocity with value delivery. Years of performance reviews based on throughput metrics create institutional inertia. The organization has not genuinely adopted the "new green" reporting framework.

How to Fix: The Product Strategist eliminates story point estimation entirely during the first six sprints. Velocity charts are removed from stakeholder reports and replaced with Chapter 9 transformation metrics. Performance evaluations are explicitly decoupled from throughput metrics during the transformation window.

Figure 13.39: Transitioning From Velocity to Capability Metrics. *Agile teams conditioned to measure success through story points must adopt transformation metrics emphasizing capability formation. Organizations eliminate velocity charts during formation windows replacing them with Chapter 9 metrics that accurately reflect AI Scrum transformation progress.*

The Solo AI Experimenter

What It Looks Like: Individual developers pursue personal AI experimentation outside the governance framework. High performers build private LLM workflows that bypass institutional memory contribution. Developers use unapproved AI tools because they're "better" than the sanctioned platform.

Why It Happens: Agile teams celebrate individual initiative and experimentation. The "fail fast" mentality encourages unsanctioned exploration. High-autonomy cultures resist governance structures that feel like micromanagement.

How to Fix: The Governance Coordinator establishes clear boundaries between sanctioned experimentation and rogue behavior. Innovation hours are scheduled where developers can explore new techniques within governance guardrails. The organization makes institutional memory contribution mandatory for all experimentation.

The Retrospective Regression

What It Looks Like: Retrospectives devolve into complaints about AI tools rather than structured improvement planning. Teams fixate on what they lost (coding speed) instead of what they're gaining (validation rigor). Action items focus on reverting to manual processes rather than refining AI workflows.

Why It Happens: Traditional retrospectives work well when teams share common framework understanding. AI transformation creates asymmetric knowledge where some members adapt faster than others. Teams have not developed vocabulary to articulate AI-specific challenges constructively.

How to Fix: The Delivery Facilitator uses structured retrospective formats specifically designed for transformation phases. The "4Ls" framework (Liked, Learned, Lacked, Longed For) replaces generic discussions. Each retrospective must produce at least one institutional memory contribution documenting lessons learned.

The Standup Bloat

[B Only] What It Looks Like: Multi-team standups balloon to 45 minutes as teams over-share AI experimentation details. Developers provide granular prompt engineering updates that are irrelevant to coordination needs. Cross-team dependencies get lost in the noise of individual AI tool discussions.

Why It Happens: Agile practitioners are conditioned to share extensively during standups. The excitement of AI discovery makes developers want to showcase their learning. Teams confuse coordination communication with knowledge transfer communication.

How to Fix: The Delivery Facilitator enforces strict time limits and formats for multi-team standups. AI technique discussions are moved to dedicated knowledge-sharing sessions. Standups focus exclusively on blockers, dependencies, and coordination needs.

The Governance Rebellion

What It Looks Like: Teams view the Governance Checkpoint as antithetical to agile principles. Developers argue that mandatory validation violates self-organization and autonomy. The Product Strategist faces resistance when enforcing sanctuary system boundaries. Teams lobby to eliminate governance protocols after Sprint 3. They claim they've "learned enough."

Why It Happens: Traditional Agile emphasizes trust and empowerment over process and control. Teams interpret governance as a reversion to waterfall command-and-control structures. Developers confuse autonomy with absence of accountability.

How to Fix: The Governance Coordinator frames validation as enabling autonomy rather than constraining it. Teams that demonstrate consistent hallucination detection earn relaxed checkpoint frequency. Executive sponsors explicitly state that AI governance is non-negotiable regardless of agile philosophy preferences.

SCENARIO C+D SPECIFIC ANTI-PATTERNS

Organizations learning Scrum and AI simultaneously face anti-patterns unique to dual learning curves. These patterns rarely emerge in agile-mature organizations because they stem from lack of agile muscle memory.

[Greenfield Organizations Only]

Organizations learning Scrum and AI simultaneously face anti-patterns that don't emerge in agile-mature environments. These teams lack the muscle memory and shared vocabulary that agile teams take for granted. Their struggles stem from building two capability sets concurrently.

The Waterfall Reversion

What It Looks Like: Teams retreat to sequential planning phases under delivery pressure. They create requirements documents before allowing any implementation work. The sprint becomes a two-week mini-waterfall with distinct analysis, coding, and testing phases.

Why It Happens: Waterfall feels psychologically safer than iterative development when teams are overwhelmed by AI complexity. Sequential workflows provide the illusion of control during chaotic formation periods. Leaders trained in traditional project management default to familiar patterns during stress.

How to Fix: The Delivery Facilitator institutes strict sprint rules that prevent waterfall behavior. All work items must demonstrate incremental progress within the first three days. The organization assigns an external agile coach to observe and intervene when waterfall patterns emerge.

The Dual Learning Overwhelm

What It Looks Like: Team members shut down cognitively because they're learning Scrum fundamentals and AI tools simultaneously. Developers cannot distinguish between Scrum confusion and AI confusion when problems arise. transformation metrics flatline across all dimensions because teams lack baseline competency in either domain.

Why It Happens: The cognitive load of dual learning exceeds human processing capacity for many team members. Organizations underestimate the difficulty of learning agile methodology while also mastering AI augmentation. Teams lack experienced mentors who can triage issues into agile versus AI categories.

How to Fix: The Delivery Facilitator slows AI adoption during Sprints 1-4 to solidify fundamental Scrum mechanics first. Teams run basic agile workflows without AI augmentation until ceremonies function reliably. The Deep Immersion Week is split into two separate training sessions separated by four weeks of practice.

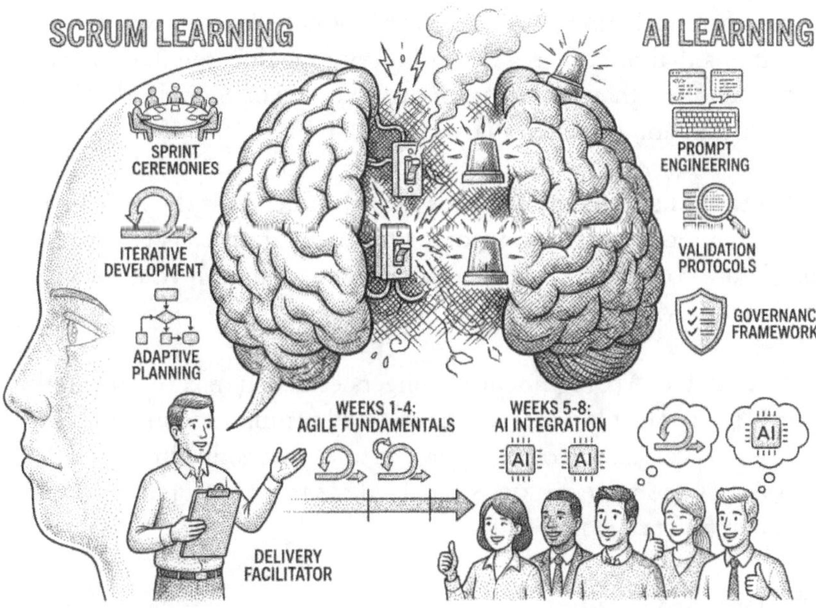

Figure 13.40: Managing Dual Learning Curve Complexity. *Non-agile teams learning Scrum and AI simultaneously experience cognitive overload exceeding human processing capacity. Organizations succeed by sequencing learning phases, solidifying fundamental agile mechanics before introducing AI augmentation, preventing overwhelming capability formation demands.*

The Agile Theater

What It Looks Like: Teams perform Scrum ceremonies without understanding their purpose. Daily standups become status reports to managers rather than peer coordination sessions. Retrospectives generate action items that nobody implements. The ceremonies exist because the playbook mandates them, not because teams find them valuable.

Why It Happens: Greenfield teams lack experiential understanding of why agile ceremonies exist. They're following a script without internalizing the principles. The Delivery Facilitator is learning alongside the team rather than bringing deep agile expertise.

How to Fix: The organization hires an external agile coach with ten-plus years of experience to mentor the Delivery Facilitator. Ceremonies are restructured to emphasize outcomes over format. Each ceremony must produce specific artifacts; standups produce updated dependency maps, retrospectives produce implemented improvements.

The Documentation Paralysis

What It Looks Like: Teams over-document everything to compensate for lack of institutional knowledge. Context blocks balloon to twenty-page specifications that nobody reads. Prompt lineage documentation becomes more burdensome than the actual coding work. The SharePoint library fills with redundant and outdated material.

Why It Happens: Greenfield teams lack confidence in their ability to remember or reconstruct knowledge. Documentation feels like insurance against future confusion. Teams confuse documentation volume with documentation value.

How to Fix: The AI Alignment Engineers establish maximum length standards for context blocks (two pages) and prompt documentation (one page). The Governance Coordinator institutes quarterly library audits where obsolete documentation is archived. Teams are taught "just enough documentation" principles.

The Governance Vacuum

What It Looks Like: Teams skip governance checkpoints because they don't understand their purpose. The Governance Coordinator lacks authority because nobody comprehends the risks being managed. Validation Leads rubber-stamp work without genuine review. Sanctuary systems are not protected because teams don't grasp the concept of risk-based AI restrictions.

Why It Happens: Greenfield teams have not experienced the consequences of ungoverned AI usage. They lack the horror stories that motivate governance discipline in agile-mature organizations. The transformation training emphasized tools and processes but didn't adequately explain the "why" behind governance protocols.

How to Fix: The Governance Coordinator presents case studies from other organizations where lack of governance led to production failures. The organization creates simulated hallucination incidents during training. Governance checkpoints include explicit education on what could go wrong if validation didn't occur.

The Role Confusion Cascade

[D Only] What It Looks Like: Multi-team greenfield organizations experience systemic role confusion that cascades across squads. Product Strategists make technical architecture decisions. Delivery Facilitators attempt governance enforcement. Developers bypass AI Alignment Engineers and create their own context blocks. Nobody understands their boundaries or accountabilities.

Why It Happens: The organization lacks existing role definitions to evolve from. The 11-person elite squad model introduces specialization that feels foreign to generalist cultures. Job descriptions were written before anyone understood how the roles would actually function.

How to Fix: The Governance Coordinator establishes a RACI matrix (Responsible, Accountable, Consulted, Informed) for every major decision type. Monthly calibration sessions bring together people in the same role across teams to align on standards. The organization accepts that role clarity will take six months to solidify.

——PART V: THE IMPLEMENTATION REFERENCE——

THE 9-MONTH PREPARATION PERIOD

All scenarios require nine months of preparation between executive approval and Sprint 1 launch. This timeline breakdown shows specific activities during each month with scenario-specific variations noted explicitly.

Overview: Months -9 to 0

[All] Organizations execute 9-month preparation period between executive approval and Sprint 1 launch. Month -9 begins when executive sponsors approve transformation budget and commit to timeline. Month 0 marks Sprint 1 Day 1 when team begins structured AI-augmented delivery under AI Scrum framework. Preparation period establishes all infrastructure, governance, training, staffing, and backlog refinement required for successful Sprint 1 launch.

Preparation divides into three phases aligned with decreasing time horizons and increasing tactical focus:

- **Phase 1 (Months -9 to -6): Strategic Alignment** - Stakeholder management, budget approval, hiring begins
- **Phase 2 (Months -6 to -3): Infrastructure and Governance** - Technical platform buildout, governance framework development
- **Phase 3 (Months -3 to 0): Tactical Readiness** - Backlog refinement, intensive training, final readiness verification

All preparation activities reference frameworks, ceremonies, and artifacts defined in Chapters 1-12. Chapter 13 provides implementation timing showing WHEN each activity occurs during preparation period rather than introducing new concepts.

Phase 1: Strategic Alignment (Months -9 to -6)

Phase 1 establishes stakeholder commitment, secures budget, initiates hiring for elite squad roles, and begins strategic infrastructure planning. Organizations complete Phase 1 activities sequentially during first three months of preparation period.

Month -9: Executive Approval and Budget Finalization

Product Strategists present complete transformation budget to executive sponsors using ROI framework from the Budget section earlier in this chapter. Financial presentation includes direct costs ($75,000-$950,000 depending on scenario), opportunity cost calculation showing velocity shortfall during transformation, and conservative ROI showing 24-34 month investment recovery timeline. Executive sponsors approve budget and commit formally to 18-month transformation timeline including 9-month preparation and 9-month capability building through sustained excellence achievement.

Organizations initiate hiring process for three elite squad roles: Product Strategist, Delivery Facilitator, and Governance Coordinator. Chapter 4 defines role responsibilities, required qualifications, and salary ranges for each position.

Organizations hiring external candidates should budget 6-12 weeks for recruitment and onboarding. Organizations promoting internal candidates complete role transitions during Month -9 enabling these individuals to lead preparation activities starting Month -8.

Months -9 to -8: J-Curve Education Campaign (Session 1-2)

Product Strategists conduct first two stakeholder education sessions during Months -9 and -8 using the J-curve education framework. Chapter 11 establishes stakeholder education philosophy and communication strategies. Session 1 (Month -9) introduces AI Scrum concepts and productivity improvement potential. Session 2 (Month -8) presents J-curve velocity pattern with scenario-specific projections. Expected velocity patterns show 40-60% during Sprints 1-4, recovery to 60-90% during Sprints 5-10, and sustained excellence at 115%+ by Sprint 19+.

Education sessions address stakeholder concerns about velocity drops and competitive timing using talking points developed earlier in this chapter under Stakeholder & Change Management section. Sessions establish realistic expectations preventing stakeholder panic when velocity trough occurs during Sprints 4-8.

Months -8 to -7: J-Curve Education Campaign (Session 3-4) and Stakeholder Contracts

Product Strategists conduct Sessions 3 and 4 during Months -8 and -7. Session 3 covers organizational impacts including resource allocation and stakeholder role changes. Session 4 finalizes scope amnesty agreements documented in stakeholder contracts.

Stakeholder contracts are completed during Month -7 following education campaign conclusion. Contracts specify velocity expectations through each transformation phase, scope amnesty periods (Sprints 1-10), reporting cadence, escalation paths, and success criteria. The Stakeholder Contract section earlier in this chapter provides complete contract components and negotiation protocols.

Product Strategists lead contract negotiations addressing stakeholder-specific concerns about customer commitments, competitive timing, or budget constraints. Executive sponsors sign contracts during Month -7 creating formal accountability for transformation support.

Month -6: Elite Squad Hiring Completion and Role Transitions

Organizations complete hiring for Product Strategist, Delivery Facilitator, and Governance Coordinator roles by Month -6. New hires complete onboarding during Month -6 preparing to lead Phase 2 infrastructure buildout. Organizations using internal promotions finalize role transitions ensuring individuals have cleared previous responsibilities to focus exclusively on transformation leadership.

Product Strategists begin preliminary tool selection research during Month -6 identifying AI coding assistants, collaboration platforms, and delivery automation tools for evaluation during Phase 2. Tool evaluation should focus on integration capabilities, security compliance, and team collaboration features.

Phase 2: Infrastructure and Governance Buildout (Months -6 to -3)

Phase 2 establishes technical infrastructure and governance frameworks required for AI Scrum execution. Technical teams provision AI Sandbox environment, configure delivery platforms, and establish institutional memory systems.

Governance Coordinators develop Governance Canvas mapping AI usage decisions across work categories and risk levels. Phase 2 represents most infrastructure-intensive preparation phase requiring significant technical effort and budget expenditure.

Months -6 to -5: Tool Selection and Procurement

Product Strategists lead tool selection process during Months -6 to -5 evaluating AI coding assistants, documentation generators, and collaboration platforms. Teams conduct proof-of-concept testing with shortlisted tools assessing prompt quality, integration capabilities, and security compliance. Organizations should evaluate tools through proof-of-concept testing with shortlisted candidates.

Organizations select primary AI tooling stack by Month -5 and initiate procurement process. Tool subscriptions begin Month -4 allowing team members to start familiarization during final preparation months. Organizations should budget tool costs for 21-27 months covering preparation period plus transformation duration through sustained excellence achievement.

[A+B] Agile teams often continue using existing GitHub Copilot, Cursor, or similar tools already deployed in organization. These teams focus tool selection on governance automation and institutional memory platforms rather than replacing established AI coding assistants.

[C+D] Non-agile teams typically select complete new tooling stack during Months -6 to -5 as existing development tools may not support AI integration or modern delivery workflows.

Months -6 to -4: AI Sandbox Provisioning

Technical teams provision AI Sandbox environment during Months -6 to -4 following standard sandbox architecture patterns for isolated AI experimentation. Sandbox provides isolated environment for AI collaboration skill development separate from production systems. Organizations maintain sandbox permanently for ongoing learning and capability experimentation beyond initial transformation period.

Sandbox configuration includes LLM endpoint integration, API security protocols, and network isolation preventing sandbox activities from affecting production systems. Standard sandbox architecture including infrastructure requirements, security controls, and access management policies.

Months -6 to -3: Delivery Platform Configuration

Teams configure delivery platforms during Months -6 to -3 to support AI collaboration patterns. Platform configuration includes version control integration, CI/CD automation, testing frameworks, and monitoring systems. The Infrastructure section of this chapter details platform requirements and configuration approaches.

[A+B] Agile teams augment existing delivery platforms with AI integration rather than building infrastructure from scratch. These teams typically have version control, CI/CD pipelines, and automated testing already operational from previous Scrum implementations. Phase 2 work focuses on integrating AI tools into existing workflows and configuring platforms to capture AI usage metrics. Agile teams typically complete Phase 2 platform work within standard 3-month window.

[C+D] Non-agile teams must build complete delivery platforms during Phase 2. Traditional project management often relies on manual deployment processes and approval-based review gates. AI Scrum requires automated platforms enabling rapid experimentation without manual bottlenecks.

Version control systems must track every code change. CI/CD pipelines automatically test and deploy code. Testing frameworks validate AI-generated outputs without human intervention for every change. The Infrastructure and Governance Setup section provides detailed guidance on platform selection, configuration, and integration patterns for teams building these capabilities from zero. Platform buildout for non-agile teams represents the most intensive Phase 2 activity. Complete platform implementation requires 8-12 weeks and consider engaging infrastructure specialists for guidance.

Months -5 to -3: Governance Canvas Development

Governance Coordinators develop initial Governance Canvas during Months -5 to -3 using framework defined in Chapter 10. Canvas maps AI usage decisions across work categories (business logic, APIs, database operations, infrastructure, etc.) and risk levels (low, medium, high). Canvas establishes which AI usage patterns teams can execute independently versus which require consultation or approval.

Initial Governance Canvas takes restrictive approach prohibiting AI usage for high-risk work categories during Sprints 1-4. The High-Risk Work Categories section earlier in this chapter lists universal prohibitions including security controls, payment processing, regulated data handling, production database modifications, and customer-facing critical paths. Canvas also establishes conditional usage rules for moderate-risk categories requiring enhanced validation during early sprints.

Governance Coordinators should involve prospective Validation Leads in canvas development ensuring validation requirements align with team capabilities during early sprints. Canvas undergoes quarterly refinement through governance checkpoint ceremonies as team capabilities improve and restrictions relax progressively.

Month -3: Institutional Memory System Establishment

Teams establish institutional memory systems during Month -3 following architecture defined in Chapter 6. Institutional memory captures architectural decisions, validated prompt patterns, coding standards, and domain knowledge enabling AI tools to maintain consistency across implementations. Organizations build institutional memory incrementally starting with basic architectural decisions and standards during preparation, expanding coverage throughout transformation sprints.

Initial memory population during Month -3 focuses on documenting existing architectural patterns, coding conventions, and design decisions governing current codebase. Teams should aim for 20-30 foundational memory entries before Sprint 1 launch covering high-level architecture, core domain concepts, and critical technical constraints. Memory expands to hundreds or thousands of entries during transformation as teams document AI collaboration patterns and validated approaches.

Phase 3: Tactical Readiness (Months -3 to 0)

Phase 3 finalizes tactical preparation including backlog refinement, intensive team training, and readiness verification. Teams complete hiring for remaining roles, execute Deep Immersion Week training, apply 3-Month Rule to backlog, and conduct final readiness certification before Sprint 1 launch.

Months -3 to -2: Remaining Role Hiring

Organizations complete hiring for remaining elite squad roles during Months -3 to -2. Chapter 4 defines complete 11-person elite squad including Product Strategist, Delivery Facilitator, Governance Coordinator (hired Month -6), plus AI Alignment Engineers, Validation Leads, and development team members. Organizations should finalize all hiring by Month -2 enabling complete team to participate in Deep Immersion Week training during Month -1.

Job descriptions for all roles are defined in Chapter 4 including required qualifications, experience levels, and compensation ranges. Organizations update job descriptions reflecting AI Scrum role responsibilities during Month -3 before posting positions. Hiring timeline should account for 6-8 week recruitment cycles ensuring new hires onboard before Deep Immersion Week.

Months -3 to -1: Backlog Cleaning and 3-Month Rule Application

Product Strategists lead backlog refinement during Months -3 to -1 applying 3-Month Rule defined in Chapter 7. Rule states that work items older than three months without advancement represent scope creep accumulation requiring purge. Teams review entire backlog removing stale items, consolidating duplicates, and restructuring remaining work for AI collaboration patterns.

Chapter 7 provides complete backlog refinement methodology. It covers work item restructuring for context clarity and acceptance criteria enhancement for AI validation. Teams format items to enable AI tools to understand requirements without excessive human explanation. Teams should complete backlog refinement by Month -1 ensuring Sprint 1 begins with clean, AI-ready backlog supporting effective capability building.

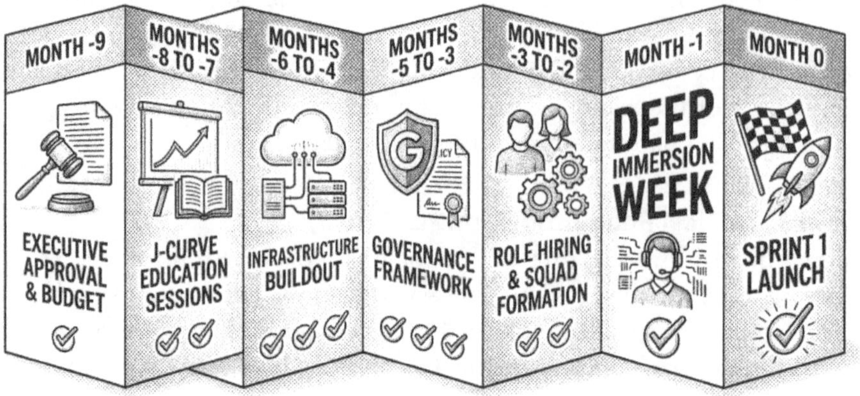

Figure 13.41: Complete Preparation Period Roadmap. *The nine-month preparation timeline progresses systematically through strategic alignment, infrastructure development, and tactical readiness. Each month builds essential foundations enabling successful Sprint 1 launch with all scenarios requiring the full duration regardless of team count or methodology.*

Month -1: Deep Immersion Week Training

Complete 11-person elite squad participates in Deep Immersion Week intensive training during Month -1. Chapter 3 defines Deep Immersion Week curriculum covering AI collaboration fundamentals, prompt engineering techniques, governance framework understanding, and role-specific skill development. Training duration is five days for agile teams, six days for non-agile teams requiring additional agile foundations coverage.

[A+B] Agile teams complete 5-day Deep Immersion Week focusing on AI augmentation skills including prompt engineering for different work categories, validation approaches for AI-generated outputs, and governance canvas navigation. Training assumes existing agile ceremony knowledge and iterative development familiarity.

[C+D] Non-agile teams complete 6-day Deep Immersion Week including one additional day covering agile principles, sprint ceremonies, and iterative development concepts. Day 1 establishes agile foundations. Days 2-6 cover AI collaboration skills matching agile team curriculum.

Organizations may use internal trainers for Deep Immersion Week if trainers possess genuine AI collaboration expertise demonstrated through practical prompt engineering skills rather than theoretical knowledge only. Organizations lacking internal expertise should engage external training providers during Month -1. Training costs are addressed in Budget section earlier in this chapter.

Month -1 to 0: Dashboard and Workflow Configuration

Delivery Facilitators configure transformation dashboards during Month -1 using frameworks defined in Chapter 9. Dashboards track AI First-Pass Acceptance Rate, Validation Lead Time, Cognitive Load Score, Sprint Outcome Achievement, and Governance Cycle Time providing visibility into transformation progress. Teams establish baseline measurements during Sprint 1 for comparison throughout transformation.

Organizations configure workflow automation during Month -1 establishing governance checkpoints, validation gates, and institutional memory updates within delivery platforms. Chapter 5 defines ceremony structures and workflow patterns supporting AI Scrum execution. Platform configurations should enforce mandatory checkpoints and prompt lineage tracking automatically reducing manual governance overhead.

Month 0: Sprint 1 Launch

Sprint 1 launches on Day 1 of Month 0 following 9-month preparation completion. Teams conduct Sprint Inception ceremony on Day 1 selecting work items from refined backlog, committing to sprint outcomes, and establishing sprint goals. Chapter 5 defines Sprint Inception ceremony structure and facilitation approaches.

Sprint 1 readiness verification occurs during the final week of Month -1. Readiness criteria include complete elite squad staffing (though defining "complete" proves surprisingly difficult in practice), Deep Immersion Week training completion, backlog refinement finalization, infrastructure operational status, and governance canvas approval. Teams discovering readiness gaps during final week should address immediately or postpone Sprint 1 launch by one week rather than launching with incomplete preparation.

Sprint 1 begins transformation journey through cognitive load spike, recovery climb, parity achievement, and sustained excellence. All preparation investments during Months -9 to 0 enable effective Sprint 1 execution and capability building throughout transformation timeline.

WHAT COMES NEXT

This playbook provides the complete framework for AI Scrum transformation. Understanding what comes after this chapter helps you sequence remaining preparation activities and leverage supporting resources effectively.

[All Scenarios]

The preceding sections provided the complete implementation playbook for AI Scrum transformation. You now possess the frameworks, timelines, budgets, and anti-patterns required to execute across all four scenario archetypes. The operational guidance ends here. What follows is deployment into your specific organizational context.

The Immediate Next Steps

[All] Organizations ready to begin transformation should initiate the nine-month prep period immediately. The clock starts when executive sponsors approve the transformation budget and commit to the timeline. Delaying execution while waiting for "perfect conditions" wastes the competitive window.

During Month -9, strategic alignment and financial modeling take priority. The Product Strategist collaborates with finance to build the complete cost model, incorporating the 26% talent premium for elite squad members. Meanwhile, executive sponsors secure stakeholder commitments for both Scope Amnesty protocols and J-Curve tolerance expectations.

The infrastructure and governance buildout phase spans Months -6 through -3. Technical teams provision the AI Sandbox while configuring delivery platforms. Concurrently, the Governance Coordinator drafts the Governance Canvas and establishes sanctuary system boundaries to protect critical functions.

Tactical readiness activities consume Months -3 through 0. Organizations formally appoint the 11-person elite squad, who then complete Deep Immersion Week training. The backlog undergoes rigorous purging via the 3-Month Rule. Teams pass Readiness Certification gates before Sprint 1 launches.

Infrastructure and Tooling Details

[All] The Infrastructure, Tooling, and Governance Setup section earlier in this chapter provides foundation for technical implementation. Organizations should revisit that section for detailed guidance on sandbox configuration and platform selection.

The AI Sandbox requires specific LLM endpoint configurations, API integrations, and security protocols. Tool evaluation should focus on integration capabilities, conduct proof-of-concept testing with shortlisted vendors, and negotiate contracts that allow flexibility as AI capabilities evolve.

Governance automation depends on platform capabilities that vary across tools like Azure DevOps, Jira, and Linear. Organizations should configure workflows that enforce the Governance Canvas automatically through mandatory checkpoints and automated prompt lineage tracking.

Multi-Team Scaling Considerations

[B+D] Multi-team organizations should reference the Wave-Based Implementation sections earlier in this chapter for scaling strategies after Wave 1 stabilizes. These sections address cross-team coordination patterns, institutional memory sharing protocols, and governance capacity management at scale.

Wave 2 launch timing depends on Wave 1 capability formation, though in practice this timing proves messier than the framework suggests. Organizations should not initiate Wave 2 until Wave 1 teams demonstrate sustained improvement in Chapter 9 metrics for at least four consecutive sprints. Organizations determine wave readiness when Wave 1 teams demonstrate consistent transformation metric improvements and stable governance patterns.

Post-Transformation Maintenance

[All] Transformation does not end when teams reach sustained excellence. Organizations must maintain governance discipline, institutional memory hygiene, and capability development indefinitely. The Governance Coordinator role transitions from intensive coaching to ongoing stewardship after Month 18.

Quarterly audits ensure the institutional memory library remains current and valuable. Obsolete prompts are archived. High-performing patterns are elevated to standard practices. Annual reassessment using Chapter 9 metrics verifies that teams haven't regressed to Traditional Agile or manual workflows.

Technology evolution requires periodic platform evaluations. LLM capabilities improve quarterly. The organization remains open to evolution while avoiding the churn of constant tool switching.

The Long-Term Vision

[All] AI Scrum represents the foundation for continuous AI capability expansion. Teams that master validation-centric workflows become equipped to adopt advanced techniques like agent-based development and automated architecture generation. The governance discipline installed during transformation provides the safety infrastructure for responsible experimentation.

Organizations that successfully transform gain competitive advantages that compound over time. Institutional memory becomes increasingly valuable as the library accumulates patterns spanning hundreds of sprints. Developer productivity continues climbing beyond the initial 115% plateau as teams discover optimization techniques.

The methodology itself will evolve as the industry gains operational experience. This playbook represents early 2026 best practices. Teams should view these frameworks as starting points requiring adaptation to specific contexts.

Figure 13.42: Achieving Sustained Excellence Together. *Successful AI Scrum transformation culminates when teams consistently exceed previous baseline productivity by fifteen percent or more while maintaining quality standards. The journey requires complete organizational commitment, systematic capability building, and genuine patience through the formation window—but delivers compounding competitive advantages.*

Your Transformation Begins Now

[All] The diagnostic phase is complete. The playbook is delivered. The decision point has arrived. Organizations commit to the full transformation or explicitly opt out. There is no middle ground where teams "try it and see."

Commit fully with proper budget, timeline, and executive support or postpone until preconditions for success are established. But don't halfheartedly attempt transformation with inadequate resources or impatient stakeholders. That path leads to wasted effort and damaged credibility.

The competitive window for managed AI adoption is finite. Organizations that transform successfully in 2026 gain advantages that laggards cannot easily replicate. The institutional memory and capability formation require time to mature. Your transformation begins now!

KEY TAKEAWAYS

The transformation playbook contains extensive detail across multiple scenarios. These key takeaways distill the most critical principles that determine transformation success regardless of your specific path.

[All Scenarios]

The preceding sections delivered the operational playbook for AI Scrum transformation across all four organizational archetypes. The following principles represent the essential insights that determine transformation success or failure.

Universal Truths Across All Scenarios

The J-Curve is mandatory, not optional. Every transformation experiences velocity decline during the formation window. Teams that bottom at 40% to 65% of baseline during Sprints 2-4 are performing normally, not failing. Organizations that panic during the wobble and demand immediate course correction guarantee transformation collapse. Executive sponsors must commit to eighteen-month timelines with genuine patience for the recovery curve.

The 11-person elite squad architecture is non-negotiable. The Product Strategist, Delivery Facilitator, and Governance Coordinator provide core leadership. The two AI Alignment Engineers and two Validation Leads provide specialized technical capabilities. Organizations attempting transformation without these dedicated roles create accountability vacuums that collapse under pressure. The 26% talent premium pays for itself within twelve months of reaching sustained excellence.

Governance exists to enable velocity, not constrain it. Teams that view the Governance Checkpoint as bureaucratic overhead fundamentally misunderstand the framework. Validation protocols prevent catastrophic hallucination failures that would destroy stakeholder trust. Teams that skip governance "just this once" create technical debt that compounds exponentially.

Institutional memory is the compound interest of AI adoption. Organizations that neglect SharePoint library contribution forfeit the primary advantage of AI Scrum. Reusable context blocks and documented prompt patterns enable teams to operate at 115%+ productivity indefinitely. The organization must make institutional memory contribution a performance evaluation criterion.

Context quality determines AI output quality. The "garbage in, garbage out" principle applies with brutal efficiency in AI contexts. High-fidelity context blocks containing architectural constraints, business rules, and edge cases enable accurate generation. Generic prompts without system dependencies produce hallucinations that waste validation cycles. The AI Alignment Engineers must certify context-ready work items before they enter sprints.

Scenario-Specific Insights

[A+B] Agile organizations face unlearning challenges that greenfield teams avoid. Traditional Agile conditions teams to optimize for velocity and individual productivity. AI Scrum requires validation rigor and collective capability formation. Teams must abandon story point estimation, velocity tracking, and personal hero cultures. The transition from creation-centric to validation-centric workflows requires genuine methodology evolution, not incremental tool adoption.

[C+D] Greenfield organizations must resist waterfall reversion under pressure. Teams learning Scrum and AI simultaneously experience intense cognitive load. The psychological safety of sequential workflows becomes tempting during stressful formation periods. Organizations must enforce strict sprint rules preventing requirements documents and stage-gate approval processes. External agile coaches provide essential intervention when waterfall patterns emerge.

[B+D] Multi-team implementations require wave-based sequencing. Governance capacity constraints limit how many teams can transform simultaneously. A single Governance Coordinator supports approximately four teams effectively during formation. Organizations that launch all teams simultaneously destroy support infrastructure. Wave 2 should not begin until Wave 1 demonstrates sustained metric improvements across Chapter 9's five dimensions.

Critical Success Factors

Scope Amnesty must be genuine and enforced. Stakeholders who demand new features during the formation window sabotage capability investment. The Product Strategist must have veto authority over non-essential work intake through Sprint 8. Organizations that allow backdoor feature requests through executive channels guarantee transformation failure. The Scope Amnesty Protocol requires board-level commitment, not departmental negotiation.

Chapter 9 transformation metrics replace velocity as the primary success metrics. Traditional throughput measurements are misleading during transformation. Teams track five new metrics instead of story points. AI First-Pass Acceptance Rate measures how often AI outputs require minimal revision. Validation Lead Time tracks hours spent reviewing AI-generated work. Cognitive Load Score assesses team stress levels. Sprint Outcome Achievement monitors delivery against commitments. Governance Cycle Time measures decision-making speed. Upward metric trends indicate successful capability formation despite velocity decline. Executives who demand velocity charts during Months 1-6 demonstrate they haven't internalized the transformation principles. This fixation on traditional metrics actively harms transformation by pressuring teams to optimize the wrong outcomes.

Early intervention prevents transformation collapse. Warning signs like Silent Revert, Prompt Hoarding, and Governance Rebellion require immediate remediation using the 4Cs Protocol. Teams that struggle without support develop learned helplessness and abandon transformation mentally before officially reverting. The Governance Coordinator must trigger diagnostic reviews when dashboard scores decline across two consecutive assessments.

Honest self-assessment determines transformation viability. Organizations must acknowledge wrong-fit scenarios rather than forcing adoption through willpower. Safety-critical systems requiring determinism should not adopt AI Scrum. Command-and-control cultures that punish failure cannot build psychological safety. Teams with 30%+ annual turnover cannot accumulate institutional memory. The Chapter 12 assessment provides diagnostic criteria; organizations must respect the results.

The Transformation Mindset

AI Scrum is a marathon, not a sprint. Organizations expecting quick wins within three months fundamentally misunderstand the framework.

Key Takeaways

Capability formation requires twelve to eighteen months before sustained excellence emerges. The structured repository of team decisions, coding standards, and prompt templates—the institutional memory system—matures over years, not quarters. Competitive advantages build incrementally as teams deepen expertise. Each sprint adds validated patterns to institutional memory. Later work leverages accumulated knowledge. This creates compounding returns through sustained commitment rather than explosive short-term gains.

Figure 13.42: Commitment Decision Point. *Organizations face explicit choice: commit fully to transformation with adequate resources and timeline or postpone until prerequisites exist.*

Adaptation beats rigid adherence. These playbooks represent starting points requiring contextual adjustment. Organizations should treat guidelines as frameworks rather than immutable specifications. The methodology will evolve as the industry gains operational experience. Teams must balance disciplined execution with intelligent adaptation to specific organizational realities.

Commitment trumps capability. Organizations with modest technical capabilities but genuine executive commitment outperform technically sophisticated teams with impatient leadership. The transformation succeeds or fails based on whether stakeholders honor the J-Curve contract during the wobble phase. Capability gaps can be addressed through training and hiring. Commitment failures destroy transformations irreversibly.

The Final Principle

There is no middle ground. Organizations either commit fully to transformation with proper budget, timeline, and support, or they explicitly opt out until preconditions improve. Halfhearted attempts with inadequate resources guarantee failure and damage organizational credibility. The competitive window for managed AI adoption is finite. Organizations that transform successfully gain advantages that laggards cannot easily replicate.

MIND MAP – THE AI SCRUM IMPLEMENTATION PLAYBOOK

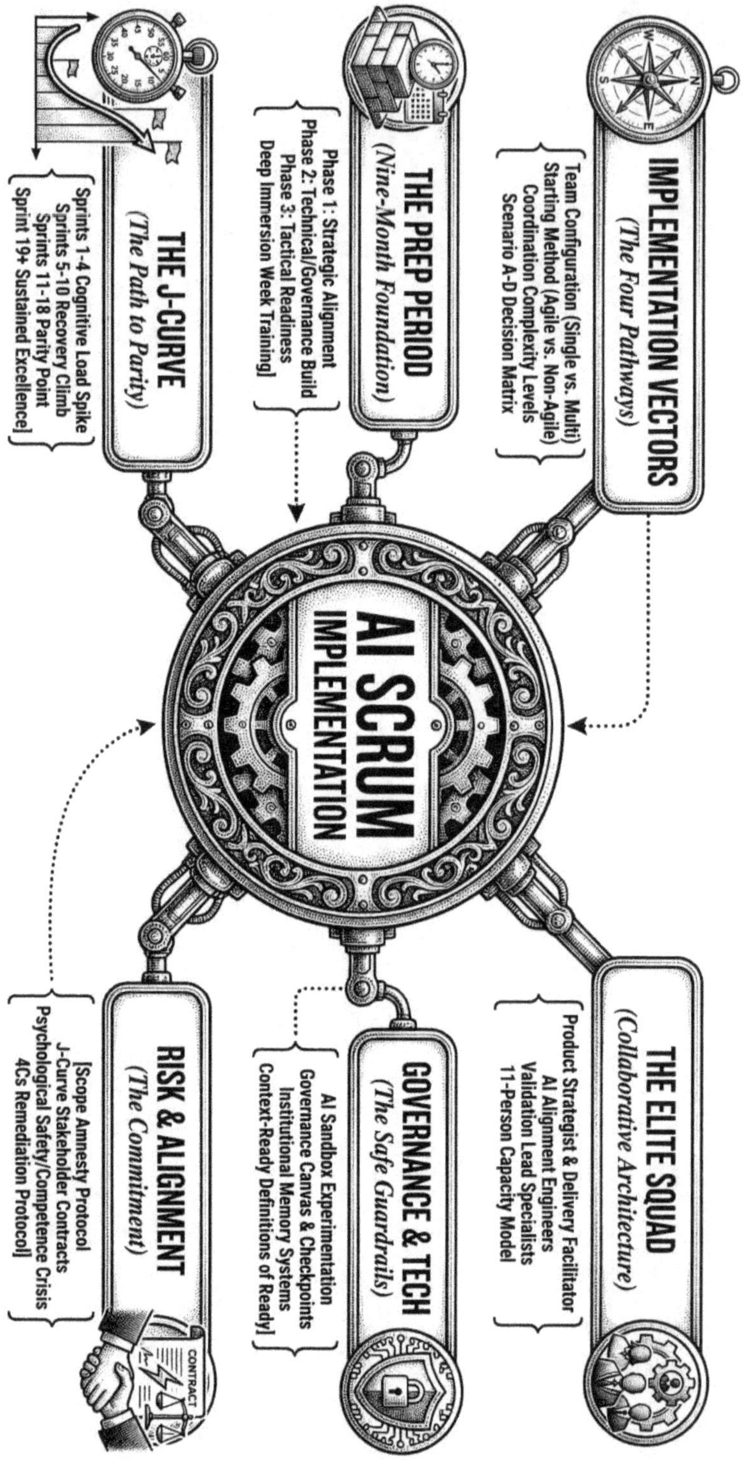

14

Organizations with existing teams face unique challenges adopting AI Scrum due to legacy technical debt, competing initiatives, and fragmented agile maturity. Success requires systematic risk assessment, phased migration planning, strategic debt management, and careful team sequencing. The approach addresses governance evolution, manages transformation fatigue, and avoids common failure patterns like big-bang replacements and analysis paralysis while maintaining production stability.

CHAPTER 14: A SUMMARY OF THE AI TOOLING FOR DIFFERENT DISCIPLINES

The Chief Technology Officer surveyed the conference room holding representatives from six disciplines. Marketing wanted Jasper for content generation. Software developers needed GitHub Copilot for coding assistance. Data scientists requested DataRobot for AutoML. Human resources sought HireVue for recruiting automation. Legal required Harvey AI for contract analysis. Compliance needed ComplyAdvantage for regulatory monitoring. Fifteen different AI subscriptions sat in various procurement stages. Nobody'd considered integration requirements or governance implications.

This pattern repeats across most organizations. Individual departments discover compelling AI capabilities in isolation. They pilot tools solving immediate problems without considering enterprise architecture. Procurement approves subscriptions based on departmental budgets rather than strategic coherence. Six months later, tool sprawl creates three devastating consequences. These'll include financial waste from redundant capabilities, integration failures preventing collaboration, and governance violations from fragmented audit trails.

The solution requires deliberate architectural thinking about AI tooling ecosystems. Not every discipline needs identical tools, but they'll need compatible approaches. Marketing teams using Jira for campaign management must coordinate smoothly with development teams using Jira for sprint planning. Legal teams using Azure DevOps for matter management must integrate seamlessly with compliance teams using Azure DevOps for regulatory tracking.

This chapter provides that architectural guidance. You'll learn which AI tools serve which disciplines across marketing, software development, data science, human resources, legal, and compliance. You'll discover how to configure Jira or Azure DevOps to support multi-discipline AI Scrum workflows. You'll understand adoption sequencing strategies preventing teams from drowning in complexity.

The frameworks from previous chapters explained AI Scrum methodology. This chapter bridges methodology to implementation reality by specifying exact tools and configurations.

One important thing to keep in mind as time unfolds is that tools will change and evolve and the same holds true concerning pricing. There may be quantity or customer discounts that have not been included. Use these costs as a general rule of thumb as things can change without warning.

Some tools will fade from view while new ones are created. The AI landscape is rapidly changing and what was available at the beginning of any given year may be replaced several months later by something more powerful and easier to use. Keep this in mind when looking over these currently available tools and choosing what is best for your organization.

WHAT THIS CHAPTER COVERS

- A summary of the AI tooling landscape organized by six categories with discipline applicability
- Marketing and creative operations tool recommendations with organization size guidance
- Software and web development tool stacks from small teams to enterprise scale
- Data science and analytics platforms for AutoML and experiment management

- Human resources AI tools for recruiting, assessment, and talent intelligence
- Legal AI capabilities for contract analysis, research, and due diligence
- Compliance tools for regulatory monitoring, AML, and fraud detection
- Jira and Azure DevOps configurations supporting multi-discipline workflows
- Cross-discipline integration patterns when multiple teams collaborate
- Security and governance considerations for enterprise AI deployments
- Three-phase adoption sequencing building capability without overwhelming teams
- Cost planning guidance with 2025 pricing for budgeting purposes

KEY TERMS AND CONCEPTS

- **Large Language Model (LLM)**: General-purpose AI system for understanding and generating human language across multiple domains. Handles tasks from requirements analysis to documentation generation to strategic thinking support for any discipline.
- **AI Code Assistant**: Development environment integration providing real-time code generation, completion, and explanation directly within IDEs. Examples include GitHub Copilot, Cursor, Tabnine, and Amazon Q Developer.
- **AutoML Platform**: Automated machine learning system that handles model selection, hyperparameter tuning, and feature engineering automatically. Reduces data science expertise requirements for building production-quality models.
- **Talent Intelligence**: AI-powered recruiting and workforce analytics platform analyzing skills, career trajectories, and organizational fit. Transforms reactive hiring into proactive talent pipeline management.
- **Contract Lifecycle Management (CLM)**: Platform managing contracts from initial drafting through execution, renewal, and termination. AI enhances CLM with clause extraction, risk analysis, and compliance checking.
- **Anti-Money Laundering (AML)**: Regulatory compliance process detecting suspicious financial transactions indicating money laundering activity. AI analyzes transaction patterns identifying anomalies faster than rule-based systems.

- **Work Item Type**: Configurable entity in Jira or Azure DevOps representing units of work. Customizable with fields, workflows, and automation rules specific to discipline requirements.
- **Tool Sprawl**: Uncontrolled proliferation of redundant AI tools across an organization without integration or governance. Creates waste, integration failures, and compliance gaps despite high subscription costs.
- **Data Residency**: Legal or regulatory requirement that data remain stored and processed within specific geographic boundaries. Critical for healthcare, finance, and government organizations subject to jurisdiction-specific regulations.
- **Prompt Library**: Organized collection of proven prompts versioned and categorized for reuse across teams. Accelerates AI adoption by sharing successful patterns rather than requiring each team to discover them independently.

THE AI TOOLING LANDSCAPE

AI tools relevant to organizational work cluster into six categories. Understanding this landscape prevents random adoption driven by vendor marketing. Each category serves specific purposes while overlapping in complementary ways.

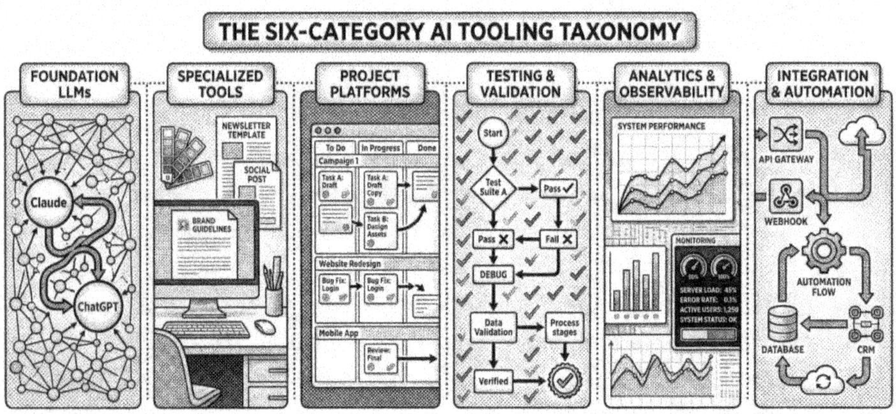

Figure 14.1: The Six-Category AI Tooling Taxonomy. *AI tools cluster into complementary categories: foundation models, discipline-specific specialization, project coordination, quality automation, intelligent monitoring, and ecosystem integration. Strategic selection balances comprehensive coverage against budget constraints while maintaining integration coherence.*

Large language models form the foundation for all disciplines. Claude from Anthropic excels at nuanced reasoning with 200,000-token context windows. ChatGPT from OpenAI offers broad capabilities with extensive plugin ecosystems. Gemini from Google provides native Workspace integration. Organizations typically standardize on one or two foundation models based on use case requirements and existing technology ecosystems.

Specialized tools address requirements that foundation models handle inadequately. Marketing needs tools optimized for brand voice consistency. Software developers need code assistants integrated into development environments. Data scientists need AutoML platforms handling full machine learning pipelines. Legal teams will need contract analysis tools trained on legal language and precedents. Compliance teams will need regulatory monitoring systems tracking rule changes across jurisdictions.

Project management platforms provide the central coordination layer. Jira with Atlassian Intelligence delivers AI capabilities within existing Agile workflows. Azure DevOps incorporates AI features across work items, pipelines, and repositories. These platforms become essential when multiple disciplines collaborate on shared initiatives. Platform choice influences integration patterns significantly.

Testing and validation tools automate quality assurance at scale. Marketing validates brand compliance before publication. Software teams automate regression testing with self-healing frameworks. Data science validates model performance across demographic segments. Legal verifies contract clause compliance. Compliance tests control effectiveness with automated evidence collection.

Analytics and observability platforms provide intelligent monitoring beyond traditional dashboards. These tools detect anomalies, predict failures, and correlate events across complex systems. Each discipline gains automated insights into performance trends and emerging issues without manual analysis.

Integration and automation capabilities connect tools into coherent ecosystems. APIs enable data exchange between platforms. Automation workflows trigger actions across systems based on events. Shared authentication provides consistent access control. Audit logging captures activities for governance requirements.

The tooling landscape evolves rapidly. The recommendations here reflect December 2025 market conditions. Treat them as starting points requiring periodic reassessment. Strategic selection balances detailed coverage against

budget constraints, deep specialization against integration complexity, and enterprise security against usability.
The following sections provide detailed tool recommendations for each discipline with organization size guidance and multi-discipline coordination scenarios.

MARKETING & CREATIVE OPERATIONS TOOLS

Marketing and creative teams face distinct challenges in AI adoption. They must balance creative quality with production velocity while maintaining brand consistency across AI-generated content. Copyright concerns and regulatory compliance add complexity that other disciplines avoid. The tooling strategy for these teams prioritizes content generation, brand compliance validation, and campaign performance analytics.

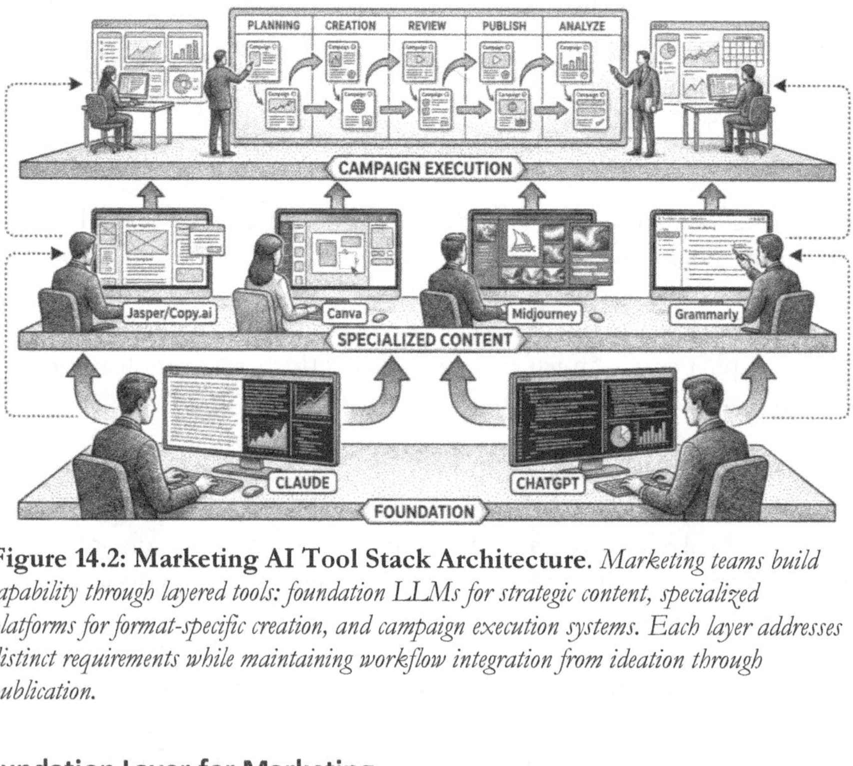

Figure 14.2: Marketing AI Tool Stack Architecture. *Marketing teams build capability through layered tools: foundation LLMs for strategic content, specialized platforms for format-specific creation, and campaign execution systems. Each layer addresses distinct requirements while maintaining workflow integration from ideation through publication.*

Foundation Layer for Marketing

Marketing teams require versatile LLMs capable of adapting to brand voice and generating diverse content formats. Claude 3.5 Sonnet excels at long-form content with superior nuance in tone and style matching. The 200,000-token context window enables analysis of complete brand guidelines and past campaign examples. ChatGPT offers faster iteration cycles for social media content and short-form copy. Gemini provides native integration with Google Workspace for teams already using Google Marketing Platform.

Most marketing teams adopt Claude as their primary tool for strategic content. Campaign briefs, white papers, and thought leadership articles benefit from its analytical depth. ChatGPT serves as the secondary tool for rapid iteration on social posts, email subject lines, and ad copy variations. This dual-model approach costs $40 per marketer monthly based on Claude Pro and ChatGPT Plus pricing as of January 2026.

Essential Marketing AI Tools

Jasper (Jasper AI) dominates marketing-specific content generation with brand voice training capabilities. Teams upload brand guidelines, past campaigns, and approved messaging. Jasper learns organizational voice patterns and generates on-brand content consistently. The platform includes templates for blog posts, social media, email campaigns, and product descriptions. According to Jasper's pricing page, Creator plan costs $49 monthly for one user. Teams plan runs $125 monthly for three users. Business plan requires custom quotes typically ranging $400-600 monthly for teams of ten.

Figure 14.3: Brand-Compliant Content Validation Flow. *Marketing workflows incorporate mandatory validation checkpoints ensuring AI-generated content maintains brand consistency and legal compliance. Sequential review stages prevent publication errors while preserving production velocity through parallel processing of multiple campaigns.*

Copy.ai provides similar capabilities at lower price points for smaller teams. The platform focuses on short-form content like social posts, ad copy, and email subject lines. Copy.ai's website lists individual plans at $49

monthly and team plans at $249 monthly for five users. Copy.ai works well for teams prioritizing speed over deep brand voice customization.
Canva AI transforms visual content creation for marketing teams without design expertise. The Magic Write feature generates copy directly within design templates. Magic Edit removes backgrounds and adjusts images with natural language commands. Text to Image creates custom graphics from descriptions. Canva Teams pricing shows $130 annually per user with AI features included, bringing annual costs to $1,300 for teams of ten.

Grammarly Business ensures brand consistency and error-free content at scale. The platform checks grammar, tone, clarity, and brand voice compliance across all content. Style guides enforce organizational terminology and messaging consistency. Teams share custom rules ensuring everyone writes on-brand. Grammarly Business pricing runs $15 monthly per user according to their enterprise sales documentation.

Recommended Marketing Tools

Several specialized tools extend marketing AI capabilities beyond foundation models. Midjourney's subscription tiers range from $30 monthly for Basic to $60 monthly for Standard plans per their pricing page. Adobe Firefly integrates AI image generation within Adobe Creative Cloud subscriptions at existing subscription rates. Synthesia's pricing page shows individual creator plans at $89 monthly. Descript handles audio and video editing with AI transcription starting at $24 monthly for individual creators according to their website. RunwayML provides advanced video editing capabilities with pricing beginning at $15 monthly for basic plans.

These tools address specific creative needs. Midjourney suits teams needing high-quality custom imagery regularly. Adobe Firefly makes sense for teams already using Creative Cloud. Synthesia works for teams producing regular video content without video production expertise. Selection depends on content formats teams produce most frequently.

Tool Configuration by Organization Size

Small Marketing Team (1-10 people, $300-600 monthly)

- Claude Pro: $20/user × 5 users = $100
- Canva Teams: $130/year per user × 10 = $1,300 annually ($108 monthly)
- Grammarly Business: $15/user × 5 users = $75
- Jasper Creator: $49
- Total: $332 monthly plus $1,300 annual Canva = $440 monthly averaged

This configuration provides full content generation and quality control. Small teams maximize value from general-purpose tools before adding specialized capabilities.

Growth Marketing Team (10-50 people, $2,000-5,000 monthly)

- Claude Team: Anthropic lists Team pricing at $25/user monthly × 20 users = $500
- Canva Enterprise: Estimated $180/year per user × 30 = $5,400 annually ($450 monthly)
- Grammarly Business: $15/user × 20 users = $300
- Jasper Business: Estimated $800-1,200 based on typical custom quotes
- Midjourney Teams: $60/user × 5 designers = $300
- Total: Approximately $2,350-2,750 monthly

Growth teams add specialized tools as content volume increases. Image generation and advanced design capabilities support larger campaign portfolios.

Enterprise Marketing (50+ people, $10,000+ monthly)

- Claude Enterprise: Estimated $40-50/user based on typical enterprise pricing × 100 users = $4,000-5,000
- Canva Enterprise: Negotiated volume discount pricing varies
- Full creative suite: Adobe Creative Cloud with Firefly, Synthesia Enterprise, advanced analytics
- Custom LLM fine-tuning for brand voice: Estimated $5,000 10,000 monthly
- Total: Approximately $15,000-25,000 monthly

Enterprise teams invest in custom solutions and volume discounts. Fine-tuned models trained on organizational content provide maximum brand consistency.

Multi-Discipline Coordination

Marketing teams frequently coordinate with other disciplines on shared initiatives. When tracking web development projects for website redesigns, marketing uses Jira boards with development teams. Shared sprint planning ensures marketing content aligns with technical delivery timelines. When requiring legal review for advertising compliance, marketing workflows integrate legal approval checkpoints. See Platform Configuration sections for cross-discipline integration patterns.

SOFTWARE & WEB DEVELOPMENT TOOLS

Software development teams approach AI adoption from positions of technical strength and methodological maturity. Most development organizations already practice Agile or Scrum making AI Scrum migration more familiar than greenfield adoption. The tooling strategy emphasizes code generation, automated testing, security scanning, and quality analysis while maintaining engineering rigor.

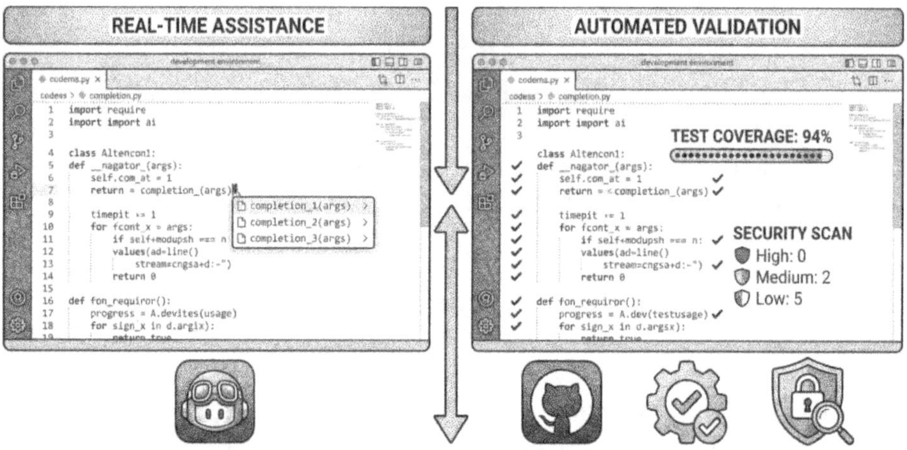

Figure 14.4: Integrated Development Environment AI Augmentation. *Modern development workflows embed AI assistance directly within IDEs, providing context-aware suggestions during coding while automated validation systems immediately assess quality, security, and test coverage. Tight integration eliminates context switching between tools.*

Foundation Layer for Software Development

Development teams benefit from LLMs capable of generating code across multiple programming languages and frameworks. Claude 3.5 Sonnet excels at architectural reasoning and complex algorithm development. Developers use Claude for system design discussions, code review analysis, and technical documentation generation. The model understands context spanning entire codebases enabling refactoring suggestions.

ChatGPT provides faster iteration for routine coding tasks and quick problem-solving. Developers use ChatGPT for syntax questions, debugging assistance, and standard implementation patterns. The plugin ecosystem extends capabilities into specialized development domains.

Most development teams adopt both models for complementary purposes. Claude handles architectural decisions and complex feature development. ChatGPT accelerates routine implementation and troubleshooting.

Combined subscriptions cost approximately $40 monthly per developer based on December 2025 pricing.

Code Generation and Completion Tools

GitHub Copilot (Microsoft/GitHub) dominates AI-assisted development with deep IDE integration and extensive language support. The tool provides context-aware code suggestions across twelve programming languages. Developers accept suggestions inline while writing code reducing time spent on boilerplate and routine implementation. Whole-function generation from natural language comments accelerates feature development. Test case generation capabilities improve code coverage without manual effort.

According to GitHub's pricing documentation, individual plans cost $10 monthly per developer. Business tier runs $19 monthly per developer. Enterprise tier provides $39 monthly per developer including security features and policy controls. Organizations with one hundred developers pay $1,000-3,900 monthly depending on tier selection.

Cursor provides AI-first IDE experience with integrated code generation and editing. The tool understands entire codebases enabling sophisticated refactoring across multiple files. Developers describe desired changes in natural language and Cursor implements modifications systematically. Pricing runs approximately $20 monthly per developer according to their website. Organizations valuing integrated experience over plugin architecture favor Cursor.

Tabnine offers enterprise-focused code completion with on-premise deployment options. Organizations with strict data residency requirements or intellectual property concerns deploy Tabnine locally. The platform learns from organizational codebases improving suggestions based on internal patterns. Starter plans cost $12 monthly per developer. Pro plans run $39 monthly per developer. Enterprise pricing requires custom quotes typically ranging $50-100 monthly per developer.

Amazon Q Developer (AWS) integrates AI assistance within AWS development workflows. Teams building cloud-native applications on AWS infrastructure gain context-aware suggestions for AWS services and best practices. The tool analyzes existing AWS deployments recommending optimization opportunities. Pricing follows usage-based models tied to AWS consumption.

Testing and Quality Assurance Automation

Testim provides AI-powered test automation with self-healing capabilities reducing test maintenance overhead. The platform identifies when applications change and adapts test scripts automatically. UI tests remain stable across cosmetic changes while detecting genuine functionality breaks. Pricing starts around $450 monthly for basic plans. Enterprise deployments range $2,000-5,000 monthly depending on test volume.

Mabl delivers intelligent test automation designed for continuous integration pipelines. The platform executes tests automatically on every commit and learns which tests reliably identify real issues. False alarm fatigue decreases significantly as Mabl distinguishes signal from noise. Pricing follows per-application models with typical costs ranging $500-2,000 monthly.

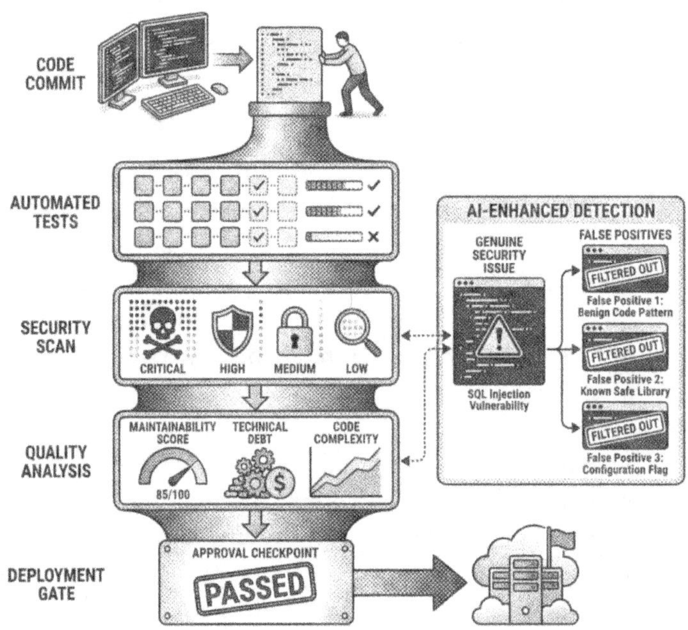

Figure 14.5: AI-Enhanced Continuous Integration Pipeline. *Development pipelines incorporate AI-powered quality gates that distinguish genuine issues from false positives. Automated testing, security scanning, and code analysis execute on every commit, preventing defects from reaching production while reducing manual review burden.*

Applitools specializes in visual testing using AI to detect UI regressions. Traditional pixel-perfect comparisons break whenever fonts or colors change slightly. Applitools understands visual hierarchy identifying genuine layout problems while ignoring insignificant variations. Pricing ranges $99-999 monthly based on test volume and features.

Security and Code Quality Tools

Snyk provides AI-enhanced vulnerability scanning beyond traditional pattern matching. The platform understands code context distinguishing genuine security issues from false positives. Integration with CI/CD pipelines catches vulnerabilities before production deployment. Free tier supports open-source projects. Team plans cost $52 monthly per developer. Enterprise pricing requires custom quotes.

SonarQube incorporates AI features for code quality analysis across multiple dimensions. The platform identifies maintainability issues, code smells, and AI Debt accumulation. Security vulnerability detection complements Snyk with different detection approaches. Community edition provides free basic scanning. Developer edition costs approximately $150 annually per developer. Enterprise edition pricing varies based on lines of code analyzed.

GitHub Advanced Security delivers security scanning integrated with GitHub repositories. Secret scanning prevents credential leaks. Dependency scanning identifies vulnerable libraries. Code scanning detects security weaknesses in custom code. Pricing bundles with GitHub Enterprise subscriptions at $21 monthly per user with Advanced Security features.

Tool Configuration by Organization Size

Small Development Team (3-10 developers, $400-800 monthly)

- GitHub Copilot Individual: $10/developer × 10 = $100
- Claude Pro: $20/developer × 5 = $100
- Testim Starter: Estimated $450
- Snyk Free Tier: $0
- Total: Approximately $650 monthly

Small teams prioritize code generation and basic test automation. Free security scanning tiers provide adequate protection for early-stage products.

Growth Development Team (10-50 developers, $3,000-8,000 monthly)

- GitHub Copilot Business: $19/developer × 30 = $570
- Cursor Pro: $20/developer × 30 = $600
- Claude Team: $25/developer × 20 = $500
- Testim Pro: Estimated $2,000
- Mabl: Estimated $1,500
- Snyk Team: $52/developer × 30 = $1,560

- SonarQube Developer: Estimated $150/developer × 30 = $4,500 annually ($375 monthly)
- Total: Approximately $7,105 monthly

Growth teams invest in comprehensive testing and security platforms. Multiple code assistants provide developers choice based on preferences and task requirements.

Enterprise Development (50+ developers, $15,000+ monthly)

- GitHub Copilot Enterprise: $39/developer × 100 = $3,900
- Cursor Enterprise: Custom pricing estimated $30/developer × 100 = $3,000
- Claude Enterprise: Estimated $50/developer × 50 = $2,500
- Testim Enterprise: Estimated $5,000
- Mabl Enterprise: Estimated $4,000
- Applitools Enterprise: Estimated $3,000
- Snyk Enterprise: Custom quote estimated $75/developer × 100 = $7,500
- GitHub Advanced Security: Bundled with GitHub Enterprise
- Total: Approximately $28,900 monthly

Enterprise teams deploy thorough security scanning and quality analysis. Custom integrations with internal development platforms require additional investment.

Multi-Discipline Coordination

Development teams frequently coordinate with other disciplines on shared initiatives. When building features for marketing campaigns, developers use shared Jira boards tracking both marketing content and technical implementation. When implementing data science models, development teams coordinate deployment pipelines with data science experimentation workflows. See Platform Configuration sections for cross-discipline integration patterns supporting collaborative development.

DATA SCIENCE & ANALYTICS TOOLS

Data science and analytics teams approach AI adoption from unique positions. They already possess deep technical expertise in machine learning and statistical modeling. Their challenge involves integrating AI augmentation into existing workflows without compromising analytical rigor. The tooling strategy focuses on AutoML platforms, experiment management, and collaborative analysis environments.

Foundation Layer for Data Science

Data science teams benefit from LLMs capable of code generation and analytical reasoning. Claude 3.5 Sonnet excels at data analysis with extended context enabling entire dataset exploration in single conversations. The model generates Python and R code for statistical analysis, creates visualizations, and explains complex analytical approaches. ChatGPT Advanced Data Analysis (formerly Code Interpreter) provides complementary capabilities with built-in code execution. Data scientists can upload datasets, run statistical analyses, and generate charts without leaving the chat interface.

Most data science teams adopt both tools for different purposes. Claude handles exploratory analysis, hypothesis development, and complex statistical reasoning. ChatGPT Advanced Data Analysis serves rapid prototyping and ad-hoc analytical requests. This dual approach costs $40 monthly per data scientist based on December 2025 pricing.

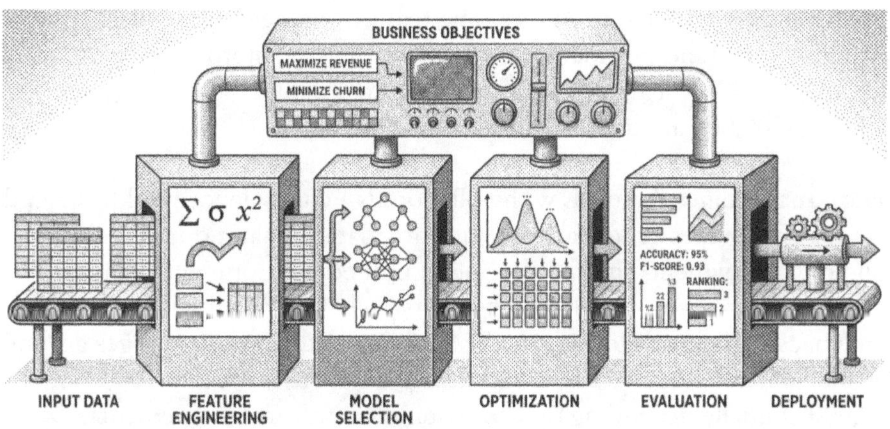

Figure 14.6: Automated Machine Learning Pipeline Architecture.
AutoML platforms automate the machine learning workflow from feature engineering through deployment. Data scientists define business objectives while the platform handles model selection, hyperparameter tuning, and performance optimization, reducing weeks of manual experimentation to hours.

AutoML Platforms

Databricks AutoML dominates enterprise data science workflows with machine learning automation. The platform handles feature engineering, model selection, hyperparameter tuning, and deployment pipelines automatically. Teams define business objectives and data sources. Databricks generates multiple candidate models, evaluates performance across metrics, and recommends optimal approaches. Integration with existing Databricks lakehouse architecture reduces friction significantly. Pricing operates on consumption basis with DBU (Databricks Unit) charges. Standard tier runs approximately $0.40 per DBU plus underlying compute costs. Typical monthly spending will range $2,000-5,000 for small teams to $20,000+ for enterprise deployments.

DataRobot provides enterprise-grade AutoML with extensive model explainability features. The platform excels at regulated industry deployments requiring model transparency and audit trails. Financial services and healthcare organizations favor DataRobot for compliance capabilities. Pricing requires custom enterprise quotes. Typical starting costs range $50,000-100,000 annually for small team deployments. Enterprise deployments with hundreds of data scientists reach $500,000+ annually.

H2O.ai offers open-source and enterprise AutoML options. H2O AutoML provides free open-source automated machine learning for teams comfortable managing infrastructure. H2O Driverless AI delivers enterprise features including automatic feature engineering, model documentation, and production deployment tools. Enterprise pricing starts around $30,000 annually for small teams. Large deployments range $100,000-300,000 annually.

Dataiku combines AutoML with collaborative data science workflows and MLOps capabilities. The platform supports data preparation, model building, deployment, and monitoring in unified environments. Business analysts, data scientists, and ML engineers collaborate within shared workspaces. Pricing follows per-user licensing with starter packages around $30,000 annually for five users. Enterprise deployments cost $100,000-500,000 annually depending on user count and feature requirements.

Specialized Analytics Tools

AWS SageMaker Autopilot provides AutoML within AWS ecosystems. Teams already using AWS infrastructure gain AutoML capabilities without platform switching. The service automatically builds, trains, and tunes machine learning models based on tabular data. Pricing follows pay-per-use model charging for underlying compute and storage resources. Typical monthly costs range $500-5,000 depending on usage patterns.

Google Vertex AI AutoML delivers similar capabilities within Google Cloud Platform. Organizations using BigQuery and Google Cloud Storage benefit from native integration. Pricing operates on prediction volume with training costs separate from inference charges. Monthly costs typically range $1,000-10,000 based on model complexity and prediction volume.

Azure Machine Learning AutoML completes major cloud provider offerings. Microsoft-centric organizations leverage existing Azure subscriptions and Active Directory integration. The platform provides automated machine learning with strong enterprise governance features. Pricing follows consumption model with costs typically ranging $1,000-8,000 monthly.

Experiment Tracking and MLOps

MLflow provides open-source experiment tracking, model registry, and deployment management. Data science teams track experiments, compare model performance, and manage model lifecycles without vendor lock-in. The tool integrates with existing Python workflows seamlessly. MLflow itself costs nothing as open-source software. Infrastructure and maintenance costs vary based on deployment scale.

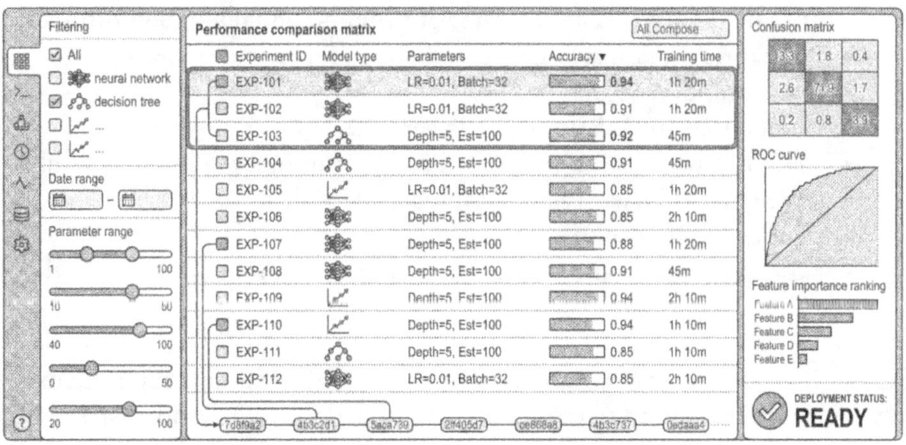

Figure 14.7: Collaborative Experiment Tracking System. *Experiment management platforms capture every model iteration with complete metadata, enabling data science teams to compare performance systematically and reproduce results reliably. Shared visibility accelerates learning by making successful patterns immediately available across the team.*

Weights & Biases offers commercial experiment tracking with superior visualization and collaboration features. Data scientists track experiments, visualize training metrics, and share results with stakeholders through intuitive dashboards. Pricing starts at $50 monthly per user for team plans. Enterprise pricing requires custom quotes typically ranging $200-500 per user annually.

Neptune.ai provides metadata management and experiment tracking optimized for team collaboration. The platform captures experiment details automatically and enables sophisticated experiment comparison. Pricing begins at $58 monthly per user for individual plans. Team plans cost custom amounts based on user count and storage requirements.

Tool Configuration by Organization Size

Small Data Team (2-5 analysts, $500-1,500 monthly)

- Claude Pro: $20/user × 5 users = $100
- ChatGPT Plus: $20/user × 3 users = $60
- H2O AutoML: Open-source (infrastructure costs only)
- MLflow: Open-source (infrastructure costs only)
- AWS/GCP/Azure credits: Approximately $500-1,000 monthly
- Total: Approximately $660-1,160 monthly

Small teams maximize open-source tools while building analytical capabilities. Cloud platform credits provide flexible compute resources.

Growth Data Team (5-20 analysts/engineers, $5,000-15,000 monthly)

- Databricks Standard: Estimated $0.40/DBU with typical consumption $3,000-5,000
- Claude Team: $25/user × 10 users = $250
- H2O Driverless AI: Estimated $30,000 annually ($2,500 monthly)
- Weights & Biases Teams: Estimated $200/user/year × 10 = $2,000 annually ($167 monthly)
- Cloud compute and storage: $3,000-6,000 monthly
- Total: Approximately $8,917-13,917 monthly

Growth teams invest in commercial AutoML platforms and experiment tracking. Scaling compute requirements drive increasing cloud costs.

Enterprise Data Science (20+ people, $25,000+ monthly)

- Databricks Enterprise: Custom pricing estimated $8,000-15,000 monthly
- DataRobot Enterprise: Estimated $500,000 annually ($41,667 monthly)
- Dataiku Enterprise: Estimated $200,000 annually ($16,667 monthly)
- Cloud infrastructure: $10,000-30,000 monthly
- Custom model monitoring and governance tools
- Total: Approximately $75,000-100,000+ monthly

Enterprise teams deploy MLOps platforms with dedicated support. Custom governance and monitoring capabilities ensure regulatory compliance.

Multi-Discipline Coordination

Data science teams frequently support other disciplines through analytical insights. When analyzing marketing campaign performance, data scientists collaborate using shared Jira boards tracking analysis requests. When building predictive models for HR talent pipelines, data teams integrate with HR systems for candidate data. See Platform Configuration sections for cross-discipline analytics integration patterns.

HUMAN RESOURCES TOOLS

Human resources teams face unique AI adoption challenges balancing efficiency gains against compliance requirements and candidate experience quality. Bias detection, data privacy, and regulatory oversight constrain tool selection more heavily than most disciplines. The tooling strategy emphasizes talent intelligence, candidate screening, and policy drafting while maintaining strict governance controls.

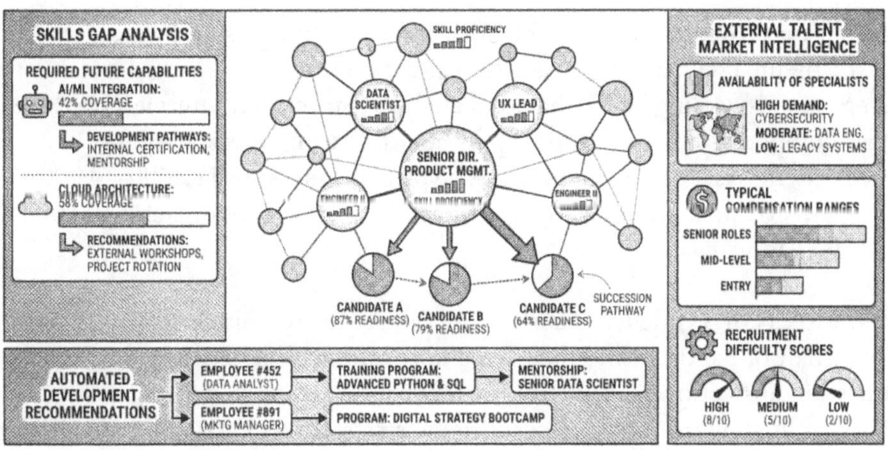

Figure 14.8: Enterprise Talent Intelligence Mapping. *AI-powered talent platforms analyze internal skill inventories, identify development pathways, and predict succession readiness. Organizations shift from reactive hiring to proactive talent pipeline management by understanding current capabilities and strategic gaps systematically.*

Foundation Layer for Human Resources

HR teams require LLMs capable of handling sensitive personnel data with appropriate privacy controls. Claude excels at policy drafting, job description generation, and nuanced communication requiring empathy. The model helps HR professionals craft offer letters, performance review templates, and employee handbook updates. ChatGPT handles rapid iteration on recruitment messaging and candidate communication templates. Both tools require enterprise tiers with data handling guarantees for any personnel information processing.

Most HR teams adopt Claude Pro at $20 monthly per HR professional for non-sensitive work. Enterprise deployments handling employee data require Claude Enterprise with custom privacy agreements. Never input actual employee names, performance data, or protected characteristics into AI tools without explicit organizational approval and legal review.

Talent Intelligence Platforms

Eightfold AI dominates enterprise talent intelligence with detailed workforce analytics and skills mapping. The platform analyzes internal talent pools, identifies skill gaps, and recommends development pathways. Recruitment modules match candidates to roles based on potential rather than just resume keywords. Pricing follows enterprise licensing models with typical costs ranging $100,000-500,000 annually depending on employee count. Organizations with 1,000-5,000 employees typically pay $150,000-250,000 annually.

HireVue provides AI-powered video interview analysis and candidate assessment. The platform evaluates candidate responses using natural language processing and behavioral analysis. Structured interviews reduce bias compared to unstructured approaches. Pricing operates on per-assessment basis or annual licenses. Typical costs range $15,000-50,000 annually for small organizations. Enterprise deployments reach $100,000-300,000 annually.

Paradox (Olivia) delivers conversational AI for recruitment automation. The chatbot handles candidate screening, interview scheduling, and basic question answering. Candidates interact through text messaging reducing friction in application processes. Pricing starts around $10,000 annually for small implementations. Mid-market deployments range $30,000-80,000 annually. Enterprise pricing reaches $150,000+ annually.

Recruitment and Sourcing Tools

Textio enhances job posting effectiveness through augmented writing analysis. The platform analyzes job descriptions for bias, clarity, and appeal across demographic groups. Real-time suggestions improve inclusive language and candidate attraction. Annual pricing starts around $10,000 for small teams. Mid-market deployments typically cost $20,000-40,000 annually.

Beamery provides talent CRM and marketing automation for proactive recruiting. HR teams build candidate pipelines and nurture relationships before roles open. The platform tracks candidate engagement and optimizes outreach timing. Pricing follows per-recruiter licensing with costs typically ranging $15,000-50,000 annually for small teams. Enterprise deployments reach $100,000-300,000 annually.

Fetcher automates candidate sourcing through AI-powered search and outreach. The platform identifies potential candidates matching role requirements across multiple sources. Automated initial outreach reduces recruiter workload significantly. Pricing ranges $20,000-60,000 annually depending on role volume and recruiter count.

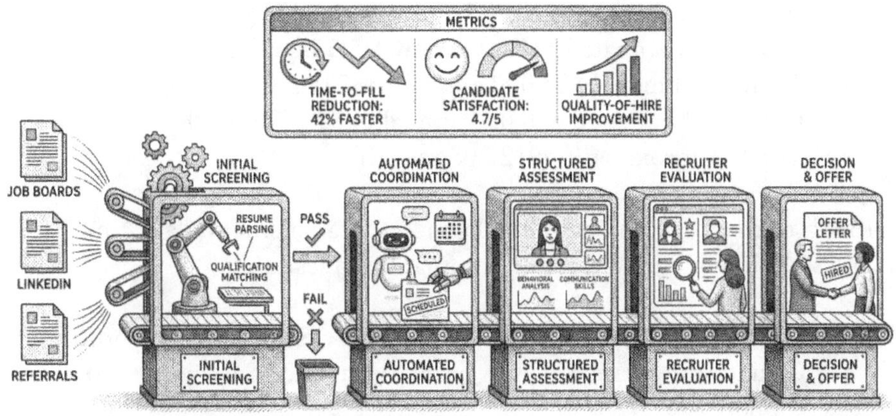

Figure 14.9: End-to-End Recruitment Automation Pipeline. *Modern HR systems automate candidate sourcing, initial screening, scheduling coordination, and assessment analysis. Recruiters focus on relationship building and final evaluation rather than administrative coordination, reducing time-to-hire while improving candidate experience and selection quality.*

Assessment and Screening Tools

Pymetrics uses neuroscience-based games and AI analysis for behavioral assessment. Candidates complete cognitive and emotional assessments providing objective hiring data. The platform reduces bias by focusing on traits rather than backgrounds. Pricing requires custom enterprise quotes typically ranging $30,000-100,000 annually.

Criteria Corp delivers pre-employment testing with AI-enhanced assessment recommendations. The platform suggests optimal test combinations based on role requirements and organizational patterns. Pricing follows per-assessment model or annual licenses starting around $5,000 annually.

Tool Configuration by Organization Size

Small HR Team (1-3 people, 100-500 employees, $500-1,000 monthly)

- Claude Pro: $20/user × 3 users = $60
- Textio: Estimated $10,000 annually ($833 monthly)
- Paradox Basic: Estimated $10,000 annually ($833 monthly)
- Total: Approximately $1,726 monthly

Small teams prioritize writing enhancement and basic automation. Manual processes remain viable for limited hiring volume.

Growth HR Team (3-10 people, 500-2,000 employees, $3,000-8,000 monthly)

- Claude Team: $25/user × 5 users = $125
- HireVue: Estimated $40,000 annually ($3,333 monthly)
- Beamery: Estimated $30,000 annually ($2,500 monthly)
- Textio: Estimated $25,000 annually ($2,083 monthly)
- Paradox: Estimated $50,000 annually ($4,167 monthly)
- Total: Approximately $12,208 monthly

Growth teams invest in talent intelligence and assessment platforms. Increased hiring volume justifies automation investment.

Enterprise HR (10+ people, 2,000+ employees, $15,000+ monthly)

- Claude Enterprise: Custom pricing estimated $40/user × 20 users = $800
- Eightfold AI: Estimated $200,000 annually ($16,667 monthly)
- HireVue Enterprise: Estimated $150,000 annually ($12,500 monthly)
- Beamery Enterprise: Estimated $100,000 annually ($8,333 monthly)
- Paradox Enterprise: Estimated $150,000 annually ($12,500 monthly)
- Detailed assessment suite: $50,000-100,000 annually
- Total: Approximately $54,167-62,500 monthly

Enterprise teams deploy talent platforms with dedicated support. Custom integrations with HRIS systems and applicant tracking systems require additional development investment.

Multi-Discipline Coordination

HR teams coordinate across multiple disciplines on talent and organizational initiatives. When HR lacks project management capabilities in HRIS platforms, teams use Jira with HR-specific workflows for recruitment campaigns and learning development rollouts. When supporting legal teams on employment policy reviews, HR workflows integrate legal approval checkpoints. See Platform Configuration sections for HR project management patterns and cross-functional integration approaches.

LEGAL TOOLS

Legal teams face the highest regulatory scrutiny and risk exposure of any discipline adopting AI tools. Attorney-client privilege, work product doctrine, and professional responsibility rules create complex constraints. Confidentiality obligations prevent sharing client information with third-party AI services without explicit consent. The tooling strategy prioritizes contract analysis, legal research, and document drafting while maintaining strict confidentiality controls.

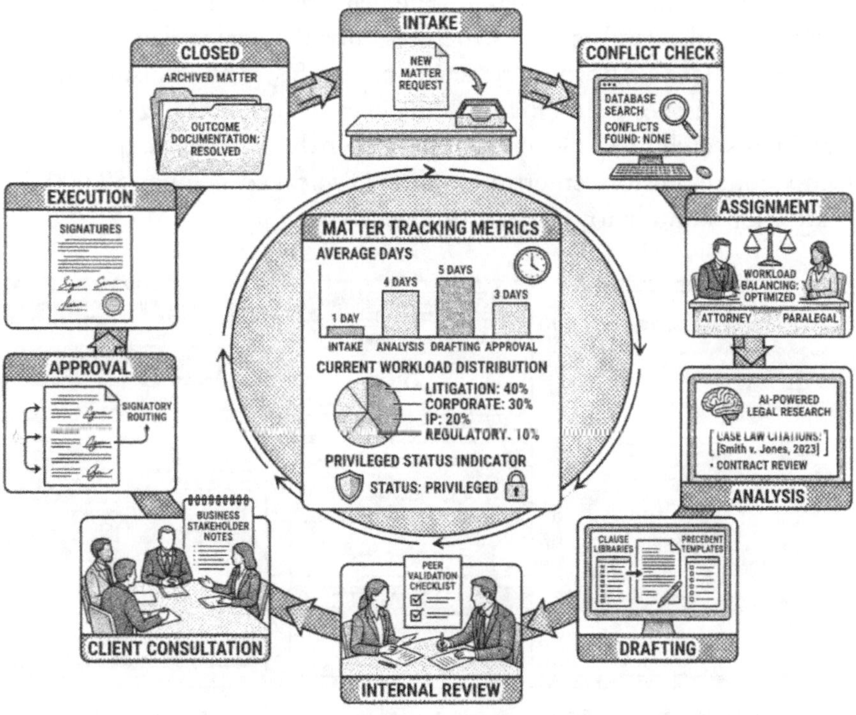

Figure 14.10: Comprehensive Legal Matter Management Lifecycle. *Legal workflows progress through standardized stages from intake through closure, with AI assistance embedded at research and drafting phases. Structured progression ensures quality control, maintains privilege protection, and provides complete audit trails for regulatory compliance.*

Foundation Layer for Legal

Legal teams require LLMs with enterprise data handling guarantees preventing training on confidential information. Claude Enterprise provides appropriate controls for attorney work product with contractual assurances against data retention. The model assists with legal research, contract drafting, and policy analysis. ChatGPT Enterprise offers similar protections with different interface preferences. Both require enterprise agreements before processing any client information or privileged communications.

Legal teams shouldn't use consumer-tier AI tools for client matters. The risk of waiving privilege or violating confidentiality obligations far exceeds any cost savings. Enterprise agreements typically cost $40-60 per attorney monthly with minimum seat commitments.

Contract Analysis and Review

Harvey AI dominates legal-specific AI with purpose-built tools for contract analysis and legal research. The platform trained on legal documents understands clause structures, jurisdictional variations, and precedent patterns. Attorneys review contracts faster by focusing attention on AI-flagged issues rather than reading every clause. Legal research capabilities search case law and statutes with natural language queries. Pricing requires custom enterprise quotes. Typical costs range $2,000-3,000 per attorney monthly for full platform access.

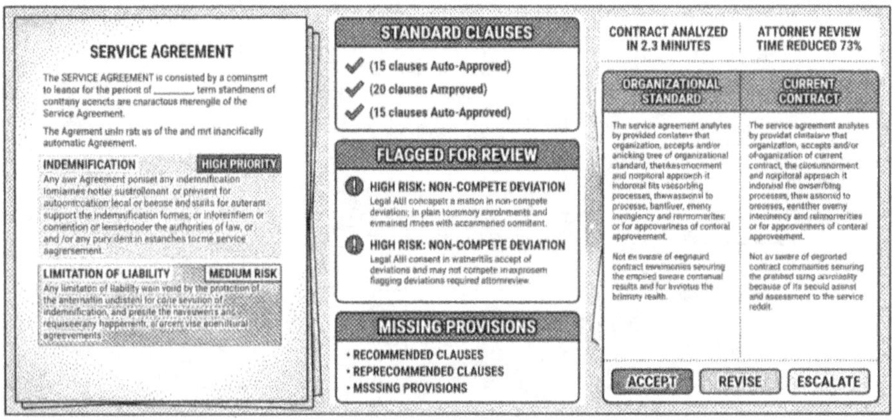

Figure 14.11: AI-Powered Contract Analysis and Risk Identification. *Legal AI platforms parse contracts against organizational playbooks, auto-approving standard language while flagging deviations for attorney review. Legal teams focus on risk assessment over clause-by-clause reading, accelerating review cycles without compromising quality.*

LawGeex provides contract review automation focused on standard agreement types. The platform analyzes NDAs, vendor agreements, and employment contracts against organizational playbooks. Approval workflows route non-standard clauses to attorneys while auto-approving standard language. Pricing follows per-contract or subscription models. Annual costs typically range $30,000-100,000 depending on contract volume.

Kira Systems specializes in contract analysis for due diligence and portfolio reviews. The platform extracts provisions from large contract sets and identifies deviations from standard terms. M&A due diligence and contract migration projects benefit from automated extraction capabilities. Pricing requires enterprise licensing with costs typically ranging $50,000-150,000 annually.

Legal Research and Intelligence

Casetext CoCounsel delivers AI-powered legal research with case law and statutory coverage. Attorneys ask research questions in natural language. The platform identifies relevant precedents, distinguishes controlling authority, and summarizes key holdings. Research time drops significantly compared to traditional Boolean keyword searching. Subscription pricing runs approximately $500-800 monthly per attorney.

Luminance provides AI for due diligence and document review in M&A transactions. The platform analyzes data rooms, identifies key documents, and flags potential issues. Legal teams review documents faster during compressed transaction timelines. Pricing follows project-based or annual licensing models with costs typically ranging $100,000-500,000 annually.

Contract Lifecycle Management

Ironclad combines contract lifecycle management with AI-powered clause extraction and analysis. The platform manages contracts from initial drafting through renewal or termination. Workflow automation routes approvals and captures metadata automatically. AI features extract key dates, obligations, and financial terms. Pricing follows per-user licensing with typical costs ranging $15,000-50,000 annually for small legal teams. Enterprise deployments reach $100,000-300,000 annually.

Onit provides enterprise contract management with AI-enhanced search and analysis. The platform supports legal operations teams managing large contract portfolios. Integration with matter management and e-billing systems creates unified legal operations platforms. Pricing requires custom enterprise quotes typically ranging $50,000-200,000 annually.

Specialized Legal AI Tools

eBrevia focuses on document analysis for lease abstraction and contract intelligence. Real estate and finance teams use eBrevia for large-scale document review projects. Pricing follows per-document or subscription models with typical costs ranging $20,000-80,000 annually.

RAVN (now part of iManage) provides intelligent contract analysis within document management systems. Teams already using iManage gain AI capabilities without platform switching. Pricing bundles with iManage subscriptions at additional cost.

Tool Configuration by Organization Size

Small Legal Team (1-3 attorneys, $2,000-5,000 monthly)

- Claude Enterprise: Estimated $50/attorney × 3 = $150
- Casetext CoCounsel: Estimated $600/attorney × 3 = $1,800
- LawGeex: Estimated $30,000 annually ($2,500 monthly)
- Total: Approximately $4,450 monthly

Small teams prioritize research efficiency and standard contract automation. Manual review remains necessary for complex matters.

Mid-Size Legal (3-10 attorneys, $10,000-25,000 monthly)

- Claude Enterprise: Estimated $50/attorney × 8 = $400
- Harvey AI: Estimated $2,500/attorney × 5 = $12,500
- Casetext CoCounsel: Estimated $600/attorney × 8 = $4,800
- LawGeex: Estimated $60,000 annually ($5,000 monthly)
- Kira Systems: Estimated $80,000 annually ($6,667 monthly)
- Total: Approximately $29,367 monthly

Mid-size teams invest in full legal AI platforms. Matter complexity and volume justify specialized tools.

Enterprise Legal (10+ attorneys, $50,000+ monthly)

- Claude Enterprise: Estimated $50/attorney × 25 = $1,250
- Harvey AI Enterprise: Custom pricing estimated $2,000/attorney × 20 = $40,000
- Ironclad Enterprise: Estimated $200,000 annually ($16,667 monthly)
- Luminance: Estimated $300,000 annually ($25,000 monthly)
- Full matter management and e-billing integration
- Total: Approximately $82,917+ monthly

Enterprise legal departments deploy full CLM suites with matter management integration. Custom AI models trained on firm precedents cost additional $10,000-30,000 monthly.

Multi-Discipline Coordination

Legal teams review materials from multiple disciplines requiring cross-functional workflows. When reviewing marketing materials for advertising compliance, legal workflows integrate with marketing approval processes. When supporting sales teams with contract negotiations, legal uses shared platforms tracking deal progress. See Platform Configuration sections for legal review integration patterns across disciplines.

COMPLIANCE TOOLS

Compliance teams operate under the strictest regulatory oversight of any discipline. Audit trail requirements, regulatory reporting obligations, and examination preparedness demands create exacting tool selection criteria. AI tools must provide explainable outputs, maintain complete activity logs, and support regulatory examination processes. The tooling strategy emphasizes regulatory monitoring, AML compliance, fraud detection, and control testing.

Figure 14.12: Real-Time Anti-Money Laundering Transaction Surveillance. *AML compliance systems analyze transaction patterns continuously, applying machine learning to identify suspicious activity while minimizing false positives disrupting legitimate business. Automated risk scoring routes transactions appropriately: routine approval for typical patterns, human review for anomalies.*

Foundation Layer for Compliance

Compliance teams require LLMs with enterprise controls ensuring audit trails and data residency compliance. Claude Enterprise provides necessary controls for regulatory analysis and policy interpretation. The model assists with regulatory gap analysis, control documentation, and compliance training content. All AI interactions require logging for examination purposes. Consumer-tier AI tools prove unsuitable for compliance work requiring documented decision processes.

Compliance professionals should implement AI usage policies before adopting any tools. Document which AI systems process what information types. Maintain logs of AI-assisted compliance decisions. Establish human review requirements for high-risk determinations. Enterprise LLM subscriptions typically cost $40-60 per compliance professional monthly.

Anti-Money Laundering and Financial Crime

ComplyAdvantage dominates AML compliance with real-time sanctions screening and adverse media monitoring. The platform monitors regulatory lists across global jurisdictions and identifies sanctioned entities automatically. Transaction monitoring rules adapt to emerging typologies using machine learning. Financial institutions use ComplyAdvantage for customer due diligence and ongoing monitoring. Pricing follows entity-based models with typical costs ranging $50,000-200,000 annually depending on transaction volume.

NICE Actimize provides thorough financial crime compliance covering AML, fraud, and conduct surveillance. The platform analyzes transactions, communications, and trading patterns for suspicious activity. Behavioral analytics detect anomalies indicating potential compliance violations. Enterprise deployments in large financial institutions cost $500,000-2,000,000 annually.

Verafin offers cloud-based BSA/AML compliance for community banks and credit unions. The platform handles suspicious activity monitoring, currency transaction reporting, and regulatory filing automation. Smaller financial institutions gain enterprise-grade compliance capabilities at accessible price points. Annual costs typically range $30,000-100,000 depending on institution size.

Communications Surveillance and Monitoring

Smarsh provides communications archiving and surveillance for regulatory compliance. The platform captures emails, instant messages, social media, and collaboration tools. AI analyzes communications for policy violations, insider trading, and market manipulation. Financial services firms use Smarsh meeting regulatory recordkeeping requirements. Pricing follows per-user models with typical costs ranging $30-60 per monitored user annually. Organizations monitoring 500 employees pay approximately $15,000-30,000 annually.

Global Relay delivers similar communications compliance capabilities with strong regulatory acceptance. The platform maintains certifications meeting SEC, FINRA, and international regulatory requirements. Annual costs typically range $40,000-150,000 depending on user count and data retention requirements.

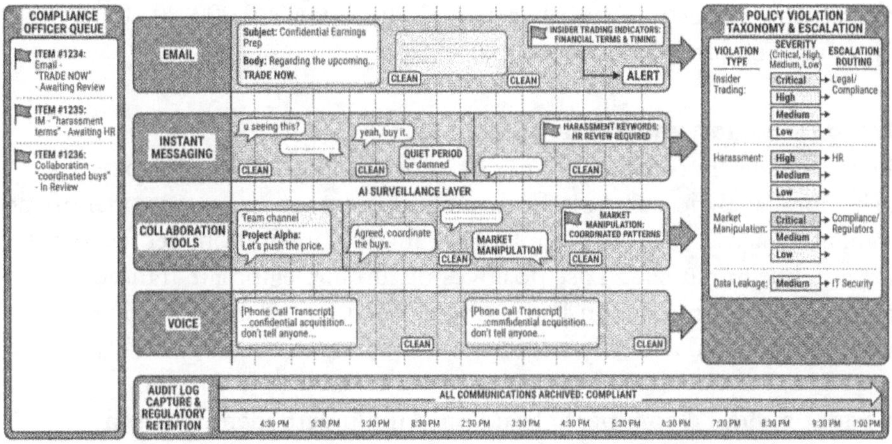

Figure 14.13: Multi-Channel Communications Compliance Surveillance. *Regulated organizations monitor employee communications across email, messaging, collaboration tools, and voice channels. AI analysis identifies policy violations and regulatory concerns while filtering routine business communications, reducing compliance review burden while maintaining comprehensive oversight.*

Governance, Risk, and Compliance Platforms

Quantivate provides GRC platform supporting policy management, risk assessment, and compliance tracking. The platform maps regulations to controls and tracks compliance obligations across frameworks. AI features suggest control improvements and identify regulatory change impacts. Pricing starts around $15,000-30,000 annually for small organizations. Enterprise deployments reach $100,000-300,000 annually.

LogicGate offers workflow automation and risk quantification for compliance programs. The platform supports custom compliance workflows without requiring coding expertise. Teams build tailored solutions for specific regulatory requirements. Annual costs typically range $25,000-100,000 depending on user count and complexity.

Regulatory Monitoring and Change Management

Comply.ai automates regulatory change monitoring across global jurisdictions. The platform tracks regulatory developments and alerts compliance teams to relevant changes. Impact analysis capabilities assess how regulatory updates affect existing programs. Pricing requires custom quotes typically ranging $30,000-100,000 annually.

RegTech Associates provides regulatory intelligence and horizon scanning services. The platform combines AI analysis with human expert validation. Compliance teams receive curated regulatory updates reducing monitoring workload. Annual costs typically range $20,000-80,000 depending on jurisdictions covered.

Fraud Detection and Prevention

Feedzai delivers real-time fraud prevention using machine learning and behavioral analytics. The platform analyzes transaction patterns identifying fraudulent activity before completion. Financial institutions reduce fraud losses while minimizing false positives disrupting legitimate transactions. Pricing follows transaction-based models with typical costs ranging $100,000-500,000 annually.

Forter provides e-commerce fraud prevention with identity-based decisioning. The platform analyzes customer behavior across merchant networks detecting fraud patterns. Retailers reduce chargebacks while improving customer experience. Pricing follows revenue-based models typically costing 0.5-1.5% of protected transaction volume.

Tool Configuration by Organization Size

Small Compliance Team (1-3 people, financial services, $3,000-8,000 monthly)

- Claude Enterprise: Estimated $50/user × 3 = $150
- Smarsh Essential: Estimated $60/user/year × 50 employees = $3,000 annually ($250 monthly)
- Quantivate Basic: Estimated $20,000 annually ($1,667 monthly)
- Verafin (community bank): Estimated $40,000 annually ($3,333 monthly)
- Total: Approximately $5,400 monthly

Small compliance teams prioritize essential surveillance and AML capabilities. Regional banks and credit unions benefit from cloud-based platforms.

Mid-Size Compliance (3-10 people, $15,000-40,000 monthly)

- Claude Enterprise: Estimated $50/user × 8 = $400
- ComplyAdvantage: Estimated $100,000 annually ($8,333 monthly)

- Smarsh: Estimated $40/user/year × 500 employees = $20,000 annually ($1,667 monthly)
- Quantivate: Estimated $75,000 annually ($6,250 monthly)
- Feedzai: Estimated $200,000 annually ($16,667 monthly)
- Comply.ai: Estimated $50,000 annually ($4,167 monthly)
- Total: Approximately $37,484 monthly

Mid-size teams deploy detailed compliance platforms. Transaction monitoring and communications surveillance become essential.

Enterprise Compliance (10+ people, multi-jurisdictional, $75,000+ monthly)

- Claude Enterprise: Estimated $50/user × 20 = $1,000
- NICE Actimize: Estimated $1,000,000 annually ($83,333 monthly)
- Smarsh Enterprise: Estimated $50/user/year × 5,000 employees = $250,000 annually ($20,833 monthly)
- Quantivate Enterprise: Estimated $200,000 annually ($16,667 monthly)
- Multiple fraud detection platforms across business lines
- Custom regulatory monitoring AI
- Total: Approximately $150,000-250,000+ monthly

Enterprise compliance organizations deploy surveillance and monitoring platforms. Multi-jurisdiction operations require global regulatory monitoring capabilities.

Multi-Discipline Coordination

Compliance teams oversee controls across all organizational disciplines. When reviewing marketing materials for regulatory advertising compliance, compliance workflows integrate with marketing approval processes. When monitoring employee communications, compliance coordinates with HR on policy enforcement. When assessing legal contract risks, compliance uses shared governance frameworks. See Platform Configuration sections for compliance integration patterns across all disciplines.

PLATFORM CONFIGURATION PATTERNS

Organizations adopting AI Scrum across multiple disciplines require deliberate platform configuration supporting both single-discipline work and cross-functional collaboration. The platform choice between Jira/Confluence and Azure DevOps matters less than ensuring consistent configuration patterns across disciplines. This section provides detailed configuration examples and adaptation guidance.

Jira/Confluence Configuration for AI Scrum

Jira and Confluence dominate Agile project management across marketing, software development, and data science teams. Organizations standardized on Atlassian tools gain AI capabilities through Atlassian Intelligence without platform migration disruption. The following configurations adapt Jira workflows for AI Scrum requirements.

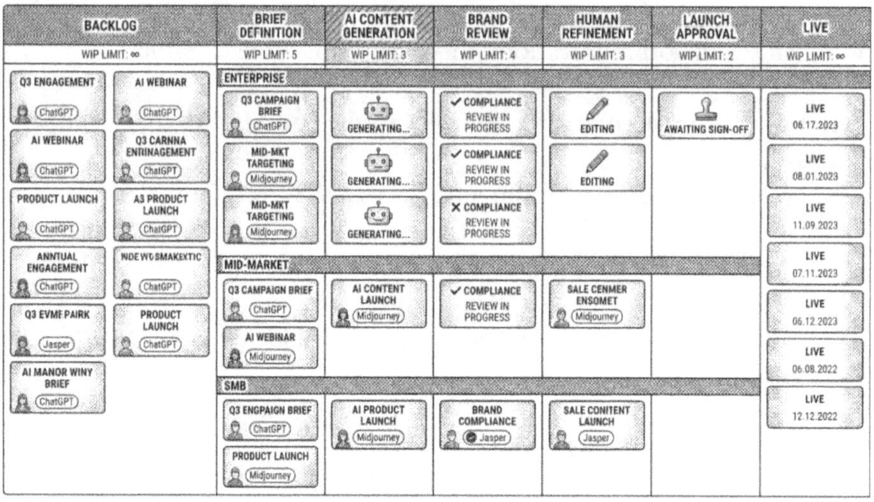

Figure 14.14: Marketing Kanban Board with AI Workflow Integration. *Jira boards configured for AI Scrum incorporate validation checkpoints as workflow columns with WIP limits preventing teams from generating content faster than validation capacity. Custom fields track AI tool usage and compliance status, ensuring governance without disrupting campaign velocity.*

Marketing Team Configuration (Detailed Example)

Marketing teams typically use Kanban workflows reflecting continuous campaign work rather than time-boxed sprints. AI Scrum principles still apply through validation-first thinking, outcome focus, and governance integration.

Custom Fields for Marketing Workflows

Marketing-specific custom fields track AI usage and brand compliance throughout campaign development. Each field serves specific tracking and governance purposes.

The AI Content Generator Used field uses multi-select dropdown format. Values include Claude, ChatGPT, Jasper, Copy.ai, Midjourney, Canva AI, None, and Multiple. This field tracks which AI tools contributed to deliverables for quality pattern analysis. The field requires completion for all work items in Content Creation category. Automation attempts auto-population from tool integration logs where possible.

The Brand Compliance Status field uses dropdown format with workflow triggers. Values include Not Reviewed, Under Review, Approved, Needs Revision, and Escalated. This mandatory checkpoint ensures brand team review of AI-generated content. Workflow rules prevent transitions to Ready for Launch without Approved status. Automation notifies brand teams when status changes to Under Review.

The Campaign Objective field captures business outcomes targeted by campaigns using text format. Validation rules require completion before moving to In Progress. Integration links objectives to outcome metrics dashboards for tracking.

The Target Audience Segment field uses multi-select dropdown with values including Enterprise, Mid-Market, SMB, Consumer, and Partner. This ensures content addresses correct audiences. Validation requires completion for all campaigns.

The Legal Review Required boolean field flags campaigns requiring legal approval before launch. When set to True, workflow rules add Legal Checkpoint states automatically. Automation creates linked legal review tasks without manual intervention.

The Performance Prediction number field captures AI-predicted campaign performance metrics. The field remains optional but proves recommended for A/B test planning. Integration compares predictions to actual results during retrospectives.

The Content Format field uses multi-select dropdown tracking content types. Values include Blog Post, Social Media, Email, Video, Infographic, White Paper, and Case Study. Reporting dashboards show content format distribution and performance patterns.

The Launch Date field requires date format specifying target publication dates. Workflow rules require completion before moving to Ready for Launch. Automation sends reminder notifications one week before scheduled launches.

Workflow States for AI-Augmented Marketing

Marketing workflow incorporates AI generation and validation checkpoints while maintaining campaign flow.

Workflow: Marketing Campaign with AI

- Backlog: Ideas and requests awaiting prioritization
- Brief Definition: Campaign objectives, audience, success metrics defined (required fields enforced)
- AI Content Generation: Using AI tools to create first draft content (AI tools field required)
- Brand Compliance Review: Brand team validates AI output (compliance status field required)
- Human Refinement: Creative team enhances AI output and adds human elements
- Legal Review (conditional): Appears only if Legal Review Required field is True
- Performance Prediction: AI analytics predict campaign results (optional prediction field)
- Launch Approval: Final stakeholder approval before publication
- Live/In-Market: Campaign actively running
- Performance Analysis: Post-campaign results compared to predictions

Board Configuration

Marketing boards use Kanban with WIP limits preventing teams from starting too many campaigns before completing validation.

Columns match workflow states with WIP limits enforced:

- Backlog (no limit)
- Brief Definition (limit 5)
- AI Content Generation (limit 3)
- Brand Review (limit 4)
- Human Refinement (limit 3)
- Launch Approval (limit 2)
- Live (no limit)

Swim lanes organize campaigns by audience segment or content format. Quick filters enable viewing by AI tool used, compliance status, or launch timeframe.

Confluence Space Structure

Marketing Confluence spaces maintain campaign documentation, brand guidelines, and AI prompt libraries.

Structure:

- /Marketing-Team
    - /Campaign-Briefs (AI-generated templates)
    - /Brand-Guidelines (for AI context loading)
    - /AI-Tool-Usage-Guides (team documentation)
    - /Content-Review-Archive (approved examples)
    - /Prompts
        - /Content-Generation
        - /Image-Creation
        - /Performance-Analysis

Each campaign links to Confluence brief capturing objectives, target audience, success metrics, and approval history.

Software Development Team Configuration (Brief Example)

Software development teams use Scrum boards with two-week sprints. Key differences from marketing configuration include technical validation fields and code quality tracking.

Key Custom Fields:

Software development teams require fields tracking code quality and technical validation. The AI Code Assistant Used field tracks tools including GitHub Copilot, Cursor, Tabnine, or None. Test Coverage Percentage uses number format measuring automated test coverage. Code Review Status tracks progress through Not Started, In Review, Changes Requested, or Approved states. Architecture Review Required boolean flags significant design changes needing architectural oversight. Performance Benchmark number field compares AI-generated versus manually written code performance.

Workflow follows AI Scrum pattern through Definition, Prototyping, Validation, Governance, Integration, and Done states.

Development boards use Scrum with sprint planning, daily standups, and retrospectives. Confluence maintains architecture decision records, coding standards, and validated prompt libraries.

Adaptation Pattern for Other Disciplines

Data science, HR, legal, and compliance teams adapt the marketing or software development patterns based on workflow similarities.

Data science resembles software development with experiment tracking fields replacing code review fields. HR teams mirror marketing workflows for recruitment campaigns with candidate pipeline tracking. Legal and compliance teams require additional audit trail fields and approval routing configurations.

Core adaptation principles:

- Identify discipline-specific validation checkpoints
- Add custom fields tracking AI tool usage and output quality
- Configure workflow states matching discipline process steps
- Establish Confluence structure supporting knowledge management
- Integrate with discipline-specific tools through APIs where possible

Azure DevOps Configuration for AI Scrum

Azure DevOps serves organizations requiring strong governance, audit trails, and Microsoft ecosystem integration. Legal and compliance teams favor Azure DevOps for regulatory documentation and matter management capabilities.

Legal Team Configuration (Detailed Example)

Legal teams manage matters rather than projects, requiring customized work item types and workflows.

Work Item Type: Legal Matter

Custom work item type Legal Matter tracks cases, contracts, and regulatory matters through legal workflows.

Custom Fields for Legal Workflows:

Legal matter tracking requires specialized fields supporting confidentiality and compliance requirements.

The Matter Type dropdown categorizes work with values including Contract Review, Litigation, Regulatory Compliance, Corporate Governance, Employment, and IP. This required field drives resource allocation decisions across the legal department.

The Client/Business Unit text field identifies internal clients or business units served. Integration links matters to organizational hierarchy. Completion remains required for all matters.

The AI Legal Tool Used multi-select field tracks assistance from Harvey AI, Casetext CoCounsel, LawGeex, Kira Systems, Claude Enterprise, or None. Required completion supports quality analysis and privilege considerations.

The Privilege Status dropdown manages confidentiality with values including Attorney-Client Privileged, Work Product, Not Privileged, or Mixed. Privileged matters trigger additional access controls automatically. Required completion protects discovery obligations.

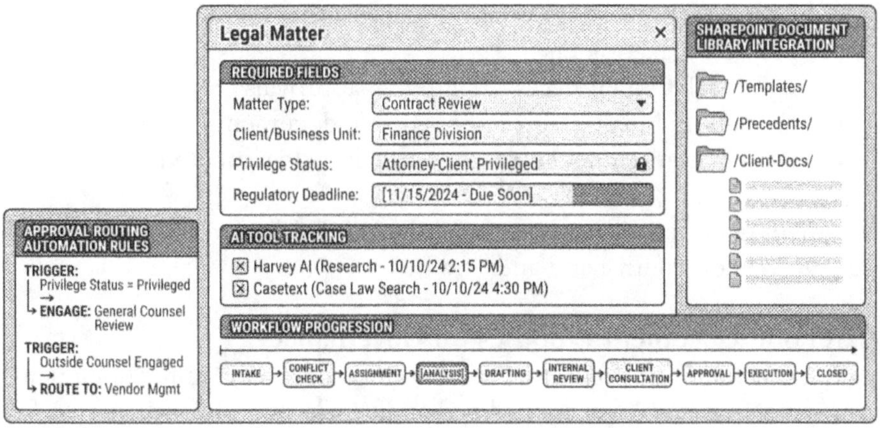

Figure 14.15: Azure DevOps Legal Matter Tracking Configuration. *Legal teams customize Azure DevOps work items to support matter lifecycle management with privilege protection, deadline tracking, and document control. Specialized fields capture regulatory requirements while workflow automation ensures proper authorization and audit trail compliance.*

The Outside Counsel Involved Boolean tracks matters requiring external law firm engagement. Selection triggers budget and invoice tracking field activation. Required completion supports cost management.

The Regulatory Deadline date field captures compliance or filing deadlines. Automated alerts escalate as deadlines approach. Validation prevents backdating when creating matters.

The Approval Authority user picker field designates attorneys authorized to approve matter resolutions. Workflow requires designation before closing significant matters. Required completion ensures proper authorization.

Workflow States for Legal Matters:
Legal matters progress through specialized workflow states reflecting legal processes rather than software development cycles.

Legal Matter Workflow includes these states:

- Intake: New matter request received
- Conflict Check: Verify no conflicts of interest exist

- Assignment: Assign to attorney and paralegal resources
- Analysis: Legal research and document review (AI tools field required)
- Drafting: Create legal documents or advice (track AI usage)
- Internal Review: Peer attorney review for quality
- Client Consultation: Discuss findings and recommendations with business stakeholders
- Approval: Obtain necessary internal approvals
- Execution: Finalize agreements or file documents
- Closed: Matter concluded with outcome documented

Document management integrates tightly with legal matter workflows to maintain version control and audit trails.

SharePoint Document Libraries for Governance
Legal teams use SharePoint integrated with Azure DevOps for document management with version control and audit trails.

Libraries:
- /Matters/{Matter-ID}/ (matter-specific documents)
- /Templates/ (contract templates and legal forms)
- /Precedents/ (approved legal positions and reasoning)
- /Policies/ (legal department policies and procedures)
- /Training/ (legal training materials)

Each library enforces metadata requirements ensuring proper classification and retention policies.

Compliance Team Configuration (Brief Example)
Compliance teams adapt legal configuration with additional regulatory tracking and control testing workflows.

Key Differences:
- Work Item Type: Compliance Initiative (instead of Legal Matter)
- Additional Fields: Regulatory Framework (GDPR, SOX, HIPAA, etc.), Control ID, Test Frequency, Finding Severity
- Enhanced Audit Trail: All field changes logged with timestamp and user
- Regulatory Documentation: SharePoint libraries organized by regulation with automatic retention policies

Compliance workflows emphasize control testing cycles, finding remediation tracking, and regulatory examination preparedness.

Cross-Discipline Integration Patterns

Multi-discipline scenarios require coordination across platform configurations without duplicating work or fragmenting governance.

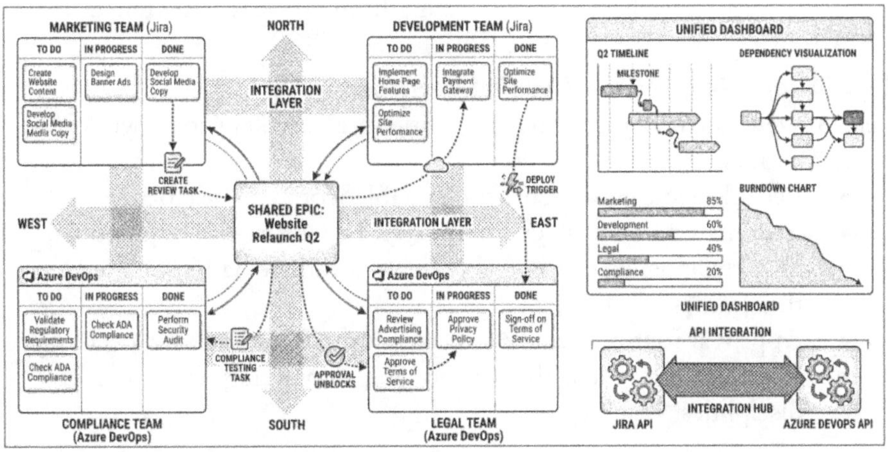

Figure 14.16: Multi-Platform Cross-Discipline Integration Architecture. *Organizations supporting both Jira and Azure DevOps build integration bridges enabling seamless collaboration across platform boundaries. Shared epics, automated task creation, and unified dashboards maintain coordination without forcing inappropriate platform standardization.*

Marketing Tracking Development Projects

Marketing website refresh initiatives require coordination between marketing content creators and software development teams.

Configuration:

- Shared Epic in Jira spanning both marketing and development boards
- Marketing stories track content creation, brand approval, and asset delivery
- Development stories track technical implementation and deployment
- Dependencies link marketing content completion to development integration tasks
- Shared sprint planning ensures alignment on delivery dates
- Unified retrospectives capture cross-functional lessons learned

HR Using Project Management Tools

HR departments lacking robust project management in HRIS platforms use Jira or Azure DevOps for recruitment campaigns and learning development initiatives.

Configuration:

- Custom HR Project board with recruitment-specific workflows
- Integration with applicant tracking systems through APIs
- Candidate pipeline tracking using Kanban workflow states
- Interview scheduling automation through calendar integrations
- Compliance checkpoints ensuring hiring process adherence

Legal Supporting Multi-Discipline Work

Legal reviews span all disciplines requiring integrated approval workflows. Configuration:

- Legal Review work item type linkable from any discipline
- Automated legal task creation when Legal Review Required flag set
- SLA tracking ensuring legal reviews complete within defined timeframes
- Approval routing based on matter complexity and risk level
- Centralized legal dashboard showing all pending reviews across disciplines

Compliance Integration Across Disciplines

Compliance oversight applies to all organizational work requiring pervasive integration.

Configuration:

- Compliance checkpoint custom field available in all work item types
- Automated compliance notifications for high-risk work categories
- Centralized compliance board with links to discipline-specific items
- Risk scoring automation triggering appropriate review levels
- Unified audit trail capturing compliance decisions across all disciplines

Automated compliance workflows enforce governance requirements without manual intervention. Organizations configure automation rules matching their risk tolerance and regulatory requirements. These rules operate consistently without requiring compliance team involvement in every transaction.

Integration Architecture and Automation

Platform integrations connect discipline-specific tools into coherent ecosystems. APIs enable data exchange between Jira and specialized tools. Automation workflows trigger cross-platform actions based on events. Shared authentication through SSO provides consistent access control. Audit logging aggregates activities across platforms for unified governance reporting.

SECURITY AND GOVERNANCE CONSIDERATIONS

AI tool adoption introduces security and governance challenges beyond traditional software procurement. Data classification requirements, vendor security assessment protocols, and audit trail obligations demand systematic evaluation before deployment.

Data Classification and Handling

Organizations must classify data before processing through AI tools. Public information requires minimal controls while confidential and regulated data demands strict protections. Marketing teams can process public campaign content through consumer AI tools. Legal teams must never input client privileged information into consumer services.

Data classification drives tool tier selection across disciplines. Consumer AI tiers train models on user inputs making them unsuitable for confidential information. Enterprise tiers provide contractual assurances against training on customer data. Teams handling regulated information require enterprise agreements regardless of cost.

Financial services organizations face particularly strict requirements. Banking regulations prohibit sharing customer financial data with third parties without consent. Healthcare providers must comply with HIPAA restrictions on protected health information. Legal teams must preserve attorney-client privilege and work product protections.

Establish data classification policies before AI tool deployment. Document which information types can use consumer versus enterprise AI tiers. Train teams on classification requirements and consequences of violations. Monitor tool usage ensuring compliance with policies. Violations of data handling policies create regulatory exposure far exceeding any cost savings.

Audit Trail Requirements

Regulated industries require audit trails documenting AI tool usage and decisions. Financial institutions must demonstrate compliance with fair lending and anti-discrimination requirements. Healthcare organizations must prove HIPAA compliance in patient care decisions. Legal departments must maintain privilege logs and work product documentation.

Enterprise AI tools provide audit logging capabilities tracking user interactions, prompts submitted, and outputs generated. Organizations configure retention periods matching regulatory requirements. Financial services firms typically retain AI interaction logs for seven years. Healthcare providers align retention with medical record requirements.

Audit trail architecture requires centralized logging aggregating activities across multiple AI platforms. Security information and event management systems collect logs from various tools. Correlation engines identify unusual patterns indicating policy violations or security incidents. Compliance teams review logs during internal audits and regulatory examinations.

Organizations should implement AI usage logging before widespread adoption. Establish baseline monitoring detecting unauthorized tool usage. Configure alerts for high-risk activities like processing regulated data. Review logs quarterly identifying training opportunities and policy gaps.

Vendor Security Assessment

AI tool vendors require thorough security assessment before organizational deployment. Traditional software vendor assessments focus on infrastructure security and data protection. AI vendor assessments must additionally evaluate model training practices, data handling policies, and intellectual property protections.

Security assessment questionnaires should cover specific AI concerns. Does the vendor train models on customer data? What data residency options exist for regulated industries? How does the vendor handle intellectual property in AI outputs? What incident response procedures exist for data breaches?

Organizations should require SOC 2 Type II attestations from AI vendors demonstrating security control effectiveness. Financial services firms may require additional certifications meeting banking regulatory standards. Healthcare organizations should verify HIPAA Business Associate Agreement compliance.

Vendor assessment extends beyond initial procurement to ongoing monitoring. Quarterly reviews track vendor security posture changes and incident disclosures. Annual reassessments verify continued compliance with organizational requirements. Vendor security incidents trigger immediate reviews determining continued usage appropriateness.

ADOPTION SEQUENCING STRATEGY

AI tool adoption requires phased implementation building organizational capability without overwhelming teams. Sequential rollout enables learning from early adopters before organization-wide deployment. Three-phase sequencing balances speed and risk management.

Phase One: Foundation Building (Weeks 1-8)

Phase one establishes AI tool infrastructure and initial capability development. Organizations select and procure foundation LLM subscriptions for pilot teams. IT teams configure enterprise authentication and audit logging. Governance teams draft initial usage policies and training materials.

Pilot team selection proves critical for successful foundation building. Choose teams with high AI enthusiasm and low regulatory constraints. Marketing teams often serve as excellent pilots given creative work tolerance for experimentation. Software development teams provide strong technical feedback on tool capabilities.

Foundation training introduces teams to AI tool capabilities and organizational policies. Four-hour workshops cover prompt engineering basics, data classification requirements, and tool-specific features. Hands-on exercises build confidence through supervised practice. Follow-up office hours provide ongoing support during initial usage.

Foundation phase success metrics include pilot team adoption rates, usage policy compliance, and initial productivity improvements. Target 80% pilot team weekly active usage by week eight. Zero data classification violations during foundation period. Measurable productivity gains in pilot use cases.

Phase Two: Role Evolution (Weeks 9-20)

Phase two expands AI tool access to additional disciplines and use cases. Organizations add specialized tools addressing discipline-specific requirements. Teams develop role-specific prompt libraries and best practices. Early adopters begin coaching broader organizational populations.

Discipline rollout follows risk-based sequencing. Low-risk disciplines like marketing and internal communications adopt before high-risk functions like legal and compliance. Each discipline receives customized training addressing specific use cases and regulatory constraints.

Role evolution requires updating job descriptions and performance expectations. Organizations document how AI augmentation changes role responsibilities. Individual contributors spend less time on execution and more on validation and strategy. Managers focus on outcome definition rather than task assignment.

Phase two introduces governance checkpoints ensuring quality before expanding adoption. Weekly reviews assess tool usage patterns and policy compliance. Monthly stakeholder updates communicate progress and address concerns. Quarterly business reviews evaluate ROI and adjust adoption plans.

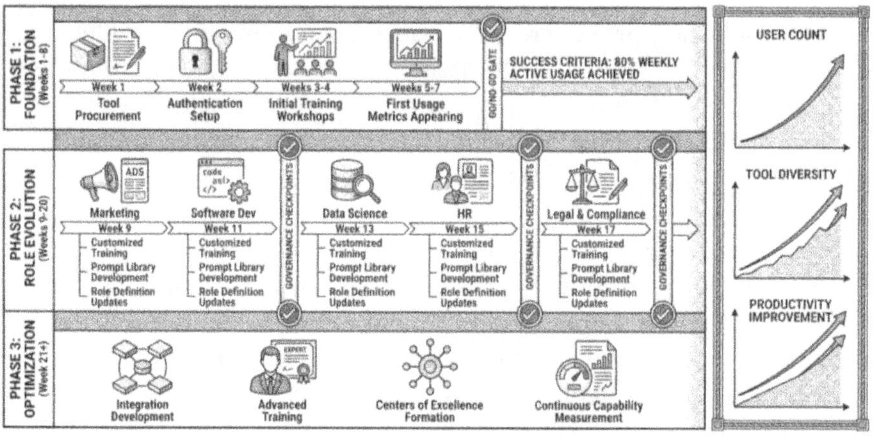

Figure 14.17: Phased AI Tool Adoption Sequencing Strategy. *Successful enterprise AI adoption follows deliberate three-phase sequencing: foundation building with pilot teams, systematic role evolution across disciplines, and ongoing optimization. Sequential rollout enables learning from early adopters before organization-wide deployment, balancing speed with risk management.*

Phase Three: Optimization (Ongoing)

Phase three focuses on continuous improvement and advanced capability development. Organizations implement sophisticated workflows integrating AI tools across disciplines. Teams develop custom solutions addressing unique organizational requirements. Centers of excellence emerge providing guidance and accelerating adoption.

Optimization phase activities include prompt library curation, integration development, and capability measurement. Teams contribute successful prompts to centralized repositories. IT teams build integrations connecting AI tools to existing workflows. Analytics teams measure productivity improvements and ROI realization.

Advanced training introduces complex use cases and multi-tool workflows. Teams learn to combine foundation LLMs with specialized tools. Cross-functional training builds collaboration capabilities for multi-discipline initiatives. Executive briefings maintain leadership engagement and secure continued investment.

COST PLANNING AND ROI

AI tool costs vary dramatically based on organizational size, discipline mix, and tool selection. Accurate cost planning requires understanding per-seat licensing, platform fees, and usage-based pricing models. ROI calculations must account for productivity improvements, quality enhancements, and strategic capabilities.

Per-Seat Cost Breakdown by Discipline

Marketing teams average $300-600 monthly per marketer for AI tooling. Foundation LLM subscriptions cost $20-40 monthly. Specialized content tools like Jasper or Copy.ai add $50-150 monthly. Visual creation tools contribute $100-200 monthly. Small marketing teams spend approximately $3,000-6,000 monthly total.

Software development teams average $150-400 monthly per developer depending on tooling depth. Foundation LLM subscriptions and code assistants cost $30-80 monthly combined. Testing automation platforms add $50-150 monthly per developer. Enterprise security scanning contributes $30-100 monthly. Development teams of twenty spend $3,000-8,000 monthly.

Data science teams require highest per-user costs given expensive AutoML platforms. Foundation tools cost $40-60 monthly per data scientist. AutoML platform costs range $200-1,000 monthly per user depending on usage intensity. Small data teams spend $5,000-15,000 monthly for five to ten people.

HR teams average $500-1,500 monthly per HR professional given specialized talent intelligence platforms. Foundation LLM costs remain modest at $20-40 monthly. Talent intelligence and assessment platforms drive costs with $200-800 monthly per user. HR teams of five spend $2,500-7,500 monthly.

Legal teams face highest per-attorney costs given specialized legal AI pricing. Foundation tools with enterprise controls cost $40-60 monthly. Legal-specific AI like Harvey AI runs $2,000-3,000 monthly per attorney. Legal teams of five spend $10,000-15,000 monthly minimum.

Compliance teams mirror legal costs given regulatory requirements and specialized tools. Foundation enterprise tools cost $40-60 monthly. AML and surveillance platforms add $500-2,000 monthly per compliance professional. Compliance teams of five spend $8,000-20,000 monthly.

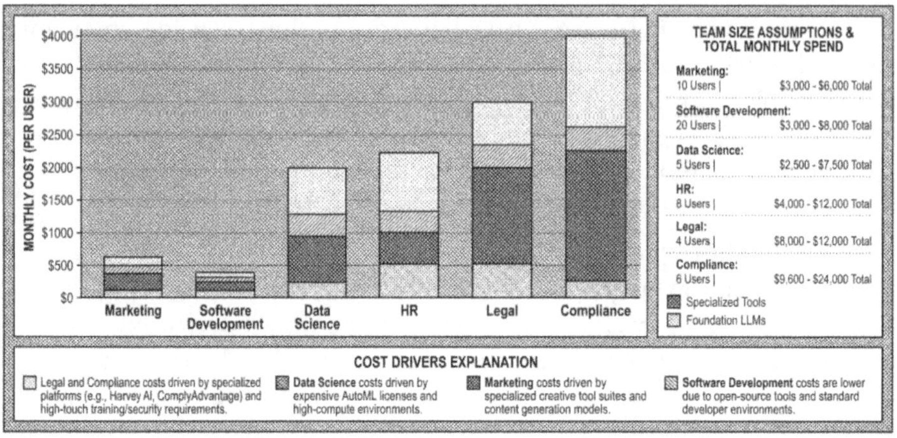

Figure 14.18: Discipline-Specific AI Tool Cost Structures. *Per-user AI tool costs vary dramatically across disciplines based on specialized platform requirements. Legal and compliance teams average $2,000-4,000 monthly per professional due to enterprise legal AI and regulatory monitoring platforms, while marketing and development teams average $150-600 monthly with lighter tooling needs.*

Platform Costs

Jira and Confluence subscriptions add platform costs beyond AI tools. Standard tier costs $7.75 monthly per user. Premium tier runs $15.25 monthly per user. Enterprise tier requires custom quotes typically $20-30 monthly per user. Organizations with one hundred users pay approximately $1,500-3,000 monthly for platform access.

Azure DevOps pricing follows similar per-user models. Basic plans provide free access for up to five users. Basic plus Test Plans costs $52 monthly per user. Enterprise agreements provide volume discounts for large deployments. Organizations should budget $10-30 monthly per user for Azure DevOps access.

Atlassian Intelligence and Microsoft Copilot add incremental costs to platform subscriptions. Atlassian Intelligence pricing bundles with premium and enterprise Jira tiers. Microsoft Copilot costs $30 monthly per user as add-on to Microsoft 365 subscriptions. Platform AI features reduce specialized tool requirements for some use cases.

Budget Planning Templates

Organizations should develop discipline-specific budget templates accounting for team size and tool selection. Templates include foundation LLM costs, specialized tools, platform fees, training expenses, and implementation support.

Small organizations with one hundred total employees and fifty AI tool users require approximately $14,500 monthly.

- Foundation LLMs: $20/user × 50 = $1,000 monthly
- Discipline tools (blended average): $200/user × 50 = $10,000 monthly
- Platform costs: $15/user × 100 = $1,500 monthly
- Training and support: $2,000 monthly
- Total: Approximately $14,500 monthly or $174,000 annually

Mid-size organizations with five hundred employees and two hundred AI users require approximately $83,000 monthly.

- Foundation LLMs: $25/user × 200 = $5,000 monthly
- Discipline tools (blended average): $300/user × 200 = $60,000 monthly
- Platform costs: $20/user × 500 = $10,000 monthly
- Training and support: $8,000 monthly
- Total: Approximately $83,000 monthly or $996,000 annually

Enterprise organizations with five thousand employees and fifteen hundred AI users require approximately $835,000 monthly.

- Foundation LLMs: $40/user × 1,500 = $60,000 monthly
- Discipline tools (blended): $400/user × 1,500 = $600,000 monthly
- Platform costs: $25/user × 5,000 = $125,000 monthly
- Training, support, integration: $50,000 monthly
- Total: Approximately $835,000 monthly or $10,020,000 annually

ROI calculation should account for productivity improvements, quality enhancements, and capability expansion. Marketing teams reducing content production time by forty percent realize significant cost savings. Development teams reducing defect rates improve customer satisfaction and reduce maintenance costs. Legal teams completing contract reviews faster enable faster deal closure.

Conservative ROI estimates assume twenty to thirty percent productivity improvement in AI-augmented tasks. Tasks consuming fifty percent of role time see ten to fifteen percent overall productivity gains. You'll want to measure baseline productivity before AI adoption and track improvements quarterly. Payback periods typically range six to eighteen months depending on implementation quality and organizational change management effectiveness.

FROM THE TRENCHES: THE PLATFORM MIGRATION THAT ALMOST WASN'T

A financial services firm with two thousand employees attempted multi-discipline AI Scrum adoption across six departments. Marketing, software development, data analytics, HR, legal, and compliance teams all required AI tooling. The organization made a critical early decision that nearly derailed the entire transformation.

IT leadership mandated Azure DevOps as the universal platform across all disciplines. The decision seemed logical given existing Microsoft enterprise agreements and security team familiarity with Azure infrastructure. Legal and compliance teams adapted successfully given their governance requirements and SharePoint integration needs. Software development teams migrated from aging on-premise tools without major disruption.

Marketing teams revolted within three weeks. Creative workflows requiring visual collaboration found Azure DevOps interfaces clunky compared to Jira's Kanban boards. Campaign planning requiring rapid iteration suffered from Azure DevOps page load times. Marketing leadership threatened abandoning AI Scrum entirely rather than accepting reduced productivity.

HR teams encountered similar friction. Recruitment campaign tracking in Azure DevOps felt forced compared to applicant tracking system workflows. Learning development initiatives lacked natural templates in Azure DevOps. HR directors began lobbying executives for Jira access or exemption from AI Scrum requirements.

The organization faced a choice. Force universal Azure DevOps adoption risking marketing and HR resistance. Allow dual platforms creating integration complexity and governance fragmentation. Abandon AI Scrum for resistant disciplines accepting competitive disadvantage.

The CTO made a pragmatic decision. Marketing and HR teams received Jira/Confluence licenses. Software development, legal, and compliance teams remained on Azure DevOps. Data analytics teams chose based on primary collaboration patterns. IT built integration bridges between platforms using APIs and automation workflows.

Cross-discipline initiatives required careful coordination. Marketing campaigns requiring legal review spanned both platforms through linked work items. HR policies requiring legal approval used Azure DevOps as governance platform with Jira tracking HR execution. Development projects requiring marketing content coordination used shared epics visible in both systems.

The solution cost more than universal platform adoption. Dual platform licensing added twenty percent to tool budgets. Integration development consumed three months of IT capacity. Training materials required platform-specific versions. Governance teams monitored both platforms for compliance.

The solution worked better than forced standardization would've. Marketing teams achieved thirty percent productivity improvements within six months. HR teams successfully launched AI-augmented recruitment campaigns. Legal and compliance teams maintained regulatory compliance without compromise. Software development teams delivered faster with fewer defects.

The lesson proved simple yet powerful. Platform choice matters less than platform fit for discipline workflows. Architectural coherence matters more than vendor standardization. Integration investment proves worthwhile when it enables rather than constrains adoption.

Organizations shouldn't pursue premature platform standardization. Evaluate platforms based on discipline needs not vendor relationships. Build integration capabilities supporting multi-platform environments. Accept modestly higher costs enabling significantly higher adoption.

KEY TAKEAWAYS

- **Multi-discipline AI adoption requires discipline-specific tool selection rather than one-size-fits-all approaches.** Marketing teams will need content generation tools with brand compliance features. Legal teams require contract analysis platforms with privilege protection. Compliance teams demand surveillance and monitoring tools with audit trails. Data science teams benefit from AutoML platforms handling full ML pipelines. HR teams use talent intelligence platforms for recruiting and assessment. Software development teams integrate code assistants with testing automation.

- **Platform configuration determines adoption success more than tool selection determines productivity gains.** Jira and Azure DevOps require extensive customization supporting AI Scrum workflows. Custom fields track AI tool usage, validation status, and governance approvals. Workflow states incorporate AI generation, validation, and governance checkpoints. Board configurations enforce WIP limits preventing teams from overwhelming validation capacity. Cross-discipline integration patterns enable coordination without duplication.
- **Phased adoption sequencing builds organizational capability without overwhelming teams or creating governance gaps.** Foundation phase establishes infrastructure and pilot team capabilities over eight weeks. Role evolution phase expands adoption across disciplines with customized training over twelve weeks. Optimization phase drives continuous improvement and advanced capabilities indefinitely. Each phase includes specific success metrics and intervention thresholds.
- **Security and governance requirements differ dramatically across disciplines and drive enterprise tier adoption.** Marketing teams can use consumer AI tiers for public content creation. Legal teams require enterprise agreements protecting attorney-client privilege. Compliance teams will need audit logging meeting regulatory examination requirements. Financial services organizations face banking regulations prohibiting customer data sharing. Healthcare providers must comply with HIPAA restrictions on protected health information.
- **Cost planning must account for per-discipline variation with marketing and development averaging lower costs than legal and compliance.** Small organizations budget approximately $15,000 monthly for fifty AI tool users across disciplines. Mid-size organizations budget approximately $80,000 monthly for two hundred users. Enterprise organizations budget approximately $800,000 monthly for fifteen hundred users. Discipline mix significantly impacts total costs given legal and compliance premium pricing.
- **ROI realization requires measuring productivity improvements and quality enhancements rather than focusing solely on cost reduction.** Marketing teams reducing content production time by forty percent enable more campaigns without proportional headcount increases. Development teams reducing defect rates improve customer satisfaction and reduce maintenance

overhead. Legal teams completing contract reviews faster accelerate deal closure and revenue recognition. Conservative estimates assume twenty to thirty percent productivity improvement in AI-augmented tasks with payback periods ranging six to eighteen months.

FINAL THOUGHTS

You began this book with a simple observation: velocity metrics were lying. Teams hitting their story point targets sprint after sprint while actual delivery to customers slowed to a crawl. The numbers looked right, but the reality on the ground told a different story.

That observation led you through fourteen chapters of systematic transformation. You discovered that the bottleneck in software delivery has permanently shifted. For twenty-five years, from the Agile Manifesto through the public release of ChatGPT, writing code was expensive and slow. Every ceremony, every metric, every organizational structure optimized around that single constraint: protecting scarce developer capacity.

AI demolished that assumption. Code generation collapsed from days to minutes. But the constraint didn't disappear—it moved. Validation, governance, integration, and outcome definition emerged as the new binding constraints. Traditional Scrum, optimized for a world where writing code was hard, became actively counterproductive in a world where validating AI-generated code became the critical path.

This realization forced a complete methodological inversion. You learned that planning must shift from estimating development effort to establishing validation criteria before a single line of code generates. You discovered that Daily Standups focused on task completion miss the systemic validation bottlenecks that actually control delivery timelines. You recognized that Sprint Reviews celebrating velocity increases create false confidence while governance debt accumulates silently in the background. The framework you now possess represents more than process changes. It represents a fundamental reconception of what software delivery means when creation becomes nearly free and validation becomes expensive.

You learned the six core principles that distinguish AI Scrum from traditional Agile: validation-first development that prevents quality debt rather than detecting it late; outcome-oriented metrics that measure business impact rather than development activity; probabilistic thinking that acknowledges AI's non-deterministic nature; integrated governance that becomes an enabler rather than a bottleneck; continuous learning systems that prevent cognitive atrophy; and human-AI partnership patterns that multiply capability rather than simply automating tasks.

You mapped the new organizational structure where the Validation Lead becomes the power center, the AI Alignment Engineer translates ambiguous business requirements into precise AI specifications, and the Product Owner shifts from feature management to outcome definition. These aren't rebranded versions of traditional roles. They're fundamentally different positions addressing fundamentally different constraints.

You reimagined every ceremony for the validation-constrained world. Sprint Inception expands from two hours to four hours, establishing validation criteria and governance requirements before implementation begins. Daily Synchronization focuses on validation progress and governance blockers rather than task completion. Sprint Review evaluates outcome achievement against predefined success criteria rather than celebrating velocity increases. Sprint Retrospective analyzes AI collaboration effectiveness and validation bottlenecks rather than development friction.

You established partnership multipliers that produce three-to-ten-times better results than treating AI as a simple tool or attempting full automation. You learned the Junior Developer Test for deciding between automation and augmentation. You built cognitive atrophy defenses that preserve deep thinking capabilities while leveraging AI's strengths.

You mastered outcome-based planning that specifies measurable results to achieve rather than features to build, opening solution space rather than constraining approaches. You adopted validation-first development sequences that plan detailed quality criteria before generating code. You implemented outcome metrics that track business impact, user satisfaction, and problem-solving effectiveness instead of misleading output proxies like velocity and story points.

You built AI governance frameworks that address non-deterministic behavior, emergent properties, data dependencies, model drift, and novel adversarial vulnerabilities through tiered review processes, sprint-synchronized decision-making, and diverse expertise across technical, business, legal, ethics, and security domains.

You navigated change management complexities unique to AI transformation, addressing identity threats where developers fear becoming "operators" and Product Owners fear losing control. You learned that resistance stems not from fear but from genuine belief that AI Scrum sacrifices software quality for speed—a belief that demands engagement rather than dismissal.

You assessed organizational maturity across eight capability dimensions, identifying catalyst actions required to progress from Initial chaos through Structured, Optimized, and Adaptive levels toward Transformative industry leadership. You learned that activity does not equal capability and that overall risk is determined by the lowest maturity dimension—high AI speed with low governance creates catastrophic exposure.

You selected implementation paths for four distinct organizational scenarios: single-team traditional Scrum migration, multi-team traditional Scrum transformation, single-team non-Agile adoption, and multi-team non-Agile implementation. Each path addresses different starting constraints, different resistance patterns, and different organizational dynamics.

You built the tooling foundation that enables AI Scrum at scale, understanding that multi-discipline adoption requires discipline-specific tool selection rather than one-size-fits-all approaches, that platform configuration determines adoption success more than tool selection determines productivity gains, and that phased adoption sequencing builds organizational capability without overwhelming teams or creating governance gaps.

The framework is complete. You possess the methodology, the organizational structure, the ceremony designs, the validation approaches, the governance frameworks, the change management strategies, the maturity assessment tools, the implementation playbooks, and the tooling foundation required to transform your organization from traditional Agile to AI-native delivery.

But possessing the framework and implementing it successfully are different challenges entirely. The gap between understanding and execution is where most transformations fail. Not because the methodology is flawed, but because organizations underestimate the human dynamics, the psychological resistance, the political complexity, and the sustained leadership commitment transformation demands.

The constraint shift is permanent. AI will only improve at creation, making validation even more critical. This framework is built for that structural reality, not the current state of AI technology. Organizations that continue optimizing for development velocity while ignoring validation capacity will watch their delivery timelines paradoxically lengthen as their code generation accelerates. Organizations that embrace the constraint inversion and restructure around validation, governance, and outcome definition will realize order-of-magnitude improvements in actual delivery to customers.

The choice is no longer whether to adopt AI. The choice is whether to adopt it systematically with appropriate methodology and governance, or haphazardly through Shadow AI and organizational drift. The former creates competitive advantage. The latter creates governance debt, quality risk, and eventual crisis.

You now understand The Agile Inversion. The question is whether your organization will act on that understanding before competitors force the decision through market pressure.

The appendices that follow provide practical reference materials—role comparison tables, workflow diagrams, migration templates, and implementation checklists. These supporting materials translate the methodology into actionable guidance for teams beginning their transformation journey.

The framework is yours. The transformation is yours to lead!

DECISION MATRIX – AI TOOL SELECTION BY DISCIPLINE

	Code Generation	Testing & Validation	Observability & Metrics	Governance & Compliance	Planning & Estimation	Collaboration & Docs	LLM / Prompt Management
Software Development	●	●	●	○	◐	◐	◐
Web Development	●	●	◐	○	◐	◐	○
Data Science	●	●	◐	○	◐	◐	●
Legal Projects	○	○	○	●	◐	◐	●
HR Projects	○	○	○	◐	●	●	●
General / Cross-Functional	◐	◐	◐	◐	●	●	●

Legend:
● = Primary/Essential
◐ = Situational/Recommended
○ = Rarely Applicable

APPENDICIES

The following appendices are provided as baseline information and tools for you to use in your practice. Just like Agile is a mindset and AI Scrum, Scrum, Kanban, and others are frameworks, these templates can also change and be tweaked to best assist you in your success.

Information such as costs and salaries are clearly a line in the sand and over time will undoubtedly change but you are provided some initial numbers to help your organization plan. Much effort has been expelled to achieve high degrees of accuracy throughout the book including these appendices. Depending upon where your organization is based, these numbers may appear to be too low or high. Plug in the most up to date information you gather for your discipline.

These templates are there to stoke your imagination, enhance your communication, and to help accelerate your path to sustained excellence. Use them all or just a few and add to them as you expand your knowledge over time. A little grease on the cogs of the machine helps make things run smoother. This is grease for your cogs.

APPENDIX A: ROLE COMPARISON TABLE

TRADITIONAL SCRUM VS. AI SCRUM

This table maps the evolution from traditional Scrum roles to AI-native AI Scrum roles, highlighting key responsibilities, skill requirements, and the fundamental shifts in focus driven by AI capabilities. The table demonstrates how roles have evolved from output-focused activities to outcome-focused, validation-centric responsibilities.

Role Dimension	Traditional Scrum	AI Scrum
Product Owner ➤ Product Strategist	**Primary Focus:** Feature backlog management, story prioritization, acceptance criteria definition **Key Activities:** Writing user stories, grooming backlog, attending sprint ceremonies, accepting/rejecting deliverables **Success Metrics:** Story points completed, velocity trends, backlog health	**Primary Focus:** Business outcome ownership, strategic alignment, stakeholder orchestration **Key Activities:** Defining outcome metrics, leveraging AI for market analysis, validating outcome achievement, predictive value modeling **Success Metrics:** Business value delivered, outcome achievement rate, stakeholder satisfaction, strategic alignment score
Business Analyst ➤ AI Alignment Engineer	**Primary Focus:** Requirements gathering, documentation, business-to-technical translation **Key Activities:** Stakeholder interviews, requirements documents, use case modeling, acceptance criteria **Success Metrics:** Requirements clarity, stakeholder approval, minimal rework	**Primary Focus:** Translating business intent to AI-compatible specifications, prompt engineering excellence **Key Activities:** Prompt library curation, AI collaboration optimization, maintaining AI Collaboration Log, training team on effective AI interaction **Success Metrics:** AI output acceptance rate, prompt iteration efficiency, team AI proficiency growth

QA Engineer ➤ Validation Lead	**Primary Focus:** Test case creation, manual testing, bug reporting **Key Activities:** Writing test plans, executing tests, logging defects, regression testing **Success Metrics:** Test coverage percentage, defects found, testing cycle time	**Primary Focus:** Full validation strategy, red-teaming, AI-specific testing, deployment readiness **Key Activities:** Defining Validation Canvas, adversarial testing, AI-assisted test generation, security validation, performance benchmarking **Success Metrics:** Validation completeness, escaped defect rate, deployment success rate, time-to-validation
Governance Coordinator *(NEW)*	***No Direct Equivalent*** Traditional Scrum had no dedicated governance role. Compliance and risk management were handled ad hoc or by separate teams outside sprint workflows.	**Primary Focus:** AI governance integration, regulatory compliance, risk management, audit readiness **Key Activities:** Completing Governance Canvas, coordinating with AI Governance Board, maintaining Governance Record, ensuring EU AI Act/NIST compliance **Success Metrics:** Governance cycle time, compliance adherence, audit findings, governance approval rate
System Architect ➤ Integration Architect	**Primary Focus:** High-level system design, technology selection, architectural patterns **Key Activities:** Creating architecture diagrams, defining standards, technology evaluation, design reviews **Success Metrics:** Architectural coherence, technology adoption, AI Debt levels	**Primary Focus:** AI system integration, dependency management, observability architecture, rapid prototyping architecture **Key Activities:** Maintaining Integration Map, AI-specific integration patterns, monitoring / telemetry design, managing AI-generated component integration **Success Metrics:** Integration success rate, dependency resolution time, system observability coverage, deployment reliability

Developer ➤ AI-Augmented Developer	**Primary Focus:** Writing code, fixing bugs, implementing features **Key Activities:** Coding, code reviews, unit testing, debugging, refactoring **Success Metrics:** Lines of code, features completed, code quality, bug fix rate	**Primary Focus:** AI-assisted implementation, validation of AI outputs, human-AI collaboration optimization **Key Activities:** Prompt engineering for code generation, validating AI-generated code, multi-alternative evaluation, refinement and integration **Success Metrics:** AI output acceptance rate, outcome delivery speed, code quality of final integrated work, prompt effectiveness
Scrum Master ➤ Delivery Facilitator	**Primary Focus:** Ceremony facilitation, process adherence, impediment removal **Key Activities:** Running ceremonies, coaching on Scrum, removing blockers, protecting team from distractions **Success Metrics:** Velocity consistency, ceremony attendance, impediment resolution time	**Primary Focus:** AI-native workflow optimization, validation/governance/alignment impediment removal, AI collaboration coaching **Key Activities:** Facilitating AI Scrum ceremonies, monitoring Sprint Telemetry Dashboard, coaching AI collaboration, removing validation and governance blockers **Success Metrics:** Outcome achievement rate, team health metrics, impediment resolution speed, AI collaboration effectiveness

KEY INSIGHTS FROM ROLE EVOLUTION

The Constraint Shift

Traditional Scrum roles optimized for development velocity as the primary constraint. AI Scrum roles address the new constraint hierarchy: validation, alignment, governance, integration, and safety. This fundamental shift explains why every role evolves significantly—they're optimizing for different bottlenecks.

Key Insights from Role Evolution

From Outputs to Outcomes

Traditional roles measured success through outputs (story points, code written, tests executed). AI Scrum roles measure success through outcomes (business value delivered, validation completeness, governance approval, integration reliability). This outcome orientation aligns team incentives with organizational objectives.

New Capabilities Required

Every AI Scrum role requires new capabilities:

- Prompt engineering and AI collaboration
- Validation of non-deterministic AI outputs
- Risk assessment and governance frameworks
- Telemetry and observability practices
- Outcome-centric thinking and measurement

The Emergence of New Roles

The Governance Coordinator represents entirely new territory for Scrum teams. Traditional Scrum assumed governance happened outside sprint workflows, creating downstream bottlenecks. AI Scrum integrates governance directly into sprints, requiring dedicated coordination. Similarly, the AI Alignment Engineer addresses the critical challenge of translating business intent into AI-compatible specifications—a skill that didn't exist in traditional development.

Career Path Implications

Every Organizations implementing AI Scrum must address career development for evolved roles:

- Training pathways for traditional practitioners transitioning to AI Scrum roles
- Compensation structures reflecting new skill requirements
- Career progression frameworks for specialized roles (AI Alignment Engineers, Governance Coordinators)
- Communities of practice enabling knowledge sharing across teams
- Certification and credentialing programs validating AI Scrum capabilities

APPENDIX B: TRADITIONAL SCRUM WORKFLOW

CURRENT STATE: DEVELOPMENT-CONSTRAINED WORKFLOW

This diagram illustrates the traditional Scrum workflow optimized for development velocity as the primary constraint. The 14-day sprint cadence assumes that development activities consume the majority of available time, with validation and integration occurring primarily at sprint boundaries.

PHASE 1: SPRINT PLANNING (DAY 1)

Sprint planning consumes approximately 4 hours at the beginning of the sprint. The Product Owner presents prioritized user stories from the backlog. The development team estimates complexity using story points (typically on a Fibonacci scale: 1, 2, 3, 5, 8, 13). The team commits to a sprint goal and a set of stories they believe achievable based on historical velocity. Stories are decomposed into technical tasks, and team members self-assign work. Generally, if something reaches an 8 or higher, the work should be evaluated and broken into smaller work items, if possible.

Critical Limitation: Planning focuses on development capacity allocation. Validation requirements, governance needs, and integration complexity receive minimal explicit attention. The assumption is that if developers have time to code, other activities will fit within remaining capacity.

PHASE 2: DEVELOPMENT PHASE (DAYS 2-7)

The core development phase occupies approximately 80% of sprint time. Developers write code, conduct peer code reviews, and implement unit tests. Daily standups provide coordination and surface blocking issues. A mid-sprint backlog refinement session prepares upcoming work.

Why This Made Sense: In traditional development, writing code was genuinely the slowest activity. Implementing a user story required days of careful coding, debugging, and refinement. The sprint timeline allocated time proportional to this dominant constraint.

Governance Gap: No systematic governance integration exists. Security reviews, architecture reviews, and compliance validations happen outside sprints, creating downstream bottlenecks. Teams often face 'done but not deployable' situations where completed work awaits governance approval.

PHASE 3: TESTING AND BUG FIXES (DAYS 8-9)

QA engineers receive completed features and execute test plans. Bugs are identified, triaged by severity, and critical issues are fixed before sprint end. Integration testing verifies component interactions. Regression testing ensures existing functionality remains intact.

Chronic Time Pressure: Testing phases are frequently compressed as development runs long. The phrase 'testing is always the bottleneck' became an agile cliché. However, the real issue wasn't testing duration—it was that development consumed so much time that validation received inadequate attention.

PHASE 4: REVIEW AND RETROSPECTIVE (DAY 10)

Sprint review demonstrates working software to stakeholders. The Product Owner accepts or rejects work based on acceptance criteria. The team calculates velocity for planning future sprints. Retrospective examines process effectiveness and identifies improvements. Release to production occurs and data is verified for accuracy.

Velocity-Centric Metrics: Success is measured by story points completed and velocity trends. These output metrics don't necessarily correlate with business value delivery. Teams can achieve high velocity while failing to solve actual customer problems.

Critical Characteristics of Traditional Workflow

Development-Constrained: The workflow optimizes for development velocity. The implicit assumption: if we maximize coding time, we maximize value delivery.

Sequential Phases: Development largely completes before testing begins. Integration occurs late. This sequential approach made sense when development dominated time but created knowledge silos.

Validation as Afterthought: Validation criteria are defined during planning but validation execution is deferred until late in the sprint. This delay means that validation failures require either accepting AI Debt or extending sprint scope.

Governance External: No governance activities occur within sprints. Architecture review boards, security teams, and compliance functions operate on separate schedules, creating unpredictable approval delays.

Output-Focused: Metrics track developer output (velocity, story points) rather than business outcomes. This creates misalignment between team incentives and organizational objectives.

Traditional Scrum Two-week Sprint
DAY 1: Sprint Planning (4 hours – 2 hours per sprint week)
• Review prioritized backlog • Select stories for sprint • Estimate story points • Commit to sprint goal • Decompose stories into tasks
DAYS 2-7: Development (6 ½ Days – 80% of the sprint)
Primary Activities: • Developers write code (Main Constraint) • Code reviews within team • Unit testing as development proceeds • Integration work ongoing Daily Standup (15 mins): • What did you do yesterday? • What will you complete today? • Any blockers? Backlog Refinement (Mid-sprint, 2 hours): • Groom upcoming stories • Add acceptance criteria Re-estimate if needed
DAYS 8-9: Testing & Bug Fixes (2 days)
• QA testing begins • Bug identification and triage • Critical bug fixes • Regression testing • Integration testing ⚠ *Often rushed due to time pressure*
DAY 10: Review, Release, & Retrospective (4 hours)
Sprint Review (2 hours): • Demonstrate completed features to stakeholders • Stakeholder feedback • Calculate velocity Retrospective (1 hour): • What went well? • Where do we need to improve? Release and Confirm: Release work to production and confirm accuracy

APPENDIX C: AI SCRUM WORKFLOW

FUTURE STATE: VALIDATION-CONSTRAINED WORKFLOW

This diagram illustrates the AI Scrum workflow optimized for validation, governance, and integration as primary constraints. The same 14-day sprint cadence is maintained for organizational consistency, but the distribution of activities reflects AI-accelerated development and the new constraint hierarchy.

Phase 1: Sprint Inception (Day 1)

Sprint Inception expands traditional sprint planning from 2-4 hours to 4 hours, reflecting the critical importance of upfront definition. Unlike traditional planning that focuses on capacity allocation, Sprint Inception establishes frameworks for validation, governance, and integration before any generation begins.

The Product Strategist leads objective alignment, ensuring the team understands not just what to build but why and what success looks like. The Validation Lead completes the Validation Canvas, defining all acceptance criteria, performance benchmarks, security requirements, and quality gates. The Governance Coordinator prepares the Governance Canvas, classifying risk levels and identifying compliance obligations. The Integration Architect reviews the Integration Map, planning how new capabilities will connect with existing systems.

Critical Difference: This upfront investment prevents downstream rework. Teams that shortchange Sprint Inception discover validation failures, governance rejections, or integration problems late in sprints when remediation becomes expensive and time-consuming.

Phase 2: Rapid Prototyping (Day 2)

With validation and governance frameworks established, the AI Alignment Engineer crafts prompts that translate business requirements into AI-compatible specifications. AI-Augmented Developers work with AI to generate 3-5 implementation alternatives, each exploring different approaches, architectural patterns, or technology choices.

This multi-alternative generation surfaces trade-offs invisible when pursuing single implementations. One alternative might optimize for performance, another for maintainability, a third for minimal dependencies. The team conducts preliminary technical evaluation, eliminating approaches with obvious flaws while preserving promising candidates for deeper analysis.

The AI Acceleration: What would have consumed 10 days of development in traditional Scrum now requires hours. This acceleration doesn't reduce work—it shifts where teams invest time. Instead of writing code, teams evaluate alternatives, validate outputs, and ensure alignment.

Phase 3: Validation and Evaluation (Days 3-5)

Validation represents the new critical path. AI Show & Tell #1 on Day 3 presents alternatives to stakeholders, gathering rapid feedback that informs validation priorities. The Validation Lead coordinates testing including automated test suite generation, red-team adversarial testing, security scanning, performance benchmarking, edge case identification, and AI-specific validation for bias and safety concerns.

Validation findings drive iterative refinement. Developers re-prompt AI to address identified issues. The AI Collaboration Log documents the evolution of prompts and the rationale for selection decisions. This validation-refinement cycle typically requires 2-4 iterations before implementations satisfy all validation criteria.

Why This Consumes Time: AI-generated code requires validation that human-written code doesn't. Non-deterministic behaviors demand probabilistic testing. Adversarial vulnerabilities need red-teaming. Bias and fairness require explicit evaluation. This validation ensures AI-generated implementations meet quality, security, and ethical standards.

Phase 4: Governance and Final Alignment (Day 6)

The Governance Checkpoint on Day 7 provides the AI Governance Board formal review opportunity. The Governance Coordinator presents the completed Governance Canvas, validation evidence, and risk assessment. The board evaluates model selection rationale, data governance, security controls, fairness considerations, regulatory compliance, and monitoring plans.

The board issues one of three decisions: full approval authorizing deployment, conditional approval specifying remediation requirements, or rejection with concerns documented. This mid-sprint governance review enables course correction while teams still have time to address concerns.

AI Show & Tell #2 follows the Governance Checkpoint, presenting refined implementations with governance approval to stakeholders. This ceremony confirms final alignment before integration work begins.

Phase 5: Integration and Deployment Prep (Days 7-8)

With validated and governance-approved implementations, the Integration Architect coordinates system integration. This phase addresses integration testing, end-to-end validation, performance verification in staging environments, monitoring and observability setup, and deployment automation configuration.

Deployment preparation establishes operational readiness including rollback procedures, incident response protocols, monitoring dashboards, and documentation. Teams deploy to staging environments for final validation before production deployment authorization.

Phase 6: Review, Retrospective, and release (Days 9-10)

Sprint Review emphasizes validation evidence and governance approval, not merely feature demonstration. The Product Strategist presents outcome achievement evidence, validation metrics, and governance approval documentation. Stakeholders evaluate whether delivered capabilities achieve committed outcomes, making deployment authorization decisions based on evidence rather than feature completeness alone.

The Retrospective incorporates AI-generated insights from sprint telemetry. Machine learning analyzes outcome metrics, validation efficiency, governance cycle times, and AI collaboration patterns, identifying improvement opportunities humans might miss. The team commits to specific process improvements for subsequent sprints.

CRITICAL CHARACTERISTICS OF AI SCRUM WORKFLOW

Validation-Constrained: The workflow optimizes for validation quality, not development velocity. All-inclusive validation consumes the majority of sprint time, reflecting its criticality for AI-generated outputs.

Parallel Activities: Validation, refinement, and stakeholder alignment occur concurrently rather than sequentially. This parallelism accelerates learning cycles while maintaining quality standards.

Governance Integrated: Governance occurs within sprint workflows on Day 7, not as external approval process. This integration eliminates downstream bottlenecks while maintaining rigorous oversight.

Continuous Stakeholder Engagement: Two AI Show & Tell ceremonies (Days 3 and 7) provide frequent stakeholder touchpoints, preventing misalignment and enabling rapid course correction.

Outcome-Focused: Metrics track outcome achievement, validation completeness, governance approval, and stakeholder satisfaction—not story points or velocity. This outcome orientation aligns team incentives with business objectives.

Evidence-Based Decisions: All major decisions—alternative selection, governance approval, deployment authorization—rely on detailed evidence from validation results, performance data, and risk assessments rather than intuition alone.

AI Scrum Two-week Sprint
DAY 1: Sprint Planning (4 hours – 2 hours per sprint week)
• Product Strategist leads objective alignment • Validation Lead completes Validation Canvas • Define acceptance criteria & performance benchmarks • Governance Coordinator prepares Governance Canvas • Classify risk levels & identify compliance obligations • Integration Architect reviews Integration Map • Plan how new capabilities connect with existing systems
DAY 2: Prototyping (1 ½ Days)
• AI Alignment Engineer crafts prompts • Translate business requirements into AI specifications • Generate 3-5 implementation alternatives with AI • Explore different approaches & architectural patterns • Surface trade-offs across alternatives • Conduct preliminary technical evaluation • Eliminate approaches with obvious flaws • Preserve promising candidates for deeper analysis
DAYS 3-5: Validation & Evaluation (3 Days)
• AI Show & Tell #1 – Present alternatives to stakeholders • Automated test suite generation • Adversarial testing • Security scanning & performance benchmarking • Edge case identification • Ai-specific validation for bias & safety • Iterative refinement based on validation findings • Document prompt evolution in AI Collaboration Log • Typically 2-4 validation-refinement iterations
DAY 6: Governance & Final Alignment (1 Day)
Governance Checkpoint: • AI Governance Board formal review • Present Governance Canvas & validation evidence • Evaluate model selection rationale • Review data governance & security controls • Assess fairness considerations & regulatory compliance • Examine monitoring plans Board Decisions: • Full approval – authorize deployment • Conditional approval – specify remediation requirements • Rejection – document concerns AI Show & Tell #2: • Present refined implementations with governance approval • Confirm final alignment before integration begins
DAYS 7-8: Integration & Deployment Prep (2 Days)
System Integration:

- Integration Architect coordinates system integration
- Integration testing
- End-to-end validation
- Performance verification in staging environments
- Monitoring and observability setup
- Deployment automation configuration

Deployment Preparation:

- Establish rollback procedures
- Define incident response protocols
- Configure monitoring dashboards
- Complete documentation
- Deploy to staging for final validation
- Obtain production deployment authorization

DAYS 9-10: Review, Retrospective, and Release (2 Days)

Sprint Review:

- Product Strategist presents outcome achievement evidence
- Review validation metrics
- Present governance approval documentation
- Stakeholders evaluate outcome achievement
- Make deployment authorization decisions
- Base decisions on evidence, not just features

Retrospective:

- Incorporate AI-generated insights from sprint telemetry
- Machine learning analyzes outcome metrics
- Review validation efficiency & governance cycle times
- Analyze AI collaboration patterns
- Identify improvement opportunities
- Commit to specific process improvements

Release:

- Deploy to production
- Validate

APPENDIX D: SALARY MATRIX COMPARISON

The salary matrix provides a detailed comparison showing:

- **Traditional Scrum Positions**: Product Owner, Business Analyst, QA Engineer, System Architect, Software Developer, Scrum Master
- **AI Scrum Positions**: Product Strategist, AI Alignment Engineer, Validation Lead, Governance Coordinator (NEW), Integration Architect, AI-Augmented Developer, Delivery Facilitator
- Salary ranges broken down by experience level (Entry/Mid/Senior)
- National averages based on January 2026 data from multiple reputable sources

Key Finding: AI Scrum roles command a 10-37% premium over traditional equivalents, with total team cost increase of approximately 26% ($152K annually), reflecting market scarcity of AI-specific skills. Salary ranges will vary from city to city and region to region.

Each description is tailored to reflect the unique requirements and market positioning of traditional vs. AI-augmented roles.

Salary data compiled from Glassdoor, Salary.com, Indeed, Built In, PayScale, and ZipRecruiter as of January 2026 and represent National U.S. Averages.

TRADITIONAL SCRUM VS AI SCRUM
SALARY COMPARISON MATRIX
January 2026
National U.S. Averages

TRADITIONAL SCRUM POSITIONS

Position: **PRODUCT OWNER**
National Salary Range: $98,000 - $155,000
Average Salary: $120,000
Entry Level (0-2 years): $93,000 - $108,000
Mid-Level (3-6 years): $108,000 - $140,000
Senior Level (7+ years): $140,000 - $184,000
Source: Glassdoor, Salary.com, Indeed (Jan 2026)

Position: **BUSINESS ANALYST**
National Salary Range: $70,000 - $135,000
Average Salary: $95,000
Entry Level (0-2 years): $65,000 - $78,000
Mid-Level (3-6 years): $85,000 - $110,000
Senior Level (7+ years): $115,000 - $140,000
Source: Glassdoor, Salary.com, PayScale (Jan 2026)

Position: **QA ENGINEER**
National Salary Range: $75,000 - $130,000
Average Salary: $95,000
Entry Level (0-2 years): $68,000 - $84,000
Mid-Level (3-6 years): $84,000 - $105,000
Senior Level (7+ years): $105,000 - $135,000
Source: Glassdoor, ZipRecruiter, PayScale (Jan 2026)

Position: **SYSTEM ARCHITECT**
National Salary Range: $125,000 - $225,000
Average Salary: $170,000
Entry Level (0-2 years): $90,000 - $125,000
Mid-Level (3-6 years): $125,000 - $180,000
Senior Level (7+ years): $180,000 - $280,000
Source: Glassdoor, Salary.com, Indeed (Jan 2026)

Position: **SOFTWARE DEVELOPER**
National Salary Range: $90,000 - $155,000
Average Salary: $120,000
Entry Level (0-2 years): $70,000 - $95,000
Mid-Level (3-6 years): $95,000 - $130,000
Senior Level (7+ years): $130,000 - $190,000
Source: Glassdoor, Built In, ZipRecruiter (Jan 2026)

Position: **SCRUM MASTER**
National Salary Range: $99,000 - $161,000
Average Salary: $118,000
Entry Level (0-2 years): $77,000 - $99,000
Mid-Level (3-6 years): $99,000 - $130,000
Senior Level (7+ years): $130,000 - $180,000
Source: Glassdoor, PayScale, Indeed (Jan 2026)

AI SCRUM POSITIONS

Position: **PRODUCT STRATEGIST**
(Evolved from Product Owner)
National Salary Range: $115,000 - $195,000
Average Salary: $145,000
Entry Level (0-2 years): $105,000 - $130,000
Mid-Level (3-6 years): $130,000 - $165,000
Senior Level (7+ years): $165,000 - $210,000
Rationale: +20-25% premium for AI strategy, predictive analytics, and outcome-based leadership capabilities beyond traditional product ownership.

Position: **AI ALIGNMENT ENGINEER**
(Evolved from Business Analyst)
National Salary Range: $105,000 - $165,000
Average Salary: $130,000
Entry Level (0-2 years): $90,000 - $115,000
Mid-Level (3-6 years): $115,000 - $145,000
Senior Level (7+ years): $145,000 - $180,000
Rationale: +30-37% premium for prompt engineering expertise, AI collaboration optimization, and technical-business translation for AI systems.

Position: **VALIDATION LEAD**
(Evolved from QA Engineer)
National Salary Range: $95,000 - $155,000
Average Salary: $118,000
Entry Level (0-2 years): $82,000 - $105,000
Mid-Level (3-6 years): $105,000 - $130,000
Senior Level (7+ years): $130,000 - $170,000
Rationale: +24% premium for AI-specific testing (adversarial, bias, security), red-teaming, and detailed validation strategy leadership.

Position: **GOVERNANCE COORDINATOR**
(NEW ROLE - No Direct Traditional Equivalent)
National Salary Range: $100,000 - $170,000
Average Salary: $125,000
Entry Level (0-2 years): $88,000 - $110,000
Mid-Level (3-6 years): $110,000 - $140,000
Senior Level (7+ years): $140,000 - $185,000
Rationale: Positioned between Compliance Analyst ($75K-$130K) and Compliance Manager ($110K-$160K) with AI governance specialization premium. Reflects growing regulatory demands (EU AI Act, NIST AI RMF).

Position: **INTEGRATION ARCHITECT**
(Evolved from System Architect)
National Salary Range: $140,000 - $245,000
Average Salary: $185,000
Entry Level (0-2 years): $105,000 - $145,000
Mid-Level (3-6 years): $145,000 - $195,000
Senior Level (7+ years): $195,000 - $295,000
Rationale: +9% premium for AI system integration patterns, observability architecture for AI systems, and rapid prototyping architecture expertise.

Position: **AI-AUGMENTED DEVELOPER**
(Evolved from Developer)
National Salary Range: $110,000 - $180,000
Average Salary: $138,000
Entry Level (0-2 years): $88,000 - $118,000
Mid-Level (3-6 years): $118,000 - $152,000
Senior Level (7+ years): $152,000 - $210,000
Rationale: +15% premium for prompt engineering for code generation, AI output validation, multi-alternative evaluation, and human-AI collaboration optimization skills.

Position: **DELIVERY FACILITATOR**
(Evolved from Scrum Master)
National Salary Range: $108,000 - $178,000
Average Salary: $130,000
Entry Level (0-2 years): $85,000 - $112,000
Mid-Level (3-6 years): $112,000 - $145,000
Senior Level (7+ years): $145,000 - $195,000
Rationale: +10% premium for AI-native workflow optimization,

validation/governance facilitation, and AI collaboration coaching expertise.

SALARY COMPARISON SUMMARY

Traditional Scrum Total Compensation Range: $535,000 - $1,046,000 (all 6 positions)

AI Scrum Total Compensation Range: $675,000 - $1,288,000 (all 7 positions)

Average Team Cost Increase: 26% (+$152,000 annually)
Premium Value: Enhanced AI governance, accelerated delivery, comprehensive validation

KEY INSIGHTS:

- AI Scrum roles command 10-37% premium over traditional equivalents
- Governance Coordinator is net-new cost but essential for regulatory compliance
- Greatest premiums: AI Alignment Engineer (37%), Validation Lead (24%),
  Product Strategist (21%)
- Premiums reflect market scarcity of AI-specific skills and regulatory demands
- Geographic variations apply: Tech hubs (SF, NYC, Seattle) add 20-40% to base ranges

SPECIAL NOTES:

- Salaries reflect full-time W-2 employment
- Additional compensation (bonuses, stock, benefits) not included in base ranges
- Certifications can add 5-15% to base salary
- Remote work may adjust salaries ±10% based on location
- Contract/1099 rates typically 30-50% higher than W-2 base

APPENDIX E: JOB DESCRIPTIONS WITH CALCULATED NATIONAL SALARY

Professional one-page descriptions for all 13 positions including:

- Position summary
- Key responsibilities
- Required skills & qualifications
- Preferred qualifications
- National salary ranges by experience level

Each description is tailored to reflect the unique requirements and market positioning of traditional vs. AI-augmented roles.
Salary data compiled from Glassdoor, Salary.com, Indeed, Built In, PayScale, and ZipRecruiter as of January 2026 and represent National U.S. Averages.

PRODUCT OWNER (TRADITIONAL SCRUM TEAM)

POSITION SUMMARY

The Product Owner is responsible for maximizing the value of the product resulting from work of the Development Team. The Product Owner is the sole person responsible for managing the Product Backlog, ensuring clear articulation of product backlog items, ordering items to best achieve goals and missions, and ensuring the Development Team understands items in the Product Backlog.

KEY RESPONSIBILITIES

• Define and prioritize product backlog items based on business value and stakeholder needs
• Clearly articulate user stories with detailed acceptance criteria
• Collaborate with stakeholders to gather requirements and validate solutions
• Make decisions regarding product features, release planning, and sprint commitments
• Participate in all Scrum ceremonies including sprint planning, daily standups, sprint reviews, and retrospectives
• Ensure product backlog is visible, transparent, and clear to all stakeholders
• Accept or reject work results based on defined acceptance criteria
• Monitor and report on product velocity and team performance metrics

REQUIRED SKILLS & QUALIFICATIONS

- Bachelor's degree in Business Administration, Computer Science, or related field
- 3-5 years of experience in product management or business analysis
- Strong understanding of Agile/Scrum methodologies
- Excellent communication and stakeholder management skills
- Proven ability to write clear user stories and acceptance criteria
- Experience with backlog management tools (JIRA, Azure DevOps, etc.)
- Strong analytical and problem-solving capabilities
- Ability to make data-driven decisions

PREFERRED QUALIFICATIONS

- Certified Scrum Product Owner (CSPO) or similar certification
- Technical background or experience in software development
- Experience with predictive analytics or business intelligence tools
- Domain expertise in relevant industry vertical

SALARY RANGE

National Average: $120,000 annually - Entry Level (0-2 years): $93,000 - $108,000 Mid-Level (3-6 years): $108,000 - $140,000 Senior Level (7+ years): $140,000 - $184,000 Note: Salaries vary by location, with tech hubs (San Francisco, NYC, Seattle) typically 20-40% above national average. Additional compensation may include bonuses (15-25% of base) and equity.

BUSINESS ANALYST (TRADITIONAL SCRUM TEAM)

POSITION SUMMARY

The Business Analyst serves as a liaison between business stakeholders and technical teams, gathering requirements, documenting business processes, and ensuring solutions align with organizational needs. This role translates business requirements into technical specifications and validates that delivered solutions meet stakeholder expectations.

KEY RESPONSIBILITIES

- Conduct stakeholder interviews and workshops to gather business requirements
- Document business processes, workflows, and system requirements
- Create detailed functional specifications and use case models
- Translate business requirements into clear user stories and acceptance criteria
- Facilitate requirement validation sessions with stakeholders
- Perform gap analysis between current and desired business states
- Support User Acceptance Testing (UAT) coordination and execution

• Maintain requirements traceability throughout project lifecycle

REQUIRED SKILLS & QUALIFICATIONS

• Bachelor's degree in Business Administration, Information Systems, or related field
• 3-5 years of business analysis experience
• Strong requirements gathering and documentation skills
• Proficiency in requirements management tools
• Excellent written and verbal communication abilities
• Experience with process modeling techniques (BPMN, flowcharting)
• Understanding of software development lifecycle
• Strong analytical and problem-solving skills

PREFERRED QUALIFICATIONS

• CBAP, PMI-PBA, or IIBA ECBA certification
• Experience with Agile/Scrum methodologies
• Knowledge of SQL and database concepts
• Familiarity with data analysis tools (Excel, Tableau, Power BI)
• Domain expertise in relevant industry

SALARY RANGE

National Average: $95,000 annually - Entry Level (0-2 years): $65,000 - $78,000 Mid-Level (3-6 years): $85,000 - $110,000 Senior Level (7+ years): $115,000 - $140,000 Note: Geographic location significantly impacts salary. Tech hubs and financial centers offer 15-30% premium. Benefits typically include health insurance, 401(k) matching, and professional development allowances.

QA ENGINEER (TRADITIONAL SCRUM TEAM)

POSITION SUMMARY

The QA Engineer ensures software quality through thorough testing strategies, test case development, and defect identification. This role designs and executes test plans, performs both manual and automated testing, and collaborates with development teams to maintain high-quality standards throughout the software development lifecycle.

KEY RESPONSIBILITIES

• Design thorough test plans and test cases based on requirements
• Execute manual and automated testing across multiple platforms
• Identify, document, and track software defects using bug tracking tools
• Perform regression testing to ensure existing functionality remains intact
• Conduct integration testing and end-to-end system testing

• Collaborate with developers to reproduce and resolve defects
• Develop and maintain automated test scripts and frameworks
• Participate in requirement reviews to identify testability issues

REQUIRED SKILLS & QUALIFICATIONS
• Bachelor's degree in Computer Science, Engineering, or related field
• 2-4 years of software quality assurance experience
• Strong knowledge of software testing methodologies and best practices
• Experience with test management tools (TestRail, Zephyr, qTest)
• Proficiency in defect tracking systems (JIRA, Bugzilla)
• Understanding of software development lifecycle and Agile methodologies
• Strong analytical and troubleshooting skills
• Excellent attention to detail and organizational abilities

PREFERRED QUALIFICATIONS
• ISTQB certification or equivalent
• Experience with test automation tools (Selenium, Cypress, Playwright)
• Programming knowledge (Python, Java, JavaScript)
• API testing experience (Postman, SoapUI)
• Performance testing familiarity (JMeter, LoadRunner)
• Mobile testing experience (iOS, Android)

SALARY RANGE
National Average: $95,000 annually - Entry Level (0-2 years): $68,000 - $84,000 Mid-Level (3-6 years): $84,000 - $105,000 Senior Level (7+ years): $105,000 - $135,000 Note: Automation skills and certifications can increase compensation by 10-20%. Major metro areas offer higher salaries to account for cost of living.

SYSTEM ARCHITECT (TRADITIONAL SCRUM TEAM)

POSITION SUMMARY
The System Architect designs high-level system architecture, defines technology standards, and ensures technical solutions align with business objectives. This role provides technical leadership, evaluates emerging technologies, and creates architectural blueprints that guide development teams in building scalable, maintainable systems.

KEY RESPONSIBILITIES
• Design system architecture and technology stack
• Define architectural standards, patterns, and best practices
• Evaluate and recommend technologies, frameworks, and tools
• Create detailed architectural documentation and diagrams

• Conduct architecture reviews and provide technical guidance
• Ensure system scalability, performance, and security
• Collaborate with stakeholders to align technical solutions with business needs
• Lead technical discussions and resolve architectural challenges

REQUIRED SKILLS & QUALIFICATIONS

• Bachelor's or Master's degree in Computer Science or related field
• 7-10 years of software development and architecture experience
• Deep expertise in system architecture patterns and design principles
• Strong knowledge of cloud platforms (AWS, Azure, GCP)
• Experience with microservices, distributed systems, and APIs
• Proficiency in multiple programming languages and frameworks
• Excellent problem-solving and analytical capabilities
• Strong communication and stakeholder management skills

PREFERRED QUALIFICATIONS

• Architecture certifications (TOGAF, AWS/Azure/GCP Architect)
• Experience with container technologies (Docker, Kubernetes)
• Knowledge of DevOps practices and CI/CD pipelines
• Security architecture expertise
• Domain-specific architectural experience
• Published technical articles or conference speaking experience

SALARY RANGE

National Average: $170,000 annually Entry Level (0-2 years): $90,000 - $125,000 Mid-Level (3-6 years): $125,000 - $180,000 Senior Level (7+ years): $180,000 - $280,000 Note: Highly experienced architects at FAANG companies can earn $250K-$400K+ including equity. Consulting architects may command premium rates.

SOFTWARE DEVELOPER (TRADITIONAL SCRUM TEAM)

POSITION SUMMARY

The Software Developer designs, develops, tests, and maintains software applications. This role writes clean, efficient code, participates in code reviews, implements features based on specifications, and collaborates with cross-functional teams to deliver high-quality software solutions.

KEY RESPONSIBILITIES

• Write clean, maintainable, and efficient code following best practices
• Develop new features and functionality based on requirements
• Participate in code reviews and provide constructive feedback

• Debug and resolve software defects and technical issues
• Collaborate with team members in Agile/Scrum ceremonies
• Write unit tests and ensure adequate code coverage
• Maintain and improve existing codebase through refactoring
• Document code, APIs, and technical processes

REQUIRED SKILLS & QUALIFICATIONS
• Bachelor's degree in Computer Science, Software Engineering, or related field
• 2-4 years of professional software development experience
• Proficiency in one or more programming languages (Java, Python, C#, JavaScript)
• Strong understanding of data structures, algorithms, and OOP concepts
• Experience with version control systems (Git)
• Knowledge of software development lifecycle and methodologies
• Problem-solving skills and ability to learn new technologies
• Good communication and teamwork abilities

PREFERRED QUALIFICATIONS
• Experience with modern frameworks (React, Angular, Spring Boot, .NET)
• Database knowledge (SQL and NoSQL)
• Cloud platform familiarity (AWS, Azure, GCP)
• DevOps and CI/CD pipeline experience
• Mobile development experience
• Open source contributions or personal projects portfolio

SALARY RANGE
National Average: $120,000 annually - Entry Level (0-2 years): $70,000 - $95,000 Mid-Level (3-6 years): $95,000 - $130,000 Senior Level (7+ years): $130,000 - $190,000 Note: Specialized skills (AI/ML, blockchain, security) command premium compensation. FAANG and high-growth startups offer significantly higher packages including equity.

SCRUM MASTER (TRADITIONAL SCRUM TEAM)

POSITION SUMMARY
The Scrum Master facilitates Scrum practices, removes impediments, and coaches teams on Agile principles. This role ensures the team follows Scrum framework, facilitates ceremonies, protects the team from external disruptions, and fosters an environment of continuous improvement and collaboration.

KEY RESPONSIBILITIES

- Facilitate all Scrum ceremonies (sprint planning, daily standups, reviews, retrospectives)
- Remove impediments and blockers that hinder team progress
- Coach team members on Agile and Scrum best practices
- Protect the team from external interruptions and scope changes
- Track and report team velocity and sprint metrics
- Foster culture of continuous improvement and innovation
- Facilitate collaboration between Product Owner and Development Team
- Identify and address team dynamics and conflict resolution

REQUIRED SKILLS & QUALIFICATIONS

- Bachelor's degree in Computer Science, Business, or related field
- 3-5 years of Scrum Master or Agile coaching experience
- Certified ScrumMaster (CSM) or equivalent certification
- Deep understanding of Scrum framework and Agile principles
- Experience with Agile project management tools (JIRA, Rally, Azure DevOps)
- Excellent facilitation and communication skills
- Strong problem-solving and conflict resolution abilities
- Servant leadership mindset

PREFERRED QUALIFICATIONS

- Scrum certifications (CSM, A-CSM, CSP-SM)
- Experience with scaled Agile frameworks (SAFe, LeSS)
- Technical background or software development experience
- Multiple team/program-level Scrum experience
- Change management and organizational coaching experience

SALARY RANGE

National Average: $118,000 annually - Entry Level (0-2 years): $77,000 - $99,000 Mid-Level (3-6 years): $99,000 - $130,000 Senior Level (7+ years): $130,000 - $180,000 Note: Enterprise Scrum Masters and Agile Coaches at large organizations can earn $150K-$200K+. Scaled Agile expertise increases compensation by 15-25%.

PRODUCT STRATEGIST (AI SCRUM TEAM)

Evolved from Traditional Product Owner

POSITION SUMMARY

The Product Strategist owns business outcomes and strategic alignment for AI-augmented product development. Beyond traditional backlog management, this role leverages AI for market analysis, predictive value modeling, and outcome validation. The Product Strategist orchestrates stakeholders around measurable business value rather than feature completion.

KEY RESPONSIBILITIES

• Define and measure business outcome metrics aligned with strategic objectives
• Leverage AI tools for market analysis, competitive intelligence, and trend prediction
• Create predictive value models to forecast feature impact on business outcomes
• Orchestrate stakeholder alignment around outcome achievement vs. feature delivery
• Validate delivered capabilities against outcome metrics, not just acceptance criteria
• Use AI-assisted analytics to identify highest-value opportunities
• Balance innovation velocity with governance and validation requirements
• Translate business strategy into AI-compatible specifications for development teams

REQUIRED SKILLS & QUALIFICATIONS

• Bachelor's degree in Business, Computer Science, Data Science, or related field
• 5-7 years of product management experience, including AI/ML products
• Deep understanding of AI capabilities, limitations, and ethical considerations
• Expertise in outcome-based product management and OKR frameworks
• Proficiency with AI-powered analytics and business intelligence tools
• Strong data analysis skills and comfort with predictive modeling
• Experience with Agile methodologies adapted for AI development
• Exceptional stakeholder management and strategic communication skills

PREFERRED QUALIFICATIONS

• Advanced Product Owner certifications plus AI/ML coursework
• Experience with AI governance frameworks and ethical AI principles
• Background in data science or machine learning
• Familiarity with prompt engineering and AI tool ecosystems
• Track record of successful AI product launches
• Experience in regulated industries with AI compliance requirements

SALARY RANGE

National Average: $145,000 annually Entry Level (0-2 years AI experience): $105,000 - $130,000 Mid-Level (3-6 years AI experience): $130,000 - $165,000 Senior Level (7+ years AI experience): $165,000 - $210,000 Note: 20-25% premium over traditional Product Owner reflects AI strategy expertise and outcome-based leadership. Tech hubs and AI-focused companies offer significant equity packages.

AI ALIGNMENT ENGINEER (AI SCRUM TEAM)

Evolved from Traditional Business Analyst

POSITION SUMMARY

The AI Alignment Engineer translates business intent into AI-compatible specifications through advanced prompt engineering and AI collaboration optimization. This role bridges business requirements and AI capabilities, maintains prompt libraries, documents AI collaboration patterns, and trains teams on effective human-AI interaction for optimal output quality.

KEY RESPONSIBILITIES

• Translate complex business requirements into precise AI-compatible prompts
• Curate and maintain organization-wide prompt library with versioning
• Optimize AI collaboration patterns for maximum output quality and efficiency
• Maintain AI Collaboration Log documenting prompt evolution
• Train team members on effective prompt engineering techniques
• Evaluate AI model capabilities and match them to business needs
• Design multi-alternative generation strategies for complex requirements
• Measure and improve AI output acceptance rates and iteration efficiency

REQUIRED SKILLS & QUALIFICATIONS

• Bachelor's degree in Business, Computer Science, Linguistics, Cognitive Science, or related field
• 4-6 years combining business analysis with AI/ML implementation
• Expert-level prompt engineering skills across multiple AI platforms
• Deep understanding of LLM capabilities, limitations, and behavior patterns
• Experience creating and maintaining technical documentation systems
• Strong grasp of both business processes and AI system architectures
• Excellent communication skills for translating between business and technical domains
• Proficiency with AI development tools and APIs

PREFERRED QUALIFICATIONS

• Master's degree in AI, NLP, or related technical field
• Certifications in Prompt Engineering, AI/ML (DeepLearning.AI, Google AI, AWS ML)
• Experience with multiple LLM platforms (GPT, Claude, Gemini, etc.)
• Background in computational linguistics or natural language processing
• Published research or articles on prompt engineering
• Experience training others on AI tools and methodologies

SALARY RANGE

National Average: $130,000 annually - Entry Level (0-2 years AI specialization): $90,000 - $115,000 Mid-Level (3-6 years AI specialization): $115,000 - $145,000 Senior Level (7+ years AI specialization): $145,000 - $180,000 Note: 30-37% premium over traditional Business Analyst due to specialized AI skills scarcity. High demand for prompt engineering expertise drives competitive compensation.

VALIDATION LEAD (AI SCRUM TEAM)

Evolved from Traditional QA Engineer

POSITION SUMMARY

The Validation Lead develops thorough validation strategies for AI-generated outputs, including adversarial testing, bias detection, security scanning, and performance benchmarking. This role coordinates red-team testing, ensures deployment readiness, and validates AI systems against quality, security, and ethical standards beyond traditional software testing.

KEY RESPONSIBILITIES

- Design and execute validation strategies for AI outputs
- Define and maintain Validation Canvas covering all acceptance criteria
- Conduct adversarial red-team testing to identify AI vulnerabilities
- Implement AI-specific testing for bias, fairness, and safety concerns
- Coordinate automated test suite generation leveraging AI tools
- Perform security scanning and penetration testing on AI systems
- Execute performance benchmarking and edge case identification
- Lead validation-refinement iteration cycles (typically 2-4 per sprint)
- Document validation findings and collaborate on remediation strategies

REQUIRED SKILLS & QUALIFICATIONS

- Bachelor's degree in Computer Science, Cybersecurity, or related field
- 5-7 years of software testing with 2+ years in AI/ML validation
- Expertise in adversarial testing and red-teaming methodologies
- Deep knowledge of AI bias detection and fairness evaluation techniques
- Experience with security testing tools and vulnerability assessment
- Understanding of AI model behavior, edge cases, and failure modes
- Strong analytical skills for complex problem diagnosis
- Proficiency in test automation frameworks and scripting

PREFERRED QUALIFICATIONS

- ISTQB Advanced certification plus AI testing specialization
- Offensive security certifications (CEH, OSCP)
- Background in data science or machine learning
- Experience with AI safety frameworks and responsible AI practices
- Familiarity with regulatory compliance (EU AI Act, NIST AI RMF)
- Published work on AI testing methodologies

SALARY RANGE
National Average: $118,000 annually - Entry Level (0-2 years AI testing): $82,000 - $105,000 Mid-Level (3-6 years AI testing): $105,000 - $130,000 Senior Level (7+ years AI testing): $130,000 - $170,000 Note: 24% premium over traditional QA reflects specialized AI validation skills. Organizations with strict AI compliance needs offer higher compensation.

GOVERNANCE COORDINATOR (AI SCRUM TEAM)

NEW ROLE - No Traditional Scrum Equivalent

POSITION SUMMARY
The Governance Coordinator integrates AI governance into sprint workflows, ensuring regulatory compliance, managing risk, and maintaining audit readiness. This role coordinates with AI Governance Boards, completes Governance Canvases, ensures adherence to frameworks like EU AI Act and NIST AI RMF, and maintains detailed governance records for all AI implementations.

KEY RESPONSIBILITIES
- Complete and maintain Governance Canvas for all AI initiatives
- Coordinate with AI Governance Board for mid-sprint reviews
- Ensure compliance with AI regulations (EU AI Act, NIST AI RMF, industry-specific)
- Classify AI systems by risk level and determine governance requirements
- Document model selection rationale, data governance, and security controls
- Maintain detailed Governance Record with audit trail
- Facilitate governance checkpoints and remediation workflows
- Monitor emerging AI regulations and update governance processes
- Train teams on governance requirements and best practices

REQUIRED SKILLS & QUALIFICATIONS
- Bachelor's degree in Law, Compliance, Computer Science, or related field
- 4-6 years in compliance, governance, or risk management with AI/tech focus
- Deep knowledge of AI governance frameworks and emerging regulations
- Understanding of AI model lifecycle, data governance, and algorithmic accountability
- Experience with audit processes and documentation requirements
- Strong project coordination and stakeholder management skills
- Excellent written communication for governance documentation
- Ability to translate complex regulatory requirements into actionable processes

PREFERRED QUALIFICATIONS
- Certifications in compliance (CIPP, CRISC) or AI ethics
- Legal background or JD with focus on technology law

• Experience with EU AI Act, GDPR, CCPA, or sector-specific regulations
• Technical background enabling understanding of AI systems
• Experience in regulated industries (finance, healthcare, government)
• Familiarity with responsible AI frameworks and algorithmic impact assessments

SALARY RANGE
National Average: $125,000 annually - Entry Level (0-2 years AI governance): $88,000 - $110,000 Mid-Level (3-6 years AI governance): $110,000 - $140,000 Senior Level (7+ years AI governance): $140,000 - $185,000 Note: New specialized role positioned between Compliance Analyst and Manager levels. Growing regulatory landscape (EU AI Act) driving demand. Organizations facing AI audits offer premium compensation.

INTEGRATION ARCHITECT (AI SCRUM TEAM)

Evolved from Traditional System Architect

POSITION SUMMARY
The Integration Architect coordinates AI system integration, manages dependencies, designs observability architecture, and enables rapid prototyping. This role maintains Integration Maps, implements AI-specific integration patterns, designs monitoring/telemetry for AI systems, and manages integration of AI-generated components into existing architectures.

KEY RESPONSIBILITIES
• Maintain detailed Integration Map for all AI system dependencies
• Design and implement AI-specific integration patterns and best practices
• Coordinate rapid prototyping architecture enabling quick AI alternative evaluation
• Design monitoring, observability, and telemetry systems for AI components
• Manage integration of AI-generated code into production systems
• Resolve dependency conflicts and ensure system compatibility
• Define integration testing strategies and deployment pipelines
• Plan migration paths from legacy to AI-augmented architectures

REQUIRED SKILLS & QUALIFICATIONS
• Bachelor's or Master's degree in Computer Science, Software Engineering, or related field
• 8-10 years of system architecture with 3+ years integrating AI/ML systems
• Deep expertise in microservices, APIs, and distributed system patterns
• Experience with AI model deployment, serving, and monitoring
• Proficiency in cloud platforms and containerization (K8s, Docker)
• Strong knowledge of observability tools (Prometheus, Grafana, ELK)

• Understanding of AI system-specific challenges (model drift, latency, versioning)
• Excellent system design and technical communication skills

PREFERRED QUALIFICATIONS
• Advanced architecture certifications with AI specialization
• Experience with MLOps platforms and model management systems
• Background in both traditional software and AI/ML engineering
• Knowledge of feature stores, model registries, and experiment tracking
• Familiarity with AI explainability and interpretability tools
• Published architectural patterns for AI system integration

SALARY RANGE
National Average: $185,000 annually Entry Level (0-2 years AI architecture): $105,000 - $145,000 Mid-Level (3-6 years AI architecture): $145,000 - $195,000 Senior Level (7+ years AI architecture): $195,000 - $295,000 Note: 9% premium over traditional System Architect for AI integration expertise. High-scale AI implementations and architecture innovation roles command top-tier compensation. Principal-level roles at tech companies can exceed $350K.

AI-AUGMENTED DEVELOPER (AI SCRUM TEAM)

Evolved from Traditional Software Developer

POSITION SUMMARY
The AI-Augmented Developer masters human-AI collaboration for code generation, validates AI outputs for quality and security, evaluates multiple implementation alternatives, and optimizes prompt engineering for development tasks. This role shifts focus from writing all code manually to orchestrating AI generation, validation, and refinement for maximum development velocity.

KEY RESPONSIBILITIES
• Engineer effective prompts for AI code generation across multiple alternatives
• Validate AI-generated code for correctness, security, and maintainability
• Evaluate 3-5 implementation alternatives to identify optimal approaches
• Refine and integrate AI-generated code into production systems
• Optimize human-AI collaboration workflows for team efficiency
• Identify and document patterns for effective AI-assisted development
• Measure AI output acceptance rates and iterate on prompt effectiveness
• Mentor team members on AI tool usage and best practices

REQUIRED SKILLS & QUALIFICATIONS
• Bachelor's degree in Computer Science, Software Engineering, or related field

• 3-5 years of software development with 1+ years using AI coding assistants
• Strong programming fundamentals enabling code quality evaluation and prompt engineering
• Experience with AI development tools (GitHub Copilot, Cursor, Replit, etc.)
• Understanding of code security, performance, and maintainability principles
• Ability to rapidly evaluate and compare multiple implementation approaches
• Knowledge of AI model capabilities and limitations for coding tasks
• Strong debugging and code review skills

PREFERRED QUALIFICATIONS
• Advanced programming certifications plus AI/ML coursework
• Certified Prompt Engineer
• Experience contributing to prompt engineering communities
• Background in multiple programming paradigms and languages
• Familiarity with AI model fine-tuning or custom model development
• Published examples of effective AI-assisted development workflows
• Experience training others on AI development tools

SALARY RANGE
National Average: $138,000 annually Entry Level (0-2 years AI development): $88,000 - $118,000 Mid-Level (3-6 years AI development): $118,000 - $152,000 Senior Level (7+ years AI development): $152,000 - $210,000 Note: 15% premium over traditional Developer for AI collaboration skills. High productivity gains from AI augmentation make these developers highly valuable. Top performers with strong AI prompt engineering can command significant premiums.

DELIVERY FACILITATOR (AI SCRUM TEAM)
Evolved from Traditional Scrum Master

POSITION SUMMARY
The Delivery Facilitator optimizes AI-native workflows, removes validation and governance impediments, coaches teams on AI collaboration effectiveness, and monitors sprint telemetry dashboards. This role facilitates AI Scrum ceremonies adapted for AI development, ensuring teams achieve outcome-based goals while navigating new AI-specific challenges.

KEY RESPONSIBILITIES
• Facilitate AI-adapted Scrum ceremonies (Sprint Inception, AI Show & Tell, Governance Checkpoints)
• Remove impediments related to validation, governance, and AI alignment

• Monitor Sprint Telemetry Dashboard and identify improvement opportunities
• Coach team members on effective AI collaboration and prompt engineering
• Optimize AI-native workflows for maximum outcome achievement
• Facilitate resolution of validation failures and governance concerns
• Track outcome achievement rates and team health metrics
• Drive continuous improvement in AI development practices

REQUIRED SKILLS & QUALIFICATIONS
• Bachelor's degree in Computer Science, Business, or related field
• 4-6 years of Scrum Master experience with 1+ years in AI development environments
• Advanced Scrum certifications (A-CSM or CSP-SM)
• Understanding of AI development lifecycle and unique challenges
• Experience facilitating governance and validation workflows
• Strong prompt engineering skills
• Strong coaching skills adapted for AI tool adoption
• Proficiency with sprint metrics and data-driven improvement
• Excellent facilitation and conflict resolution abilities

PREFERRED QUALIFICATIONS
• Experience with AI governance frameworks and compliance workflows
• Technical background enabling understanding of AI system challenges
• Certification in Organizational Change Management and Prompt Engineering
• Familiarity with AI development tools and platforms
• Experience scaling AI development across multiple teams
• Background in DevOps or MLOps practices

SALARY RANGE
National Average: $130,000 annually Entry Level (0-2 years AI facilitation): $85,000 - $112,000 Mid-Level (3-6 years AI facilitation): $112,000 - $145,000 Senior Level (7+ years AI facilitation): $145,000 - $195,000 Note: 10% premium over traditional Scrum Master for AI workflow optimization expertise. Organizations undergoing AI transformation place high value on effective change facilitation. Agile Coaches specializing in AI adoption can command higher compensation.

APPENDIX F: TEMPLATES, CHECKLISTS, AND MISCELLANEOUS

#	Template Title	Chapter	Description
1	**Outcome Backlog Template**	7	A structured format for defining and prioritizing outcomes instead of features. Includes fields for: outcome statement, success metrics, validation criteria, business value, dependencies, and constraints. Helps teams shift from "build X feature" to "achieve Y outcome."
2	**Sprint Inception Agenda & Checklist**	5	A 4-hour meeting template with time-boxed sections for: Outcome Alignment (60 min), Validation Definition (60 min), Governance Framing (45 min), Capacity Planning (45 min), and AI Collaboration Protocol Review (30 min). Includes facilitator notes and decision checkpoints.
3	**Validation Canvas Template**	3, 5	A visual framework for defining validation criteria before implementation begins. Includes sections for: functional requirements, performance benchmarks, security requirements, compliance checkpoints, acceptance thresholds, and evidence needed. Maps directly to the "Validation-First" principle.
4	**Context Specification Template**	6	A structured prompt format for human-AI collaboration. Fields include: task objective, architectural constraints, business rules, edge cases to handle, what NOT to do, expected output format, and validation approach. Ensures AI receives complete context for better outputs.
5	**AI Collaboration Log**	6, 9	A daily tracking template documenting: prompts used, AI tool/model, outputs generated, time saved vs. traditional approach, validation issues found, and lessons learned. Enables teams to build institutional knowledge about effective AI collaboration patterns.
6	**Daily Synchronization Question Framework**	5	A structured replacement for traditional standup questions focused on validation progress: (1) What validation did we complete? (2) What validation are we pursuing today? (3) What validation blockers exist? (4) What AI collaboration worked/didn't work?

7	**AI Show & Tell Session Structure**	5	A mid-sprint stakeholder feedback template for Days 3 and 7. Includes: agenda (demo preparation, prototype presentation, feedback capture), decision framework (accept/refine/pivot), and next steps documentation. Prevents late-stage rework through early course correction.
8	**Governance Checkpoint Checklist**	5, 10	A mid-sprint governance review template covering: risk tier validation, compliance documentation review, data privacy verification, security assessment, ethical considerations, and decision log. Integrates governance throughout the sprint rather than as an end gate.
9	**Risk Assessment & Classification Template**	10	A four-tier decision framework (Delegated Authority, Standard Review, Board Review, Prohibited) for classifying AI work. Includes risk factors to evaluate: data sensitivity, regulatory impact, user exposure, system criticality, and AI hallucination potential.
10	**Role Transition Playbook**	4	A practical guide for evolving traditional Scrum roles to AI Scrum roles. For each role (Product Owner → Product Strategist, QA → Validation Lead, etc.), includes: capability gaps to close, training needed, timeline expectations (2-6 months), and success indicators.
11	**Test Strategy Template for AI-Generated Code**	8	A detailed testing approach covering: unit test generation checklist, integration test scenarios, adversarial testing cases, drift monitoring setup, and test maintenance procedures. Addresses the unique validation challenges of probabilistic AI outputs.
12	**Metrics Dashboard Blueprint**	9	A visual dashboard template tracking AI Scrum metrics: Validation Lead Time, First-Pass Acceptance Rate, Governance Cycle Time, AI Debt Index, Outcome Achievement Rate, Token Efficiency, and Defect Escape Ratio. Designed for 5-second comprehension with traffic light indicators.
13	**AI Scrum RACI Matrix**	4	A responsibility assignment matrix showing Responsible, Accountable, Consulted, and Informed roles across AI Scrum activities. Covers 15+ key activities from outcome definition through deployment, mapped to all seven AI Scrum roles.

14	**Maturity Assessment Scorecard**	12	A self-assessment tool evaluating organizational readiness across 8 dimensions: Agile Process Maturity, AI Integration Depth, Governance & Risk Management, Team Capabilities, Tooling & Infrastructure, Metrics & Observability, Change Management, and Strategic Alignment. 5-level scoring (1=Reactive to 5=Transformative).
15	**Sprint Retrospective Framework with AI Metrics**	5, 9	An enhanced retrospective template incorporating: traditional "What Went Well/What Needs Improvement" plus AI-specific reflection on: prompt effectiveness, validation quality, governance friction, tool performance, and learning velocity. Includes objective telemetry review.
16	**Center of Excellence (CoE) Charter**	4	Formalize organizational learning and standardization by establishing Centers of Excellence that prevent duplicate problem-solving across teams and accelerate AI capability development.
17	**Governance Canvas Template**	4	Provide a visual, comprehensive framework for planning and documenting governance requirements for AI work before implementation begins, enabling teams to identify compliance obligations, assess risks, and establish appropriate oversight levels during Sprint Inception.
18	**European Union AI Act Compliance Checklist**	10	Ensure AI systems comply with the European Union AI Act through comprehensive step-by-step guidance, enabling proactive governance and reducing regulatory risk for any organization that may serve EU customers or operate in regulated domains.
19	**Budget Planning Template for AI Scrum Adoption**	4, 14, Apdx D, Apdx E	Provide a simple, complete budgeting framework for AI Scrum adoption covering both team salaries (the largest cost) and tools/technology costs.
20	**AI Scrum ROI Calculator – Example and Template**	4	Provide a simple, complete ROI calculator for AI Scrum adoption.

TEMPLATE IMPLEMENTATION NOTES

Prioritization for Initial Adoption

If teams can only implement a subset initially, prioritize in this order:

1. **Validation Canvas Template** (Template #3) - Core to the entire framework
2. **Outcome Backlog Template** (Template #1) - Shifts mindset from features to outcomes
3. **Sprint Inception Agenda** (Template #2) - Establishes new ceremony structure
4. **Context Specification Template** (Template #4) - Improves AI collaboration quality
5. **AI Collaboration Log** (Template #5) - Builds institutional learning
6. **Center of Excellence Template** (Template #16) – Reduces waste

Customization Guidance

- All templates should be customized to organizational context
- Start with the provided structure, then adapt based on team feedback
- Digital formats (Confluence, Notion, Miro) recommended over static documents
- Templates should evolve through retrospective insights

Integration with Existing Tools

- Templates can be implemented in existing tools (Azure DevOps, Jira, etc.)
- Consider creating custom work item types or dashboard views
- Automation opportunities exist for logging and metrics collection
- Version control recommended for context specifications and prompt libraries

OUTCOME BACKLOG

Purpose
This template shifts teams from feature-centric planning to outcome-oriented delivery. Instead of specifying what to build, define what business result to achieve and let AI-augmented teams determine optimal implementations.

How to Use This Template

1. Complete one outcome card for each desired business result
2. Prioritize outcomes based on business value and strategic alignment
3. Refine outcomes during backlog grooming before Sprint Inception
4. Validate that success metrics are measurable and time-bound
5. Review dependencies and constraints before sprint commitment

---

OUTCOME BACKLOG ITEM
Outcome ID: [AUTO-GENERATED OR MANUAL ID]
Status:
☐ Draft
☐ Refined
☐ Ready
☐ In Progress
☐ Validated
☐ Delivered

OUTCOME STATEMENT
What business result are we trying to achieve?
[Write a clear, concise statement of the desired outcome. Focus on the "why" and "what," not the "how."]

Example: Reduce customer support escalation rate from 15% to 10% within 60 days.

SUCCESS METRICS
How will we measure achievement of this outcome?

Metric Name	Current Baseline	Target Value	Measurement Method	Review Frequency
[Metric 1]	[Current state]	[Desired state]	[How to measure]	[Daily/Weekly/Sprint]
[Metric 2]				
[Metric 3]				

Primary Success Indicator: [Which single metric best indicates success?]

VALIDATION CRITERIA
What evidence proves this outcome was achieved?

- **Functional Validation:** [What must work correctly?]
- **Performance Validation:** [What benchmarks must be met?]
- **User Acceptance:** [How do we know users are satisfied?]
- **Business Impact:** [What business metrics must improve?]
- **Compliance/Security:** [What governance requirements must be satisfied?]

Definition of Done for Outcome: [Clear statement of what "complete" means for this outcome]

BUSINESS VALUE & PRIORITY
Business Value Score: [1-10, with 10 being highest value]
Strategic Alignment:

- Revenue Growth
- Cost Reduction
- Customer Satisfaction
- Risk Mitigation
- Competitive Advantage
- Regulatory Compliance
- Other: ___________

Priority Ranking: [P0 (Critical) | P1 (High) | P2 (Medium) | P3 (Low)]
Rationale for Priority: [Explain why this outcome matters now and why it's prioritized at this level]

CONSTRAINTS & BOUNDARIES

Must Have (Non-Negotiable):

1. [Requirement that cannot be compromised]
2.
3.

Must NOT Have (Explicit Exclusions):

1. [What should specifically NOT be included]
2.
3.

Technical Constraints:

- Platform/Technology: [Any required or prohibited technologies]
- Integration Points: [Systems that must be integrated]
- Performance Requirements: [Response time, throughput, etc.]
- Data Constraints: [Privacy, retention, access requirements]

Budget/Resource Constraints:

- Estimated Effort: [Story points or effort estimate]
- Required Skills: [Specialized expertise needed]
- External Dependencies: [Third-party vendors, APIs, etc.]
- Timeline Constraints: [Hard deadlines or milestones]

DEPENDENCIES

Upstream Dependencies (What must be completed first):

Dependency	Owner	Status	Risk Level	Mitigation Plan
[Item 1]	[Team /Person]	[Status]	[High/Med/Low]	[If blocked, what's plan B?]

Downstream Dependencies (What depends on this):

Dependent Outcome	Impact if Delayed	Owner
[Item 1]	[Consequence]	[Team]

Cross-Team Coordination Required: [List teams that must be involved and nature of coordination]

RISK ASSESSMENT

Risk Tier: ☐ Delegated Authority ☐ Standard Review ☐ Board Review ☐ Prohibited

Identified Risks:

Risk Description	Probability	Impact	Mitigation Strategy	Owner
[Risk 1]	[High/Med/Low]	[High/Med/Low]	[How to address]	[Person]

Governance Requirements:

- Legal Review Required
- Security Review Required
- Privacy/GDPR Impact Assessment
- Compliance Documentation
- Ethical AI Review
- Accessibility Standards

STAKEHOLDER INFORMATION

Outcome Owner: [Person accountable for business result]

Key Stakeholders:

Name	Role	Interest	Engagement Level	Contact
			[Daily/Weekly/As Needed]	

Decision Authority: [Who has final say on scope/acceptance?]

AI COLLABORATION NOTES

AI Suitability Assessment:

- High - AI can generate most implementation
- Medium - AI can assist with portions
- Low - Primarily human-driven work

Recommended AI Approach: [Context specifications, prompt patterns, or AI collaboration strategies that might be effective]

Known AI Limitations: [Areas where AI assistance may not be suitable or requires extra validation]

SOLUTION SPACE (Optional - Completed During Sprint Inception)

Potential Approaches (Generated with AI assistance):

1. [Approach A: Brief description, pros/cons]
2. [Approach B: Brief description, pros/cons]

3. [Approach C: Brief description, pros/cons]

Selected Approach: [Chosen after evaluation during Sprint Inception]
Rationale: [Why this approach best satisfies outcome criteria]

ACCEPTANCE THRESHOLD

Minimum Viable Outcome: [What is the minimum acceptable result that still delivers business value?]
Ideal Outcome: [What would constitute exceptional achievement?]
Acceptance Decision Criteria: [Under what conditions do we accept the outcome as complete vs. requiring additional work?]

TRACKING & HISTORY

Created By: [Name]
Date: [Date]
Last Modified By: [Name]
Date: [Date]
Sprint Assignment: [Sprint number/name when committed]

Change Log:

Date	Change	Reason	Modified By

Retrospective Notes: [After completion, what did we learn about pursuing this outcome?]

TEMPLATE USAGE TIPS

For Product Strategists:

- Focus on business outcomes, not technical solutions
- Ensure success metrics are measurable and time-bound
- Validate stakeholder alignment before marking "Ready"
- Use AI to generate potential approaches during refinement

For AI Alignment Engineers:

- Review constraints to identify prompt requirements
- Flag outcomes with unclear intent for refinement
- Suggest validation criteria based on similar past outcomes
- Prepare context specifications during backlog grooming

For Delivery Facilitators:

- Use dependency mapping to prevent sprint disruption
- Identify cross-team coordination needs early
- Track risk tier to plan appropriate governance touchpoints

- Ensure outcome statement clarity before sprint commitment

Common Pitfalls to Avoid:

- Writing outcomes as technical features ("Build REST API")
- Success metrics that can't be measured objectively
- Vague validation criteria that enable interpretation
- Missing critical constraints discovered mid-sprint
- Underestimating dependencies on external teams

Template Version: 1.0
Last Updated: [Date]

AI SPRINT INCEPTION AGENDA & CHECKLIST

Purpose

Sprint Inception replaces traditional Sprint Planning in AI Scrum. This 4-hour ceremony establishes validation frameworks, governance requirements, and AI collaboration protocols before any implementation begins. Front-loading this work prevents expensive mid-sprint rework.

SPRINT INCEPTION MEETING DETAILS

Sprint Number: _______
Sprint Duration: _______ days
Date: _________
Time: _________ to _________
Location/Video Link:_________________________

Required Attendees:

☐ Product Strategist
☐ AI Alignment Engineer
☐ Validation Lead
☐ Governance Coordinator
☐ Integration Architect
☐ AI-Augmented Developers (all)
☐ Delivery Facilitator

Optional Attendees:

☐ Key Stakeholders (for Outcome Alignment)
☐ Security Representative (for high-risk outcomes)
☐ Legal/Compliance (as needed)

PRE-MEETING PREPARATION (Complete 24 Hours Before)

Product Strategist:

☐ Refined Outcome Backlog items are marked "Ready"
☐ Success metrics defined for each proposed outcome
☐ Stakeholder alignment confirmed on priorities
☐ Business value justification documented

AI Alignment Engineer:

☐ Reviewed outcomes for AI suitability
☐ Identified ambiguous requirements needing clarification
☐ Prepared questions about intent and constraints

☐ Collected relevant context from previous sprints

Validation Lead:
☐ Reviewed similar past outcomes for validation patterns
☐ Identified required test environments and data
☐ Checked availability of security/compliance resources
☐ Prepared validation complexity estimates

Governance Coordinator:
☐ Completed preliminary risk tier assessment
☐ Identified regulatory requirements for proposed outcomes
☐ Confirmed governance board member availability (if needed)
☐ Reviewed compliance checklist for sprint

Delivery Facilitator:
☐ Sent calendar invites with pre-reading materials
☐ Prepared collaboration tools (Miro, virtual whiteboard)
☐ Reviewed previous sprint velocity and lessons learned
☐ Set up sprint tracking artifacts

PHASE 1: OUTCOME ALIGNMENT (60 MINUTES)
Objective: Establish shared understanding of business outcomes and prioritize sprint commitments.

Time Block 1: Outcome Review (30 min)
Facilitator: Product Strategist

Activities:

1. **Review Sprint Goal (5 min)**
   ☐ Present overarching sprint objective
   ☐ Explain how outcomes align to strategic priorities
   ☐ Clarify how success will be measured

2. **Outcome Presentations (20 min, ~5 min each)** For each proposed outcome:
   ☐ State outcome clearly ("Achieve X by improving Y")
   ☐ Explain business value and urgency
   ☐ Present success metrics and targets
   ☐ Share stakeholder context

   Outcome #1: ______________________________
   - Business Value Score: ___/10
   - Key Metric: ______________________________

- Questions/Clarifications: ____________________

Outcome #2: ______________________________

- Business Value Score: ___/10
- Key Metric: ______________________________
- Questions/Clarifications: ____________________

Outcome #3: ______________________________

- Business Value Score: ___/10
- Key Metric: ______________________________
- Questions/Clarifications: ____________________

3. **Alignment Check (5 min)**
   - ☐ Team consensus on outcome importance
   - ☐ All questions about business context answered
   - ☐ No fundamental disagreements on priorities

Red Flag Check:

- ☐ Any outcome lacks clear success metrics? → STOP, refine before proceeding
- ☐ Any outcome lacks stakeholder alignment? → STOP, confirm before proceeding
- ☐ Any outcome conflicts with another? → STOP, clarify priorities

---

Time Block 2: AI Suitability Assessment (15 min)
Facilitator: AI Alignment Engineer

Activities:

1. **Evaluate Each Outcome for AI Collaboration (10 min)**

Outcome	AI Suitability	AI Approach	Human Oversight Needs
#1	High/Med/Low	[Strategy]	[What requires human judgment]
#2			
#3			

2. **Identify Intent Ambiguities (5 min)**
   - ☐ Flag vague requirements that will cause AI generation problems
   - ☐ Note where additional context specifications are needed
   - ☐ Highlight outcomes requiring prompt engineering research

Output:

- Each outcome rated for AI suitability
- Known AI limitations documented
- Intent clarification needs identified

Time Block 3: Capacity-Based Prioritization (15 min)
Facilitator: Delivery Facilitator

Activities:

1. **Review Sprint Capacity (5 min)**
   - Team Availability: ___% (account for PTO, meetings, etc.)
   - Historical Velocity: ___ points
   - Adjusted Capacity: ___ points (accounting for unknowns)
2. **Rough Sizing (5 min)**
   - Outcome #1: ___ points (focus on validation complexity)
   - Outcome #2: ___ points
   - Outcome #3: ___ points
   - Total: ___ points
3. **Prioritize & Commit (5 min)**
   - ☐ Select outcomes that fit capacity
   - ☐ Identify stretch goals (attempt if capacity allows)
   - ☐ Park lower-priority outcomes for future sprints

Sprint Commitment:

Outcome	Priority	Points	Committed?
	P0/P1/P2		Yes/Stretch/No

PHASE 2: VALIDATION DEFINITION (60 MINUTES)

Objective: Define how we will know each outcome succeeded before generating any implementations.

Facilitator: Validation Lead
For Each Committed Outcome:
Outcome Name: ______________________________

Validation Canvas Construction (15 min per outcome)
1. Functional Requirements (What must work?)

- ☐ Core capability: ______________________________
- ☐ Edge cases: ______________________________
- ☐ Error handling: ______________________________
- ☐ User interactions: ______________________________

2. Performance Benchmarks (How fast/efficient?)

- ☐ Response time: < _____ ms/sec
- ☐ Throughput: _____ requests/sec
- ☐ Resource usage: < _____ CPU/memory
- ☐ Scalability: support _____ concurrent users

3. Security Requirements

- ☐ Authentication/authorization: __________________
- ☐ Data encryption: __________________
- ☐ Access controls: __________________
- ☐ Audit logging: __________________

4. Compliance Checkpoints

- ☐ GDPR/privacy requirements: __________________
- ☐ Industry regulations: __________________
- ☐ Internal policies: __________________
- ☐ Documentation requirements: __________________

5. Acceptance Thresholds

- **Minimum Viable:** [What's the floor for acceptable?]
- **Target State:** [What's the goal?]
- **Ideal State:** [What would be exceptional?]

6. Evidence Requirements

What proof do we need to show validation?

- ☐ Automated test results: __________________
- ☐ Performance benchmarks: __________________
- ☐ Security scan reports: __________________
- ☐ Stakeholder sign-off: __________________
- ☐ Compliance documentation: __________________

7. Validation Approach

- ☐ Unit tests (AI-generated + human-verified)
- ☐ Integration tests
- ☐ End-to-end tests
- ☐ Adversarial testing / red-teaming
- ☐ User acceptance testing
- ☐ Performance testing
- ☐ Security testing

Validation Complexity Score: ___/10 (1=simple, 10=complex)

Red Flag Check:

- ☐ Validation criteria vague or subjective? → REFINE NOW
- ☐ Missing security/compliance requirements? → ADD NOW
- ☐ No clear acceptance threshold? → DEFINE NOW

PHASE 3: GOVERNANCE FRAMING (45 MINUTES)

Objective: Classify risks and establish governance oversight before implementation.

Facilitator: Governance Coordinator

Governance Framework Application (10 min per outcome)

Outcome Name: ________________________________

Risk Tier Classification
Risk Assessment Factors:

Factor	Score (1-5)	Notes
Data Sensitivity (PII, financial, health)		
Regulatory Impact (industry regulations)		
User Exposure (# of users affected)		
System Criticality (importance to ops)		
AI Hallucination Potential		
Reversibility (can we undo easily?)		

Total Risk Score: ___/30

Risk Tier Assignment:

- ☐ **Delegated Authority** (Score 6-12): Team decides, document rationale
- ☐ **Standard Review** (Score 13-18): Governance Coordinator reviews, documents
- ☐ **Board Review** (Score 19-24): Governance Board approval required
- ☐ **Prohibited** (Score 25-30): Do not proceed without executive exception

Governance Requirements
For This Outcome:

- ☐ Legal review required: Yes/No | Due: _______
- ☐ Security assessment required: Yes/No | Due: _______
- ☐ Privacy impact assessment: Yes/No | Due: _______
- ☐ Compliance documentation: Yes/No | Due: _______
- ☐ Ethical AI review: Yes/No | Due: _______

Governance Checkpoint Scheduled:

- Date/Time: _____________ (typically Day 5-6 of sprint)
- Required Attendees: ________________________________
- Materials Due: 24 hours before checkpoint

Compliance Checklist
Regulatory Requirements:

- ☐ EU AI Act compliance (if applicable)

- ☐ GDPR/privacy regulations
- ☐ Industry-specific regulations: _______________
- ☐ Accessibility standards (WCAG, etc.)
- ☐ Data retention policies

Documentation Requirements:

- ☐ Decision log (why this approach chosen)
- ☐ Risk assessment record
- ☐ Validation evidence
- ☐ AI collaboration log entries
- ☐ Post-deployment monitoring plan

PHASE 4: CAPACITY & DEPENDENCY PLANNING (45 MINUTES)

Objective: Identify constraints and dependencies that could derail sprint execution.
Facilitator: Delivery Facilitator

Validation Capacity Assessment (20 min)
Question: How much can we validate this sprint?

Resource Availability

Resource Type	Available This Sprint	Constraint Level
Validation Lead capacity	___%	High/Med/Low
Security expertise	___%	High/Med/Low
Test environments	___% uptime	High/Med/Low
Stakeholder availability	___ hours	High/Med/Low
Compliance reviewers	___ hours	High/Med/Low

Skill Gaps

- ☐ Specialized skills needed: _________________________
- ☐ Team members with skills: _________________________
- ☐ Training/pairing plan: ___________________________

External Dependencies (15 min)
Critical Path Analysis:

Dependency	Owner	Status	Needed By	Mitigation Plan
API access from Team X		Green/Yellow/Red	Sprint Day ___	[Plan B if blocked]
Test data from Team Y				
Security review				
Stakeholder approval				

Coordination Needs:

- ☐ Daily sync with: ____________________________________
- ☐ Weekly sync with: ____________________________________
- ☐ Integration points requiring coordination: __________________

Environment & Infrastructure (10 min)
Required Infrastructure:

- ☐ Development environments: Ready/Pending
- ☐ Test environments: Ready/Pending
- ☐ Staging environments: Ready/Pending
- ☐ AI tool access: Ready/Pending
- ☐ Data pipelines: Ready/Pending

Infrastructure Risks:

- ☐ Known instability: ____________________________________
- ☐ Planned maintenance: ____________________________________
- ☐ Access restrictions: ____________________________________

PHASE 5: AI COLLABORATION PROTOCOL REVIEW (30 MINUTES)

Objective: Establish how the team will work with AI during this sprint.
Facilitator: AI Alignment Engineer

Prompt Library Review (10 min)
Existing Prompts from Previous Sprints:

- ☐ Reviewed prompt library for relevant patterns
- ☐ Identified prompts to reuse
- ☐ Flagged prompts needing updates
- ☐ Noted prompts with high success rates

New Prompts Needed:

- ☐ Outcome #1 requires: ____________________
- ☐ Outcome #2 requires: ____________________
- ☐ Research/experimentation needed: ____________

Context Specification Planning (10 min)
For Each Outcome:

Outcome	Context Spec Required?	Complexity	Owner	Due Before
#1	Yes/No	High/Med/Low	[Person]	Day ___
#2				

Context Elements to Prepare:

- ☐ Architectural constraints documentation
- ☐ Business rules reference
- ☐ Edge cases list
- ☐ Integration requirements
- ☐ Code standards/patterns

AI Collaboration Logging (5 min)
Logging Protocol This Sprint:

- ☐ AI Collaboration Log location: ____________
- ☐ Update frequency: Daily/Per-task
- ☐ Required fields: ____________________
- ☐ Log review schedule: ____________________

What to Capture:

- ☐ Prompts used (successful and unsuccessful)
- ☐ Time saved vs. traditional approach
- ☐ Validation issues discovered in AI outputs
- ☐ Lessons learned about effective collaboration

AI Show & Tell Scheduling (5 min)
Mid-Sprint Stakeholder Sessions:
Session 1 (Day 3):

- Date/Time: ____________
- Stakeholders invited: ____________________
- Outcomes to demo: ____________________
- Preparation owner: ____________

Session 2 (Day 7):

- Date/Time: ____________

- Stakeholders invited: ________________________________
- Outcomes to demo: ________________________________
- Preparation owner: ________________

SPRINT COMMITMENT SUMMARY

Sprint Goal

[One clear statement of what this sprint aims to achieve]

Committed Outcomes

1. **Outcome:** ________________________________
    - **Success Metric:** ________________________________
    - **Validation Complexity:** ___/10
    - **Risk Tier:** ________________
    - **Points:** ___
2. **Outcome:** ________________________________
    - **Success Metric:** ________________________________
    - **Validation Complexity:** ___/10
    - **Risk Tier:** ________________
    - **Points:** ___

Total Committed Points: ___

Stretch Goals (Attempt if Capacity Allows)

- ☐ ________________________________

Governance Milestones

Milestone	Date	Owner	Status
Governance Checkpoint	Sprint Day ___	Governance Coordinator	Scheduled
Legal Review (if needed)			
Security Assessment (if needed)			

Key Risks & Mitigation

Risk	Probability	Impact	Mitigation Owner
	High/Med/Low	High/Med/Low	

Dependencies Requiring Tracking

- ☐ ________________________________
- ☐ ________________________________

POST-INCEPTION ACTIONS

Delivery Facilitator:

- ☐ Publish Sprint Commitment Summary to team/stakeholders
- ☐ Create sprint tracking artifacts (board, burndown, etc.)
- ☐ Schedule AI Show & Tell sessions
- ☐ Set up dependency tracking notifications

AI Alignment Engineer:

- ☐ Create context specifications for outcomes
- ☐ Prepare prompt templates
- ☐ Set up AI collaboration logging

Validation Lead:

- ☐ Set up test environments
- ☐ Prepare validation canvas artifacts
- ☐ Schedule validation checkpoints

Governance Coordinator:

- ☐ Notify governance board (if Board Review needed)
- ☐ Prepare compliance documentation templates
- ☐ Schedule Governance Checkpoint

Product Strategist:

- ☐ Communicate sprint commitments to stakeholders
- ☐ Confirm AI Show & Tell availability
- ☐ Prepare acceptance criteria clarifications

MEETING EFFECTIVENESS CHECKLIST

At the End of Sprint Inception, Verify:

- ☐ Every committed outcome has clear success metrics
- ☐ Validation criteria defined before implementation
- ☐ Risk tiers assigned and governance plan established
- ☐ Dependencies identified and mitigation plans created
- ☐ Team capacity realistic (not over-committed)
- ☐ AI collaboration protocols established
- ☐ All team members understand their responsibilities
- ☐ Sprint commitment published and visible

Warning Signs of Poor Sprint Inception:

- Team jumps to solution discussion before defining validation
- Outcomes lack measurable success criteria
- Governance requirements discovered mid-sprint
- Dependencies identified during execution, not planning
- Team over-commits beyond realistic capacity
- AI collaboration is ad hoc, not systematic

FACILITATION TIPS

For Delivery Facilitators:

- Time-box ruthlessly; each phase has a purpose
- Use parking lot for off-topic discussions
- Capture action items and owners in real-time
- Insist on specificity; reject vague commitments
- End with explicit sprint commitments

For Product Strategists:

- Come prepared with refined outcomes
- Be ready to defend business value
- Accept that not everything fits in one sprint
- Prioritize ruthlessly based on value

For AI Alignment Engineers:

- Challenge vague requirements early
- Surface AI limitations upfront
- Share relevant lessons from previous sprints
- Prevent "AI will figure it out" assumptions

For Validation Leads:

- Define criteria before seeing any implementation
- Estimate validation complexity honestly
- Flag resource constraints immediately
- Refuse to accept subjective validation criteria

Template Version: 1.0
Last Updated: [Date]

VALIDATION CANVAS

Purpose
The Validation Canvas defines success criteria before any implementation begins. This prevents the expensive waste that AI's speed can amplify by ensuring teams validate the right things in the right way.

VALIDATION CANVAS OVERVIEW
Outcome Name: ______________________________
Outcome ID: ________
Sprint: ________
Canvas Owner: _______________ (Validation Lead)
Created: ________
Last Updated: ________

Validation Philosophy:
Define success criteria first. Generate implementations second. Validate against predetermined standards third.

SECTION 1: FUNCTIONAL VALIDATION
What Must Work Correctly?

Core Functionality Requirements
Primary Capability: [What is the main function this outcome must perform?]
Success Looks Like:

- Capability 1: ______________________________
    - Validation method: ____________________
    - Pass criteria: ________________________
- Capability 2: ______________________________
    - Validation method: ____________________
    - Pass criteria: ________________________
- Capability 3: ______________________________
    - Validation method: ____________________
    - Pass criteria: ________________________

User Interaction Flows
Critical User Paths:

1. **Path:** [Describe user journey]
    - **Expected Behavior:** __________________
    - **Test Approach:** ______________________

   - **Pass/Fail Criteria:** ____________________
2. **Path:** [Describe user journey]
   - **Expected Behavior:** ____________________
   - **Test Approach:** ____________________
   - **Pass/Fail Criteria:** ____________________

Edge Cases & Error Handling
What happens when things go wrong?

Edge Case	Expected Behavior	Validation Method	Priority
[Scenario 1]	[How should system respond?]	[How to test?]	P0/P1/P2
Invalid input			
Missing data			
Network failure			
Timeout			
Concurrent access			

Integration Points
Where does this outcome connect with other systems?

Integration Point	Expected Interaction	Validation Approach	Owner
[System/API 1]	[What data flows?]	[How to verify?]	[Team]

SECTION 2: PERFORMANCE BENCHMARKS
How Fast and Efficient Must It Be?

Response Time Requirements

Operation	Target	Maximum Acceptable	Measurement Method
Page load	< ___ ms	< ___ ms	[Tool/approach]
API response	< ___ ms	< ___ ms	[Tool/approach]
Query execution	< ___ ms	< ___ ms	[Tool/approach]
Batch processing	< ___ min	< ___ min	[Tool/approach]

Throughput Requirements

- **Concurrent Users:** Must support _____ users simultaneously
- **Requests Per Second:** Must handle _____ req/sec at peak
- **Data Volume:** Must process _____ records/transactions per _____
- **Peak Load Capacity:** Must maintain performance at _____% above normal

Resource Utilization

- **CPU Usage:** Must remain below _____% under normal load
- **Memory Usage:** Must not exceed _____ GB/MB
- **Storage:** Must not exceed _____ GB growth per _____
- **Network Bandwidth:** Must not exceed _____ Mbps

Scalability Validation

- Tested at 1x normal load: Pass/Fail
- Tested at 2x normal load: Pass/Fail
- Tested at 5x normal load: Pass/Fail
- Degradation is graceful (not catastrophic): Pass/Fail

Performance Testing Tools:

- Load testing tool: ______________________________
- Monitoring/observability: ______________________________
- Baseline measurements captured: Yes/No

SECTION 3: SECURITY VALIDATION

How Do We Keep This Safe?

Authentication & Authorization

- **Authentication Required:** Yes/No
    - Method: ______________________________
    - Validation: ______________________________
- **Authorization Rules:**

User Role	Allowed Actions	Forbidden Actions	Test Method
[Role 1]			
[Role 2]			

Data Protection

- **Data in Transit:**
    - Encryption: TLS _____
    - Validation: ______________________________
- **Data at Rest:**
    - Encryption: ______________________________
    - Validation: ______________________________

- **Sensitive Data Handling:**
  - PII present: Yes/No
  - Masking/tokenization: ____________________________
  - Access controls: __________________________________

Security Testing Checklist

- **Input Validation:** Tested against injection attacks
- **Access Controls:** Verified role-based permissions
- **Secrets Management:** No hardcoded credentials
- **Encryption:** All sensitive data encrypted
- **Audit Logging:** Security events logged
- **Error Messages:** No sensitive data leakage
- **Session Management:** Proper timeout and invalidation
- **OWASP Top 10:** Tested against current vulnerabilities

Security Scan Results Required:

- Static Analysis (SAST): ______________________________
- Dynamic Analysis (DAST): ____________________________
- Dependency Scanning: ______________________________
- Secrets Detection: ___________________________________

Acceptable Vulnerability Threshold:

- Critical: 0 allowed
- High: ___ allowed (with mitigation plan)
- Medium: ___ allowed
- Low: ___ allowed

SECTION 4: COMPLIANCE & GOVERNANCE

What Regulatory Requirements Must Be Met?

Privacy & Data Protection

GDPR/Privacy Requirements (if applicable):

- Lawful basis for processing: ______________________________
- Data minimization: Only collecting necessary data
- Purpose limitation: Using data only for stated purpose
- Retention policy: Data deleted after _____ days/years
- Right to erasure: Deletion mechanism implemented
- Data portability: Export functionality available
- Privacy notice: Users informed of data usage

Data Classification:

Data Element	Classification	Handling Requirements	Validated
[Field 1]	Public / Internal / Confidential / Restricted		☐

Industry-Specific Regulations
Applicable Regulations:

- HIPAA (Healthcare): ______________________________
- PCI DSS (Payments): ______________________________
- SOX (Financial): ______________________________
- FERPA (Education): ______________________________
- Other: ______________________________

Compliance Validation:

Requirement	Implementation	Evidence	Validated
[Regulation requirement 1]	[How we comply]	[Proof of compliance]	☐

Accessibility Standards

- **WCAG 2.1 Level:** A / AA / AAA
- **Screen reader compatible:** Tested with ________________
- **Keyboard navigation:** All functions accessible via keyboard
- **Color contrast:** Meets minimum ratios
- **Alternative text:** Images have descriptive alt text

SECTION 5: AI-SPECIFIC VALIDATION
How Do We Validate Probabilistic Outputs?
AI Behavior Verification
AI Component: [Name/description of AI functionality]
Deterministic vs. Probabilistic:

- Deterministic (same input = same output)
- Probabilistic (same input = variable output)

If Probabilistic:

- **Acceptable Variation Range:** ________________________
- **Consistency Testing:** Generate _____ times, _____% must match criteria
- **Drift Monitoring:** Plan for detecting behavior changes over time

Hallucination Prevention

AI Hallucination Risks:

Risk Area	Prevention Strategy	Validation Method	Acceptable Threshold
Factual accuracy	[How we prevent]	[How we test]	_____% accuracy
Code correctness			
Data fabrication			
Logic errors			

Human Verification Points:

- AI output 1: ___________________ → Human reviews: ___________________
- AI output 2: ___________________ → Human reviews: ___________________

Model Performance Metrics

If using ML models:

- **Accuracy Target:** _____%
- **Precision Target:** _____%
- **Recall Target:** _____%
- **F1 Score Target:** _____%
- **False Positive Rate:** < _____%
- **False Negative Rate:** < _____%

Baseline Comparison:

- Current baseline: ___________________________________
- Target improvement: _______________________________
- Measurement method: _____________________________

Adversarial Testing

Red Team Testing Plan:

- Prompt injection attempts: ___________________________
- Boundary condition tests: ____________________________
- Adversarial inputs: ___________________________________
- Unexpected context: __________________________________

Robustness Criteria:

- Graceful degradation under adversarial conditions
- No unauthorized data access
- No catastrophic failures
- Error messages are appropriate

SECTION 6: ACCEPTANCE THRESHOLDS

What Level of Quality Is "Good Enough"?
Three-Tier Acceptance Framework

MINIMUM VIABLE (Floor): [Below this threshold, outcome is not acceptable]
Functional:

- Must have: ________________________________
- Performance: ________________________________
- Security: ________________________________

TARGET STATE (Goal): [This is what we aim to achieve]
Functional:

- Should have: ________________________________
- Performance: ________________________________
- Security: ________________________________

IDEAL STATE (Exceptional): [Exceeding expectations]
Functional:

- Could have: ________________________________
- Performance: ________________________________
- Security: ________________________________

Go/No-Go Decision Criteria
Outcome is ACCEPTABLE if:

- All Minimum Viable criteria met
- All P0 validation tests pass
- No critical security vulnerabilities
- No compliance blockers
- Performance within acceptable range
- Stakeholder sign-off obtained

Outcome REQUIRES ADDITIONAL WORK if:

- Missing Minimum Viable criteria
- P0 tests failing
- Critical security issues
- Compliance violations
- Performance below acceptable threshold
- Stakeholders request significant changes

Outcome is REJECTED if:

- Fundamentally wrong approach
- Cannot meet minimum security standards
- Compliance impossible with current design
- Performance issues unfixable without redesign

SECTION 7: EVIDENCE REQUIREMENTS

What Proof Must We Provide?
Automated Test Results

Required Test Coverage:

- **Unit Tests:** _____% code coverage minimum
- **Integration Tests:** All integration points covered
- **End-to-End Tests:** All critical user paths covered
- **Performance Tests:** All benchmark scenarios covered

Test Execution Evidence:

- Test suite name: ______________________________
- Pass rate required: _____%
- Evidence location: ______________________________
- Automated in CI/CD: Yes/No

Manual Verification Evidence
What Requires Human Review:

Item	Reviewer Role	Evidence Artifact	Status
[UX/UI quality]	[Stakeholder/Designer]	[Screenshots/recording]	☐
[Business logic]	[Subject matter expert]	[Walkthrough notes]	☐
[Security]	[Security team]	[Security assessment]	☐

Documentation Requirements
Required Documentation:

- Test results report: ______________________________
- Performance benchmark report: ______________________
- Security scan results: ______________________________
- Compliance checklist: ______________________________
- Stakeholder sign-off: ______________________________
- AI collaboration log entries: _________________________
- Known limitations documented: ______________________

Stakeholder Acceptance
Acceptance Sign-Off Required From:

Stakeholder	Role	Acceptance Criteria	Sign-Off	Date
[Name]	[Title]	[What they validate]	☐	

Demonstration Required:

- Live demo to stakeholders: Date _________
- Recorded demo acceptable: Yes/No
- Demo scenarios: ______________________________

SECTION 8: VALIDATION EXECUTION PLAN

How and When Will We Validate?

Validation Timeline

Sprint Day 1-2: Setup

- Test environments ready
- Test data prepared
- Validation tools configured
- Team members assigned

Sprint Day 3: Initial Validation (AI Show & Tell)

- Functional validation of early prototype
- Stakeholder feedback on direction
- Course correction if needed

Sprint Day 4-6: Continuous Validation

- Daily automated test execution
- Integration testing
- Performance benchmarking
- Security scanning

Sprint Day 5-6: Governance Checkpoint

- Compliance verification
- Risk assessment review
- Documentation review

Sprint Day 7: Mid-Sprint Validation (AI Show & Tell)

- Comprehensive validation review
- Stakeholder acceptance testing
- Identify any gaps

Sprint Day 8-9: Final Validation

- All validation criteria verified
- Evidence collected
- Acceptance thresholds confirmed

Sprint Day 10: Sprint Review

- Demonstrate validated outcome
- Present evidence
- Obtain final sign-off

Validation Resources

People:

Role	Time Allocation	Availability Confirmed
Validation Lead	____%	☐
AI-Augmented Developers	____%	☐
Security Specialist	____ hours	☐
Stakeholders	____ hours	☐

Tools & Environments:

- Test environment: ________________________________
- Performance testing tool: ____________________________
- Security scanning tool: ______________________________
- Test data: _________________________________

Validation Risks

Risk	Mitigation Plan	Owner
Test environment unstable	[Backup plan]	[Person]
Stakeholder unavailable	[Alternative reviewer]	[Person]
Test data insufficient	[Data generation plan]	[Person]
Performance tools fail	[Manual testing approach]	[Person]

SECTION 9: VALIDATION OUTCOMES

After Validation, What Did We Learn?

[Complete during/after sprint execution]

Validation Results Summary

Overall Status: Pass / Pass with Conditions / Fail

Validation Metrics:

Validation Area	Target	Actual	Status
Functional tests pass rate	____%	____%	✓/✗
Performance benchmarks			
Security scan			
Compliance checks			
Stakeholder acceptance			

Gaps Identified

What didn't meet acceptance criteria?

Gap	Severity	Action Required	Owner	Due Date
[Issue 1]	[Critical / High / Med / Low]	[Fix / Document / Defer]		

Lessons Learned

What worked well in validation?

- ______________________________
- ______________________________

What validation approach should we improve?

- ______________________________
- ______________________________

What would we do differently next time?

- ______________________________
- ______________________________

Validation Efficiency Metrics

Time Spent:

- Setup/preparation: _____ hours
- Automated testing: _____ hours
- Manual testing: _____ hours
- Stakeholder review: _____ hours
- **Total validation effort:** _____ hours

Defects Found:

- During development: _____
- During validation: _____
- After deployment: _____ (track in future)

Validation ROI:

- Issues prevented: _____
- Estimated cost of issues if deployed: $_____
- Validation cost: $_____

CANVAS EFFECTIVENESS CHECKLIST

A Good Validation Canvas Has:

- Specific, measurable pass/fail criteria (not subjective)
- Acceptance thresholds defined at three levels (minimum/target/ideal)
- AI-specific validation for probabilistic outputs
- Security and compliance requirements explicit
- Evidence requirements clearly stated
- Validation timeline aligned with sprint ceremonies

- Resources identified and confirmed available

Warning Signs of Poor Validation Canvas:

- Vague criteria ("should work well")
- No acceptance thresholds (no definition of "good enough")
- Missing AI-specific validation
- Security/compliance as afterthought
- No evidence requirements
- Validation deferred to end of sprint
- Resource availability assumed, not confirmed

USAGE GUIDELINES

When to Create:

- During Sprint Inception, as part of Validation Definition phase
- Before any implementation begins
- One canvas per committed outcome

Who Creates:

- Primary: Validation Lead
- Contributors: AI Alignment Engineer, Product Strategist, Governance Coordinator
- Reviewers: Entire team

How to Use:

- Reference during implementation to ensure alignment
- Update as validation proceeds
- Use in Sprint Review to demonstrate outcome achievement
- Review in Retrospective to improve validation approach

Integration with Other Artifacts:

- Links to: Outcome Backlog Item
- Feeds into: AI Collaboration Log, Sprint Review
- Referenced by: Governance Checkpoint, Test Automation

Template Version: 1.0
Last Updated: [Date]

CONTEXT SPECIFICATION

Purpose
Context Specifications provide AI with complete, precise instructions for generation tasks. This template ensures AI receives sufficient context to produce high-quality outputs that require minimal human correction.

CONTEXT SPECIFICATION METADATA
Context Spec ID: ________
Created By: ________________
Date: ________
Related Outcome: ________________________________
Sprint: ________
Task/Story ID: ________
AI Tool/Model:
☐ GPT-x ☐ Claude ☐ Copilot ☐ Gemini ☐ Other: ________

SECTION 1: TASK OBJECTIVE
What Are We Asking AI to Do?
Primary Objective: [One clear sentence describing what you want AI to create/generate]
Example: Generate a REST API endpoint that retrieves customer order history with pagination support.

Success Looks Like: [Describe the ideal output in concrete terms]
Example: A working API endpoint that returns JSON-formatted order data, handles errors gracefully, includes unit tests, and follows our team's code standards.

Scope Boundaries: Include:

- ☐ ________________________________
- ☐ ________________________________
- ☐ ________________________________

Exclude (explicitly OUT of scope):

- ☐ ________________________________
- ☐ ________________________________
- ☐ ________________________________

SECTION 2: ARCHITECTURAL CONSTRAINTS

What Technical Guardrails Must AI Respect?

Technology Stack
Required Technologies:

- **Language/Framework:** ____________________
- **Version:** _________________________________
- **Libraries/Packages:** _______________________
- **Prohibited Libraries:** ______________________

Example:

```
- Language: Python 3.11+
- Framework: FastAPI 0.104+
- Database: PostgreSQL 15
- ORM: SQLAlchemy 2.0
- DO NOT USE: Django, Flask (we're
standardized on FastAPI)
```

Architectural Patterns
Design Patterns to Follow:

- ☐ Repository pattern
- ☐ Service layer pattern
- ☐ Dependency injection
- ☐ Factory pattern
- ☐ Other: _________________________________

Patterns to Avoid:

- ☐ Singleton (causes testing issues)
- ☐ God objects
- ☐ Other: _________________________________

Code Structure Requirements
File Organization:

```
[Describe expected file structure]

Example:
/src
  /api
    /endpoints
      orders.py  ← Generate this
  /services
    order_service.py  ← Generate this
  /models
    order.py  ← Use existing
```

```
    /tests
      test_orders.py  ← Generate this
```

Naming Conventions:

- **Classes:** ______________________________
- **Functions:** ______________________________
- **Variables:** ______________________________
- **Constants:** ______________________________

Example:

```
- Classes: PascalCase (OrderService)
- Functions: snake_case (get_customer_orders)
- Variables: snake_case (order_list)
- Constants: SCREAMING_SNAKE_CASE
(MAX_PAGE_SIZE)
```

Integration Points
Existing Systems to Integrate:

System	Integration Method	Authentication	Data Format
[System name]	REST API/GraphQL/DB	API Key/OAuth	JSON/XML

APIs/Services to Call:

```
[Provide example API calls, authentication
methods, expected responses]

Example:
GET /api/v1/customers/{id}/orders
Authorization: Bearer {token}
Response: { "orders": [...], "total": 42,
"page": 1 }
```

SECTION 3: BUSINESS RULES & LOGIC
What Business Logic Must AI Implement?

Core Business Rules
Rule 1:

- **Description:** ______________________________
- **Implementation:** ______________________________
- **Example:** ______________________________

Example:

```
Rule: Pagination must never exceed 100 items
per page
Implementation: If requested page_size > 100,
return error 400
Example: GET /orders?page_size=150 → returns
{"error": "page_size must be ≤ 100"}
```

Rule 2:

- **Description:** ______________________________
- **Implementation:** ______________________________
- **Example:** ______________________________

Rule 3:

- **Description:** ______________________________
- **Implementation:** ______________________________
- **Example:** ______________________________

Validation Rules

Input Validation:

Field	Required	Type	Constraints	Error Message
[field_name]	Yes/No	String/Int/etc	[Rules]	[Message]

Example:

```
| customer_id | Yes | UUID | Must be valid
UUID v4 | "Invalid customer_id format" |
| page | No | Integer | 1-1000, default=1 |
"Page must be between 1-1000" |
| page_size | No | Integer | 1-100, default=20
| "Page size must be 1-100" |
```

Calculation Logic

If calculations are involved:

```
[Provide formulas, examples, edge cases]

Example:
Total Price = (Item Price × Quantity) + Tax +
Shipping
Tax = Subtotal × Tax Rate (varies by state)
Shipping = Free if Subtotal > $50, else $5.99
```

State Management

Valid States:

```
[Define state machine if applicable]
Example:
Order States: PENDING → CONFIRMED →
PROCESSING → SHIPPED → DELIVERED
               ↓
          CANCELLED (allowed from PENDING,
CONFIRMED, PROCESSING)
```

SECTION 4: DATA CONTRACTS & SCHEMAS

What Data Structures Must AI Use?

Input Schema

Request Format:

```
json
[Provide exact JSON/YAML structure]

Example:
{
  "customer_id": "uuid-string",
  "page": 1,
  "page_size": 20,
  "start_date": "2024-01-01",  // optional,
ISO 8601
  "end_date": "2024-12-31",    // optional,
ISO 8601
  "status": "DELIVERED"         // optional,
filter by status
}
```

Output Schema

Response Format:

```
json
[Provide exact JSON structure]

Example:
{
  "orders": [
    {
      "order_id": "uuid-string",
      "order_date": "2024-01-15T10:30:00Z",
      "status": "DELIVERED",
```

```
        "total": 125.99,
        "items": [...]
      }
    ],
    "pagination": {
      "page": 1,
      "page_size": 20,
      "total_pages": 5,
      "total_items": 87
    }
  }
```

Database Schema
Relevant Tables/Collections:

```
sql
[Provide schema for tables AI needs to query]

Example:
CREATE TABLE orders (
  order_id UUID PRIMARY KEY,
  customer_id UUID NOT NULL,
  order_date TIMESTAMP NOT NULL,
  status VARCHAR(20) NOT NULL,
  total_amount DECIMAL(10,2) NOT NULL,
  FOREIGN KEY (customer_id) REFERENCES
customers(customer_id)
);

CREATE INDEX idx_orders_customer ON
orders(customer_id);
CREATE INDEX idx_orders_date ON
orders(order_date);
```

Error Response Format
Standard Error Structure:

```
json
{
  "error": {
    "code": "ERROR_CODE",
    "message": "Human-readable error message",
    "details": {
      "field": "field_name",
      "value": "invalid_value"
```

```
        }
    }
}
```

Common Error Codes:

- INVALID_INPUT - Validation failed
- NOT_FOUND - Resource doesn't exist
- UNAUTHORIZED - Authentication failed
- FORBIDDEN - Insufficient permissions
- INTERNAL_ERROR - Server-side error

SECTION 5: EDGE CASES & ERROR HANDLING

What Could Go Wrong?

Edge Cases to Handle

Edge Case 1: [Describe scenario]

- **Expected Behavior:** ______________________________
- **Example:** ______________________________
- **Test Case:** ______________________________

Example:

```
Edge Case: Customer has no orders
Expected: Return empty array with 200 status
(not 404)
Example: GET /customers/abc-123/orders →
{"orders": [], "pagination": {...}}
Test: Verify new customer with zero orders
returns empty list
```

Edge Case 2: ______________________________

Edge Case 3: ______________________________

Edge Case 4: ______________________________

Error Scenarios

What happens when:

Error Scenario	HTTP Status	Response	Handling Logic
Invalid customer_id format	400	Error message	Validate before query
Customer doesn't exist	404	Not found	Check existence first
Database unavailable	503	Service unavailable	Retry logic + circuit breaker
Timeout	504	Gateway timeout	Set reasonable timeout
Unauthorized access	401	Unauthorized	Validate token/session

Performance Edge Cases

Large Data Sets:

- What if customer has 10,000 orders? → ____________________
- What if page_size=100 with complex queries? → _____________

Concurrent Requests:

- What if 1000 users hit endpoint simultaneously? → __________
- Race conditions to consider: ______________________________

SECTION 6: QUALITY & STANDARDS

What Code Quality Standards Apply?

Code Style Guidelines

Formatting:

- ☐ Follow [Black/Prettier/ESLint] configuration
- ☐ Line length: ___ characters maximum
- ☐ Indentation: ___ spaces (not tabs)
- ☐ Comments: Required for complex logic, not obvious code

Documentation:

- ☐ Docstrings required for all public functions
- ☐ Type hints required for all function parameters and returns
- ☐ Inline comments for non-obvious logic
- ☐ README section explaining usage

Example:

```
python
def get_customer_orders(
    customer_id: UUID,
    page: int = 1,
    page_size: int = 20,
    start_date: Optional[datetime] = None
) -> OrderResponse:
    """
    Retrieve paginated order history for a
customer.

    Args:
        customer_id: UUID of the customer
        page: Page number (1-indexed)
        page_size: Items per page (1-100)
        start_date: Optional filter for orders
after this date

    Returns:
        OrderResponse containing orders and
pagination metadata

    Raises:
        ValidationError: If inputs are invalid
        NotFoundError: If customer doesn't
exist
    """
```

Testing Requirements

Unit Tests:

- Coverage target: ____%
- Test file naming: test_*.py
- Test function naming: test_[functionality]_[scenario]
- Fixtures location: conftest.py

Test Scenarios to Cover:

- Happy path (success case)
- All error cases from error handling section
- All edge cases listed above
- Boundary conditions (page=1, page=max, page_size=1, page_size=100)
- Performance (can handle 100 items efficiently)

Example Test Structure:

```
python
def test_get_customer_orders_success():
    """Should return orders when customer
exists"""
    # Arrange
    customer_id = create_test_customer()
    create_test_orders(customer_id, count=5)

    # Act
    response =
get_customer_orders(customer_id)

    # Assert
    assert response.status_code == 200
    assert len(response.orders) == 5
```

Security Requirements

- ☐ No hardcoded credentials
- ☐ Input sanitization against SQL injection
- ☐ Output escaping to prevent XSS
- ☐ Rate limiting on endpoint: ___ requests per minute
- ☐ Authentication: Required/Not Required
- ☐ Authorization: Check user has access to customer data

Performance Requirements

- ☐ Response time: < ___ ms (95th percentile)
- ☐ Database queries: Maximum ___ queries per request
- ☐ Query optimization: Use indexes on [fields]
- ☐ Caching: Cache results for ___ seconds (if applicable)

SECTION 7: EXAMPLES & REFERENCES

Concrete Examples to Guide AI

Working Example

Input:

```
[Provide actual example request]

GET /api/v1/customers/550e8400-e29b-41d4-a716-
446655440000/orders?page=1&page_size=10
Authorization: Bearer
eyJhbGciOiJIUzI1NiIsInR5cCI6IkpXVCJ9...
```

Expected Output:

```
json
[Provide actual example response]

{
  "orders": [
    {
      "order_id": "660e8400-e29b-41d4-a716-
446655440001",
      "order_date": "2024-01-15T14:30:00Z",
      "status": "DELIVERED",
      "total": 125.99,
      "items_count": 3
    }
  ],
  "pagination": {
    "page": 1,
    "page_size": 10,
    "total_pages": 1,
    "total_items": 1
  }
}
```

Similar Code Reference
Existing Code to Use as Pattern:

```
[Point AI to similar existing code it can use
as reference]

Example:
See /src/api/endpoints/customers.py for
similar endpoint pattern
See /src/services/product_service.py for
pagination implementation
Follow error handling pattern from
/src/utils/errors.py
```

Anti-Patterns to Avoid
Do NOT do this:

```
python
[Show what NOT to do]

# DON'T: Query all orders and paginate in
memory
all_orders = db.query(Order).all()
return all_orders[start:end]  # Memory
intensive!

# DO: Let database handle pagination
query =
db.query(Order).limit(page_size).offset(offset
)
```

SECTION 8: WHAT NOT TO DO
Explicit Prohibitions

Code Patterns to Avoid:

- ☐ No global variables
- ☐ No circular imports
- ☐ No print() statements (use proper logging)
- ☐ No hardcoded configuration values
- ☐ No bare except: clauses (catch specific exceptions)

Technologies to Avoid:

- ☐ Do not use [specific library/tool]: ________________
- ☐ Do not implement [specific pattern]: ________________

Assumptions Not to Make:

- ☐ Don't assume inputs are valid (validate everything)
- ☐ Don't assume database is always available (handle errors)
- ☐ Don't assume unlimited resources (implement pagination)
- ☐ Don't assume single-threaded execution (handle concurrency)

SECTION 9: VALIDATION CHECKLIST
How Will We Verify AI Output?

Code Review Checklist:

- ☐ Implements all business rules from Section 3
- ☐ Handles all edge cases from Section 5
- ☐ Follows code style from Section 6

- ☐ Includes detailed tests
- ☐ No security vulnerabilities
- ☐ Meets performance requirements
- ☐ Documentation is complete and accurate

Functional Testing:

- ☐ Happy path works as expected
- ☐ Error cases handled correctly
- ☐ Edge cases produce correct behavior
- ☐ Integration with existing systems works
- ☐ Performance benchmarks met

Human Review Required For:

- ☐ Business logic correctness
- ☐ Security implications
- ☐ Architectural alignment
- ☐ User experience considerations
- ☐ Regulatory compliance (if applicable)

SECTION 10: AI GENERATION NOTES

Instructions Specific to AI Tool

Generation Approach:

[Specific instructions for how AI should approach this task]

Example:

1. First generate the API endpoint function
2. Then generate the service layer business logic
3. Then generate database query logic
4. Then generate comprehensive unit tests
5. Finally generate integration tests

Generate complete, working code - not pseudocode or TODO comments.

Iteration Strategy:

[How to handle multi-pass generation]

Example:

Pass 1: Generate basic implementation with happy path
Pass 2: Add error handling and edge cases
Pass 3: Add thorough tests
Pass 4: Optimize for performance

Output Format Preference:
[How you want the code delivered]
☐ Single file with all code
☐ Multiple files (specify structure)
☐ Code with inline explanations
☐ Code with separate documentation

SECTION 11: POST-GENERATION

After AI Generates Code

Human Review Process:

1. Verify all requirements from sections 1-7 met
2. Run all generated tests
3. Perform manual testing of edge cases
4. Security review
5. Performance testing
6. Code style check
7. Integration testing with existing systems

Known AI Limitations for This Task:

- AI may hallucinate library functions → Verify all imports exist
- AI may overcomplicate → Look for simpler approaches
- AI may miss edge cases → Cross-check with Section 5
- AI may generate inefficient queries → Review database calls

Refinement Prompts if Needed:
[Prepare follow-up prompts for common issues]

Example:
"The code is too complex. Simplify by removing [X feature] and using [Y approach]"
"Add error handling for [specific scenario]"
"Optimize the database query using [specific technique]"

USAGE TIPS

For AI Alignment Engineers:

- Complete this template during Sprint Inception
- Collaborate with developers to understand constraints
- Be specific - vague specifications produce vague code
- Include examples - AI learns from concrete patterns

- Explicitly state what NOT to do

For AI-Augmented Developers:

- Review the entire context spec before prompting AI
- Use the spec as your prompt foundation
- Reference specific sections when iterating
- Validate AI output against checklist in Section 9
- Update the spec if you discover missing requirements

Common Mistakes to Avoid:

- Vague objective ("make it better")
- Missing constraints (AI will make assumptions)
- No examples (AI has no pattern to follow)
- Assuming AI knows your codebase structure
- Not specifying what NOT to do
- No validation criteria

Template Version: 1.0
Last Updated: [Date]

AI COLLABORATION LOG

Purpose

The AI Collaboration Log captures how teams work with AI tools, documenting prompts, outputs, time saved, and lessons learned. This builds institutional knowledge and enables continuous improvement of AI collaboration patterns.

LOG METADATA

Sprint: _________

Week Of: _________

Team: ________________________________

Log Maintained By: _________________ (rotate responsibility weekly)

Log Location: ________________________________

Last Updated: _________

DAILY COLLABORATION ENTRIES

Date: _________

Day of Sprint: ___

ENTRY #1

Task/Story: ______________________________ (Story ID: _______)

Team Member: _________________

AI Tool Used: ☐ GPT-4 ☐ Claude ☐ Copilot ☐ Cursor ☐ Other: _____

What Was Generated

Generation Type:

☐ Code (implementation)

☐ Tests (unit/integration/e2e)

☐ Documentation

☐ Data analysis

☐ Architecture/design

☐ Refactoring

☐ Bug fix

☐ Other: _________

Artifact Generated: [Brief description of what AI created]
Lines of Code/Content: _____
Complexity: ☐ Simple ☐ Moderate ☐ Complex

Prompt Used
Prompt Strategy:

☐ Zero-shot (no examples)
☐ Few-shot (with examples)
☐ Chain-of-thought
☐ Context specification-based
☐ Iterative refinement
☐ Other: _________

Initial Prompt:
[Paste the actual prompt used]
Context Provided:

☐ None
☐ Existing code snippets
☐ API documentation
☐ Business requirements
☐ Architecture diagrams
☐ Error messages
☐ Other: _________

Prompt Effectiveness:

☐ Excellent
☐ Good
☐ Fair
☐ Poor

Iteration Details
Number of Iterations: _____
Iteration 1:

- **Refinement Needed:** ___________________________________
- **Follow-up Prompt:** ___________________________________
- **Result:** ☐ Better ☐ Worse ☐ No Change

Iteration 2:

- **Refinement Needed:** ___________________________________
- **Follow-up Prompt:** ___________________________________

- **Result:** ☐ Better ☐ Worse ☐ No Change

Iteration 3:

- **Refinement Needed:** ________________________________
- **Follow-up Prompt:** ________________________________
- **Result:** ☐ Better ☐ Worse ☐ No Change

Final State: ☐ Accepted as-is ☐ Modified by human ☐ Discarded

Time Metrics
AI Generation Time: _____ minutes
Human Review/Edit Time: _____ minutes
Total Time (Generation + Review): _____ minutes
Traditional Approach Would Take: _____ minutes/hours
Time Saved: _____ minutes
Time Saved %: ____%
Efficiency Calculation:
Efficiency = (Traditional Time - AI Time) / Traditional Time × 100
Example: (120 min - 30 min) / 120 min × 100 = 75% time saved

Quality Assessment
First-Pass Acceptance:

☐ Yes - Used without modification

☐ Partial - Minor edits needed

☐ No - Significant rework required

☐ Rejected - Completely rewritten by human

Issues Found in AI Output:

- Logic errors
- Syntax errors
- Missing edge cases
- Inefficient implementation
- Security vulnerabilities
- Style violations
- Incorrect assumptions
- Hallucinated functions/libraries
- Other: ________________________________

Severity of Issues:

☐ None

☐ Minor (quick fix)

☐ Moderate (30+ min to fix)

☐ Major (faster to rewrite)

Validation Approach:

- Code review by human
- Automated tests passed
- Manual testing performed
- Peer review completed
- Security scan run
- Performance benchmarked

Validation Outcome:

☐ Passed all checks

☐ Passed with minor fixes

☐ Required significant fixes

☐ Failed validation

Human Intervention
What Did Human Add/Change?
Category of Change:

☐ Bug fixes

☐ Performance optimization

☐ Security hardening

☐ Edge case handling

☐ Code style/formatting

☐ Documentation

☐ Error handling

☐ Other: _________

Specific Changes: [Describe what was modified and why]
Why Was Change Necessary?

☐ AI misunderstood requirement

☐ AI lacked context

☐ AI hallucinated capability

☐ AI optimized for wrong criteria

☐ Prompt was ambiguous

☐ Edge case not in prompt

☐ Security/compliance requirement

☐ Other: _________

Lessons Learned
What Worked Well:

- ______
- ______

What Didn't Work:

- ______
- ______

Prompt Improvements for Future: [What would you change in the prompt next time?]

Context Spec Gaps: [What should have been in the Context Specification but wasn't?]

Knowledge Captured:

☐ Added to team prompt library

☐ Updated Context Specification template

☐ Documented as anti-pattern

☐ Shared in team channel

☐ None - one-off learning

Cost Tracking (Optional)
If tracking AI API costs:
Model Used: ______________________________
Tokens Used:

- Input tokens: _____
- Output tokens: _____
- Total tokens: _____

Estimated Cost: $_____
Cost per Outcome: $_____
ROI Calculation:
ROI = (Time Saved × Hourly Rate - AI Cost) / AI Cost × 100
Example: (2 hours × $100/hr - $1.50) / $1.50 × 100 = 13,233% ROI

WEEKLY SUMMARY
Complete at end of each week
Week: ________ | **Sprint Days: -**
Aggregate Metrics
Total Entries This Week: _____

Generation Types:

- Code: _____ entries
- Tests: _____ entries
- Documentation: _____ entries
- Other: _____ entries

AI Tools Used:

- Tool 1: _____ times
- Tool 2: _____ times
- Tool 3: _____ times

Time Saved This Week: _____ hours
Average Time Saved per Entry: _____ minutes
First-Pass Acceptance Rate: _____%
Calculation: (Accepted as-is + Minor edits) / Total entries × 100

Quality Insights
Most Common Issues:

1. _________________ (occurred _____ times)
2. _________________ (occurred _____ times)
3. _________________ (occurred _____ times)

Issue Trends:

☐ Issues decreasing (team learning)

☐ Issues steady

☐ Issues increasing (need intervention)

Validation Pass Rate: _____%

Learning & Improvement
Top 3 Effective Prompts:

1. [Prompt pattern that worked well]
2. [Prompt pattern that worked well]
3. [Prompt pattern that worked well]

Top 3 Ineffective Prompts:

1. [Prompt pattern that failed]
2. [Prompt pattern that failed]
3. [Prompt pattern that failed]

Prompt Library Updates:

- Added _____ new prompts
- Refined _____ existing prompts
- Deprecated _____ prompts

Context Spec Improvements:

- Identified _____ missing constraints
- Added _____ new examples
- Clarified _____ ambiguities

Team Patterns
Who Used AI Most Effectively: [Team member name]: _____ successful generations

What Made Them Effective:

- ______________________________
- ______________________________

Who Struggled Most: [Team member name]: _____ failed generations
Support Needed:

- Training on prompt engineering
- Pairing with effective AI user
- Better Context Specifications
- Other: ______________________________

Knowledge Sharing:

- Team shared learnings in Daily Sync
- Retrospective discussion scheduled
- Prompt library updated
- Best practices documented

Recommendations for Next Week
Process Improvements:

1. ______________________________
2. ______________________________
3. ______________________________

Tool/Technique to Experiment:

- ______________________________

Training Needs Identified:

- ______________________________

SPRINT RETROSPECTIVE INSIGHTS
Complete at end of sprint
Sprint: _________
Sprint-Level Metrics
Total AI Collaboration Entries: _____
Total Time Saved: _____ hours
Estimated Value of Time Saved: $_____ (time × avg hourly rate)
Average First-Pass Acceptance Rate: _____%
Average Validation Pass Rate: _____%
Average Time Saved per Task: _____ minutes

Trend Compared to Last Sprint:

☐ Improving (acceptance rate up)

☐ Stable

☐ Declining (need intervention)

Sprint Learnings

Most Valuable AI Use Cases This Sprint:

1. ______
2. ______
3. ______

Least Effective AI Use Cases:

1. ______
2. ______
3. ______

Breakthrough Moments: [Describe any significant learning or improvement in AI collaboration]
Persistent Challenges: [What issues keep recurring despite iteration?]

Prompt Library Evolution
Prompts Added: _____
Prompts Refined: _____
Prompts Deprecated: _____
Total Library Size: _____

Most Reused Prompt: [Name/description of most valuable prompt]
Usage Count: _____ times
Success Rate: _____%

Impact on Sprint Outcomes
Outcomes Enhanced by AI:

- Outcome 1: ________________ (impact: ____________)
- Outcome 2: ________________ (impact: ____________)
- Outcome 3: ________________ (impact: ____________)

AI Contribution to Velocity:

- Without AI: Estimated _____ points deliverable
- With AI: Actual _____ points delivered
- AI contribution: _____ points (____% increase)

Validation Bottleneck Impact:

- Time spent on AI generation: _____ hours
- Time spent on AI output validation: _____ hours
- Validation/Generation ratio: _____:1

Ideal Ratio: 2:1 (twice as much validation as generation)

Current Performance: ☐ Above ideal ☐ At ideal ☐ Below ideal

Recommendations for Next Sprint

Continue:

- ______________________________
- ______________________________

Start:

- ______________________________
- ______________________________

Stop:

- ______________________________
- ______________________________

Action Items:

- Action 1: _______________ | Owner: _______ | Due: _______
- Action 2: _______________ | Owner: _______ | Due: _______
- Action 3: _______________ | Owner: _______ | Due: _______

ORGANIZATIONAL LEARNING NETWORK

Share insights across teams

Knowledge Worth Sharing

Discoveries to Share with Other Teams:

1. ______________________________
2. ______________________________
3. ______________________________

Prompt Patterns with >80% Success Rate:

- Pattern 1: ____________________________________
- Pattern 2: ____________________________________
- Pattern 3: ____________________________________

Context Specification Improvements: [What additions to Context Spec template helped most?]

USAGE GUIDELINES

Daily Logging

When to Log:

- Immediately after AI generation completes
- Before end of day (don't let entries accumulate)
- Minimum: Any AI generation taking >15 minutes

Who Logs:

- Person who performed AI collaboration
- Rotate log review responsibility weekly

How Detailed:

- Daily entries: 3-5 minutes to complete
- Focus on facts, not opinions
- Capture actual prompts and outcomes

Weekly Review

When: End of each week (Friday afternoon)
Who: Entire team + Delivery Facilitator
Duration: 15-30 minutes
Agenda:

1. Review aggregate metrics
2. Identify top effective/ineffective prompts
3. Update prompt library
4. Identify training needs
5. Plan next week improvements

Sprint Retrospective

When: During Sprint Retrospective ceremony
Who: Entire team
Duration: 20-30 minutes (portion of retro)
Focus:

- Trends over sprint
- Impact on outcomes
- Learning velocity
- Recommendations for next sprint

ANTI-PATTERNS TO AVOID

Don't:

- Log only successes (log failures too - they teach more)
- Wait until end of day/week (details forgotten)
- Copy prompts without context (future you won't understand)
- Skip "lessons learned" (critical for improvement)
- Let log become pure compliance exercise (it's a learning tool)
- Hoard knowledge (share discoveries with team)

Do:

- Log both successes and failures
- Capture prompts exactly as used
- Document why changes were needed
- Share insights immediately
- Review and discuss patterns regularly
- Update prompt library continuously

INTEGRATION WITH OTHER ARTIFACTS

Feeds Into:

- Sprint Retrospective discussion
- Metrics Dashboard (aggregated data)
- Prompt Library (effective patterns)
- Context Specification Template (improvements)
- Training needs assessment

References:

- Context Specification (what prompted AI)
- Outcome Backlog (what was being built)
- Validation Canvas (what was validated)

Ownership:

- Primary: AI Alignment Engineer (maintains structure)
- Contributors: All team members (daily entries)
- Reviewers: Delivery Facilitator (weekly synthesis)

Template Version: 1.0
Last Updated: [Date]

DAILY SYNCHRONIZATION QUESTION FRAMEWORK

Purpose
Daily Synchronization replaces traditional standup questions with validation-focused coordination. When AI eliminates the development bottleneck, teams need visibility into validation progress, not task completion.

MEETING DETAILS
Date: _________
Sprint Day: ___ of ___
Time: _________
Duration: 15 minutes maximum
Format: ☐ Synchronous (live meeting) ☐ Asynchronous (Slack/Teams)
Facilitator: _________________ (Delivery Facilitator)
Attendees:

- Product Strategist
- AI Alignment Engineer
- Validation Lead
- Governance Coordinator
- Integration Architect
- AI-Augmented Developers (all)
- Delivery Facilitator

THE FOUR VALIDATION-FOCUSED QUESTIONS
Replace traditional standup questions with these validation-oriented alternatives:

QUESTION 1: What Validation Did We Complete Yesterday?
Purpose: Report validation criteria satisfied, not tasks completed.

Format per Team Member:
Name: _________________
Validation Completed:

- ☐ **Validation Criteria:** __________________________________
  - **Evidence:** ________________________________
  - **Status:** ☐ Passed ☐ Passed with Conditions ☐ Failed

- ☐ **Validation Criteria:** ______________________________
    - **Evidence:** ______________________________
    - **Status:** ☐ Passed ☐ Passed with Conditions ☐ Failed

Example (Good):
"We validated performance benchmarks for the order API. Database queries return results under 100ms. Evidence: JMeter test results in CI/CD pipeline."

Example (Poor - Don't do this):
"I finished implementing the order API" ← This is task completion, not validation

---

Collective Team Progress:

Outcome	Validation Criteria Met	Evidence Accumulated	% Complete
Outcome 1	/	[Evidence type]	____%
Outcome 2	/	[Evidence type]	____%
Outcome 3	/	[Evidence type]	____%

Validation Trend:

☐ Ahead of schedule

☐ On track

☐ Behind schedule (discuss after meeting)

QUESTION 2: What Validation Are We Pursuing Today?
Purpose: Coordinate validation activities and identify collaboration opportunities.

Format per Team Member:
Name: ________________
Planned Validation Activities:

- ☐ **Validation Target:** ______________________________
    - **Success Criteria:** ______________________________
    - **Dependencies:** ______________________________
    - **Estimated Completion:** Today / Tomorrow / This Week

Collaboration Needed:

- Need pairing with: ____________ for ____________
- Awaiting input from: __________ on ____________
- Stakeholder review scheduled: ________________

Example:
"Today I'm validating security requirements for the payment flow. Success means passing OWASP Top 10 scans with zero critical findings. I need 30 minutes from the Validation Lead to review the security test plan."

Team Validation Coordination:

Team Member	Validation Focus	Collaboration Points	Risks
[Name]	[What validating]	[Who need/helping]	[Concerns]

Swarming Opportunities: [Identify if multiple people should collaborate on complex validation]

QUESTION 3: What Validation Blockers Exist?
Purpose: Surface obstacles preventing validation progress for immediate resolution.
Format per Team Member:
Name: ________________
Blockers:

☐ No blockers

☐ **Blocker Type:** Missing Test Data / Environment Issue / Stakeholder Unavailable / Dependency Delay / Other
Description: ________________________________
Impact:

☐ Critical (stops all work)

☐ High (delays outcome)

☐ Medium (slows progress)
Owner: ________________ (who can resolve?)
Mitigation Plan: ________________________________
Status:

☐ New

☐ In Progress

☐ Escalated

Common Blocker Types:

Blocker Category	Example	Resolution Approach
Test Environment	Staging database down	Escalate to infrastructure team
Missing Data	No test PII for validation	Generate synthetic data, or get approval for anonymized production data
Stakeholder Access	Product owner unavailable for UAT	Defer to AI Show & Tell, or identify backup approver
Governance Delay	Waiting on legal review	Governance Coordinator follows up, considers parallel path
Integration Failure	Third-party API returning errors	Contact vendor, implement mock for testing
Tool/License Issue	Security scanner license expired	Escalate to IT/procurement

---

Active Blockers Summary:

Blocker	Owner	Days Blocked	Resolution Plan	ETA
[Description]	[Person]	___	[Next steps]	[Date]

Escalation Needed:

- Blocker requires management intervention
- Blocker requires cross-team coordination
- Blocker threatens sprint commitment

QUESTION 4: What AI Collaboration Worked Well or Poorly?
Purpose: Capture real-time learnings about effective AI interactions.

Format per Team Member:
Name: ________________
AI Collaboration Highlights:
What Worked:

- **Prompt/Approach:** ________________________________
- **Result:** ______________________________
- **Reusable Pattern:** ☐ Yes (add to library) ☐ No (one-off)

What Didn't Work:

- **Prompt/Approach:** ________________________________
- **Issue:** ________________________________
- **Lesson:** ________________________________

Example (Good):
"Using the Context Specification template for API generation worked perfectly. First-pass acceptance. Adding this prompt pattern to our library."

Example (Poor):
"AI hallucinated a library function again. We need better prompts." ← Too vague, not actionable

---

Team Learning Capture:
Today's Discoveries:

- Pattern to share: ________________________________
- Anti-pattern to avoid: ________________________________
- Prompt library update needed: Yes / No

AI Collaboration Health Check:

- First-pass acceptance rate today: _____%
- Time saved vs. traditional approach: _____ hours
- Validation issues from AI outputs: _____ (count)

ASYNCHRONOUS ALTERNATIVE FORMAT
If team uses asynchronous daily sync (Slack/Teams):
Channel: #daily-sync
Post By: 9:00 AM daily

Template:
Sprint Day X - [Date]

VALIDATION COMPLETED YESTERDAY:
[Your validation achievements with evidence]

VALIDATION PURSUING TODAY:
[What you're validating, dependencies, collaboration needs]

BLOCKERS:
[Any obstacles, or "None"]

AI LEARNINGS:
[What worked/didn't work with AI collaboration]

% Complete on My Outcomes: [X]%

Review Pattern:

- All posts by 9:00 AM
- Delivery Facilitator aggregates by 9:30 AM
- Live sync only if blockers need real-time discussion (< 2× per week)

FACILITATION GUIDELINES

For Delivery Facilitators

Before Meeting:

- Review yesterday's progress from tracking tools
- Check for blocker updates in channels
- Identify any outcomes falling behind
- Prepare to facilitate focused discussion

During Meeting:

- Time-box ruthlessly (15 min max)
- Focus on validation progress, not activity
- Capture blockers for follow-up
- Note collaboration opportunities
- Update visual board in real-time

After Meeting:

- Update sprint tracking artifacts
- Follow up on blockers with owners
- Share learnings in appropriate channels
- Escalate critical issues if needed

Red Flags to Watch:

- Same blockers mentioned 3+ days in row → Escalate
- Validation progress <50% by sprint day 7 → Team discussion needed
- Multiple team members blocked by same issue → Immediate resolution required
- No AI learnings shared for 3+ days → Check if team using AI effectively

VISUAL DASHBOARD INTEGRATION

Validation Board Updates:

After each daily sync, update the Validation-First Board:

Column	Status
Validation Defined	Criteria established, ready to pursue
Validation In Progress	Actively being validated
Validation Complete	Evidence collected, criteria met
Validation Blocked	Blocker preventing progress

Sprint Burndown - Validation Focus:

- Y-axis: Validation criteria remaining
- X-axis: Sprint days
- Track validated vs. remaining (not tasks completed)

COMPARISON: TRADITIONAL VS. AI SCRUM DAILY SYNC

Traditional Standup	AI Scrum Daily Sync
"What did you do yesterday?"	"What validation did we complete yesterday?"
"What will you do today?"	"What validation are we pursuing today?"
"What blockers do you face?"	"What validation blockers exist?"
[No AI-specific question]	"What AI collaboration worked well or poorly?"
Focus: Task completion	**Focus:** Validation progress
Success: Closed tickets	**Success:** Evidence accumulated
Metric: Velocity	**Metric:** Validation criteria satisfied

EFFECTIVENESS METRICS

Track weekly:

- Average meeting duration: _____ minutes
- Blockers identified: _____
- Blockers resolved within 24 hours: _____%
- AI learnings captured: _____
- Validation criteria met vs. planned: _____%

Health Indicators:

☐ Meeting stays under 15 minutes

☐ All team members participate

☐ Blockers get resolved quickly

☐ Validation progress is transparent

☐ AI learnings are captured daily

Warning Signs:

☐ Meetings consistently run long (>20 min)

☐ Same people never speak

☐ Blockers linger for days

☐ No concrete validation evidence shared

☐ No AI collaboration insights

COMMON MISTAKES TO AVOID

Don't:

- Discuss implementation details (save for later)
- Turn into status report to management (it's team coordination)
- Skip when remote workers are unavailable (use async format)
- Let one person dominate time (time-box per person)
- Focus on activity vs. outcomes ("I was busy" doesn't matter)

Do:

- Focus on validation evidence
- Surface blockers immediately
- Coordinate collaboration
- Capture AI learnings
- Keep it short and focused

Template Version: 1.0
Last Updated: [Date]

AI SHOW & TELL SESSION STRUCTURE

Purpose

AI Show & Tell enables mid-sprint stakeholder feedback on AI-generated prototypes, preventing late-stage rework through early course correction. This ceremony occurs twice per sprint (Days 3 and 7).

SESSION METADATA

Session: ☐ #1 (Day 3) ☐ #2 (Day 7)
Date: _________
Time: _________
Duration: 60 minutes maximum
Sprint: _________
Sprint Day: ___
Location/Link: ___________________________________

REQUIRED ATTENDEES

Team:

- Product Strategist
- AI-Augmented Developers (presenting outcomes)
- Validation Lead
- Delivery Facilitator

Stakeholders:

- __________________ (Primary decision-maker)
- __________________ (Domain expert)
- __________________ (End user representative)
- __________________ (Other: _________)

PRE-SESSION PREPARATION (24 Hours Before)

Team Preparation Checklist

For Each Outcome:

- Working prototype ready (functional, not polished)
- Demo script prepared
- Known limitations documented
- Feedback questions prepared
- Demo environment tested

Materials to Share:

- Outcome statement & success metrics
- Current validation status
- Known issues/limitations

- Specific questions for stakeholders

Expectation Setting:

- Stakeholders notified this is early prototype review
- "Direction validation, not production sign-off" communicated
- Feedback method explained (accept/refine/pivot)

SESSION AGENDA

Part 1: Context Setting (5 minutes)

Facilitator: Product Strategist

What to Cover:

1. **Sprint Goal Reminder:** [Restate overarching sprint objective]
2. **Today's Focus:** [Which outcomes being demonstrated]
3. **Stage of Development:**
    - Day 3: Early prototype, validating direction
    - Day 7: More complete, validating acceptance criteria
4. **How Feedback Works:**
    - Accept: Continue on current path
    - Refine: Adjust based on feedback
    - Pivot: Fundamentally change direction

Part 2: Outcome Demonstrations (40 minutes total, ~10-15 min each)

For Each Outcome:

Outcome Name: ________________________________

Presenter: ________________

Demo Structure (10-15 minutes):

A. Outcome Recap (2 min)

- **What we're trying to achieve:** ________________________________
- **Success metric:** ________________________________
- **Where we are:** ___% complete

B. Prototype Demonstration (5-8 min)

Show:

- Core functionality working
- User flow example
- Key integration points
- Performance characteristics (if relevant)

Don't:

- Apologize for incomplete features
- Explain code implementation

- Demo edge cases not yet working

C. Validation Status (2 min)

Validation Area	Status	Evidence
Functional	☐ Complete ☐ In Progress ☐ Not Started	[Brief description]
Performance	☐ Complete ☐ In Progress ☐ Not Started	[Brief description]
Security	☐ Complete ☐ In Progress ☐ Not Started	[Brief description]

D. Known Limitations (1 min)

- **Not yet implemented:** ________________________________
- **Known issues:** ________________________________
- **Planned for Days 8-10:** ________________________________

E. Specific Questions (2 min) [Ask targeted questions to guide feedback]

1. Is this the right approach to achieve [outcome]?
2. Does the user flow make sense for [use case]?
3. Is [specific feature] solving the right problem?

Part 3: Stakeholder Feedback (10 minutes)
Facilitator: Delivery Facilitator

Structured Feedback Collection:

Outcome 1: ________________________________
Direction Validation:

☐ **Accept** - Continue on current path, looks good

☐ **Refine** - Right direction, but needs adjustments

☐ **Pivot** - Wrong approach, need to change direction
If REFINE, what needs to change?

- Priority 1: ________________________________
- Priority 2: ________________________________
- Priority 3: ________________________________

If PIVOT, what's the new direction?

- New approach: ________________________________
- Reason for change: ________________________________

Acceptance Criteria Validation:

- Current approach will meet criteria: Yes / No / Unsure
- If No: What's missing? ________________________________

User Experience Feedback:

- Works well: ________________________________
- Needs improvement: ________________________________

Value Confirmation:

- Still worth pursuing: Yes / Maybe / No
- If No: Should we stop work? ____________________________

Outcome 2: ________________________________

[Repeat feedback structure]

Part 4: Action Items & Next Steps (5 minutes)

Facilitator: Delivery Facilitator

Decisions Made:

Outcome	Decision	Action Required	Owner	Due
Outcome 1	Accept/Refine/Pivot	[What changes]	[Who]	[When]
Outcome 2				

Course Corrections:

- **High Priority Change:** ________________ (Do immediately)
- **Medium Priority:** ________________ (Do this sprint)
- **Low Priority:** ________________ (Consider for future)

Follow-Up Meetings Needed:

- Additional stakeholder session: Date _______
- Subject matter expert consultation: _______
- User testing session: _______

Next AI Show & Tell:

- Date/Time: _________ (Day 7 or Sprint Review)
- What we'll demonstrate: ________________________________

POST-SESSION ACTIVITIES

Immediate Follow-Up (Within 2 Hours)

Delivery Facilitator:

- Distribute session notes to all attendees
- Update Outcome Backlog with feedback
- Create action items in sprint tracking tool

- Schedule any follow-up meetings

Product Strategist:

- Clarify any ambiguous feedback with stakeholders
- Reprioritize work based on pivot/refine decisions
- Update acceptance criteria if needed

AI-Augmented Developers:

- Implement high-priority refinements
- Re-prompt AI if pivot required
- Update validation criteria if scope changed

Validation Lead:

- Adjust validation plan for refinements
- Identify new validation needs from feedback

Update Artifacts
Outcome Backlog:

- Update outcome statements if pivoted
- Adjust success metrics if refined
- Document stakeholder decisions

Validation Canvas:

- Add new acceptance criteria from feedback
- Remove criteria for deprioritized features
- Update evidence requirements

Sprint Tracking:

- Adjust sprint commitments if major pivot
- Update burndown if scope changed
- Flag risks from course corrections

DAY 3 VS. DAY 7 SESSION DIFFERENCES
Day 3 Session (Early Prototype)

Purpose: Direction validation **Completeness:** 30-50% functional **Stakeholder Mindset:** "Are we building the right thing?"

Team Should Show:

- Core functionality working (even if rough)
- Basic user flow demonstrable
- Key integration points mocked/working

Team Should NOT:

- Polish UI/UX
- Implement all edge cases
- Optimize performance (unless it's the outcome)

Typical Outcome:

- Accept: 70% of outcomes
- Refine: 25% of outcomes
- Pivot: 5% of outcomes

Day 7 Session (Nearly Complete)
Purpose: Acceptance criteria validation **Completeness:** 70-90% functional **Stakeholder Mindset:** "Are we ready to ship?"

Team Should Show:

- Comprehensive functionality
- Polished user experience
- Real integration (not mocked)
- Performance benchmarks met
- Security/compliance validated

Team Should NOT:

- Introduce new features
- Make major architectural changes
- Promise future enhancements

Typical Outcome:

- Accept (ready for Sprint Review): 80%
- Refine (minor adjustments): 15%
- Pivot (rare at this stage): 5%

DECISION FRAMEWORK

Accept Decision

Criteria:

- Direction aligns with stakeholder expectations
- Approach will achieve desired outcome
- No fundamental concerns raised
- Team can continue current implementation

Next Steps:

- Continue implementation
- No scope changes
- Proceed to next validation phase

Refine Decision

Criteria:

- Right direction, but needs adjustments
- Changes are additive or subtractive, not fundamental

- Refinements fit within sprint timeline
- Team agrees changes are achievable

Change Categories:

Minor Refinements (Continue sprint):
- UI/UX adjustments
- Additional validation rules
- Performance tuning
- Enhanced error messages

Moderate Refinements (May extend timeline):
- Additional features within scope
- Integration changes
- Significant validation additions
- Security enhancements

Major Refinements (Reassess sprint commitment):
- Scope expansion requiring discussion
- Architectural modifications
- New dependencies

Next Steps:
- Document specific changes
- Estimate effort impact
- Adjust sprint plan if needed
- May require additional stakeholder review

Pivot Decision

Criteria:
- Current approach will not achieve outcome
- Fundamental misalignment with stakeholder expectations
- Better approach identified
- Salvaging current work not viable

Pivot Types:

Mild Pivot (Same outcome, different approach):
- Change technical approach
- Different user flow
- Alternative architecture

Moderate Pivot (Adjust outcome definition):
- Revise success metrics
- Narrow/broaden scope
- Change target user

Major Pivot (Abandon outcome):
- Stop work on outcome
- Remove from sprint

- Replace with different outcome

Next Steps:

- Document new direction clearly
- Re-prompt AI with new context
- Reset validation criteria
- Communicate to broader team
- May require Sprint Inception-like session

Escalation:

- Pivot impacts sprint goal → Notify management
- Pivot requires additional resources → Request approval
- Pivot suggests systemic issue → Schedule retrospective

FACILITATION TIPS

For Delivery Facilitators:

Before Session:

- Set expectations that prototypes won't be perfect
- Remind team to focus on substance, not polish
- Prepare stakeholders for "work in progress" viewing

During Session:

- Keep demos moving (time-box strictly)
- Prevent deep technical discussions (parking lot)
- Focus stakeholders on direction, not details
- Capture feedback clearly

After Session:

- Ensure decisions are unambiguous
- Follow up on unclear feedback within 2 hours
- Update all artifacts same day

For Presenters:

Do:

- Show working functionality first
- Acknowledge known limitations upfront
- Ask specific questions
- Demo real scenarios
- Show evidence of validation

Don't:

- Apologize excessively
- Explain technical implementation
- Promise features not in scope
- Argue with feedback
- Hide problems (they'll surface later)

SUCCESS METRICS

Effective AI Show & Tell Has:

- <20% pivot rate (most directions correct)
- Refinements captured in <5 minutes
- Action items assigned with owners
- Stakeholder attendance >80%
- Session completes within 60 minutes
- Team leaves with clear next steps

Warning Signs:

- High pivot rate (>30%) → Requirements process broken
- Vague feedback → Need more specific questions
- No decisions made → Need decision authority present
- Stakeholders never available → Escalate priority
- Team defensive about feedback → Culture issue

EXAMPLE SCENARIOS

Scenario 1: Successful Accept
Outcome: Reduce customer support escalation rate
Demo: Chatbot prototype with ML-powered response suggestions
Stakeholder Reaction: "This is exactly what we need"
Decision: Accept - continue implementation
Action: None needed, proceed to comprehensive validation

Scenario 2: Productive Refine
Outcome: Improve order processing speed
Demo: Automated workflow with AI prioritization
Stakeholder Feedback: "Love the automation, but prioritization logic is backwards"
Decision: Refine - adjust AI prompt for prioritization
Action: AI Alignment Engineer updates context specification, regenerate logic
Impact: 4 hours of rework, still within sprint capacity

Scenario 3: Early Pivot (Day 3)
Outcome: Streamline inventory management
Demo: Real-time dashboard with AI predictions
Stakeholder Reaction: "This solves the wrong problem - we need batch processing, not real-time"
Decision: Pivot - redesign for batch workflow
Action: Re-prompt AI with new context, redesign validation

Impact: Essentially starting over, but Day 3 pivot better than Day 10 failure

---

Template Version: 1.0
Last Updated: [Date]

GOVERNANCE CHECKPOINT CHECKLIST

Purpose
The Governance Checkpoint integrates compliance review mid-sprint (typically Day 5-6) rather than as an end-of-sprint gate. This enables course correction while changes remain inexpensive.

CHECKPOINT METADATA
Sprint: ________
Checkpoint Date: ________ (Day 5-6)
Duration: 60 minutes maximum
Facilitator: Governance Coordinator

ATTENDEES
Required:

- Governance Coordinator
- Product Strategist
- Validation Lead
- AI-Augmented Developers (outcome owners)

As Needed (Based on Risk Tier):

- Legal representative (for Board Review outcomes)
- Security specialist (for high-risk data outcomes)
- Compliance officer (for regulated industry)
- Ethics officer (for sensitive AI applications)

PRE-CHECKPOINT PREPARATION (48 Hours Before)
For Each Outcome Being Reviewed:
Outcome Name: ________________________________
Risk Tier:

☐ Delegated

☐ Standard Review

☐ Board Review

☐ Prohibited

Documentation Ready:

- Outcome description & business justification
- AI collaboration approach documented
- Validation evidence collected to date
- Risk assessment completed

- Compliance checklist filled
- Decision log started

SECTION 1: RISK TIER VALIDATION

Confirm Initial Risk Assessment

For each outcome, verify the risk tier assigned during Sprint Inception remains accurate:

Outcome 1: ______________________________

Original Risk Tier: ________ (from Sprint Inception)

Risk Factor Reassessment:

Factor	Original Score (1-5)	Current Score	Changed?
Data Sensitivity			Y/N
Regulatory Impact			Y/N
User Exposure			Y/N
System Criticality			Y/N
AI Hallucination Potential			Y/N
Reversibility			Y/N

Total Risk Score: _____/30 (Original: _____/30)

Risk Tier Status:

☐ Confirmed - No change

☐ Escalated - Higher risk than expected

☐ Downgraded - Lower risk than expected

If Changed:

- New Risk Tier: ______________________________
- Reason for Change: ______________________________
- Additional Requirements: ______________________________

SECTION 2: COMPLIANCE DOCUMENTATION REVIEW

Privacy & Data Protection

GDPR/CCPA Compliance (if applicable):

- **Lawful Basis Documented:**
  - Basis:

    ☐ Consent

    ☐ Contract

- ☐ Legal Obligation
- ☐ Legitimate Interest
  - Documentation: ____________________________________
- **Data Processing:**
  - What data: ____________________________________
  - Why collected: ____________________________________
  - Who accesses: ____________________________________
  - How long retained: ____________________________________
  - Data subject rights: ☐ Implemented
- **Privacy Notice:**
  - Users informed: Yes/No
  - Location: ____________________________________
  - Last updated: __________

PII Handling:

Data Element	Classification	Protection Method	Validated
[Customer email]	PII	Encrypted at rest/transit	☐

Compliance Status:

☐ Pass

☐ Conditional Pass

☐ Fail

Issues Found: ____________________________________

Security Assessment

Security Checklist:

- **Authentication & Authorization:**
  - Method documented: ____________________________________
  - Role-based access implemented: Yes/No
  - Tested: Yes/No
- **Data Protection:**
  - Encryption in transit: ☐ TLS 1.2+
  - Encryption at rest: ☐ AES-256 or equivalent
  - Secrets management: ☐ No hardcoded credentials
- **Vulnerability Assessment:**
  - SAST scan completed: Date ________
  - DAST scan completed: Date ________
  - Dependency scan: Date ________

- Critical vulnerabilities: _____ (must be 0)
- High vulnerabilities: _____ (threshold: ___)

Security Scan Results:

Scan Type	Date	Critical	High	Medium	Low	Pass/Fail
SAST		0				
DAST		0				
Dependency		0				

Penetration Testing (if required):

- Scheduled: Date _______
- Completed: Date _______
- Findings remediated: ☐

Compliance Status: ☐ Pass ☐ Conditional Pass ☐ Fail
Issues Found: ______________________________
Industry-Specific Regulations
Applicable Regulations:

☐ HIPAA (Healthcare)

☐ PCI DSS (Payments)

☐ SOX (Financial)

☐ FERPA (Education)

☐ EU AI Act

☐ Other: ____

For Each Applicable Regulation:
Regulation Name: ______________________________
Requirements:

Requirement	Implementation	Evidence	Status
[Specific req 1]	[How compliant]	[Proof]	☐

Documentation:

- Compliance documentation complete
- Required approvals obtained
- Audit trail maintained

Compliance Status:

☐ Pass

☐ Conditional Pass

☐ Fail

Issues Found: ______________________________

SECTION 3: AI-SPECIFIC GOVERNANCE

AI Behavior Verification

Model/AI Component: ______________________________

Explainability & Transparency:

- AI involvement disclosed to users: Yes/No
- How AI decisions are made: Documented/Not Documented
- Users can understand AI outputs: Yes/No/Partial

Fairness & Bias Assessment:

- Training data evaluated for bias: Yes/No/N/A
- Output tested across demographics: Yes/No/N/A
- Disparate impact analysis: Completed/In Progress/N/A

Bias Test Results:

Demographic	Metric	Acceptable Range	Actual	Pass/Fail
[Group 1]	[Accuracy/etc]	[Range]	[Value]	☐

Hallucination Prevention:

- Grounding mechanism: ______________________________
- Human-in-loop verification: ______________________________
- Hallucination rate: __% **(target: <_%)**

AI Compliance Status: ☐ Pass ☐ Conditional Pass ☐ Fail

Issues Found: ______________________________

Ethical Considerations

Ethical AI Checklist:

- **Human Autonomy:**
  - AI supports humans, doesn't replace decision-making: Yes/No
  - Override mechanism available: Yes/No
- **Harm Prevention:**
  - Potential harms identified: ______________________________
  - Mitigation strategies: ______________________________
  - Monitoring plan: ______________________________
- **Accountability:**
  - Clear ownership of AI decisions: Person _______
  - Audit trail maintained: Yes/No
  - Incident response plan: Documented/Not Documented

Ethical Review Status: ☐ Pass ☐ Conditional Pass ☐ Fail
Issues Found: ____________________________

SECTION 4: DECISION LOG & RATIONALE
Document Key Decisions

For this outcome, document:
Decision 1:
- **What:** ____________________________
- **Why:** ____________________________
- **Alternatives Considered:** ______________
- **Rationale for Choice:** ________________
- **Risk Accepted:** ____________________

Decision 2:
- **What:** ____________________________
- **Why:** ____________________________
- **Alternatives Considered:** ______________
- **Rationale for Choice:** ________________
- **Risk Accepted:** ____________________

AI Generation Decisions:
- **Prompts Used:** See AI Collaboration Log entry _____
- **Why This Approach:** ____________________________
- **Alternatives Considered:** __________________________
- **Validation Applied:** ____________________________

SECTION 5: EVIDENCE REVIEW
Documentation Completeness Check
Required Evidence Present:
- **Functional Evidence:**
  - Test results: Location ______________________
  - Coverage metrics: _____%
- **Security Evidence:**
  - Scan reports: Location ______________________
  - Vulnerability summary: ______________________
- **Compliance Evidence:**
  - Compliance checklist: Completed/Incomplete
  - Required approvals: All Obtained/Missing _____
  - Audit documentation: Complete/Incomplete
- **AI-Specific Evidence:**
  - AI Collaboration Log entries: Count _____
  - Context Specifications used: Count _____

- Validation evidence for AI outputs: Present/Missing

Evidence Gaps: [List any missing evidence and plan to obtain]

SECTION 6: CHECKPOINT DECISION

Overall Governance Status

For Each Outcome Reviewed:

Outcome: ______________________________

Checkpoint Decision:

☐ **Approved** - Proceed to completion

☐ **Conditional Approval** - Proceed with remediation required

☐ **Not Approved** - Stop work until issues resolved

☐ **Escalate** - Requires Board Review

If CONDITIONAL APPROVAL:

Required Remediation:

Issue	Severity	Remediation Plan	Owner	Due Date
[Issue 1]	Critical/High/Med	[Action needed]	[Person]	[Date]

Conditional Approval Terms:

- Work may continue on unaffected areas
- Work must stop until resolution on: _____________________
- Re-review required: Yes/No | Date: ______

If NOT APPROVED:

Blocking Issues:

1. ______________________________
2. ______________________________
3. ______________________________

Required Actions Before Work Can Resume:

- Action 1: ____________ | Owner: ______ | ETA: ______
- Action 2: ____________ | Owner: ______ | ETA: ______

Re-Checkpoint Scheduled: Date ______

Impact on Sprint:

- Sprint commitment at risk
- Management notification required
- Outcome may need to be dropped

If ESCALATE to Board Review:
Escalation Reason:

☐ Risk tier requires Board approval

☐ Novel use case needs Board guidance

☐ Regulatory uncertainty needs legal review

☐ Significant business risk identified

Board Review Request:

- Request submitted: Date _______
- Supporting materials prepared: _________________
- Board review scheduled: Date _______
- Outcome work: ☐ Continue ☐ Pause ☐ Stop

Temporary Approval (while awaiting Board):

☐ Yes, work may continue at team's risk

☐ No, work must pause

SECTION 7: POST-CHECKPOINT ACTIONS

Immediate Follow-Up (Same Day)

Governance Coordinator:

- Distribute checkpoint results to team
- Document decisions in governance repository
- Schedule re-review if needed
- Escalate to Board if required
- Update risk register

Product Strategist:

- Communicate status to stakeholders
- Adjust sprint plan if work stopped
- Prioritize remediation work

Validation Lead:

- Address validation gaps identified
- Collect missing evidence
- Schedule additional testing if needed

AI-Augmented Developers:

- Remediate flagged issues
- Update documentation
- Implement required controls

Artifact Updates
Update:

- Outcome Backlog (status, blockers)
- Validation Canvas (new requirements)
- Sprint Tracking (risks, blockers)
- Risk Register (new/updated risks)
- Compliance Documentation (findings, approvals)

GOVERNANCE EFFECTIVENESS METRICS

Track Over Time:

- Checkpoint duration: _____ minutes (target: <60)
- Approval rate: ____% (target: >80% approved or conditional)
- Critical issues found: _____ (lower is better)
- Remediation time: _____ days (target: <2 days)
- Re-review rate: ____% (target: <20%)

Health Indicators:

☐ Most outcomes approved or conditionally approved
☐ Issues found are minor, not critical
☐ Remediation happens quickly
☐ Rare escalations to Board
☐ Team prepared with documentation

Warning Signs:

☐ High not-approved rate (>30%)
☐ Same issues recurring across sprints
☐ Frequent Board escalations
☐ Missing documentation consistently
☐ Last-minute checkpoint prep

COMMON ISSUES & RESOLUTIONS

Issue: Missing Compliance Documentation
Resolution:

- Identify specific gaps
- Assign owners for each document
- Set 24-hour deadline
- Schedule mini-review when complete

Issue: Security Scans Not Run
Resolution:

- Run scans immediately (during checkpoint if possible)
- Conditional approval with scan completion deadline
- If critical vulnerabilities found, stop work

Issue: AI Behavior Concerns
Resolution:

- Document specific concern
- Increase human validation checkpoints
- Consider additional testing
- May require architecture change

Issue: Regulatory Uncertainty
Resolution:

- Pause work on uncertain areas
- Escalate to legal/compliance
- Continue work on compliant portions
- Document conservative interpretation

BOARD REVIEW PREPARATION

If escalation to Board Review required, prepare:

Executive Summary (1 page):

- Outcome description
- Business value
- Risk assessment
- Governance concern
- Recommended decision
- Impact of approval/denial

Supporting Materials:

- Detailed risk analysis
- Compliance assessment
- Technical documentation
- Validation evidence
- Cost-benefit analysis

Board Presentation (if needed):

- 10-minute overview
- Q&A preparation
- Decision options
- Recommendation

Template Version: 1.0
Last Updated: [Date]

RISK ASSESSMENT & CLASSIFICATION

Purpose
The four-tier risk framework classifies AI work to determine appropriate governance oversight. This template ensures teams apply consistent risk assessment across all outcomes.

ASSESSMENT METADATA
Outcome Name: ______________________________
Outcome ID: ________
Sprint: ________
Assessed By: _______________ **Date:** ________
Review Date: ________ (reassess mid-sprint at Governance Checkpoint)

RISK SCORING FRAMEWORK
Rate Each Factor (1-5 Scale)
1 = Minimal Risk | 3 = Moderate Risk | 5 = Severe Risk

Factor 1: Data Sensitivity
Score: ☐ 1 ☐ 2 ☐ 3 ☐ 4 ☐ 5
Scoring Guide:

- **1:** Public data only, no sensitivity
- **2:** Internal business data, low sensitivity
- **3:** Confidential business data
- **4:** PII present (names, emails, addresses)
- **5:** Highly sensitive (health records, financial accounts, SSNs)

Data Involved: [List data elements this outcome will process, store, or access]
Justification for Score: ____________________________

Factor 2: Regulatory Impact
Score: ☐ 1 ☐ 2 ☐ 3 ☐ 4 ☐ 5
Scoring Guide:

- **1:** No regulatory requirements
- **2:** General data protection laws apply
- **3:** Industry-specific regulations (minor impact)
- **4:** Heavy regulatory oversight (HIPAA, PCI DSS, SOX)
- **5:** Multiple jurisdictions, complex compliance (EU AI Act, GDPR + others)

Applicable Regulations:

☐ GDPR

☐ CCPA

☐ HIPAA

☐ PCI DSS

☐ SOX

☐ FERPA

☐ EU AI Act

☐ Other: ____________________

Justification for Score: __________________________________

Factor 3: User Exposure

Score: ☐ 1 ☐ 2 ☐ 3 ☐ 4 ☐ 5

Scoring Guide:

- **1:** Internal tool, <10 users
- **2:** Internal tool, <100 users
- **3:** Internal tool, enterprise-wide OR external tool, limited users
- **4:** External-facing, thousands of users
- **5:** Public-facing, millions of users

User Base:

- Number of users: _____
- User type: ☐ Internal ☐ External ☐ Public
- User vulnerability: ☐ General ☐ Vulnerable Population (children, elderly, etc.)

Justification for Score: __________________________________

Factor 4: System Criticality

Score: ☐ 1 ☐ 2 ☐ 3 ☐ 4 ☐ 5

Scoring Guide:

- **1:** Non-essential, failure has no business impact
- **2:** Low priority, workarounds available
- **3:** Important but not critical, temporary disruption acceptable
- **4:** Business-critical, failure causes significant disruption
- **5:** Mission-critical, failure causes catastrophic business impact

Business Impact of Failure:

- Revenue impact:

  ☐ None

  ☐ Minor

  ☐ Moderate

  ☐ Significant

☐ Severe

- Operational impact:

☐ None

☐ Minor

☐ Moderate

☐ Significant

☐ Severe

- Reputational risk:

☐ None

☐ Minor

☐ Moderate

☐ Significant

☐ Severe

Justification for Score: ______________________________

Factor 5: AI Hallucination Potential

Score: ☐ 1 ☐ 2 ☐ 3 ☐ 4 ☐ 5

Scoring Guide:

- **1:** No AI generation, deterministic code only
- **2:** AI-assisted but fully validated by humans
- **3:** AI generates outputs with human review
- **4:** AI generates user-facing content with limited validation
- **5:** AI makes autonomous decisions affecting users/business

AI Role:

- Type:

☐ No AI

☐ Code Generation

☐ Content Generation

☐ Decision Making

- Human oversight:

☐ Every output

☐ Sampling

☐ Exception-based

☐ Minimal

- Consequence of error:

☐ Low

☐ Medium

☐ High

☐ Severe

Justification for Score: ________________________________

Factor 6: Reversibility

Score: ☐ 1 ☐ 2 ☐ 3 ☐ 4 ☐ 5

Scoring Guide:

- **1:** Easily reversible, instant rollback possible
- **2:** Reversible with some effort, <1 hour to undo
- **3:** Reversible but requires significant work, <1 day
- **4:** Partially reversible, some permanent effects
- **5:** Irreversible actions (delete data, send communications, financial transactions)

Reversibility Analysis:

- Can we roll back?

  ☐ Yes, instantly

  ☐ Yes, with effort

  ☐ Partially

  ☐ No
- Permanent effects: ________________________________
- Recovery time: ________________________________

Justification for Score: ________________________________

TOTAL RISK SCORE

Calculate Total

Factor	Score
Data Sensitivity	___
Regulatory Impact	___
User Exposure	___
System Criticality	___
AI Hallucination Potential	___
Reversibility	___
TOTAL	**___/30**

RISK TIER ASSIGNMENT

Based on total score, assign tier:

☐ TIER 1: Delegated Authority (Score 6-12)

Definition: Low risk, team can proceed with standard practices

Governance Requirements:

- Team documents decisions
- Standard validation practices
- No special approvals needed
- Report in Sprint Review

Decision Authority: Team self-governs
Examples:

- Internal tool with no PII
- AI-assisted code generation with full testing
- Non-critical feature improvements

☐ **TIER 2: Standard Review (Score 13-18)**
Definition: Moderate risk, Governance Coordinator reviews
Governance Requirements:

- Governance Coordinator reviews at Checkpoint
- Enhanced documentation required
- Security scan mandatory
- Compliance checklist completed

Decision Authority: Governance Coordinator
Examples:

- Internal tool processing confidential data
- Customer-facing feature with AI assistance
- Integration with regulated systems

☐ **TIER 3: Board Review (Score 19-24)**
Definition: High risk, governance board approval required
Governance Requirements:

- Formal risk assessment documented
- Legal/compliance review
- Executive summary prepared
- Board presentation (if requested)
- Enhanced monitoring post-deployment

Decision Authority: Governance Board
Examples:

- Public-facing AI decision-making
- Processing of sensitive PII (health, financial)
- Mission-critical system changes
- Novel AI application

☐ **TIER 4: Prohibited (Score 25-30)**
Definition: Unacceptable risk, do not proceed without executive exception
Governance Requirements:

- Work must stop immediately
- Executive escalation required

- Risk mitigation plan mandatory before any approval
- May require external audit/review

Decision Authority: Executive leadership only

Examples:

- Autonomous AI decisions with irreversible financial impact
- AI processing highly sensitive data (children's health records)
- Unproven AI in life-safety applications
- Regulatory non-compliance inevitable

Exception Process: If executive believes exception warranted, require:

- Comprehensive risk mitigation plan
- External expert review
- Enhanced monitoring
- Legal opinion
- Board approval

RISK MITIGATION PLAN

For Tier 2 and above, document mitigation strategies:

Identified Risks

Risk 1:

- **Description:** ________________________________
- **Likelihood:** ☐ Low ☐ Medium ☐ High
- **Impact:** ☐ Low ☐ Medium ☐ High
- **Mitigation:** ________________________________
- **Owner:** ________________
- **Residual Risk:** ☐ Low ☐ Medium ☐ High

Risk 2: [Repeat structure]

CONTINGENCY PLANNING

If This Goes Wrong:

Incident Response:

- Who to notify: ________________________________
- Escalation path: ________________________________
- Response time target: ________________________________

Rollback Plan:

- Can we rollback? Yes/No
- Rollback procedure: ________________________________
- Rollback time: ________________________________
- Data recovery: ________________________________

Communication Plan:

- Internal notification: ________________________________
- Customer notification: ☐ Required ☐ Not Required

- Regulatory notification: ☐ Required ☐ Not Required

MONITORING & REVIEW

Ongoing Monitoring:

- Metrics defined: ________________________________
- Alert thresholds set: ________________________________
- Review frequency: Daily/Weekly/Monthly
- Owner: _________________

Review Triggers: [When to reassess risk tier]

- Scope change
- User base expansion
- Regulatory change
- Security incident
- AI behavior anomaly

Post-Deployment Review:

- Scheduled: _____ days after launch
- Owner: _________________
- Focus: Validate risk assessment accuracy

Template Version: 1.0
Last Updated: [Date]

ROLE TRANSITION PLAYBOOK

Purpose
Guide for evolving traditional Scrum roles to AI Scrum roles, with capability gaps, training needs, timelines, and success indicators for each transition.

TRANSITION 1: Product Owner → Product Strategist
Current State Assessment
Product Owner Capabilities:
- Backlog prioritization
- Stakeholder management
- User story writing
- Sprint planning participation
- Acceptance criteria definition

Experience Level:
- Years as PO: _____
- Team size managed: _____
- Domain expertise: ______________________________

Capability Gaps to Close
New Skills Required:
1. **Outcome-Oriented Thinking**
   - Gap: Feature lists → Business outcomes
   - Training: Outcome definition workshop
   - Practice: Convert 5 user stories to outcomes
2. **AI Strategy & Suitability Assessment**
   - Gap: Manual development assumption → AI-first thinking
   - Training: AI capabilities course
   - Practice: Assess 10 backlog items for AI suitability
3. **Predictive Analytics**
   - Gap: Historical data usage → Predictive modeling
   - Training: Data analysis for POs
   - Practice: Build outcome forecast model
4. **Economic Value Measurement**
   - Gap: Velocity → ROI/business impact
   - Training: Business metrics for product
   - Practice: Define value metrics for 3 outcomes

Transition Timeline: 3-4 Months

Month 1: Foundation

- Complete AI fundamentals training (8 hours)
- Shadow AI Alignment Engineer (2 weeks)
- Convert 5 user stories to outcome format
- Attend outcome definition workshop

Month 2: Practice

- Lead outcome alignment for 1 outcome
- Define success metrics for all sprint outcomes
- Learn prompt engineering basics
- Practice AI suitability assessment

Month 3: Integration

- Own entire Outcome Backlog
- Lead Sprint Inception ceremony
- Use AI for backlog analysis
- Mentor other Product Owners

Month 4: Mastery

- Predictive outcome forecasting
- Advanced stakeholder management
- AI strategy consultation
- Full Product Strategist responsibilities

Success Indicators

Week 4:

☐ Can articulate difference between features and outcomes

☐ Has converted personal backlog to outcomes

☐ Understands AI capabilities and limitations

Week 8:

☐ Leads outcome definition sessions confidently

☐ Asks right questions about AI approach

☐ Defines measurable success metrics

Week 12:

☐ Stakeholders embrace outcome focus

☐ Team delivers business value, not just features

☐ Uses data to prioritize effectively

Month 4:

☐ Recognized as Product Strategist by team/stakeholders

☐ Influences roadmap with predictive insights

☐ Mentors others in outcome thinking

TRANSITION 2: QA Engineer → Validation Lead

Current State Assessment

QA Engineer Capabilities:

- Manual testing
- Test automation
- Bug reporting
- Test plan creation
- Regression testing

Testing Focus:

- Manual: ____%
- Automated: ____%
- Specialization: ________________________________

Capability Gaps to Close

New Skills Required:

1. **Validation-First Mindset**
    - Gap: Post-development testing → Pre-implementation validation definition
    - Training: Validation Canvas workshop
    - Practice: Create 5 validation canvases
2. **AI-Specific Testing**
    - Gap: Deterministic testing → Probabilistic validation
    - Training: AI testing methodologies
    - Practice: Test AI-generated code
3. **Adversarial Testing & Red-Teaming**
    - Gap: Happy path focus → Hostile input scenarios
    - Training: Security testing fundamentals
    - Practice: Red-team 3 AI outputs
4. **Validation Strategy Leadership**
    - Gap: Execute tests → Design validation approach
    - Training: Strategic thinking for QA
    - Practice: Lead validation planning

Transition Timeline: 2-3 Months

Month 1: Foundation

- AI testing fundamentals (8 hours)
- Shadow Validation Lead (if available)
- Create first Validation Canvas
- Learn adversarial testing basics

Month 2: Practice

- Lead validation definition for 2 outcomes

- Conduct first red-team session
- Implement AI output validation
- Present validation strategy

Month 3: Integration

- Own all sprint validation planning
- Mentor developers on validation
- Build validation automation
- Full Validation Lead responsibilities

Success Indicators

Week 4:

☐ Creates validation criteria before seeing code

☐ Understands AI-specific testing needs

☐ Can explain validation-first approach

Week 8:

☐ Team validates during development, not after

☐ Red-teams AI outputs effectively

☐ Validation prevents rework

Week 12:

☐ Validation canvas drives sprint work

☐ Team trusts validation leadership

☐ AI outputs validated comprehensively

TRANSITION 3: Scrum Master → Delivery Facilitator

Capability Gaps to Close

1. **AI Workflow Optimization**
    - Gap: Human workflow → Human-AI collaboration patterns
    - Training: AI augmentation workflows
    - Practice: Optimize 3 team processes
2. **Governance Facilitation**
    - Gap: Remove impediments → Navigate governance
    - Training: Governance framework understanding
    - Practice: Facilitate 2 Governance Checkpoints
3. **Validation Orchestration**
    - Gap: Development coordination → Validation coordination
    - Training: Validation-first ceremonies
    - Practice: Lead Daily Sync with validation focus

4. **AI Metrics & Observability**
    - Gap: Velocity tracking → Outcome + AI metrics
    - Training: AI Scrum metrics
    - Practice: Build validation-focused dashboard

Transition Timeline: 2-3 Months
[Similar structure to above]

TRANSITION 4: Business Analyst → AI Alignment Engineer
Capability Gaps to Close

1. **Prompt Engineering Mastery**
    - Gap: Requirements → AI instructions
    - Training: Advanced prompt engineering certification
    - Practice: Build prompt library (50+ prompts)
2. **Context Specification Creation**
    - Gap: User stories → Context specifications
    - Training: Context spec workshop
    - Practice: Create 10 context specifications
3. **AI Collaboration Optimization**
    - Gap: Human stakeholders → Human-AI interface
    - Training: AI collaboration patterns
    - Practice: Optimize team AI usage

Transition Timeline: 4-6 Months
[Similar structure]

TRANSITION 5: Developer → AI-Augmented Developer
Capability Gaps to Close

1. **Prompt Engineering for Code**
    - Gap: Write code → Instruct AI to write code
    - Training: Developer-focused prompt engineering
    - Practice: Generate 20 implementations with AI
2. **AI Output Validation**
    - Gap: Trust own code → Validate AI code
    - Training: AI code review techniques
    - Practice: Review 50 AI-generated implementations
3. **Architecture for AI Integration**
    - Gap: Traditional patterns → AI-friendly architectures
    - Training: Designing for AI

  - Practice: Refactor 3 systems

Transition Timeline: 2-4 Months
[Similar structure]

GENERAL TRANSITION SUPPORT
For All Roles:
Psychological Safety:

- Growth framed as opportunity, not obsolescence
- Mistakes during transition expected and accepted
- Pairing with experienced practitioners
- Regular check-ins with manager

Resources Provided:

- Training budget: $_____
- Conference attendance
- Certification support
- Book/course allowance

Success Metrics:

- Transition completion rate: Target 90%
- Role effectiveness at 6 months: Target 80% proficiency
- Team satisfaction with new structure: Target 75%+

Template Version: 1.0
Last Updated: [Date]

TEST STRATEGY FOR AI-GENERATED CODE

Purpose
Comprehensive testing approach for AI-generated implementations, addressing unique validation challenges of probabilistic outputs.

Test Strategy Components

1. Unit Test Generation
Pre-Generation Checklist:
- Context Specification defines test requirements
- Validation Canvas specifies coverage targets
- Edge cases documented

AI Prompt for Unit Tests:
Generate comprehensive unit tests for [function/class].

Requirements:
- Test framework: [Jest/PyTest/JUnit]
- Coverage target: 90%+
- Include: happy path, edge cases, error conditions
- Test data: [specify]
- Assertion style: [specify]

Edge cases to cover:
- [List specific scenarios]

Do NOT:
- Skip error handling tests
- Use magic numbers
- Hardcode test data

Human Validation Required:
- All edge cases from Validation Canvas covered
- Tests actually test logic (not just syntax)
- Assertions meaningful, not trivial
- No false positives (tests pass when they shouldn't)
- No test interdependencies

Coverage Metrics:
- Line coverage: Target ≥90%
- Branch coverage: Target ≥85%
- Function coverage: Target 100%

2. Integration Test Scenarios

Integration Points Matrix:

System A	System B	Interface	Test Scenario	Priority
[Service 1]	[Service 2]	REST API	[Scenario]	P0/P1/P2

AI-Assisted Integration Test Generation:

Generate integration tests for [component A] → [component B].

Interface contract:
[Paste API spec/schema]

Test scenarios:
1. Normal flow: [describe]
2. Error flow: [describe]
3. Timeout: [describe]
4. Data mismatch: [describe]

Validation:

- Response format correct
- Data transformations accurate
- Error handling graceful

Failure Scenario Testing:

- Network timeout
- Service unavailable
- Partial failure
- Data corruption
- Race conditions

---

3. End-to-End Test Automation

User Workflow Mapping:

Workflow 1: ______________________________

- Steps: 1) ______ 2) ______ 3) ______
- Expected outcome: ____________________
- Test data needed: ____________________

AI-Generated E2E Tests:

- Tool: Playwright/Selenium/Cypress
- Realistic user scenarios
- Test data generation
- Visual regression checks

E2E Test Validation:

- Tests reflect actual user behavior

- No brittle selectors
- Proper wait strategies
- Cleanup after tests

---

4. Adversarial Testing (AI-Specific)
Adversarial Test Categories:
A. Prompt Injection Attempts:

- Direct injection: "Ignore previous instructions..."
- Indirect injection: Via user input fields
- Nested prompts
- Context manipulation

B. Hallucination Detection:

- Factual accuracy verification
- Citation checking (if applicable)
- Consistency across runs
- Known truth validation

C. Boundary Testing:

- Token limit approaches
- Unusual input formats
- Multilingual inputs
- Special characters/encoding

D. Bias Testing:

- Demographic variations
- Cultural sensitivity
- Fairness metrics

Red Team Checklist:

- 50+ adversarial inputs tested
- All P0 edge cases pass
- Graceful degradation verified
- No data leakage

---

5. Test Maintenance Strategy
AI-Assisted Test Maintenance:
Detect Obsolete Tests:
Analyze test suite for:

- Tests that always pass (code changed, test didn't)
- Tests covering removed functionality
- Duplicate test coverage

Update Tests for Code Changes:
Given code change: [describe]

Update affected tests:

- Modified functions: [list]
- New edge cases: [list]
- Changed interfaces: [list]

Refactor Brittle Tests:

- Identify flaky tests (pass/fail inconsistently)
- Root cause analysis
- AI-suggested improvements
- Human validation of fixes

Test Suite Health Metrics:

- Test execution time: Target <5 min for unit tests
- Flaky test rate: Target <2%
- Test coverage trend: Target stable or increasing
- Maintenance overhead: Track time spent fixing tests

Testing Workflow

Phase 1: Definition (During Sprint Inception)

- Validation Canvas defines test requirements
- Test strategy selected per outcome
- Test data requirements identified

Phase 2: Generation (During Development)

- AI generates tests alongside code
- Human validates test quality
- Red team testing planned

Phase 3: Execution (Continuous)

- Automated tests run on every commit
- Integration tests run on PR
- E2E tests run on staging deploy

Phase 4: Maintenance (Ongoing)

- Weekly test suite health check
- Quarterly test refactoring
- Continuous test evolution

Template Version: 1.0
Last Updated: [Date]

METRICS DASHBOARD BLUEPRINT

Purpose
Visual dashboard for AI Scrum metrics, designed for 5-second comprehension with traffic light indicators.

Dashboard Structure
Section 1: Outcome Achievement (Top Priority)
Committed Outcomes This Sprint: _____
Outcome Achievement Rate:

- Delivered outcomes: /
- Target: 80%+
- Status: 🟢 ≥80% | 🟡 60-79% | 🔴 <60% *(Green, Yellow, Red)*

Visual: Horizontal bar showing completed vs. remaining outcomes

Metric Details:

Outcome	Status	Validation	Evidence	% Complete
[Name]	🟢 🔴	Pass/Fail	[Type]	___%

Section 2: Validation Metrics
Validation Lead Time:

- Average: _____ hours (time from code-complete to validated)
- Target: <24 hours
- Trend: ↗ ↘ → (compared to last sprint)
- Status: 🟢 <24h | 🟡 24-48h | 🔴 >48h

First-Pass Acceptance Rate:

- Rate: _____%
- Target: ≥70%
- Calculation: (Accepted without changes) / Total validations
- Status: 🟢 ≥70% | 🟡 50-69% | 🔴 <50%

Validation Criteria Satisfied:

- This sprint: / criteria met
- By category: Functional ___%, Performance ___%, Security ___%, Compliance ____%

Section 3: Governance & Risk
Governance Cycle Time:

- Average review time: _____ days
- Target: <3 days

- Status: ● <3d | ● 3-5d | ● >5d

Risk Distribution:

Tier	Count	%
Delegated	__	__%
Standard Review	__	__%
Board Review	__	__%
Prohibited	__	__%

Compliance Status:

- Compliant outcomes: /
- Open compliance issues: _____
- Status: ● All compliant | ● Minor issues | ● Critical issues

---

Section 4: AI Collaboration Effectiveness

AI Generation Efficiency:

- Time saved this sprint: _____ hours
- Average time saved per task: _____ minutes
- ROI: ____% (time saved vs. AI cost)

AI Output Quality:

- First-pass acceptance: _____%
- Refinement iterations avg: _____
- Rejection rate: _____%

Prompt Library Health:

- Total prompts: _____
- Reused this sprint: _____
- New prompts added: _____
- Success rate of library prompts: _____%

Section 5: Velocity & Capacity (De-emphasized but tracked)

Story Points (Outcome-Focused):

- Committed: _____ points
- Completed: _____ points
- Completion rate: _____%

Capacity Utilization:

- Validation capacity used: _____%
- Development capacity used: _____%
- Ideal ratio (2:1 validation:development)

Section 6: Quality & Defects
Defect Escape Ratio:

- Production defects: _____
- Pre-production defects: _____
- Ratio: _____ (lower is better)
- Target: <5%
- Status: 🟢 <5% | 🟡 5-10% | 🔴 >10%

AI Debt Index:

- Unvalidated AI outputs: _____
- Uncertain AI behaviors: _____
- Technical debt from AI: _____ story points
- Trend: ↗ ↘ →

Dashboard Design Principles
5-Second Comprehension Rule:

- Green/Yellow/Red traffic lights visible immediately
- Arrows show trends (↗ improving, ↘ declining, → stable)
- Numbers secondary to visual indicators

Information Hierarchy:

1. Outcome achievement (most important)
2. Validation metrics
3. Governance/risk
4. AI collaboration
5. Traditional velocity (least important)

Update Frequency:

- Real-time: Outcome status
- Daily: Validation metrics
- Weekly: Sprint trends
- Monthly: Historical comparisons

Template Version: 1.0
Last Updated: [Date]

AI SCRUM RACI MATRIX

Purpose

Clarify responsibility, accountability, consultation, and information flow across AI Scrum activities for all seven roles.

RACI Legend

- **R (Responsible):** Does the work
- **A (Accountable):** Final decision authority, one A per activity
- **C (Consulted):** Provides input before decision
- **I (Informed):** Notified after decision

AI SCRUM RACI MATRIX

Activity	Product Strategist	AI Alignment Engineer	Validation Lead	Governance Coordinator	Integration Architect	AI-Augmented Developer	Delivery Facilitator
PLANNING & DEFINITION							
Outcome Backlog Management	A/R	C	C	C	C	I	I
Sprint Inception Facilitation	R	R	R	R	R	R	A
Outcome Alignment	A	R	C	C	I	I	R
Validation Definition	C	R	A/R	C	C	C	I
Governance Framing	C	I	I	A/R	C	I	R
IMPLEMENTATION							
Context Specification Creation	C	A/R	C	I	C	R	I
AI Prompt Engineering	C	A/R	I	I	I	R	I
Code Generation (AI-assisted)	I	C	I	I	C	A/R	I
Code Review	C	R	R	I	R	A	I
VALIDATION & QUALITY							
Test Strategy Design	C	R	A/R	C	C	R	I
Test Execution	I	R	A	I	C	R	I
Adversarial Testing	C	R	A/R	C	I	R	I
Performance Validation	C	I	A/R	I	R	R	I
Security Validation	I	I	R	A	R	R	C
GOVERNANCE & COMPLIANCE							
Risk Assessment	C	C	C	A/R	C	C	R
Governance Checkpoint	R	R	R	A	R	R	R

Compliance Documentation	C	I	C	A/R	I	I	I
Board Review Preparation	A	C	C	R	C	I	R
CEREMONIES							
Daily Synchronization	R	R	R	R	R	R	A
AI Show & Tell	A	R	R	I	R	R	R
Sprint Review	A	R	R	R	R	R	R
Sprint Retrospective	R	R	R	R	R	R	A
COLLABORATION & LEARNING							
AI Collaboration Log Maintenance	I	A	C	I	I	R	R
Prompt Library Management	C	A/R	I	I	I	R	I
Knowledge Sharing	R	R	R	R	R	R	A
INTEGRATION & DEPLOYMENT							
Integration Design	C	C	C	C	A/R	R	I
Deployment Planning	C	C	R	C	R	A	R
Production Monitoring	C	I	R	C	R	A	I

Key Collaboration Interfaces
Product Strategist ↔ AI Alignment Engineer
Primary Interface: Outcome → AI-Compatible Specification
Weekly touchpoints: 3-5 times **Critical for:** Intent clarification, AI approach selection

---

AI Alignment Engineer ↔ AI-Augmented Developer
Primary Interface: Prompt Strategy → Implementation
Daily touchpoints: Multiple **Critical for:** Effective AI collaboration, prompt debugging

---

Validation Lead ↔ AI-Augmented Developer
Primary Interface: Validation Strategy → Implementation + Testing
Daily touchpoints: 2-3 times **Critical for:** Build quality in, not inspect after

---

Governance Coordinator ↔ All Roles
Primary Interface: Risk Management → Compliant Delivery
Touchpoints: Checkpoint + ad-hoc **Critical for:** Navigate governance without delays

---

Template Version: 1.0
Last Updated: [Date]

MATURITY ASSESSMENT SCORECARD

Purpose
Self-assessment tool for organizational readiness across 8 AI Scrum capability dimensions.

Maturity Levels
Level 1 - Reactive: Chaotic, ad hoc, unpredictable
Level 2 - Developing: Some consistency, high variation
Level 3 - Operational: Standardized, repeatable, systematic
Level 4 - Optimizing: Data-driven improvement, refinement
Level 5 - Transformative: Industry leadership, innovation

DIMENSION 1: Agile Process Maturity
Assessment Questions
Select ONE level that best describes your organization:

☐ **Level 1 - Reactive**
- Teams use chaotic Agile or Waterfall
- Velocity untracked or manipulated
- No outcome focus

☐ **Level 2 - Developing**
- AI Scrum ceremonies adopted but mechanical
- Teams track outputs, not outcomes
- Forms without substance

☐ **Level 3 - Operational**
- Outcome Alignment replaces Sprint Planning
- Validation-first is standard
- AI Collaboration Log maintained

☐ **Level 4 - Optimizing**
- Process bottlenecks identified via data
- AI compresses specific phases measurably
- Continuous refinement

☐ **Level 5 - Transformative**
- Organization invents new methodologies
- Industry benchmark for AI-native development
- Methodology contributions published

Score: _____/5

DIMENSION 2: AI Integration Depth

☐ **Level 1** - Experimental (sporadic AI use, manual paste)

☐ **Level 2** - Tool Deployment (Copilot but as autocomplete)

☐ **Level 3** - Systematic Integration (Context Specs, RAG, AI as teammate)

☐ **Level 4** - Partnership Optimization (Custom agents, fine-tuning)

☐ **Level 5** - Partnership Revolution (Proprietary models, competitive advantage)

Score: _____/5

DIMENSION 3: Governance & Risk Management

☐ **Level 1** - Governance Vacuum ("move fast, break things")

☐ **Level 2** - Policy Establishment (PDFs exist, not enforced)

☐ **Level 3** - Embedded Governance (Checkpoints in workflow)

☐ **Level 4** - Predictive Governance (Risk-based automation)

☐ **Level 5** - Governance Innovation (Shape regulatory standards)

Score: _____/5

DIMENSION 4: Team Capabilities

☐ **Level 1** - Traditional Skills (No AI expertise)

☐ **Level 2** - Basic Training (Intro courses completed)

☐ **Level 3** - Applied Proficiency (Prompt engineering practiced)

☐ **Level 4** - Advanced Expertise (Specialists, mentors)

☐ **Level 5** - Industry Leadership (Publish, speak, teach)

Score: _____/5

DIMENSION 5: Tooling & Infrastructure

☐ **Level 1** - No AI Tools (Blocked or unavailable)

☐ **Level 2** - Basic Access (Personal accounts, no enterprise)

☐ **Level 3** - Enterprise Deployment (Secured, logged, governed)

☐ **Level 4** - Optimized Stack (Integrated, automated, monitored)

☐ **Level 5** - Custom Platform (Proprietary capabilities)

Score: _____/5

DIMENSION 6: Metrics & Observability

☐ **Level 1** - No Metrics (Flying blind)
☐ **Level 2** - Basic Tracking (Velocity only)
☐ **Level 3** - Outcome Metrics (Validation, AI effectiveness)
☐ **Level 4** - Automated Insights (AI-powered anomaly detection)
☐ **Level 5** - Predictive Analytics (Forecasting, optimization)

Score: _____/5

DIMENSION 7: Change Management

☐ **Level 1** - Resistance (Active pushback, no adoption)
☐ **Level 2** - Awareness (Training started, skepticism remains)
☐ **Level 3** - Adoption (Teams using AI, growing confidence)
☐ **Level 4** - Advocacy (Champions emerging, organic spread)
☐ **Level 5** - Cultural Norm (AI-first thinking default)

Score: _____/5

DIMENSION 8: Strategic Alignment

☐ **Level 1** - No Strategy (AI is incidental)
☐ **Level 2** - Experimental Budget (Small pilots)
☐ **Level 3** - Tactical Integration (AI in roadmap)
☐ **Level 4** - Strategic Priority (Executive sponsorship)
☐ **Level 5** - Competitive Differentiator (Core to business model)

Score: _____/5

MATURITY CALCULATION
Total Score: _____/40

Overall Maturity Level:

- 8-16 points: **Level 1-2** (Early Stage)
- 17-24 points: **Level 3** (Operational)
- 25-32 points: **Level 4** (Optimizing)
- 33-40 points: **Level 5** (Transformative)

Maturity Profile:

Dimension	Score	Gap to Level 3	Priority
Agile Process	__/5	__	High/Med/Low
AI Integration	__/5	__	High/Med/Low
Governance	__/5	__	High/Med/Low
Team Capabilities	__/5	__	High/Med/Low
Tooling	__/5	__	High/Med/Low
Metrics	__/5	__	High/Med/Low
Change Management	__/5	__	High/Med/Low
Strategy	__/5	__	High/Med/Low

Structural Tension: [Identify dimensions with largest gaps]
Recommended Focus: [Which 2-3 dimensions to prioritize]

Template Version: 1.0
Last Updated: [Date]

SPRINT RETROSPECTIVE WITH AI METRICS

Purpose
Enhanced retrospective incorporating AI-specific reflection on prompt effectiveness, validation quality, governance friction, and learning velocity.

Retrospective Structure
Sprint: ________
Date: ________
Facilitator: ________________
Duration: 90 minutes

SECTION 1: Traditional Retrospective (30 min)
What Went Well

- ________
- ________
- ________

What Needs Improvement

- ________
- ________
- ________

Action Items from Last Sprint

Action	Owner	Status	If incomplete, why?
[Action 1]	[Person]	☐ Done ☐ In Progress ☐ Blocked	

SECTION 2: AI Collaboration Review (20 min)
Prompt Effectiveness Analysis
Most Effective Prompts This Sprint:

1. **Prompt Pattern:** ________________________________
   - **Success Rate:** _____%
   - **Why it worked:** ________________________________
   - **Reuse recommendation:** Add to library / Refine / Document
2. **Prompt Pattern:** ________________________________
   [Repeat structure]

Least Effective Prompts:

1. **Prompt Pattern:** ______________________________
   - **Failure Mode:** ______________________________
   - **Root Cause:** ☐ Vague prompt ☐ Missing context ☐ AI limitation ☐ Wrong tool
   - **Learning:** ______________________________

Context Specification Quality:

- Used Context Specs: _____ times
- First-pass acceptance with Context Specs: _____%
- First-pass acceptance without: _____%
- **Insight:** ______________________________

AI Tool Performance

Tools Used This Sprint:

Tool	Usage Count	Success Rate	Cost	Keep/Replace/Add
[GPT-x]	__	__%	$__	Keep/Replace/Add
[Claude]	__	__%	$__	Keep/Replace/Add

Tool Switching:

- Did we use the right tool for each task? Yes/No
- Examples of tool mismatch: ______________________________

AI Collaboration Log Insights

Log Completeness: ____% of AI work logged **Average time saved per entry:** _____ minutes **Total time saved this sprint:** _____ hours

Patterns Identified:

- ______________________________
- ______________________________

Prompt Library Growth:

- Prompts added: _____
- Prompts refined: _____
- Library size: ___

SECTION 3: Validation Quality Review (15 min)

Validation Canvas Effectiveness

Validation Criteria Clarity:

- Were criteria clear enough before implementation? ☐ Always ☐ Usually ☐ Sometimes ☐ Rarely

- Times we discovered missing criteria mid-sprint: _____

Validation Timeline:

- Average validation lead time: _____ hours
- Target: <24 hours
- Gap analysis: ________________________________

Validation Bottlenecks:

- What slowed validation? (Rank 1-5)
    - Missing test data (_____)
    - Environment issues (_____)
    - Stakeholder availability (_____)
    - AI output quality (_____)
    - Security scanning delays (_____)
    - Other: ________________ (_____)

First-Pass Acceptance Rate:

- This sprint: _____%
- Last sprint: _____%
- Trend: ↗ ↘ →
- **Analysis:** ________________________________

SECTION 4: Governance Friction Analysis (10 min)

Governance Checkpoint Review

Checkpoint Effectiveness:

- Held on schedule: Yes/No
- Duration: _____ minutes (target: 60)
- Outcomes reviewed: _____
- Issues found: _____ Critical, _____ High, _____ Medium

Governance as Enabler or Blocker? ☐ Enabler - Helped us move fast safely ☐ Neutral - Didn't help or hinder ☐ Blocker - Slowed us down unnecessarily

If Blocker, why?

- Late discovery of requirements
- Unclear governance criteria
- Slow turnaround on reviews
- Risk assessment inconsistent
- Other: ________________________________

Improvement Actions:

- ____
- ____

SECTION 5: Outcome Achievement Analysis (10 min)
Committed vs. Delivered
Outcomes Committed: _____
Outcomes Delivered: _____
Completion Rate: _____%

For Incomplete Outcomes:

Outcome	% Complete	Why incomplete?	Carry to next sprint?
[Name]	__%	[Reason]	Yes/No

Root Cause Analysis:

- Underestimated validation complexity
- Requirements changed mid-sprint
- AI pivot required more time
- Dependencies blocked us
- Governance issues
- Team capacity issue
- Other: ______________________________

SECTION 6: Objective Telemetry Review (5 min)
Sprint Metrics
Present Objective Data (Not opinions):
Velocity & Capacity:

- Story points committed: _____
- Story points completed: _____
- Validation/Development ratio: _____:1 (ideal: 2:1)

Quality:

- Defects found in sprint: _____
- Defects escaped to next stage: _____
- Defect escape ratio: _____%

AI Effectiveness:

- Time saved via AI: _____ hours
- AI output rework time: _____ hours
- Net efficiency: _____%

Governance:

- Governance cycle time: _____ days (target: <3)
- Compliance issues: _____
- Escalations: _____

Trends: [Show 3-sprint trend lines for key metrics]

SECTION 7: Action Planning (10 min)
Actions for Next Sprint
Category: Process

- **Action:** ______________________________
  - **Owner:** _______________
  - **Due:** Sprint Day ___
  - **Success Criteria:** ______________________________

Category: AI Collaboration

- **Action:** ______________________________
  - **Owner:** _______________
  - **Due:** Sprint Day ___
  - **Success Criteria:** ______________________________

Category: Validation

- **Action:** ______________________________
  - **Owner:** _______________
  - **Due:** Sprint Day ___
  - **Success Criteria:** ______________________________

Category: Governance

- **Action:** ______________________________
  - **Owner:** _______________
  - **Due:** Sprint Day ___
  - **Success Criteria:** ______________________________

Experiments to Try:

- ____________________________________
- ____________________________________

SECTION 8: Team Health Check (5 min)
Anonymous Team Sentiment:

How do you feel about AI collaboration?

☐ Energized

☐ Neutral

☐ Frustrated

How confident are you in our validation approach?

☐ Very

☐ Somewhat

☐ Not confident

Is governance helping or hindering?

☐ Helping

☐ Neutral

☐ Hindering

Do you feel psychological safety to make mistakes while learning AI?

☐ Yes

☐ Sometimes

☐ No

Red Flags to Address: [Facilitator notes any concerning patterns]
RETROSPECTIVE EFFECTIVENESS
Action Item Completion Rate (from last retro): _____% **Team Satisfaction with Retro:**

☐ Valuable

☐ Okay

☐ Waste of time

Attendance: ____
Meta-Retrospective (quarterly): Are our retrospectives leading to real improvement? Yes/No What should we change about how we retrospect?

Template Version: 1.0
Last Updated: [Date]

CENTER OF EXCELLENCE (COE) CHARTER TEMPLATE

Primary Purpose: Why this CoE exists: [Example for AI Alignment CoE: To centralize prompt engineering strategies, model selection standards, and Context Specification templates, preventing redundant problem-solving across teams and accelerating organizational AI capability.]

[Example for Governance & Risk CoE: To interpret regulations (EU AI Act, NIST), define data privacy standards, approve new AI tools, and ensure all AI implementations meet legal, ethical, and security requirements.]

Mission Statement
[One sentence describing the CoE's aspirational goal]
Example: "To establish [Organization Name] as an industry leader in safe, effective, and scalable AI-augmented development practices."

---

1. SCOPE & AUTHORITY

In Scope

The CoE is responsible for:

- Standardization of [prompts/governance processes/tools]
- Knowledge management and dissemination
- Capability building across teams
- Tool evaluation and approval
- Best practice documentation
- Training curriculum development
- Quality standards enforcement
- [Additional scope items]

Out of Scope

The CoE is NOT responsible for:

- Day-to-day sprint execution
- Individual team performance management
- Budget ownership for team tooling
- Direct project delivery
- [Additional out-of-scope items]

Decision Authority

Delegated Authority (CoE can decide independently):

- Standard template updates
- Internal training schedules
- Prompt library additions (pending validation)
- Best practice recommendations

Recommended Authority (CoE recommends, leadership approves):

- New AI tool adoption
- Governance policy changes
- Budget requests over $[threshold]
- Organizational process changes

Stop Authority (CoE can halt work): For Governance & Risk CoE:

- Deployment of non-compliant AI implementations
- Use of prohibited AI tools
- Data privacy violations
- [Additional stop conditions]

---

2. COMPOSITION & ROLES

Leadership Structure

CoE Lead: [Name & Title]

- **Commitment:** [X hours/week]
- **Responsibilities:**
    - Chair CoE meetings
    - Set strategic direction
    - Report to executive sponsors
    - Break tie votes
    - Approve charter changes

Deputy Lead (optional): [Name & Title]

- **Commitment:** [X hours/week]
- **Responsibilities:** Support CoE Lead, act as backup

Core Membership

Name	Role/Team	Time Commitment	Term Length
[Name]	[AI Alignment Engineer, Team Alpha]	[4 hrs/week]	[6 months]
[Name]	[Governance Coordinator, Team Beta]	[3 hrs/week]	[6 months]
[Name]	[Integration Architect, Team Gamma]	[2 hrs/week]	[6 months]
...	...	...	...

Membership Criteria:

- Must be active practitioner in relevant role (AI Alignment Engineer, Governance Coordinator, etc.)
- Demonstrated expertise in [domain area]
- Commitment to knowledge sharing
- Executive sponsor approval

Rotation Policy:

- Core members serve [6-12 month] terms
- Stagger rotations to maintain continuity (max 30% turnover per quarter)
- Former members become "CoE Alumni" with advisory status

Extended Network
CoE Contributors (ad-hoc participation):

- Subject matter experts invited for specific topics
- Pilot team representatives
- External consultants
- Vendor representatives (when appropriate)

---

3. OPERATIONS & CADENCE

Meeting Schedule

Regular Meetings:

- **Frequency:** [Bi-weekly / Monthly]
- **Duration:** [60 / 90 minutes]
- **Day/Time:** [Every other Tuesday, 2:00-3:00 PM EST]
- **Required Attendance:** [Minimum 70% of core members]

Meeting Format:

1. **Review (15 min):** Actions from previous meeting
2. **Knowledge Share (20 min):** Teams present learnings, challenges, solutions
3. **Standard Work (15 min):** Update templates, prompt library, governance canvas
4. **Strategic Topics (10 min):** New tools, regulatory changes, capability gaps
5. **Action Planning (10 min):** Assign ownership for follow-up items

Ad-Hoc Meetings: Triggered by:

- New regulation announcement
- Security incident
- Tool evaluation urgency
- Executive request

Communication Channels

- **Primary:** [Slack channel #coe-ai-alignment / Microsoft Teams]
- **Documentation:** [Confluence / SharePoint / Internal Wiki]
- **Announcements:** [Monthly newsletter to all teams]
- **Escalations:** [Email to executive sponsor]

Decision-Making Process
Consensus Model:

- Strive for unanimous agreement
- If consensus fails after two discussions, move to majority vote
- CoE Lead breaks ties
- Dissenting opinions documented in meeting notes

Quorum: [60%] of core members required for binding decisions

---

4. KEY DELIVERABLES & OUTPUTS

Primary Artifacts

For AI Alignment CoE:

Deliverable	Owner	Update Frequency	Distribution
Prompt Library	[Name]	Bi-weekly	All teams
Context Specification Templates	[Name]	Monthly	All teams
Model Selection Guide	[Name]	Quarterly	Product Strategists, Architects
AI Collaboration Best Practices	[Name]	Monthly	All teams
Approved Tools List	[Name]	As needed	All teams
Training Curriculum	[Name]	Quarterly	HR, Team Leads

For Governance & Risk CoE:

Deliverable	Owner	Update Frequency	Distribution
Governance Canvas Template	[Name]	Quarterly	All teams
Approved Tools List	[Name]	Monthly	All teams
Risk Classification Framework	[Name]	Semi-annually	Product Strategists, Governance Coordinators
Compliance Checklist	[Name]	As regulations change	All teams
Data Privacy Guidelines	[Name]	Quarterly	All teams
Incident Response Playbook	[Name]	Semi-annually	Security, Legal, Teams

Success Metrics

Efficiency Metrics:

- Time to onboard new team member to AI proficiency: [Target: <2 weeks]
- Prompt reuse rate across teams: [Target: >60%]
- Duplicate problem-solving incidents: [Target: <5% of all issues]

Quality Metrics:

- First-pass acceptance rate of AI-generated code: [Target: >75%]
- Governance review cycle time: [Target: <48 hours]
- Policy compliance rate: [Target: 100%]

Adoption Metrics:

- Prompt library usage: [Target: >80% of teams]
- Training completion rate: [Target: >90% within 3 months of hire]
- CoE meeting attendance: [Target: >80%]

---

5. BUDGET & RESOURCES

Annual Budget (if applicable)

Total: $[Amount]

Breakdown:

- Training & Development: $[Amount]
- Tool Licenses (for CoE testing): $[Amount]
- External Expertise/Consulting: $[Amount]
- Conference Attendance: $[Amount]
- Documentation Tools: $[Amount]

Budget Authority: [CoE Lead can approve up to $X without additional approval]

Resource Allocation

Time Investment:

- Core members: [2-5 hours/week average]
- CoE Lead: [8-12 hours/week]
- Executive sponsor: [1-2 hours/month]

Infrastructure:

- Dedicated collaboration space: [Yes/No]
- Shared repository access: [GitHub/GitLab/Confluence]
- Dashboard/Metrics platform: [Azure DevOps/Jira/Custom]

---

6. MATURITY ROADMAP

Crawl Phase (Months 1-4)

Goal: Establish foundational processes and build trust

Activities:

- Identify enthusiasts from each team
- Create dedicated communication channel
- Begin informal knowledge sharing
- Draft initial version of primary deliverables
- Document first 10 lessons learned

Success Criteria:

- 100% of teams have representative engaged
- First version of [Prompt Library/Governance Canvas] published
- Executive awareness established

Walk Phase (Months 5-12)

Goal: Formalize structure and demonstrate value

Activities:

- Appoint official CoE Lead
- Establish regular meeting cadence
- Publish comprehensive deliverables
- Launch first training program
- Measure and report metrics

Success Criteria:

- 70% meeting attendance
- Measurable time savings from reused solutions
- Positive feedback from teams (>4/5 satisfaction)
- Executive sponsorship secured

Run Phase (Month 13+)

Goal: Achieve operational excellence and strategic influence

Activities:

- Grant CoE budget authority
- Certify prompts/processes for production use
- Influence organizational hiring/training strategy
- Represent organization in industry forums
- Contribute to open-source AI governance frameworks

Success Criteria:

- CoE recognized as strategic asset by leadership
- Demonstrable ROI (time saved, risk avoided)
- Industry recognition (presentations, publications)
- Sustainable funding model

7. GOVERNANCE & OVERSIGHT

Executive Sponsorship

Primary Sponsor: [Name, Title]

- **Responsibilities:**
  - Remove organizational blockers
  - Secure budget and resources
  - Champion CoE in leadership meetings
  - Approve major policy changes

Frequency of Engagement: [Monthly review meetings]

Reporting Structure

Monthly Report (to Executive Sponsor) includes:

1. Key decisions made
2. Deliverables published
3. Adoption metrics
4. Budget status
5. Risks and escalations
6. Requests for leadership action

Quarterly Business Review (to broader leadership):

- Strategic impact summary
- ROI calculation
- Capability maturity assessment
- Roadmap for next quarter

Charter Review & Amendment Process

Review Cycle: Quarterly (minimum annually)

Amendment Process:

1. Any core member can propose amendment
2. CoE discusses in regular meeting
3. If majority supports, draft updated charter
4. Executive sponsor approves final version
5. Communicate changes to all stakeholders

Version Control:

- All charter versions archived
- Changes tracked in revision history
- Stakeholders notified of material changes

---

8. SUCCESS INDICATORS & RISKS

Indicators of CoE Health

Positive Signs:

- Teams proactively contribute solutions to library
- New hires ask "Where's the CoE documentation?"

- Leadership cites CoE standards in strategic decisions
- Cross-team collaboration increases
- Redundant work measurably decreases

Warning Signs:

- Meeting attendance drops below 60%
- Deliverables not updated on schedule
- Teams bypass CoE processes
- Executive sponsor disengaged
- No measurable impact after 6 months

Risk Mitigation

Risk	Probability	Impact	Mitigation Strategy
Member burnout due to time commitment	Medium	High	Rotate responsibilities, limit meetings, executive support for time allocation
CoE becomes bureaucratic blocker	Medium	High	Maintain "enabler not gatekeeper" culture, measure approval cycle times
Lack of executive support	Low	Critical	Demonstrate ROI early, regular sponsor engagement, align to business goals
Knowledge hoarding vs. sharing	Medium	Medium	Incentivize contributions, recognize sharing in performance reviews
Tool/tech changes make deliverables obsolete	High	Medium	Quarterly review cycle, version control, AI-assisted updates

---

9. APPENDICES

Appendix A: CoE Member Roles & Responsibilities

[Detailed RACI for CoE activities]

Appendix B: Template Library Index

[Links to all CoE-maintained templates]

Appendix C: Training Curriculum

[Onboarding path for new team members]

Appendix D: Metrics Dashboard

[Link to live metrics tracking CoE performance]

Appendix E: Meeting Minutes Archive

[Historical record of CoE decisions]

Appendix F: Contact Directory

[Contact information for all CoE members and stakeholders]

SIGN-OFF

CoE Lead:

Name: ____________________

Signature: ____________________

Date: ____________________

Executive Sponsor:

Name: ____________________

Signature: ____________________

Date: ____________________

Additional Approvers (if required):

[CIO, Chief Risk Officer, etc.]

REVISION HISTORY

Version	Date	Author	Changes Summary
1.0	[Date]	[Name]	Initial charter

End of Charter

IMPLEMENTATION NOTES

Quick Start Guide for Using This Template

1. **Choose Your CoE Type** (or create custom):
    - AI Alignment CoE (focus: prompts, models, collaboration)
    - Governance & Risk CoE (focus: compliance, security, ethics)
    - Custom hybrid or specialized CoE
2. **Complete All Bracketed [Sections]**:
    - Replace all placeholder text
    - Delete options that don't apply
    - Customize based on organizational needs
3. **Tailor the Maturity Roadmap**:
    - Adjust timeline based on organizational readiness

    - Set realistic success criteria
    - Align milestones to fiscal calendar if needed
4. **Socialize Before Finalizing**:
    - Share draft with potential core members
    - Get executive sponsor feedback
    - Incorporate team input before approval
5. **Treat as Living Document**:
    - Schedule first quarterly review
    - Adjust based on early learnings
    - Version control all changes

Common Mistakes to Avoid

- **Making the CoE a "Department of No"** - Always frame as enabler, not blocker
- **Insufficient executive sponsorship** - Without top cover, CoE will struggle
- **Overcommitting member time** - 2-5 hours/week max for core members
- **Too many deliverables too fast** - Start with 2-3 critical artifacts, expand over time
- **Ignoring the "Crawl, Walk, Run" model** - Trying to formalize too early kills momentum
- **Forgetting to measure impact** - Without metrics, CoE looks like overhead
- **No rotation policy** - Prevents burnout and fresh perspectives

---

Template Version: 1.0
Last Updated: [Date]

GOVERNANCE CANVAS TEMPLATE

Purpose: Provide a visual, comprehensive framework for planning and documenting governance requirements for AI work before implementation begins, enabling teams to identify compliance obligations, assess risks, and establish appropriate oversight levels during Sprint Inception.

When to Use:

- During Sprint Inception (Governance Framing phase - 45 minutes)
- When evaluating new AI initiatives
- Before deploying AI features to production
- During governance checkpoint reviews
- When regulations or policies change

Owner: Governance Coordinator (with input from entire team)
Review Frequency: Every sprint, updated as needed

GOVERNANCE CANVAS FRAMEWORK
SECTION 1: INITIATIVE OVERVIEW

Initiative Name: ______________________________________
Sprint/Release: ______________________________________
Business Outcome: ____________________________________

Technical Summary (1-2 sentences):

____

____

Stakeholders:

- Product Strategist: ______________________________
- Governance Coordinator: __________________________
- Integration Architect: ____________________________
- Other: ______________________________________

SECTION 2: REGULATORY LANDSCAPE

A. Applicable Regulations

Check all that apply and note specific requirements:

Horizontal AI Regulations:

- **EU AI Act**
  - Risk Classification:
    ☐ Unacceptable
    ☐ High
    ☐ Limited
    ☐ Minimal
  - Specific Requirements: ___________________________
- **NIST AI Risk Management Framework**
  - Application:
    ☐ Voluntary adoption
    ☐ Contractual requirement
    ☐ Government mandate
  - Focus Areas:
    ☐ Govern
    ☐ Map
    ☐ Measure
    ☐ Manage
- **Other AI-Specific Regulations:**
  - Name: ____
  - Requirements: _________________________________

Industry-Specific Regulations:

- **HIPAA** (Healthcare)
  - PHI Involved: ☐ Yes ☐ No
  - Safeguards Required: _____________________________
- **GDPR/CCPA** (Data Privacy)
  - Personal Data: ☐ Yes ☐ No
  - Data Subject Rights:
    ☐ Access
    ☐ Deletion
    ☐ Portability
    ☐ Other: _______
- **Financial Regulations** (SOX, Dodd-Frank, EMIR, etc.)
  - Type: ____
  - Audit Requirements: ______________________________
- **FDA** (Medical Devices/Clinical Decision Support)
  - Classification: ____________________________________

- Validation Requirements: ______________________________
- **Industry Standards** (ISO, NIST, etc.)
    - Standard: ______________________________________
- Compliance Level: ____________________________________
- **Other:**
    - Regulation: ___________________________________
    - Requirements: _________________________________

B. Regulatory Deadlines

Any time-sensitive compliance requirements:

Requirement Deadline Owner Status: _______________________

SECTION 3: RISK ASSESSMENT

A. Risk Tier Classification

Based on Template #9 (Risk Assessment & Classification):

Risk Tier:

☐ Delegated Authority

☐ Standard Review

☐ Board Review

☐ Prohibited

Justification: ____

B. Risk Factors Analysis

Data Sensitivity:

- No personal data
- Non-sensitive personal data
- Sensitive personal data (health, financial, biometric)
- Proprietary/confidential business data
- Classified/controlled information

Data Details:

- Types of data: ____
- Data volume: ____
- Data sources: ____
- Data retention requirements: ______________________________

User Exposure:

- Scope:

  ☐ Internal only

  ☐ Limited external

  ☐ Public-facing

  ☐ Critical infrastructure

- User count: ________________________________
- User demographic: ___________________________
- Potential impact of failure:

  ☐ Low

  ☐ Medium

  ☐ High

  ☐ Critical

System Criticality:

- Non-critical support function
- Important business process
- Revenue-impacting system
- Safety-critical system
- Life-or-death decisions

AI Hallucination Potential:

- Risk of harmful hallucinations: ☐ Low ☐ Medium ☐ High
- Mitigation strategy: ______________________________
- Human-in-the-loop validation: ☐ Yes ☐ No
- Explanation: ________________________________

Regulatory Impact:

- Penalty for non-compliance: $______________________
- Regulatory body: ___________________________
- Previous audit findings: _________________________

SECTION 4: COMPLIANCE REQUIREMENTS

A. Data Privacy & Security

Data Privacy Controls:

- Data minimization implemented
- Purpose limitation documented
- Consent mechanisms in place
- Data subject rights supported (access, deletion, portability)
- Privacy impact assessment completed
- Data Processing Agreement (DPA) required: ☐ Yes ☐ No

Security Requirements:

- Encryption at rest
- Encryption in transit
- Access controls (role-based)
- Authentication (MFA, SSO)
- Audit logging
- Penetration testing required
- Security review completed: ☐ Yes ☐ No ☐ Scheduled: _______

Data Residency:

- Geographic restrictions: ___________________________
- Cross-border transfer mechanisms: _____________________

B. AI-Specific Compliance

Transparency & Explainability:

- Users informed of AI involvement
- Decisions are explainable
- Model documentation maintained
- Disclosure statements required: ______________________

Fairness & Bias:

- Bias assessment conducted
- Protected classes identified: __________________________
- Fairness metrics defined: ____________________________
- Mitigation strategies: ________________________________

Human Oversight:

- Human-in-the-loop required
- Override mechanism available
- Escalation path defined
- Decision authority: ______________________________

Model Governance:

- Model versioning tracked
- Model performance monitored
- Drift detection implemented
- Retraining policy defined
- Model card/documentation complete

C. Documentation Requirements

Required Documentation:

- Technical design document
- Risk assessment report
- Privacy impact assessment
- Security assessment
- Bias/fairness analysis
- Model validation report
- User acceptance testing results
- Compliance checklist
- Incident response plan
- Training materials
- Other: ____________________________________

Documentation Owner: _____________________________

Documentation Due Date: ____________________________

Storage Location: __________________________________

SECTION 5: APPROVAL & OVERSIGHT

A. Approval Authority

Based on Risk Tier, this initiative requires:

Delegated Authority (Low Risk):

- Team self-approval with documentation
- Approver: ________________________________ (Governance Coordinator)
- Approval Date: ________________________________

Standard Review (Medium Risk):

- Governance Coordinator approval required
- Security team review required
- Legal review required
- Review timeline: ________________________________
- Approvers: ________________________________

Board Review (High Risk):

- Governance Board approval required
- Executive sponsor approval required
- External audit required
- Board meeting date: ________________________________
- Board members: ________________________________

Prohibited (Unacceptable Risk):

- Work cannot proceed
- Escalation path: ________________________________
- Alternative approaches: ________________________________

B. Governance Checkpoints

Mid-Sprint Checkpoint (Day 5-6):

- Scheduled date/time: ________________________________
- Attendees: ________________________________
- Review focus: ________________________________
- Decision needed: ☐ Continue ☐ Modify ☐ Stop

Pre-Deployment Review:

- Scheduled date/time: ________________________________
- Final approvers: ________________________________
- Go/No-Go criteria: ________________________________

Post-Deployment Review:

- Scheduled date/time: ________________________________
- Metrics to review: ________________________________

C. Ongoing Monitoring

Compliance Monitoring:

- Monitoring frequency: ☐ Daily ☐ Weekly ☐ Monthly ☐ Quarterly

- Key metrics: ___________________________________
- Alert thresholds: ______________________________
- Responsible party: _____________________________

Audit Schedule:

- Next audit date: _______________________________
- Audit scope: ___________________________________
- Auditor: ☐ Internal ☐ External ☐ Both

SECTION 6: ETHICAL CONSIDERATIONS

Ethical Framework Applied:

- Organizational AI Ethics Guidelines
- Industry standards (e.g., IEEE, ACM)
- Professional codes of conduct
- Other: ____

Ethical Risk Assessment:

Potential Harms:

- Privacy invasion
- Discrimination/bias
- Manipulation/deception
- Autonomy reduction
- Safety risks
- Environmental impact
- Other: ______________________________________

Harm Mitigation: ________________________________

Stakeholder Impact Analysis:

- Who benefits: _______________________________
- Who bears risks: _____________________________
- Are impacts equitable: ☐ Yes ☐ No ☐ Uncertain
- Justification: ________________________________

Ethics Review Required:

- Yes - Scheduled: ____________________________
- No - Justification: ___________________________

SECTION 7: INCIDENT RESPONSE & CONTINGENCY

Incident Types to Plan For:

- Data breach
- Privacy violation
- Bias/discrimination incident
- AI hallucination causing harm

- Regulatory violation
- Service disruption
- Other: ____

Incident Response Plan:

- Escalation contact: ______________________________
- Notification requirements: ________________________
- Regulatory reporting timeline: _____________________
- Communication plan: _____________________________
- Rollback procedure: ______________________________

Contingency Planning:

- Backup system available: ☐ Yes ☐ No
- Manual override process: _________________________
- Maximum acceptable downtime: ____________________
- Recovery time objective (RTO): ____________________

SECTION 8: APPROVED TOOLS & TECHNOLOGIES

AI Tools/Models Being Used:

Tool/Model	Version	Vendor	Approved?	Approval Date	Restrictions
			☐ Yes ☐ No		
			☐ Yes ☐ No		
			☐ Yes ☐ No		

Tool Approval Process:

- Approval authority: ______________________________
- Security review: ☐ Complete ☐ Pending ☐ Not required
- Legal review: ☐ Complete ☐ Pending ☐ Not required
- Data residency verified: ☐ Yes ☐ No ☐ N/A

Prohibited Tools: (List any tools explicitly banned by policy) __________

SECTION 9: TRAINING & AWARENESS

Team Training Requirements:

- AI ethics training completed
- Data privacy training completed
- Security awareness training completed
- Bias mitigation training completed
- Incident response training completed
- Role-specific compliance training completed

Training Records:

- Training completion tracked: ☐ Yes ☐ No

- Location: ___________________________________
- Last updated: ________________________________

End User Training:

- User documentation required
- Training materials required
- Disclosure/consent process defined
- Training owner: ______________________________

SECTION 10: SUCCESS CRITERIA & METRICS

Governance Success Metrics:

Metric	Target	Current	Status
Compliance checklist completion	100%		
Documentation completeness	100%		
Approval cycle time	<48 hours		
Policy violations	0		
Audit findings	0 critical		
Training completion	100%		

Compliance Debt:

- Known gaps: _________________________________
- Remediation plan: ____________________________
- Target completion: ____________________________

SECTION 11: SIGN-OFF & ACCOUNTABILITY

Canvas Completed By:

Name: ____
Role: ____
Date: ____

Governance Coordinator Review:

Name: ____
Signature: ____________________________________
Date: ____
Status: ☐ Approved ☐ Approved with conditions ☐ Rejected

Conditions/Notes: _____________________________

Product Strategist Acknowledgment:
Name: ____
Signature: ____
Date: ____
Additional Approvals (if required):
Legal: ____
Date: _______
Security: ____
Date: _______
Privacy Officer: _______________________________________
Date: _______
Executive Sponsor: ____________________________________
Date: _______

APPENDICES

Appendix A: Quick Reference - Risk Tier Decision Tree

```
START
│
├─ Does this involve prohibited use cases? (social scoring, real-time
biometric surveillance)
│   ├─ YES → PROHIBITED - Stop work
│   └─ NO → Continue
│
├─ High-risk domains? (employment, education, law enforcement,
critical infrastructure, biometrics)
│   ├─ YES → BOARD REVIEW required
│   └─ NO → Continue
│
├─ Sensitive data OR significant user exposure OR potential for harm?
│   ├─ YES → STANDARD REVIEW required
│   └─ NO → Continue
└─ Internal only, non-sensitive, low impact?
    └─ YES → DELEGATED AUTHORITY (team self-approval)
```

Appendix B: Compliance Checklist Quick View

Pre-Sprint:

- Governance Canvas completed
- Risk tier assigned
- Regulatory requirements identified
- Approval authority determined
- Checkpoints scheduled

During Sprint:
- Mid-sprint governance checkpoint completed
- Documentation in progress
- No new risks identified

Pre-Deployment:
- All documentation complete
- All approvals obtained
- Training materials ready
- Incident response plan tested
- Monitoring configured

Post-Deployment:
- Compliance monitoring active
- Audit schedule established
- Governance Canvas archived

Appendix C: Escalation Matrix

Issue Type	First Contact	Escalation 1	Escalation 2
Policy violation	Governance Coordinator	Governance Board	Legal/Compliance
Security incident	Security Team	CISO	Executive Team
Privacy breach	Privacy Officer	Legal	Regulatory Authority
Regulatory question	Governance Coordinator	Legal	External Counsel
Ethical concern	Governance Coordinator	Ethics Committee	Executive Team

Appendix D: Document Repository Links
- Governance policies: ___________________________
- Risk assessment templates: ___________________________
- Compliance checklists: ___________________________
- Training materials: ___________________________
- Incident response procedures: ___________________________
- Audit reports: ___________________________

USAGE NOTES

How to Complete This Canvas

1. **During Sprint Inception (Governance Framing - 45 minutes):**
    - Start with Section 1 (Initiative Overview)

   - Complete Section 2 (Regulatory Landscape) - identify all applicable regulations
   - Work through Section 3 (Risk Assessment) - determine risk tier
   - Based on risk tier, complete relevant sections
   - Schedule checkpoints in Section 5
2. **High-Risk Initiatives:**
   - Complete ALL sections thoroughly
   - Engage Legal and Security early
   - Budget extra time for Board Review process
3. **Low-Risk Initiatives:**
   - Focus on Sections 1, 3, 5, and 11
   - Other sections can be abbreviated or marked N/A
   - Still document decisions for audit trail
4. **Living Document:**
   - Update as regulations change
   - Revise after checkpoint reviews
   - Version control all changes
   - Archive completed canvases for audit

Integration with Other Templates

- **Template #2 (Sprint Inception):** Use during Governance Framing phase
- **Template #8 (Governance Checkpoint):** Reference canvas during mid-sprint review
- **Template #9 (Risk Assessment):** Feeds Section 3 of this canvas
- **Template #16 (CoE Charter):** Governance & Risk CoE maintains this template

Common Mistakes to Avoid

- Completing canvas as afterthought rather than during Sprint Inception
- Skipping sections thinking they don't apply (document N/A instead)
- Not updating canvas when requirements change mid-sprint
- Treating as checkbox exercise rather than risk identification tool
- Not involving Legal/Security early enough for high-risk work
- Forgetting to schedule governance checkpoints

Success Indicators

- Canvas completed before implementation begins
- All stakeholders identified and engaged
- Regulatory requirements clearly documented

- Risk tier justified with evidence
- Approval authority clear and agreed
- Checkpoints scheduled on team calendar
- Team understands compliance obligations

Template Version: 1.0
Last Updated: [Date]

EU AI ACT COMPLIANCE CHECKLIST

Purpose: Provide teams with a comprehensive, step-by-step checklist for ensuring AI systems comply with the European Union AI Act, enabling proactive governance and reducing regulatory risk even for organizations outside the EU.

When to Use:

- When developing AI systems that may serve EU customers
- During Sprint Inception for any AI initiative
- When planning market expansion to Europe
- As part of pre-deployment governance review
- During product compliance audits
- When designing AI systems in regulated domains (HR, healthcare, finance, law enforcement, critical infrastructure)

Owner: Governance Coordinator (with support from Legal, Product Strategist, and Integration Architect)

Reference: Regulation (EU) 2024/1689 - EU AI Act (entered into force August 1, 2024)

Penalty for Non-Compliance: Up to 7% of global annual revenue or €35 million (whichever is higher)

⚠ **CRITICAL DISCLAIMER:** This checklist is based on the EU AI Act regulation text (EU) 2024/1689. However, the EU AI Act has staggered implementation dates, and many implementation mechanisms (registration database, detailed guidance documents, harmonized standards, national authority designations) are still being developed by the European Commission and Member States.

Organizations using this checklist must:

1. Verify current implementation status with official EU sources (see Appendix D)
2. Consult legal counsel familiar with EU AI Act
3. Monitor EU AI Office announcements for implementation updates
4. Do not rely solely on this template for compliance decisions

Official EU Source for Current Status: https://digital-strategy.ec.europa.eu/en/policies/ai-office

SECTION 1: RISK CLASSIFICATION DECISION TREE
Step 1: Identify the Risk Tier

CRITICAL QUESTION 1: Is this a prohibited AI practice?
If YES to ANY of the following, STOP - this work is

UNACCEPTABLE RISK and cannot proceed:

- Social scoring by public authorities (evaluating trustworthiness based on behavior/characteristics)
- Real-time remote biometric identification in publicly accessible spaces for law enforcement (with limited exceptions)
- AI systems that manipulate human behavior to circumvent free will
- AI systems that exploit vulnerabilities of specific groups (age, disability, socioeconomic status)
- Biometric categorization to infer sensitive characteristics (race, political opinions, trade union membership, religious beliefs, sex life, sexual orientation)
- Scraping facial images from internet or CCTV to create facial recognition databases
- Emotion recognition in workplace or educational institutions (with exceptions)
- Predictive policing based solely on profiling or personality traits

⚠️ **If ANY box is checked → PROHIBITED - Escalate to Legal immediately, work cannot proceed**

CRITICAL QUESTION 2: Is this a high-risk AI system?
Check if your AI system falls into ANY of the following categories:

High-Risk Category 1: AI in Products Covered by EU Safety Legislation

- Medical devices (diagnosis, treatment, monitoring)
- In vitro diagnostic devices
- Safety components of civil aviation
- Automotive safety systems
- Machinery safety components
- Toys with AI learning capabilities
- Recreational watercraft
- Elevators and cableways
- Equipment for explosive atmospheres
- Radio equipment

- Pressure equipment
- Marine equipment
- Rail systems

High-Risk Category 2: Specific AI Use Cases (Annex III)
Biometric Identification & Categorization:

- Remote biometric identification systems
- Biometric categorization systems (not prohibited)
- Emotion recognition systems (outside prohibited contexts)

Critical Infrastructure:

- AI managing or operating critical infrastructure (water, gas, electricity, heating)
- AI controlling transportation infrastructure
- AI managing internet infrastructure

Education & Vocational Training:

- Determining access to educational institutions
- Assigning students to institutions
- Evaluating learning outcomes or performance
- Monitoring detection of prohibited behavior (exam proctoring)
- Assessing appropriate level of education

Employment, Workers Management & Access to Self-Employment:

- Recruitment or selection (CV screening, interview analysis)
- Promotion or termination decisions
- Task allocation or performance monitoring
- Affecting terms of work-related relationships

Access to Essential Private/Public Services & Benefits:

- Creditworthiness assessment (except fraud detection)
- Credit scoring
- Eligibility for public benefits/services
- Emergency first responder dispatch/prioritization
- Risk assessment for health/life insurance pricing

Law Enforcement:

- Individual risk assessment (recidivism, offense likelihood)
- Polygraph/lie detection
- Evaluating reliability of evidence
- Crime analytics (victim prediction, location, occurrence)
- Profiling during investigation/prosecution
- Assessing personality traits or characteristics

Migration, Asylum & Border Control:

- Assessing risk from irregular entry
- Assisting authorities in examining applications (asylum, visa, residence permits)
- Detecting/identifying document authenticity

- Lie detection for visa/asylum purposes

Administration of Justice & Democratic Processes:

- Assisting judicial authorities in researching/interpreting facts and law
- Applying law to specific facts

⚠ **If ANY box is checked → HIGH-RISK SYSTEM - Complete all sections below**

CRITICAL QUESTION 3: Does the system involve transparency obligations?

- Chatbot or conversational AI (users may think they're interacting with humans)
- Emotion recognition system (in permitted contexts)
- Biometric categorization system (in permitted contexts)
- AI-generated or manipulated content (deepfakes, synthetic media)
- AI-generated text, image, audio, or video content made public

If ANY box checked → LIMITED RISK - Complete Section 8 (Transparency Requirements)

If NONE of the above apply → MINIMAL RISK - Basic compliance only (Section 9)

Risk Tier Summary

Your AI System is Classified as:

- **UNACCEPTABLE RISK** (Prohibited) - Work must stop
- **HIGH RISK** - Full EU AI Act compliance required (Sections 2-7 mandatory)
- **LIMITED RISK** - Transparency obligations only (Section 8)
- **MINIMAL RISK** - No specific requirements beyond general laws (Section 9)

Rationale for Classification:

____

____

Classification Reviewed By: ______________________________

Date: ____

SECTION 2: HIGH-RISK SYSTEMS - REQUIREMENTS OVERVIEW

⚠ Complete this section ONLY if classified as HIGH-RISK

High-risk AI systems must comply with ALL of the following requirements before deployment:

Compliance Requirements Checklist

- **Risk Management System** (Article 9) - Section 3
- **Data Governance** (Article 10) - Section 4
- **Technical Documentation** (Article 11) - Section 5
- **Record-Keeping & Logging** (Article 12) - Section 5
- **Transparency & Information to Deployers** (Article 13) - Section 8
- **Human Oversight** (Article 14) - Section 6
- **Accuracy, Robustness & Cybersecurity** (Article 15) - Section 7
- **Quality Management System** (Article 17) - Section 7
- **Conformity Assessment** (Article 43) - Section 7
- **Post-Market Monitoring** (Article 72) - Section 7
- **Fundamental Rights Impact Assessment** (Article 27) - Section 6

Estimated Compliance Timeline: ___ weeks/months
Compliance Owner: ___________________________
Target Completion Date: ___________________________

SECTION 3: RISK MANAGEMENT SYSTEM (Article 9)

Requirement: Establish, implement, document, and maintain a risk management system throughout the AI system's lifecycle.

3.1 Risk Identification

Identify and analyze known and foreseeable risks to:

Health:

- Physical harm potential documented
- Mental health impact assessed
- Safety risks identified
- Medical decision risks evaluated

Safety:

- System failure scenarios documented
- Misuse scenarios identified
- Security vulnerabilities assessed
- Emergency response procedures defined

Fundamental Rights:

- Privacy risks identified
- Discrimination/bias risks assessed
- Freedom of expression impacts evaluated
- Right to fair trial (if applicable) considered
- Other fundamental rights impacts documented

Risk Register Location: ______________________________

3.2 Risk Estimation & Evaluation

For each identified risk:

Risk ID	Description	Probability	Impact	Risk Level	Mitigation Strategy
		☐ Low ☐ Med ☐ High	☐ Low ☐ Med ☐ High		
		☐ Low ☐ Med ☐ High	☐ Low ☐ Med ☐ High		
		☐ Low ☐ Med ☐ High	☐ Low ☐ Med ☐ High		

Risk Acceptance Criteria Defined: ☐ Yes ☐ No
Criteria: ____

3.3 Risk Mitigation Measures

Implemented Mitigations:

- Most appropriate design and development techniques applied
- Adequate risk management measures implemented
- Adequate information to deployers provided (instructions for use)
- Appropriate training provided to deployers

Residual Risk Assessment:

- All residual risks have been identified
- Residual risks are below defined acceptance thresholds
- Residual risks have been communicated to deployers

Mitigation Evidence Location: ________________________________

3.4 Testing & Validation

Risk management system tested:

- Using real-world data or test data representative of deployment setting
- Throughout system lifecycle (development, production, post-deployment)
- Test results documented

Testing Documentation: ______________________________
Last Test Date: ______________________________
Next Scheduled Test: ______________________________

SECTION 4: DATA GOVERNANCE (Article 10)

Requirement: Training, validation, and testing data must be subject to appropriate data governance and management practices.

4.1 Data Quality Requirements

Training Data:

- Relevant, sufficiently representative, and free of errors
- Completeness assessed and documented
- Appropriate statistical properties documented (balance, distribution)
- Design choices documented (training methodology, data collection, labeling)

Data Quality Metrics:

- Completeness: ___% (Target: ___%)
- Accuracy: ___% (Target: ___%)
- Balance across categories: ☐ Yes ☐ No
- Representativeness assessment: ☐ Complete ☐ In progress

4.2 Data Examination

Biases Examined:

- Selection bias assessed
- Measurement bias assessed
- Historical bias assessed
- Representation bias assessed
- Other biases: ______________________________

Bias Mitigation Applied:

- Dataset rebalancing
- Data augmentation
- Synthetic data generation
- Algorithmic debiasing
- Other: ______________________________

Gaps/Shortcomings Identified:

____

____

Mitigation Strategy for Gaps:

____

____

4.3 Data Processing Operations
Data Processing Documentation:

- Data cleaning procedures documented
- Data transformation procedures documented
- Labeling procedures documented
- Data augmentation methods documented
- Assumptions made about data documented

Data Processing Log Location: ______________________

4.4 Personal Data Processing
If processing personal data:
Legal Basis (GDPR Article 6):

- Consent
- Contract performance
- Legal obligation
- Vital interests
- Public task
- Legitimate interests

Special Category Data (GDPR Article 9):

- Not applicable
- Processing special category data with legal basis: __________

Data Protection Measures:

- Data minimization applied
- Purpose limitation documented
- Storage limitation defined
- Privacy by design implemented
- Data Protection Impact Assessment (DPIA) completed
- DPIA Location: ______________________

Data Subject Rights Supported:

- Right to access
- Right to rectification
- Right to erasure
- Right to restriction of processing
- Right to data portability
- Right to object
- Rights related to automated decision-making

SECTION 5: DOCUMENTATION & RECORD-KEEPING (Articles 11 & 12)

5.1 Technical Documentation (Article 11)

Required Documentation Checklist:

General Description:

- Intended purpose and use cases documented
- Deployer profile and expected users described
- Geographic, behavioral, functional deployment scope defined
- Hardware and software requirements specified
- Lifespan and maintenance requirements documented

Detailed Description:

- Methods and steps for AI system development documented
- Design specifications documented
- General logic of AI system explained
- Key design choices and rationale documented
- Description of system architecture
- Computational resources used documented

Datasets:

- Training datasets described (provenance, scope, characteristics)
- Validation datasets described
- Testing datasets described
- Information on data governance and management
- Data preparation and processing steps documented
- Assumptions and limitations of datasets documented

Training & Performance:

- Training methodology documented
- Training techniques and parameters specified
- Model selection process documented
- Model architecture details provided
- Performance metrics defined and results documented
- Accuracy, robustness, cybersecurity measures described

Risk Management:

- Risk management system documentation (see Section 3)
- Changes made through lifecycle documented
- Validation and testing procedures documented

Quality Management:

- Quality management system documentation (see Section 7.4)
- Post-market monitoring system described (see Section 7.5)

Documentation Location: ______________________________

Last Updated: ____

Version Control: ☐ Yes ☐ No

5.2 Record-Keeping & Logging (Article 12)

Automatic Logging Capabilities:

- Logging functionality designed and implemented
- Logging cannot be disabled by deployer
- Logs stored for appropriate duration
- Log retention period: ___ months/years

Information Logged:

- Duration of each use of the system
- Reference database against which input data was checked
- Input data for which the search generated a match
- Identification of natural persons involved in verification of results
- Other relevant information: ____________________________

Log Security:

- Logs protected against tampering
- Access controls implemented
- Audit trail of log access maintained

Log Storage Location: _____________________________

SECTION 6: HUMAN OVERSIGHT & FUNDAMENTAL RIGHTS (Articles 14 & 27)

6.1 Human Oversight Measures (Article 14)

Requirement: High-risk AI systems must be designed to enable effective oversight by natural persons.

Oversight Capabilities Implemented:

Understanding:

- System capabilities and limitations fully documented for overseers
- System outputs can be properly interpreted by overseers
- Training provided to oversight personnel
- Training documentation: __________________________

Monitoring:

- Real-time monitoring interface provided
- Anomaly detection mechanisms in place
- Alert system for problematic outputs
- Dashboard for oversight metrics: ____________________

Intervention:

- Ability to interrupt system operation (emergency stop)
- Ability to override automated decisions
- Ability to ignore, override, or reverse system outputs

- Intervention mechanism testing completed: ☐ Yes ☐ No

Human-in-the-Loop Design:

- System requires human approval for high-stakes decisions
- Human review checkpoints identified
- Frequency of human review: ___________________________
- Specific decisions requiring human approval: ____________

Oversight Personnel:

- Minimum qualifications: ___________________________
- Required training: ________________________________
- Number of oversight personnel: _____________________
- Escalation path: __________________________________

6.2 Fundamental Rights Impact Assessment (Article 27)

⚠️ Required BEFORE deploying high-risk AI systems (except for narrow exception in Article 6(3))

Assessment Completed: ☐ Yes ☐ No ☐ In Progress

Impact Assessment Contents:

1. Deployer & Purpose:

- Deployer identity: _____________________________
- Contact information: ___________________________
- Intended purpose: ______________________________
- Deployment context: ____________________________

2. Affected Persons:

- Categories of persons affected: ____________________
- Number of potentially affected persons: ______________
- Vulnerable groups involved: ☐ Yes ☐ No
- If yes, specify: __________________________________

3. Fundamental Rights Impacts:

Assess impact on each relevant right:

Fundamental Right	Impact Level	Description	Mitigation
Privacy & Data Protection	☐ None ☐ Low ☐ Medium ☐ High		
Non-discrimination & Equality	☐ None ☐ Low		

	☐ Medium ☐ High		
Freedom of Expression	☐ None ☐ Low ☐ Medium ☐ High		
Freedom of Assembly	☐ None ☐ Low ☐ Medium ☐ High		
Human Dignity	☐ None ☐ Low ☐ Medium ☐ High		
Rights of the Child	☐ None ☐ Low ☐ Medium ☐ High		
Right to Fair Trial	☐ None ☐ Low ☐ Medium ☐ High		
Consumer Protection	☐ None ☐ Low ☐ Medium ☐ High		
Right to Work	☐ None ☐ Low ☐ Medium ☐ High		
Other: ___________	☐ None ☐ Low ☐ Medium ☐ High		

4. Risk Mitigation Measures:

____

____

5. Assessment Results & Actions:

- No significant risks identified - deployment may proceed
- Risks identified - mitigation measures implemented
- Significant unmitigable risks - deployment should not proceed
- Consultation with affected stakeholders conducted
- Assessment shared with market surveillance authority (if required)

Assessment Documentation: ______________________________
Assessment Date: ______________________________________
Next Review Date: _____________________________________
Reviewer: ____

SECTION 7: ACCURACY, ROBUSTNESS & QUALITY MANAGEMENT (Articles 15, 17, 43, 72)

7.1 Accuracy, Robustness & Cybersecurity (Article 15)

Accuracy:

- Appropriate accuracy level defined based on intended purpose
- Accuracy metrics established
- Target accuracy: ___%
- Actual accuracy: ___%
- Accuracy testing completed across relevant subgroups
- Accuracy maintained throughout lifecycle

Robustness:

- System resilient to errors, faults, inconsistencies
- Behavior under unexpected conditions tested
- Robustness against adversarial attacks assessed
- Edge case handling validated
- Degradation gracefully under stress

Cybersecurity:

- Technical vulnerabilities identified and mitigated
- Resilience against attempts to alter use/performance
- Security testing completed
- Last security assessment: __________________________
- Vulnerability management process defined

Performance Monitoring:

- Ongoing performance monitoring implemented
- Drift detection mechanisms in place
- Performance degradation thresholds defined
- Alerting for performance issues configured

7.2 Instructions for Use (Article 13)
User Documentation Must Include:

- Identity and contact details of provider
- Characteristics, capabilities, limitations of AI system
- Performance level (accuracy, robustness, cybersecurity)
- Changes in performance due to specific conditions
- Reasonably foreseeable unintended outcomes and risks
- Training data characteristics affecting performance
- Human oversight requirements and measures
- Expected lifetime and maintenance requirements
- Installation and operation instructions
- Where applicable, conformity assessment information

Instructions Location: ______________________________
Last Updated: ______________________________
Version: ______________________________

7.3 Conformity Assessment (Article 43)
⚠ **Required before placing high-risk AI system on EU market**

Conformity Assessment Procedure:
Option 1: Internal Control (Annex VI):

- Applicable for most high-risk AI systems
- Technical documentation prepared (Section 5.1)
- Quality management system implemented (Section 7.4)
- EU declaration of conformity drafted
- CE marking affixed

Option 2: Quality Management + Conformity Assessment (Annex VII):

- Required for biometric identification/categorization systems
- Required for systems in Annex IX (critical infrastructure, law enforcement, etc.)
- Notified body involved
- Notified body: ______________________________
- Assessment certificate issued
- Certificate number: ______________________________
- Certificate expiration: ______________________________

EU Declaration of Conformity:

- Declaration drafted
- All required information included (Annex V)
- Declaration signed by authorized representative

- Declaration location: ______________________________

CE Marking:

- CE marking affixed to product/packaging/documentation
- CE marking visible, legible, indelible
- Notified body identification number included (if applicable)

Conformity Assessment Completed: ☐ Yes ☐ No
Date: ______________________________
Valid Until: ______________________________

7.4 Quality Management System (Article 17)
Requirement: Establish, document, implement, and maintain a quality management system.
Quality Management System Must Cover:

Strategy:

- Compliance with EU AI Act provisions
- Quality objectives and strategy documented
- Management commitment to quality documented

Design & Development:

- Techniques, procedures, systematic actions for design/development
- Testing procedures defined
- Version control and change management

Technical Documentation:

- Procedures for creating and maintaining technical documentation
- Documentation review and update processes
- Version control for documentation

Quality Control:

- Examination and testing procedures
- Monitoring and performance assessment
- Corrective and preventive actions

Post-Market Monitoring:

- Plan for collecting and analyzing data on performance
- Reporting system for serious incidents
- Procedures for managing corrective actions

Communication:

- Communication with authorities
- Communication with notified bodies (if applicable)
- Communication with deployers
- Communication with other stakeholders

Record-Keeping:

- All documentation and records maintained
- Records kept for at least 10 years after system placed on market

- Digital format preferred, otherwise paper format

Quality Management System Documentation: _______________
Last Internal Audit: ________________________________
Next Internal Audit: ________________________________

7.5 Post-Market Monitoring System (Article 72)
Requirement: Establish and document a post-market monitoring system.

Post-Market Monitoring Plan Must Include:

Data Collection:
- Process for collecting data on performance in real-world use
- Data sources identified
- Data collection frequency: _______________________
- Feedback mechanisms from deployers/users

Performance Monitoring:
- Actual performance vs. expected performance tracked
- Performance across different user groups monitored
- Drift detection and monitoring
- Metrics tracked: ______________________________

Incident Monitoring:
- Serious incidents tracking system
- Root cause analysis procedures
- Corrective action procedures

Reporting:
- Serious incident reporting to authorities (within 15 days)
- Malfunctioning reporting procedures
- Annual post-market monitoring reports prepared
- Reports submitted to national authorities

Review & Updates:
- Regular review of post-market data
- System updates based on findings
- Documentation of changes and rationale

Post-Market Monitoring Documentation: ________________
Last Review: ______________________________________
Next Review: ______________________________________

SECTION 8: TRANSPARENCY REQUIREMENTS (LIMITED RISK)

⚠ Complete this section if classified as LIMITED RISK or if HIGH-RISK system has transparency obligations

8.1 AI System Transparency (Article 50)
For AI systems intended to interact with natural persons:
Chatbots & Conversational AI:

- Users informed they are interacting with an AI system
- Disclosure clear, distinguishable, and easily visible
- Exceptions for obvious cases (e.g., clearly marked chatbots)
- Disclosure method: ________________________________

Emotion Recognition & Biometric Categorization:

- Users informed of operation of these systems
- Disclosure timing:

  ☐ Before use

  ☐ At start of use

  ☐ Other: ________________

- Disclosure method: __________________________________

Disclosure Examples:

- "This is an AI chatbot. Responses are generated automatically."
- "This system uses emotion recognition technology."
- "Biometric categorization is in use."

8.2 AI-Generated Content Labeling (Article 50)
For AI-generated or manipulated content:
Synthetic Media (Deepfakes):

- AI-generated/manipulated audio content disclosed
- AI-generated/manipulated image content disclosed
- AI-generated/manipulated video content disclosed
- AI-generated/manipulated text content disclosed
- Disclosure in machine-readable format (where technically feasible)
- Disclosure detectable by automated means

Labeling Methods:

- Visible watermark
- Metadata tagging
- Statement in accompanying text
- Other: ____

Exceptions:

- Authorized by law for detecting/preventing/prosecuting crime
- Free speech/artistic expression (if appropriate disclosure exists)
- Assistance for standard editing/AI-enabled improvement (not substantial alteration)

Labeling Examples:

- "This image was created with AI."
- "This video contains AI-generated content."

- "AI was used to generate this text."

Disclosure Verification:
- Disclosure testing completed
- User comprehension validated
- Accessibility for disabled users ensured

SECTION 9: MINIMAL RISK - VOLUNTARY CODES OF CONDUCT

For MINIMAL RISK systems, no specific EU AI Act requirements apply. However, voluntary adherence to best practices is recommended:
- Consider developing code of conduct (Article 95)
- Apply risk management practices voluntarily
- Implement transparency measures
- Document design decisions
- Monitor for emerging risks

Voluntary Commitments:

____

____

SECTION 10: REGISTRATION & NOTIFICATION OBLIGATIONS

10.1 EU Database Registration (Article 71)

⚠ HIGH-RISK systems must be registered in EU database before market placement

Registration Information Required:
- Provider name, address, contact details
- Authorized representative (if applicable)
- AI system name and additional identifiers
- Intended purpose
- Risk classification (Annex III reference)
- High-level summary of training data
- Member states where system is available
- CE marking information
- Link to instructions for use
- Electronic link to EU declaration of conformity

Registration Status:
- Not yet registered

- Registration in progress
- Registered - Date: ___________________________
- Registration number: _________________________

Database Information:

- **Status:** The EU AI Act database is being developed by the European Commission
- **Expected Launch:** Database will be operational when high-risk system registration requirements take effect (2026-2027 timeline)
- **Updates:** Check the official EU AI Office page for database launch announcements: https://digital-strategy.ec.europa.eu/en/policies/ai-office
- **Current Action:** Monitor official EU sources for final database URL and registration procedures

10.2 Serious Incident Reporting (Article 73)

Serious Incident Definition: Incident that leads to:

- Death or serious harm to health
- Serious and irreversible disruption of critical infrastructure
- Breach of fundamental rights obligations

Incident Reporting Procedure:

- Incident detection and logging process defined
- Reporting timeline: Within 15 days of awareness
- Reporting format and content defined
- Market surveillance authority contact: ___________________
- Escalation procedure documented

Incident Log: ____

SECTION 11: COMPLIANCE VERIFICATION & AUDIT TRAIL

11.1 Pre-Deployment Verification

Final Compliance Checklist:

- Risk classification completed and documented
- All applicable requirements sections completed
- Technical documentation complete and current
- Quality management system operational
- Risk management system operational
- Data governance practices implemented
- Human oversight measures functional
- Conformity assessment completed (if high-risk)
- EU declaration of conformity signed (if high-risk)
- CE marking affixed (if high-risk)
- System registered in EU database (if high-risk)

- Post-market monitoring system ready
- Instructions for use provided to deployers
- Training for deployers completed
- All documentation archived and accessible

Pre-Deployment Review Date: ____________________
Reviewed By: ________________________________

Deployment Approval:

☐ Approved

☐ Approved with Conditions

☐ Rejected

Conditions/Notes:

____

____

11.2 Ongoing Compliance Monitoring

Periodic Reviews:

- Review frequency: ☐ Monthly ☐ Quarterly ☐ Annually
- Next review date: ________________________
- Review scope: ___________________________

Change Management:

- Process for managing system changes documented
- Impact assessment required for changes
- Re-assessment triggers defined
- Change log maintained

Compliance Metrics:

Metric	Target	Current	Status
Documentation completeness	100%		
Serious incidents reported	0		
Post-market monitoring completion	100%		
Training completion (deployers)	100%		
Audit findings (critical)	0		

11.3 Audit Trail & Document Retention

Required Documentation Retention:

- Technical documentation: 10 years after last product on market
- Quality management system records: 10 years
- Logs: Duration defined in Article 12
- Post-market monitoring data: 10 years
- Incident reports: Permanent

- Conformity assessment documentation: 10 years

Document Archive Location: ____________________________
Archive Responsible Party: ______________________________
Archive Backup Procedures: ______________________________

SECTION 12: ROLES & RESPONSIBILITIES

12.1 EU AI Act Defined Roles

Provider (Article 3(3)):

- Natural/legal person placing AI system on market or putting into service
- Responsible for: Design, development, conformity assessment, documentation

Your Organization's Role:

☐ Provider

☐ Deployer

☐ Distributor

☐ Importer

☐ Other

If Provider:

- All provider obligations understood and accepted
- Authorized representative appointed (if not established in EU)
- Conformity assessment responsibilities assigned

Deployer (Article 3(4)):

- Natural/legal person using AI system under their authority
- Responsible for: Using system according to instructions, human oversight, monitoring

If Deployer:

- Instructions for use received and reviewed
- Human oversight measures implemented
- Fundamental rights impact assessment completed
- Incident reporting procedure understood

Authorized Representative (Article 22):

- Required (provider not established in EU)
- Not required
- Representative name: ______________________________
- Written mandate in place
- Tasks assigned: ____________________________________

12.2 Internal Compliance Roles

Compliance Team:

Role	Name	Responsibilities	Time Allocation
EU AI Act Compliance Lead		Overall compliance oversight	
Legal Counsel		Regulatory interpretation	
Governance Coordinator		Day-to-day compliance activities	
Integration Architect		Technical compliance measures	
Product Strategist		Business alignment	
Validation Lead		Testing & validation	
Data Protection Officer		GDPR alignment	

SECTION 13: INTEGRATION WITH OTHER TEMPLATES

This EU AI Act Compliance Checklist integrates with:

- **Template #9 (Risk Assessment & Classification):** Risk tier feeds into EU AI Act classification
- **Template #17 (Governance Canvas):** EU AI Act requirements feed Section 2 (Regulatory Landscape)
- **Template #2 (Sprint Inception):** EU compliance confirmed during Governance Framing phase
- **Template #8 (Governance Checkpoint):** EU compliance verified at mid-sprint checkpoint
- **Template #11 (Test Strategy):** Accuracy/robustness testing requirements from Section 7.1

Cross-Reference Documentation:

- Governance Canvas reference: ________________________________
- Risk Assessment reference: ________________________________
- Sprint documentation: ________________________________

APPENDICES

Appendix A: Quick Reference - EU AI Act Risk Tiers

Risk Tier	Examples	Obligations
Unacceptable	Social scoring, real-time biometric surveillance (generally), manipulation	**Prohibited** - Cannot deploy
High	Employment decisions, credit scoring, law enforcement tools, critical infrastructure	Full compliance (Sections 2-7), conformity assessment, CE marking, registration
Limited	Chatbots, emotion recognition (permitted contexts), deepfakes	Transparency obligations (Section 8)
Minimal	AI-enabled spam filters, video games	No specific requirements (voluntary codes of conduct)

Appendix B: Conformity Assessment Decision Tree

```
Is this a HIGH-RISK system?
│
├─YES → Which assessment procedure?
│  │
│  ├─ Biometric ID/categorization OR Annex IX system?
│  │   ├─ YES → Annex VII (Notified Body required)
│  │   └─ NO → Continue
│  │
│  └─ Most other high-risk systems
│      └─ Annex VI (Internal Control)
│
└─NO → No conformity assessment needed
```

Appendix C: Serious Incident Reporting Template

Incident Report Form:

1. **Incident Details:**
    - Date/time of incident: ______________________________
    - Location: ______________________________
    - AI system involved: ______________________________
    - Version: ______________________________
2. **Impact:**
    - Death
    - Serious harm to health

- Serious disruption of critical infrastructure
- Breach of fundamental rights
- Description: ______________________________

3. **Root Cause Analysis:**
    - Immediate cause: ______________________________
    - Contributing factors: ____________________________
    - System defect: ☐ Yes ☐ No
4. **Corrective Actions:**
    - Immediate actions taken: ________________________
    - Long-term corrective actions: ____________________
    - Timeline: ____________________________________
5. **Reporting:**
    - Market surveillance authority notified: ☐ Yes ☐ No
    - Date of notification: __________________________
    - Follow-up required: ☐ Yes ☐ No

Appendix D: EU AI Act Resources

Official EU Resources (Verified):

- **EU AI Act Full Text:** https://eur-lex.europa.eu/eli/reg/2024/1689/oj
- **European Commission AI Policy Page:** https://digital-strategy.ec.europa.eu/en/policies/regulatory-framework-ai
- **EU AI Office:** https://digital-strategy.ec.europa.eu/en/policies/ai-office

Database & Registration (In Development):

- **EU AI Act Database:** Not yet operational as of template creation
- **Status:** Database is being developed; launch expected when high-risk system registration requirements take effect
- **Updates:** Monitor EU AI Office page (link above) for database launch announcements
- **Action:** Bookmark official EU AI Office page for registration procedure updates

Implementation Guidance (To Be Published):

- European Commission implementation guidance documents are being developed
- Standards organizations (CEN-CENELEC) are developing harmonized standards
- Codes of practice are being drafted for general purpose AI
- **Recommended:** Subscribe to EU AI Office updates for guidance documents when published

National Competent Authorities:

- Each EU Member State will designate market surveillance authorities

- **Recommended:** Identify relevant national authority for your target markets once designated
- Lists will be published on EU AI Office website

⚠️ **CRITICAL NOTE:** The EU AI Act is a new regulation with staggered implementation dates. Many implementation mechanisms (database, detailed guidance, standards, national authorities) are still being developed.

Your organization should:

1. Monitor official EU sources regularly for updates
2. Consult legal counsel familiar with current EU AI Act implementation status
3. Join industry working groups developing codes of practice
4. Do not rely solely on this template - verify all regulatory requirements with current official sources

Appendix E: Revision History

Version	Date	Author	Changes
1.0			Initial checklist based on EU AI Act (EU) 2024/1689

FINAL SIGN-OFF

EU AI Act Compliance Status:

- **Compliant** - Ready for EU market
- **In Progress** - Compliance work ongoing (Target date: ________)
- **Non-Compliant** - Gaps identified (Remediation plan: ________)
- **Not Applicable** - System will not be placed on EU market

Compliance Lead Certification:

I certify that this EU AI Act Compliance Checklist has been completed to the best of my knowledge and that all applicable requirements have been addressed.

Name: ____
Title: ____
Signature: ____________________________________
Date: ____

Legal Review:

Name: ____
Signature: ____________________________________
Date: ____

Executive Approval (for high-risk systems):
Name: ______________________________
Title: ______________________________
Signature: ______________________________
Date: ______________________________

IMPLEMENTATION GUIDANCE
Critical Compliance Verification

⚠ **BEFORE using this checklist for actual compliance:**

1. **Verify Current Regulation Status**
   - Confirm you are working with the most recent version of the EU AI Act
   - Check for any amendments or updates since this template was created
   - Official source: https://eur-lex.europa.eu/eli/reg/2024/1689/oj
2. **Confirm Implementation Timeline**
   - The EU AI Act has different effective dates for different requirements
   - Verify which requirements are currently in force for your use case
   - Consult official EU AI Office timeline: https://digital-strategy.ec.europa.eu/en/policies/ai-office
3. **Engage Legal Counsel**
   - EU AI Act compliance requires legal expertise
   - Interpretation of risk classifications can be complex
   - Legal counsel should review your compliance approach before deployment
4. **Monitor Official Channels**
   - EU AI Office publishes implementation guidance and updates
   - Subscribe to official EU notifications
   - Join relevant industry working groups

❗ **This template is a tool to organize compliance work, not a substitute for legal advice or official guidance.**

For Teams New to EU AI Act

1. **Start with Risk Classification** (Section 1)
    - This determines everything else
    - Be conservative - if uncertain, classify as higher risk
    - Consult legal counsel for borderline cases
2. **If High-Risk, Don't Panic**
    - Break compliance into phases (design → development → deployment)
    - Build compliance into Sprint Inception (don't bolt on at end)
    - Use existing templates (Governance Canvas, Risk Assessment) as foundation
3. **Budget Adequate Time**
    - First conformity assessment: 3-6 months
    - Subsequent systems with similar use cases: 1-2 months
    - Time decreases as organizational capability matures
4. **Engage Experts Early**
    - Legal counsel familiar with EU AI Act
    - Notified body (if required for your system type)
    - Data protection officer (for GDPR alignment)
    - Consider consultants for first implementation
5. **Integrate, Don't Separate**
    - EU AI Act compliance should be part of standard development workflow
    - Use Template #2 (Sprint Inception) to address compliance upfront
    - Use Template #8 (Governance Checkpoint) for ongoing verification
    - Don't create separate "compliance team" - embed in delivery teams

Common Pitfalls to Avoid

- **Treating as legal-only issue** - This is a technical compliance framework requiring engineering involvement
- **Waiting until deployment to assess compliance** - Build compliance into design phase
- **Underestimating documentation burden** - Technical documentation is extensive for high-risk systems
- **Ignoring if not currently serving EU** - "Brussels Effect" means this becomes global standard
- **Copy-pasting from other systems** - Each system needs individual assessment
- **Forgetting post-market monitoring** - Compliance doesn't end at deployment

Success Indicators

- EU AI Act discussed during every Sprint Inception
- Risk classification completed before development starts
- Technical documentation maintained throughout (not created at end)
- Team understands which tier their work falls into
- Post-market monitoring integrated into operations
- Serious incident reporting procedure tested
- Conformity assessment on critical path for deployment

Template Version: 1.0
Last Updated: [Date]

BUDGET PLANNING TEMPLATE FOR AI SCRUM ADOPTION

Purpose: Provide a simple, complete budgeting framework for AI Scrum adoption covering both team salaries (the largest cost) and tools/technology costs.

Owner: CFO / Finance Team (with Transformation Lead)
Reference: Chapter 4 (Team Costs), Chapter 14 (Tool Costs)

ORGANIZATION PROFILE
Organization: ______________________________
Number of AI Scrum Teams: ______________________
Total Employees: ______________________________
Fiscal Year: __________________________________

PART 1: TEAM SALARY COSTS (Largest Cost Component)
Reference: Chapter 4, Table 4.1, Appendix D, Appendix E

⚠ **Note:** The roles below are shown for a **Software Development** AI Scrum team. Other disciplines (Marketing, Legal, HR, etc.) have **equivalent roles with different titles** but similar salary ranges and team structure.

Team Composition & Salaries
Per Software Development Team (9-11 people):

Role	Quantity	Salary Each	Annual Cost
Product Strategist	1	$145,000	$145,000
AI Alignment Engineer	2	$130,000	$260,000
Validation Lead	2	$118,000	$236,000
Governance Coordinator*	1	$125,000	$125,000
Integration Architect*	1	$185,000	$185,000
AI-Augmented Developer	2-3	$138,000	$345,000 (2.5 avg)
Delivery Facilitator	1	$130,000	$130,000
Total per Team	**9-11**		**$1,426,000**

*** Fractional Roles:** Governance Coordinator and Integration Architect can each support 2-3 teams. For multiple teams, you don't need 1 per team - 1 person in these roles can be shared across teams.

Role Equivalents by Discipline
Other disciplines have the same team structure with different role titles:

Software Dev Role	Marketing Equivalent	Legal Equivalent	HR Equivalent	Salary Range
Product Strategist	Campaign Strategist	Practice Leader	Talent Strategist	$140-150k
AI Alignment Engineer	Content AI Specialist	Legal AI Specialist	Recruiting AI Specialist	$125-135k
Validation Lead	Brand Quality Lead	Legal Review Lead	Compliance Lead	$115-120k
Governance Coordinator*	Governance Coordinator*	Governance Coordinator*	Governance Coordinator*	$120-130k
Integration Architect*	Marketing Systems Architect	Legal Systems Architect	HR Systems Architect	$180-190k
AI-Augmented Developer	**AI-Augmented Content Creator**	**AI-Augmented Legal Associate**	**AI-Augmented Recruiter**	**$135-145k**
Delivery Facilitator	Delivery Facilitator	Delivery Facilitator	Delivery Facilitator	$125-135k

Key Insight: Team structure remains consistent across disciplines (9-11 people), but role names change to match the discipline. **Total team cost remains approximately $1.4M-$1.5M per year regardless of discipline.**

Number of Teams: ___
Single Team Cost: $1,426,000 per year
Multi-Team Cost (with shared roles):

- 2 teams: ~$2,542,000 (save $310k by sharing 1 Gov Coord + 1 Int Architect)
- 3 teams: ~$3,658,000 (save $620k by sharing 1 Gov Coord + 1 Int Architect)
- 4+ teams: Add proportionally, sharing these roles across every 2-3 teams

Your Team Salary Cost: $_______ per year
Notes:

- Salaries are national averages (2025/2026)
- Tech hubs cost 20-40% more

- This is ~26% more than traditional Scrum teams
- Add 25-50% for benefits/overhead to get fully-loaded cost

Fully-Loaded Team Cost (with benefits): $_______ per year

PART 2: TOOLS & TECHNOLOGY COSTS
Reference: Chapter 14

2A. Foundation LLMs
Who needs these: All team members need access to foundation LLMs (Claude, ChatGPT, Gemini, etc.). Remember, each LLM has different strengths so some users may need access to multiple LLMs to be efficient. Ensure your group has the correct LLM(s) to be efficient and productive.

Cost per team member: $20-40/month (average $25)

Category	Team Members	Cost/User/Month	Monthly	Annual
Foundation LLMs (Claude, ChatGPT, etc.)	___ (9-11 per team)	$25	$_______	$_______

Example:
- 1 team with 10 members → 10 × $25 = $250/month ($3,000/year)
- 3 teams with 30 members → 30 × $25 = $750/month ($9,000/year)

2B. Discipline-Specific Tools
⚠ **Select the discipline(s) your team(s) belong to - don't add all disciplines together!**

Tool Costs by Discipline (per Chapter 14):

Discipline	Cost/User/Month	What This Includes
Marketing	$250 (avg)	Jasper/Copy.ai, Canva, Grammarly, campaign tools
Software Development	$200 (avg)	GitHub Copilot, Cursor, Sonar, test automation
Data Science	$400 (avg)	DataRobot, Databricks, AutoML, analytics platforms
HR	$150 (avg)	HireVue, Eightfold, recruiting & talent tools
Legal	$400 (avg)	Harvey AI, CoCounsel, LawGeex, contract analysis

Compliance	$300 (avg)	AML screening, privacy tools, audit automation

Your Team Discipline Breakdown:

Team #	Discipline	Team Members	Cost / User / Month	Monthly Cost	Annual Cost
Team 1	________	___	$______	$______	$______
Team 2	________	___	$______	$______	$______
Team 3	________	___	$______	$______	$______
Total Discipline Tools				**$______**	**$______**

Example:

- Team 1: Marketing team with 10 people → 10 × $250 = $2,500/month
- Team 2: Development team with 9 people → 9 × $200 = $1,800/month
- Team 3: Legal team with 8 people → 8 × $400 = $3,200/month
- **Total: $7,500/month ($90k/year)**

2C. Platform Costs

⚠ **Important:** These are **organization-wide costs**, not per-team. Platform licenses (Jira, Confluence, Slack) typically cover **all employees**, not just AI Scrum team members.

Item	Total Org Users	Cost / User / Month	Monthly	Annual
Jira / Azure DevOps	___ (all employees)	$15	$______	$______
Confluence / SharePoint	___ (all employees)	$10	$______	$______
Slack / Teams (if not included)	___ (all employees)	$8	$______	$______
Platform Subtotal			**$______**	**$______**

Example:

- 100 employees → Jira ($1,500) + Confluence ($1,000) + Slack ($800) = **$3,300/month**
- 500 employees → Jira ($7,500) + Confluence ($5,000) + Slack ($4,000) = **$16,500/month**

Note: These costs don't scale per AI Scrum team - they scale with total organization size

2D. Ongoing Support

Item	Monthly	Annual
Vendor support contracts	$______	$______
Internal IT support	$______	$______
Support Subtotal	**$______**	**$______**

Tools & Technology Total

Category	Monthly	Annual
Foundation LLMs	$______	$______
Discipline Tools	$______	$______
Platforms	$______	$______
Support	$______	$______
Total Tools Cost	**$______**	**$______**

PART 3: ONE-TIME COSTS (Year 1)

Training

⚠ **All training costs are PER PERSON**

Item	Participants	Cost Per Person	Total
Deep Immersion Week (5-6 days)	___	$3,000-8,000	$______
Certifications (CSM, CSPO, PE)	___	$1,000-2,000	$______
Training Subtotal			**$______**

Example:

- 1 team (10 people) × Deep Immersion ($2,000 avg) = **$20,000**
- 3 teams (30 people) × Deep Immersion ($2,000 avg) = **$60,000**

Note: Depending upon what training needs to be taken, group discounts may be available. Not all individuals will need certain types of training and certifications. All will benefit from prompt engineering as it is a standard business skill these days but only a couple would benefit from "scrum master" or "product owner" training.

Implementation

Item	Cost
External consulting/implementation	$_______
Platform setup/migration	$_______
Transformation Lead (Months -9 to +6, 15 months)	$_______
Change management	$_______
Implementation Subtotal	**$_______**

One-Time Costs Total

Total One-Time (Year 1 only): $_______

PART 4: TOTAL BUDGET SUMMARY

Annual Costs

Category	Year 1	Year 2+ (Ongoing)
Team Salaries (fully-loaded)	$_______	$_______
Tools & Technology	$_______	$_______
Training (ongoing learning)	$_______	$_______
One-Time Costs	$_______	$0
TOTAL	**$_______**	**$_______**

Per-Team Monthly Breakdown (for budgeting)

Category	Monthly Cost
Team Salaries	$_______
Tools & Technology	$_______
Total per Team per Month	**$_______**

PART 5: ROI CALCULATION

Reference: Chapter 14 - Conservative Assumptions

Expected Productivity Gains

Conservative Estimate: 20-30% improvement in AI-augmented tasks

Overall Productivity Gain: 10-15% (when AI tasks are 50% of role time)

Team Role	% Time AI-Augmented	Productivity Gain	Overall Impact
Product Strategist	60%	25%	15% overall
AI Alignment Engineer	70%	30%	21% overall
Validation Lead	50%	25%	12.5% overall
Developers	60%	30%	18% overall

Others	40%	20%	8% overall

Average Team Productivity Improvement: ~15%

Value Calculation

Per Team Annual Value:

Source	Calculation	Value
Time savings	Team cost × 15% productivity gain	$_______
Quality improvements	Reduced defects, faster delivery	$_______
Capability expansion	Work not possible before	$_______
Total Annual Value		**$_______**

ROI

Metric	Amount
Total Year 1 Investment	$_______
Total Annual Value	$_______
Net Benefit (Year 1)	$_______
ROI %	___%
Payback Period	___ **months**

Expected Payback: 6-18 months (per Chapter 14)

BENCHMARK COMPARISON (Chapter 14)

⚠ **CRITICAL WARNING:** The benchmarks below show **TOOLS COSTS ONLY** (not team salaries).

Your total budget = Tools (below) + Team Salaries (Part 1)

Tools-Only Benchmarks:

Size	Employees	AI Users	Monthly Tools Cost	Your Tools Cost	Match?
Small	100	50	~$14,500	$_______	
Mid-Size	500	200	~$83,000	$_______	
Enterprise	5,000	1,500	~$835,000	$_______	

Complete Budget Examples (Tools + Salaries):

Size	Teams	Team Salaries	Tools	TOTAL
Small (1-2 teams)	1-2	~$1.4M-$2.5M/year	~$174k/year	**~$1.6M-$2.7M/year**
Mid-Size (3-5 teams)	3-5	~$3.7M-$6M/year	~$996k/year	**~$4.7M-$7M/year**
Enterprise (10+ teams)	10+	~$14M+/year	~$10M/year	**~$24M+/year**

Key Insight: Team salaries are typically **70-85% of total AI Scrum budget**. Tools are important but smaller.

APPROVAL SIGNATURES

Budget Request: $_______ (Year 1) / $_______ (Ongoing)

CFO: ______________________________________

Date: _______

CTO/CIO: _______________________________

Date: _______

Executive Sponsor: _________________________

Date: _______

NOTES

Key Points:

- Team salaries are the largest cost (~70-80% of total budget)
- Tools are significant but smaller than team costs
- Year 1 includes one-time training and implementation costs
- Expected payback: 6-18 months
- Conservative ROI assumption: 15% productivity improvement

Regional Adjustments:

- Tech hubs (SF, NYC, Seattle): salaries will be higher
- Mid-tier cities: Use salaries as shown
- Lower-cost regions: salaries will be less

Scaling:

- Costs scale linearly per team
- Some infrastructure costs have economies of scale
- Platform costs per user decrease with volume

Template Version: 1.0

Last Updated: [Date]

AI SCRUM ROI CALCULATOR – EXAMPLE AND TEMPLATE

Purpose: Provide a simple, complete ROI calculator for AI Scrum adoption.

Owner: CFO / Finance Team (with Transformation Lead)
Reference: Chapter 4

Worked Example: MidWest Financial Technology — Three-Team AI Scrum Transition

Organization Profile

MidWest Financial Technology is a 340-person regional fintech company with a software development organization of approximately sixty people, organized into three Scrum teams. They build and maintain a commercial lending platform. They have been practicing traditional Scrum for four years. Sprint velocity is stable but feature delivery timelines consistently run eight to twelve weeks beyond initial estimates. Production defect rates have been creeping upward for eighteen months.

The CTO has proposed a full AI Scrum transition for all three teams over a twelve-month period. The CFO has asked for a three-year ROI projection before approving the budget.

Section 1: Investment (Cost Side)

Transition costs fall into three categories: talent, tooling, and transformation. Talent costs are recurring. Tooling costs are partially recurring. Transformation costs are front-loaded to Year 1.

Talent Premium (Annual, Recurring)

Item	Per Team	Three Teams
Annual talent premium	$152,000	$456,000

AI Tooling Costs (Annual)

Tool Category	Annual Cost
AI code generation — GitHub Copilot Enterprise, 60 seats	$38,400
AI testing and validation platform	$24,000
Observability and telemetry, AI-enhanced tier	$18,000
Governance and compliance tooling	$12,000
LLM API access and prompt management	$15,000
Total Annual Tooling	**$107,400**

Organizations carrying existing enterprise licenses in some of these categories should expect this figure to reduce by twenty to thirty percent.

Transformation Investment (Year One Only)

Item	Estimated Cost
AI Scrum training and certification, 60 people	$48,000
External coaching and facilitation, 6 months	$60,000
Transition productivity loss — see Chapter 11	$85,000
Total Transformation Investment	**$193,000**

Total Three-Year Investment

Year	Talent Premium	Tooling	Transformation	Annual Total
Year 1	$456,000	$107,400	$193,000	$756,400
Year 2	$456,000	$107,400	$0	$563,400
Year 3	$456,000	$107,400	$0	$563,400
Three-Year Total	**$1,368,000**	**$322,200**	**$193,000**	**$1,883,200**

Section 2: Return (Benefit Side)

Each benefit category below is translated into a conservative dollar figure tied to observable operational changes. Where a figure requires internal validation against your own data, that is noted explicitly.

Benefit 1: Sprint Velocity Improvement

Research from GitHub, Tabnine, and McKinsey on AI-assisted development shows productivity improvements in the forty to fifty-five percent range for individual developers on well-defined tasks. Applied to a full AI Scrum team — where validation discipline prevents the rework that typically erodes raw productivity gains — a conservative net velocity improvement of thirty percent is reasonable by end of Year 1. For three teams at an average fully-loaded cost of $1.4M per year, this represents approximately $420,000 in equivalent productivity capacity annually without adding headcount.

Item	Year 1	Year 2	Year 3
Velocity improvement value — conservative 30%, Year 1 prorated	$210,000	$420,000	$462,000

Benefit 2: Reduced Production Defect Cost

MidWest FinTech's estimated annual cost of production defects — including developer rework, QA re-engagement, and elevated support handling — is approximately $380,000. AI Scrum's validation-first model, specifically the Validation Canvas and adversarial testing practices described in Chapter 8, consistently reduces escaped defect rates in adopting organizations. A conservative forty percent reduction is applied here.

Item	Year 1	Year 2	Year 3
Defect cost reduction — 40% of $380K, Year 1 prorated	$76,000	$152,000	$152,000

Benefit 3: Time-to-Market Acceleration

MidWest FinTech's current average time from feature approval to production is fourteen weeks. Moving to ten weeks on an annual cadence of approximately twenty significant features, with an average revenue value per feature of $180,000, yields $2.16M in revenue acceleration potential at full run rate. The figure below applies a thirty percent realization rate to stay firmly in defensible territory. This is the input most dependent on your own product and finance team's revenue attribution model and should be validated internally before use in a board presentation.

Item	Year 1	Year 2	Year 3
Time-to-market acceleration — 30% realization, Year 1 prorated	$216,000	$648,000	$648,000

Benefit 4: Developer Retention

The fully-loaded replacement cost of an experienced software engineer runs between $150,000 and $250,000. MidWest FinTech experienced two senior developer departures in the prior year, both attributing their decisions to outdated tooling and practices. At $180,000 per replacement, that is $360,000 in avoidable annual cost. A conservative assumption of preventing one departure per year through AI Scrum adoption is used here.

Item	Year 1	Year 2	Year 3
Retention benefit — 1 departure prevented, Year 1 prorated	$90,000	$180,000	$180,000

Section 3: Three-Year ROI Summary

Category	Year 1	Year 2	Year 3	Three-Year Total
INVESTMENT				
Total Investment	($756,400)	($563,400)	($563,400)	($1,883,200)
RETURN				
Sprint Velocity Improvement	$210,000	$420,000	$462,000	$1,092,000
Defect Cost Reduction	$76,000	$152,000	$152,000	$380,000
Time-to-Market Acceleration	$216,000	$648,000	$648,000	$1,512,000
Developer Retention	$90,000	$180,000	$180,000	$450,000
Total Return	**$592,000**	**$1,400,000**	**$1,442,000**	**$3,434,000**
Net Position	**($164,400)**	**$836,600**	**$878,600**	**$1,550,800**
Cumulative Net	**($164,400)**	**$672,200**	**$1,550,800**	
Three-Year ROI				82%

Break-even point: approximately Month 16 from transformation start.

Section 4: What This Model Does Not Capture

Three benefit categories are real but excluded because they are too organization-specific to model generically:

Regulatory risk reduction. For financial services and other regulated industries, AI governance discipline directly reduces the probability of regulatory findings. A single material finding can cost multiples of the entire transformation investment in remediation. This requires your compliance team to estimate the probability-weighted value for your environment.

Competitive positioning. Organizations that establish AI-native delivery capability in 2025 and 2026 will operate at a structural cost and speed advantage that compounds over time. It does not appear in a three-year model but is arguably the most important long-term return on this investment.

Technical debt prevention. AI-generated code deployed without AI Scrum's validation discipline creates a new category of technical debt that is harder to detect and more expensive to remediate than traditional debt. The avoided future cost is real but not modeled here.

Blank ROI Template

Your Organization Profile

- Number of teams transitioning: ___
- Average fully-loaded annual team cost: $___
- Average revenue value per significant feature release: $___
- Current average time-to-market in weeks: ___
- Current annual production defect cost: $___
- Average cost to replace a senior developer: $___

Investment

Item	Your Figure
Annual talent premium ($152,000 × number of teams)	$__
Annual AI tooling cost	$__
Year 1 training and certification	$__
Year 1 external coaching	$__
Year 1 transition productivity loss	$__
Year 1 Total Investment	$__
Year 2–3 Annual Investment	$__

Return

Benefit Category	Conservative Assumption	Your Estimate
Sprint velocity improvement	25–35% net by end of Year 1	$__
Production defect cost reduction	35–45% reduction from baseline	$__
Time-to-market acceleration	25–35% realization of opportunity value	$__
Developer retention	1–2 departures prevented annually	$__
Regulatory risk reduction	Organization-specific	$__
Total Annual Return at Year 2 run rate		$__

Summary

	Year 1	Year 2	Year 3	Three-Year Total
Total Investment	$__	$__	$__	$__
Total Return	$__	$__	$__	$__
Net Position	$__	$__	$__	$__
Cumulative Net	$__	$__	$__	
Break-Even Point				Month ___
Three-Year ROI				__%

The value of this model is not its precision. It is the discipline of making your assumptions explicit before the transformation begins so that when results differ from projections, you know exactly which assumption to revisit.

Template Version: 1.0
Last Updated: [Date]

REFERENCES

Aaron, O. (2025, November 10). *How AI and machine learning are transforming software development in 2025*. Medium. https://medium.com/@aotema

Accenture. (2025). *Technology vision 2025: The human-machine partnership*. Accenture Research.

Agile Alliance. (2001). *Agile manifesto: Principles behind the Agile manifesto*. https://agilemanifesto.org/principles.html

Agile Mania (Kumar, N.). (2025). *Do we really need SAFe, LeSS, Spotify, or Nexus to be Agile?* Agile Mania Blog. https://agilemania.com/safe-vs-less-vs-spotify-or-nexus-to-be-agile

AI21 Labs. (2025, November). *9 key AI governance frameworks in 2025*. AI21 Knowledge Base. https://www.ai21.com/knowledge/ai-governance-frameworks

American Institute of Certified Public Accountants. (2025). *SOC 2: Trust services criteria*. https://www.aicpa.org/soc2

Anthropic. (2023). *Claude 2: Constitutional AI and improved safety*. Anthropic Research.

Anthropic. (2025a). *Claude for enterprise: Security and privacy*. Anthropic Documentation. https://www.anthropic.com/claude-enterprise

Anthropic. (2025b). *Constitutional AI: Harmlessness from AI feedback*. Research Publication.

Appinventiv. (2025, September 26). *Predictive analytics software development guide 2025*. Appinventiv Blog. https://appinventiv.com/blog/predictive-analytics-software-development

Applitools. (2025). *Visual AI testing platform*. Applitools Documentation. https://applitools.com/platform

Arya.ai. (2025, August 21). *AI maturity matters: Why enterprises need a capability roadmap*. Arya.ai Blog. https://arya.ai/blog/ai-maturity-roadmap-for-enterprises

Atlassian. (2024). *Team playbook: Agile at scale*. Atlassian Resources.

Atlassian. (2025). *Atlassian intelligence: AI-powered features for Jira and Confluence*. Atlassian Documentation. https://www.atlassian.com/software/atlassian-intelligence

Beck, K., Beedle, M., van Bennekum, A., Cockburn, A., Cunningham, W., Fowler, M., Grenning, J., Highsmith, J., Hunt, A., Jeffries, R., Kern, J., Marick, B., Martin, R. C., Mellor, S., Schwaber, K., Sutherland, J., & Thomas, D. (2001). *Manifesto for agile software development*. Agile Alliance.

Behara, G. K. (2024, November 20). *Artificial intelligence governance & alignment with enterprise governance*. Medium. https://medium.com/@gopalakrishnabehara

Bellamy, R. K. E., Dey, K., Hind, M., Hoffman, S. C., Houde, S., Kannan, K., Lohia, P., Martino, J., Mehta, S., Mojsilović, A., Nagar, S., Ramamurthy, K. N., Richards, J., Saha, D., Sattigeri, P., Singh, M., Kush, R. V., & Zhang, Y. (2019). AI Fairness 360: An extensible toolkit for detecting and mitigating algorithmic bias. *IBM Journal of Research and Development*, *63*(4/5), 4:1-4:15.

Bessen, J., Goos, M., Salomons, A., & Van den Berge, W. (2019). *Automatic reaction: What happens to workers at firms that automate?* (Law and Economics Research Paper). Boston University School of Law.

Bessemer Venture Partners. (2025, August 15). *The state of AI 2025: Agents, innovation, and investment*. BVP Atlas. https://www.bvp.com/atlas/the-state-of-ai-2025

Beyer, B., Jones, C., Petoff, J., & Murphy, N. R. (2016). *Site reliability engineering: How Google runs production systems*. O'Reilly Media.

BizTech Magazine. (2025, November). *AI leadership: Building enterprise AI strategy and governance*.

https://biztechmagazine.com/article/2025/11/ai-leadership-guide-building-ai-strategy-and-governance

Bommasani, R., Hudson, D. A., Adeli, E., Altman, R., Arora, S., von Arx, S., Bernstein, M. S., Bohg, J., Bosselut, A., Brunskill, E., Brynjolfsson, E., Buch, S., Card, D., Castellon, R., Chatterji, N., Chen, A., Creel, K., Davis, J. Q., ... & Liang, P. (2021). *On the opportunities and risks of foundation models*. Stanford Center for Research on Foundation Models. arXiv. https://arxiv.org/abs/2108.07258

Bostrom, N. (2014). *Superintelligence: Paths, dangers, strategies*. Oxford University Press.

Brown, S. (2025, February 25). *What's your company's AI maturity level?* MIT Sloan Management Review. https://mitsloan.mit.edu/ideas-made-to-matter/whats-your-companys-ai-maturity-level

Brynjolfsson, E., & McAfee, A. (2014). *The second machine age: Work, progress, and prosperity in a time of brilliant technologies*. W. W. Norton & Company.

Bryson, J. J., & Theodorou, A. (2019). How society can maintain human-centric artificial intelligence. In *Human-centered AI*. Springer.

Buenen, M. (2025, November 13). *Generative AI in quality engineering has shifted from early experimentation to strategic integration*. Capgemini Research. https://www.capgemini.com/news/press-releases/world-quality-report-2025

Capgemini, OpenText, & Sogeti. (2025, November 13). *World quality report 2025: Adapting to emerging worlds* (17th ed.). https://www.worldqualityreport.com

Chen, M., Tworek, J., Jun, H., Yuan, Q., Pinto, H. P. d. O., Kaplan, J., Edwards, H., Burda, Y., Joseph, N., Brockman, G., Ray, A., Puri, R., Krueger, G., Petrov, M., Khlaaf, H., Sastry, G., Mishkin, P., Chan, B., Gray, S., ... & Zaremba, W. (2021). *Evaluating large language models trained on code*. arXiv. https://arxiv.org/abs/2107.03374

Chui, M., Manyika, J., Miremadi, M., Henke, N., Chung, R., Nel, P., & Malhotra, S. (2018). *Notes from the AI frontier: Modeling the impact of*

AI on the world economy (McKinsey Global Institute Discussion Paper). McKinsey & Company.

Churchill, W. S. (1906, June 21). South African native races. *Hansard Parliamentary Debates*, House of Commons (Vol. 159, cols. 385–400). https://hansard.parliament.uk/

Cloud Native Computing Foundation (CNCF). (2025). *CNCF annual survey: State of cloud native development*. CNCF Reports.

Codeium. (2025). *Windsurf: The AI-native IDE*. Codeium Product Documentation. https://codeium.com/windsurf

Cohn, M. (2005). *Agile estimating and planning*. Prentice Hall.

Cohn, M. (2009). *Succeeding with agile: Software development using Scrum*. Addison-Wesley.

Cursor. (2025). *Cursor: The AI-first code editor*. Cursor Documentation. https://cursor.sh/docs

Dalmijn, M. (2025, March 2). *AI in agile product teams: Deep research insights*. Scrum.org Blog. https://www.scrum.org/resources/blog/ai-agile-product-teams

Datadog. (2025). *AI-powered monitoring and security*. Datadog Documentation. https://docs.datadoghq.com

DataTeams AI. (2025). *Top 10 best predictive analytics software for 2025*. DataTeams Blog. https://www.datateams.ai/blog/best-predictive-analytics-software

Davenport, T. H., & Ronanki, R. (2018). Artificial intelligence for the real world. *Harvard Business Review*, *96*(1), 108–116.

Deloitte. (2024). *State of AI in the enterprise, 5th edition*. Deloitte Insights.

Derby, E., & Larsen, D. (2006). *Agile retrospectives: Making good teams great*. Pragmatic Bookshelf.

DevOps.com. (2025, October). *The agentic AI-driven future of telemetry*. https://devops.com/the-agentic-ai-driven-future-of-telemetry

DevOps Institute. (2025). *Upskilling 2025: Enterprise DevOps skills report*. DevOps Institute Research.

Disciplined Agile Consortium. (2012). *Disciplined agile delivery: A practitioner's guide to agile software delivery in the enterprise*. IBM Press.

DoIT Software. (2025, October 16). *Generative AI vs predictive AI: Differences & 2025's examples*. DoIT Blog. https://doit.software/blog/generative-ai-vs-predictive-ai

Drata. (2025). *Compliance automation platform*. Drata Resources. https://drata.com/platform

Drexler, E. (2024, June 22). *Breaking software bottlenecks: AI and the future of development*. AI Prospects. https://aiprospects.substack.com/p/breaking-software-bottlenecks

Dynatrace. (2025). *AI-powered observability platform*. Dynatrace Documentation. https://www.dynatrace.com/support/help

Edmondson, A. C. (2019). *The fearless organization: Creating psychological safety in the workplace for learning, innovation, and growth*. Wiley.

Elastic. (2025). *State of observability & telemetry report 2025*. Elastic Research.

EPAM Systems, Inc. (2025, October 28). *Revolutionizing QA: EPAM's agentic QA brings human-AI synergy to software testing* [Press release]. https://www.epam.com/about/newsroom/press-releases/2025/revolutionizing-qa

European Commission. (2019). *Ethics guidelines for trustworthy AI*. High-Level Expert Group on Artificial Intelligence.

European Commission. (2024). *EU AI Act: Regulatory framework for artificial intelligence*. Official Journal of the European Union.

Floridi, L., Cowls, J., Beltrametti, M., Chatila, R., Chazerand, P., Dignum, V., Luetge, C., Madelin, R., Pagallo, U., Rossi, F., Schafer, B., Valcke, P., & Vayena, E. (2018). AI4People: An ethical framework for a good AI society. *Minds and Machines*, *28*, 689–707.

Forrester Research. (2024, October 23). *Predictions 2025: GenAI reality bites back for software developers*. Forrester Blogs. https://www.forrester.com/blogs/predictions-2025-software-development

Fountaine, T., McCarthy, B., & Saleh, T. (2019). Building the AI-powered organization. *Harvard Business Review*, *97*(4), 62–73.

Functionize. (2025). *AI-powered test automation*. Functionize Resources. https://www.functionize.com/platform

Gartner, Inc. (2021, November 22). *Gartner forecasts worldwide AI software revenue to reach $62 billion in 2022* [Press release]. https://www.gartner.com/en/newsroom/press-releases/2021-11-22-gartner-forecasts-worldwide-artificial-intelligence-software-revenue-to-reach-62-billion-in-2022

GitHub. (2023). *GitHub copilot: Your AI pair programmer*. GitHub Documentation. https://github.com/features/copilot

GitHub. (2025). *GitHub copilot documentation*. GitHub Docs. https://docs.github.com/en/copilot

Google. (2025). *Gemini for Google Workspace*. Google Workspace Documentation. https://workspace.google.com/solutions/ai

Google AI. (2025). *AI agent whitepaper: The future of autonomous systems*. Google Research.

Google DeepMind. (2022). *AlphaCode: Competitive programming with AI*. DeepMind Research.

Google PAIR. (2025). *What-if tool: Explore machine learning models*. People + AI Research. https://pair-code.github.io/what-if-tool

Grafana Labs. (2025). *The future of observability: AI/ML integration* [Interview with Tom Wilkie]. Grafana Resources.

H2O.ai. (2025). *Convergence of the world's best predictive and generative AI for private, protected data.* https://h2o.ai

Harrison, C. (2025). *LangChain documentation*. LangChain. https://python.langchain.com/docs

Harvard Business Review. (2018). *How to win with machine learning.* In *HBR guide to data science basics for managers.*

Herda, T., Pichler, V., Zhang, Z., Abrahamsson, P., & Hanssen, G. K. (2025, August 28). *AI and agile software development: From frustration to success--XP2025 workshop summary*. arXiv. https://arxiv.org/abs/2506.20159

Herschmann, J., Singhal, S., Power, R., & Swan, C. A. (2025, October 6). *Magic quadrant for AI-augmented software testing tools*. Gartner, Inc.

Hou, X., et al. (2025, May 14). *Mind the metrics: Patterns for telemetry-aware in-IDE AI application development using Model Context Protocol (MCP)*. arXiv. https://arxiv.org/html/2506.11019v1

Humble, J., & Farley, D. (2010). *Continuous delivery: Reliable software releases through build, test, and deployment automation*. Addison-Wesley.

Iansiti, M., & Lakhani, K. R. (2020). Competing in the age of AI. *Harvard Business Review*, *98*(1), 60–67.

IBM. (2025, November 17). *What is telemetry? Understanding modern observability*. IBM Think Topics. https://www.ibm.com/think/topics/telemetry

IBM Institute for Business Value. (2025). *The enterprise guide to AI governance*. IBM Think Leadership. https://www.ibm.com/thought-leadership/institute-business-value/en-us/report/ai-governance

IDC. (2022). *Worldwide spending on AI-centric systems forecast to reach $154 billion in 2023*. IDC Research.

IEEE Computer Society. (2025). *IEEE AITest 2025: Conference on AI testing, verification and validation*. IEEE CISOSE 2025. https://conf.researchr.org/track/cisose-2025/ai-test2025

IEEE Global Initiative on Ethics of Autonomous and Intelligent Systems. (2019). *Ethically aligned design: A vision for prioritizing human well-being with autonomous and intelligent systems*. IEEE.

International Organization for Standardization. (2022). *ISO/IEC 27001: Information security management systems*. https://www.iso.org/standard/27001

Intland (PTC). (2024, November 8). *SAFe or Scrum@Scale – Which framework is best for you?* Codebeamer Blog. https://content.intland.com/blog/agile/safe/safe-or-scrum-at-scale

ISACA. (2025a, September 19). *Collaboration and the new triad of AI governance*. ISACA Industry News. https://www.isaca.org/resources/news-and-trends/industry-news/2025/collaboration-and-the-new-triad-of-ai-governance

ISACA. (2025b). *Leveraging COBIT for effective AI system governance* [White paper]. ISACA Resources. https://www.isaca.org/resources/white-papers/2025/leveraging-cobit-for-effective-ai-system-governance

Jobin, A., Ienca, M., & Vayena, E. (2019). The global landscape of AI ethics guidelines. *Nature Machine Intelligence*, *1*, 389–399.

Kahneman, D. (2011). *Thinking, fast and slow*. Farrar, Straus and Giroux.

Kaplan, A., & Haenlein, M. (2019). Siri, Siri, in my hand: Who's the fairest in the land? On the interpretations, illustrations, and implications of artificial intelligence. *Business Horizons*, *62*(1), 15–25.

Kim, G., Humble, J., Debois, P., & Willis, J. (2016). *The DevOps handbook: How to create world-class agility, reliability, and security in technology organizations*. IT Revolution Press.

Kniberg, H., & Ivarsson, A. (2012). *Scaling agile @ Spotify*. Spotify Labs.

KnowledgeHut. (2021, May 11). *SAFe vs agile frameworks: Scrum scale vs LeSS vs Spotify*. KnowledgeHut Blog. https://www.knowledgehut.com/blog/agile/safe-vs-scrum-scale-vs-less-vs-spotify

Kosmyna, N., et al. (2025, June 10). *Your brain on ChatGPT: Accumulation of cognitive debt when using an AI assistant for essay writing task*. arXiv. https://arxiv.org/abs/2506.08872

Kotter, J. P. (2012). *Leading change*. Harvard Business Review Press.

Kuldeep, P. (2025, November 13). *Top 5 AI observability platforms in 2025*. DEV Community. https://dev.to/kuldeep_paul/top-5-ai-observability-platforms-in-2025-4216

Larman, C., & Vodde, B. (2016). *Large-scale Scrum: More with LeSS*. Addison-Wesley.

Larman, C., & Vodde, B. (2025). *Large-scale Scrum (LeSS) framework*. LeSS Works. https://less.works

Leffingwell, D. (2011). *Agile software requirements: Lean requirements practices for teams, programs, and the enterprise*. Addison-Wesley.

Leffingwell, D. (2023). *SAFe 6.0 distilled: Achieving business agility with the scaled agile framework*. Addison-Wesley.

Li, R., et al. (2023). *StarCoder: Open-source large language models for code*. BigCode Project.

LlamaIndex. (2025). *LlamaIndex documentation: Data framework for LLM applications*. https://docs.llamaindex.ai

LXT. (2025, January 31). *Report: The path to AI maturity--survey of 200 US organizations*. LXT Research. https://www.lxt.ai/path-ai-maturity

Majors, C., Fong-Jones, L., & Miranda, G. (2022). *Observability engineering: Achieving production excellence*. O'Reilly Media.

Management Solutions. (2023). *LLMOps: Operational lifecycle management of large language models* [Technical report].

McKinsey & Company. (2024, June). *The economic potential of generative AI: The next productivity frontier*. McKinsey Global Institute.

McKinsey & Company. (2025, November). *The state of AI in 2025: Agents, innovation, and transformation*. McKinsey Quarterly. https://www.mckinsey.com/capabilities/quantumblack/our-insights/the-state-of-ai

METR (Model Evaluation and Threat Research). (2025, July 10). *Measuring the impact of early-2025 AI on experienced open-source developer productivity*. METR Research. https://metr.org/blog/2025-07-10-early-2025-ai-experienced-os-dev-study

Microsoft. (2025a). *Azure DevOps services documentation*. Microsoft Learn. https://learn.microsoft.com/en-us/azure/devops

Microsoft. (2025b). *Counterfit: A CLI tool for assessing AI system security*. Microsoft Security. https://github.com/Azure/counterfit

Microsoft Digital. (2025, November). *Enterprise AI maturity in five steps: Our guide for IT leaders*. Inside Track Blog. https://www.microsoft.com/insidetrack/blog/enterprise-ai-maturity-in-five-steps

Microsoft Research. (2022). *GitHub copilot: Parrot or crow? An empirical study of GitHub copilot's code generation*. Microsoft Research Publication.

Middleware. (2025). *AI observability: Full-stack monitoring with autonomous remediation*. Middleware Documentation.

Milan, R. (2025, November 16). *Top AI trends in DevOps for 2025: Predictive monitoring, testing & incident management*. Medium. https://medium.com/@rammilan1610

MIT Technology Review. (2025). *AI in 2025: The year of enterprise transformation*. MIT Technology Review Insights.

MITRE Corporation. (2025). *MITRE artificial intelligence maturity model (AI MM) and assessment tool*. MITRE Technical Resources. https://aimaturitymodel.mitre.org

Mittelstadt, B. (2019). Principles alone cannot guarantee ethical AI. *Nature Machine Intelligence*, *1*, 501–507.

Monte Carlo Data. (2025, November). *The 17 best AI observability tools in October 2025*. Monte Carlo Blog. https://www.montecarlodata.com/blog-best-ai-observability-tools

National Institute of Standards and Technology (NIST). (2023, January). *AI risk management framework (AI RMF 1.0)* (NIST Special Publication 1270). https://nist.gov/itl/ai-risk-management-framework

New Relic. (2025a). *AI-powered observability*. New Relic Documentation. https://docs.newrelic.com

New Relic. (2025b). *Observability forecast 2025: Enterprise trends report*. New Relic Research.

NVIDIA. (2025). *Garak: LLM vulnerability scanner*. NVIDIA AI Red Team. https://github.com/leondz/garak

Octal Software. (2025, August 4). *Predictive analytics software development guide 2025: Trends, features, tools & strategies*. Octal Software Blog. https://www.octalsoftware.com/blog/predictive-analytics-software-development

OECD. (2019). *OECD principles on artificial intelligence*. OECD Legal Instruments.

OneAdvanced. (2025, April 8). *AI governance: A strategic guide (2025)*. OneAdvanced Resources. https://www.oneadvanced.com/resources/a-guide-to-mastering-ai-governance-for-business-success

OpenAI. (2023). *GPT-4 technical report*. arXiv. https://arxiv.org/abs/2303.08774

OpenAI. (2025). *ChatGPT enterprise: Security and privacy*. OpenAI Documentation. https://openai.com/enterprise

OpenTelemetry Community. (2025, March 12). *AI agent observability: Evolving standards and best practices*. OpenTelemetry Blog. https://opentelemetry.io/blog/2025/ai-agent-observability

O'Reilly. (2025). *AI adoption in the enterprise 2025*. O'Reilly Research Report.

OWASP Foundation. (2025). *OWASP ZAP: Zed attack proxy*. OWASP Project Documentation. https://www.zaproxy.org

Partnership on AI. (2024). *Responsible practices for synthetic media: A framework for collective action.*

Patton, J. (2014). *User story mapping: Discover the whole story, build the right product*. O'Reilly Media.

Perfecto. (2025). *Agentic AI and the future of quality assurance: From scripted automation to intelligent testing*. Perfecto Blog. https://www.perfecto.io/blog/ai-in-quality-assurance

PremierAgile. (2025). *Scrum vs SAFe: Difference between Scrum and SAFe frameworks*. PremierAgile Resources. https://premieragile.com/scrum-vs-safe-frameworks

Prosci. (2025). *ADKAR model: A goal-oriented change management model*. Prosci Methodology. https://www.prosci.com/methodology/adkar

PromptLayer. (2025). *PromptLayer: Prompt engineering platform*. PromptLayer Documentation. https://docs.promptlayer.com

PwC. (2017). *Sizing the prize: What's the real value of AI for your business and how can you capitalise?* PwC Global.

PwC. (2025). *2025 AI business predictions: Risk management and transformation*. PwC Tech Effect. https://www.pwc.com/us/en/tech-effect/ai-analytics/ai-predictions.html

Pydantic. (2025). *Langfuse: Open source LLM engineering platform*. Langfuse Documentation. https://langfuse.com/docs

Radford, A., Wu, J., Child, R., Luan, D., Amodei, D., & Sutskever, I. (2019). *Language models are unsupervised multitask learners*. OpenAI.

Raji, I. D., Mitchell, M., Buolamwini, J., Gebru, T., & Madrigal, A. (2020). *Closing the AI accountability gap: Defining an end-to-end framework for internal algorithmic auditing*. Conference on Fairness, Accountability, and Transparency.

Ransbotham, S., Gerbert, P., Reeves, M., Kiron, D., & Spira, M. (2019). *Winning with AI*. MIT Sloan Management Review and Boston Consulting Group.

Research and Markets. (2025). *AI maturity assessment 2025: Global survey analysis*. Market Research Report.

Rubin, K. S. (2012). *Essential Scrum: A practical guide to the most popular agile process*. Addison-Wesley.

Saleiro, P., Kuester, B., Stevens, A., Anisfeld, A., Hinkson, L., London, J., & Ghani, R. (2018). *Aequitas: A bias and fairness audit toolkit*. arXiv. https://arxiv.org/abs/1811.05577

Sawmills.ai. (2025). *Smart telemetry management platform: AI-powered cost optimization*. Sawmills Product Documentation. https://www.sawmills.ai

Scaled Agile, Inc. (2025). *SAFe 6.0: Scaled agile framework for lean enterprises*. Scaled Agile Framework. https://scaledagileframework.com

Scheibmeir, J., Murphy, T., & Herschmann, J. (2025, June 13). *4 essential steps to build test automation capabilities*. Gartner, Inc.

Schwaber, K. (2004). *Agile project management with Scrum*. Microsoft Press.

Schwaber, K., & Sutherland, J. (2020, November). *The Scrum guide: The definitive guide to Scrum: The rules of the game*. Scrum.org.

Scrum Alliance. (2025). *AI for product owners - Scrum Alliance microcredential course*. Scrum Alliance Training. https://www.scrumalliance.org/microcredentials/ai-for-product-owners

Scrum.org. (2020a). *Evidence-based management guide*. Scrum.org Resources.

Scrum.org. (2020b). *The Nexus guide: The definitive guide to scaling Scrum*.

Scrum.org. (2025a). *Professional Scrum product owner - AI essentials (PSPO-AI essentials)*. Course Catalog. https://www.scrum.org/courses/professional-scrum-product-owner-ai-essentials

Scrum.org. (2025b, June 26). *Scrum.org announces new AI training for product owners* [Press release]. https://www.scrum.org/resources/scrumorg-announces-new-ai-training-product-owners

Scrum.org. (2025c). *The augmented product owner: Amplifying Scrum with AI*. Scrum.org Resources. https://www.scrum.org/resources/blog/augmented-product-owner-amplifying-scrum-ai

Secureframe. (2025). *Security and compliance automation*. Secureframe Documentation. https://secureframe.com/platform

ServiceNow. (2025). *Enterprise AI maturity index 2025* [White paper].

Shi, Y., et al. (2024). *LLMOps: Large language model operations for production AI systems*. Conference on AI Systems.

SimpliAxis. (2025). *SAFe vs. LeSS: Choosing the right agile framework for your organization*. SimpliAxis Resources. https://www.simpliaxis.com/resources/safe-vs-less

Snyk. (2025). *Developer security platform*. Snyk Documentation. https://docs.snyk.io

SonarSource. (2025). *SonarQube documentation*. SonarQube Docs. https://docs.sonarqube.org/latest

Splunk. (2025). *The state of observability 2025*. Splunk Research Report.

Stanford University. (2025). *Artificial intelligence index report 2025*. Stanford Institute for Human-Centered Artificial Intelligence.

State of California Department of Justice. (2025). *California Consumer Privacy Act (CCPA)*. https://oag.ca.gov/privacy/ccpa

State of DevOps Report 2024. (2024). DORA (DevOps Research and Assessment), Google Cloud.

Sutherland, J. (2014). *Scrum: The art of doing twice the work in half the time*. Crown Business.

Sutherland, J., & Brown, A. (2025). *Scrum@Scale guide*. Scrum@Scale.

Sweller, J. (2011). Cognitive load theory. In *Educational psychology handbook* (pp. 37–76). Cambridge University Press.

Tabnine. (2024). *AI-powered code completion: Enterprise developer productivity study*. Tabnine Research.

Tarantino, J. (2017, June 27). *Calendar tetris is an antipattern*. Jack of Spades. https://jack.ofspades.com/calendar-tetris-is-an-antipattern/index.html

TechRadar. (2025, February 7). *A 2025 guide to unlocking AI's potential for QA & dev teams*. TechRadar Pro. https://www.techradar.com/pro/a-2025-guide-to-unlocking-ais-potential-for-qa-and-dev-teams

The Hacker News. (2025, November). *Governing AI agents: From enterprise risk to strategic asset*. Expert Insights. https://thehackernews.com/expert-insights/2025/11/governing-ai-agents.html

The New Stack. (2025, January 8). *Observability in 2025: OpenTelemetry and AI to fill in gaps*. https://thenewstack.io/observability-in-2025-opentelemetry-and-ai-to-fill-in-gaps

TheServerSide. (2025). *Scrum vs. SAFe: How to choose between top agile frameworks*. TheServerSide Tips. https://www.theserverside.com/tip/Scrum-vs-SAFe-How-to-choose-between-top-Agile-frameworks

ThoughtWorks. (2024, October). *Technology radar vol. 31: AI-augmented software development*.

Toivonen, T. (2025, March 26). *The impact of AI on agile product development (until 2030)*. Medium. https://teamdom.medium.com/the-impact-of-ai-on-agile-product-development-until-2030

Toptal. (2019, January 23). *The project management blueprint part 2: A comprehensive comparison of Waterfall, DAD, SAFe, LeSS, and Scrum@Scale*. Toptal Engineering Blog. https://www.toptal.com/project-managers/agile/project-management-blueprint-part-2

Traceloop. (2025). *OpenTelemetry instrumentation for LLMs and AI agents*. Traceloop Open Source.

UNESCO. (2021). *Recommendation on the ethics of artificial intelligence*.

U.S. Department of Health and Human Services. (2025). *HIPAA privacy rule*. https://www.hhs.gov/hipaa/for-professionals/privacy

USA AI Institute. (2025). *Understanding AI maturity levels: A roadmap for strategic AI adoption*. USA AI Resources. https://www.usaii.org/ai-insights/understanding-ai-maturity-levels

Vanta. (2025). *Trust management platform*. Vanta Resources. https://www.vanta.com/products

Vaswani, A., Shazeer, N., Parmar, N., Uszkoreit, J., Jones, L., Gomez, A. N., Kaiser, Ł., & Polosukhin, I. (2017). Attention is all you need. *Advances in Neural Information Processing Systems 30*.

Veritis. (2025, August 13). *AI maturity model: A CEO's guide to scaling AI for success*. Veritis Blog. https://www.veritis.com/blog/ai-maturity-model

VersionOne. (2024). *State of agile report 2024*. Digital.ai.

Visual Paradigm. (2025). *Comparison of scaling agile frameworks: Which one should you choose?* Visual Paradigm Guides. https://www.visual-paradigm.com/scrum/scaling-agile-frameworks-comparison

Weill, P., Woerner, S. L., & Sebastian, I. M. (2024). Building enterprise AI maturity. *MIT CISR Research Briefing*, *XXIV*(12). https://cisr.mit.edu/publication/2024_1201_EnterpriseAIMaturityModel_WeillWoernerSebastian

Woerner, S. L., Sebastian, I. M., Peter, W., & Káganer, E. (2025, August 21). Grow enterprise AI maturity for bottom-line impact. *MIT CISR Research Briefing*, *XXV*(8).

Wolpers, S. (2025a, June 6). *Food for agile thought #496: Meeker's AI trends*. Age of Product Newsletter. https://age-of-product.com/food-agile-thought-496-meekers-ai-trends

Wolpers, S. (2025b, September 21). *AI transformation déjà vu: Why today's failures look uncannily like yesterday's 'agile transformations.'* Age of Product. https://age-of-product.com/ai-transformation

Wolpers, S. (2025c, October 19). *The dangerous middle: Agile roles that AI will erode first*. Scrum.org Resources. https://www.scrum.org/resources/blog/dangerous-middle-agile-roles-ai-will-erode-first

Workhuman. (2025, October 9). *5 major challenges of AI in 2025 and practical solutions to overcome them*. Workhuman Blog. https://www.workhuman.com/blog/challenges-of-ai

World Economic Forum. (2019). *AI governance: A holistic approach to implement ethics into AI* [White paper].

World Economic Forum. (2025, July). *Enterprise AI is at a tipping point: Here's what comes next*. World Economic Forum Stories. https://www.weforum.org/stories/2025/07/enterprise-ai-tipping-point

World Journal of Advanced Research and Reviews. (2025). Ethical and governance challenges of AI in information systems: Toward responsible enterprise adoption. *WJARR*, *27*(2), 1744–1751.

XB Software. (2025, July 17). *Generative AI in software development: 2024 trends & 2025 predictions*. XB Software Blog. https://xbsoftware.com/blog/ai-in-software-development

Xebia Academy. (2025, August 18). *Scrum teams in 2030: The future of agile with AI*. Xebia Resources. https://academy.xebia.com/future-agile-ai

Zapier. (2025, June 11). *The 7 best predictive analytics software in 2025*. Zapier Blog. https://zapier.com/blog/predictive-analytics-software

www.ingramcontent.com/pod-product-compliance
Lightning Source LLC
LaVergne TN
LVHW082002200926
848481LV00010B/2789
9798995043805